Laser Processing and Analysis of Materials

Laser Processing and Analysis of Materials

W. W. Duley
York University
Toronto, Ontario, Canada

PLENUM PRESS • NEW YORK AND LONDON

Library of Congress Cataloging in Publication Data

Duley, W. W.
Laser processing and analysis of materials.

Includes bibliographical references and index.
1. Lasers—Industrial applications. I. Title.
TA1677.D84 1982 621.36′6 82-18611
ISBN 0-306-41067-2

A Division of Plenum Publishing Corporation
233 Spring Street, New York, N.Y. 10013

Printed in the United States of America

To
my parents

Preface

It has often been said that the laser is a solution searching for a problem. The rapid development of laser technology over the past dozen years has led to the availability of reliable, industrially rated laser sources with a wide variety of output characteristics. This, in turn, has resulted in new laser applications as the laser becomes a familiar processing and analytical tool.

The field of materials science, in particular, has become a fertile one for new laser applications. Laser annealing, alloying, cladding, and heat treating were all but unknown 10 years ago. Today, each is a separate, dynamic field of research activity with many of the early laboratory experiments resulting in the development of new industrial processing techniques using laser technology.

Ten years ago, chemical processing was in its infancy awaiting, primarily, the development of reliable tunable laser sources. Now, with tunability over the entire spectrum from the vacuum ultraviolet to the far infrared, photochemistry is undergoing revolutionary changes with several proven and many promising commercial laser processing operations as the result.

The ability of laser sources to project a probing beam of light into remote or hostile environments has led to the development of a wide variety of new analytical techniques in environmental and laboratory analysis. Many of these are reviewed in this book.

We now stand at a point in the development of laser technology at which one can see clearly a number of areas in which the laser has provided the solution to an outstanding problem. We can also see with clarity the potential areas in which further applications may occur. At present, some other possible applications are only interesting curiosities in the research laboratory.

The purpose of this book is to survey areas in materials science and analysis in which the laser has a proven or potential application. As such, the reader will find that while in one section we will discuss a well-established laser processing technique, in another we may lapse into the consideration of what may be a prototype laboratory experiment. By adopting this approach, one obtains an overview of a dynamic research field while at the same time seeing which laser applications have proven feasible in an industrial environment.

My own education in lasers and laser technology has been greatly aided by contacts with colleagues and students. Production of this book would have been impossible without the expert typing of Mrs. Gladys Hayward, the drawing skill of Sally Lakdawala, and the photographic expertise of Bryson Timmins. Thanks also to Elisa Bourdon for her careful reading of parts of the manuscript.

I am also indebted to all those who freely granted permission to quote and reproduce various parts of their work.

Finally, my thanks go to my wife, Irmgardt, and my sons, Nick and Mark, for their encouragement and support during the writing and production of this book.

Toronto Walter W. Duley

Contents

3 Laser Processing of Semiconductors

4 Chemical Processing

5 Lasers in Chemical Analysis

6 Lasers in Environmental Analysis

Chapter 1

Lasers and Laser Radiation

1.1. INTRODUCTION

This chapter surveys commercially available lasers and discusses some characteristics of laser radiation that make it useful in the processing and analysis of materials. Performance data for commercial versions of the most widely applied laser sources are outlined. The aim of this summary is to provide ready access to information that may be required in the assessment of the viability of particular laser applications. With this object in mind, the use of ancillary beam handling and measuring components is discussed from a practical point of view. The important problem of beam focusing is discussed in some detail.

Within this chapter, emphasis has been placed on providing the reader with easy access to data pertaining to the generation and use of laser radiation. Data presented are typical of commercially available systems and components, but the performance of laboratory prototype systems may be significantly better. However, it is hoped that by providing performance data representative of commercially available systems the reader will be able to assess the applicability of a particular laser system to the solution of a specific problem.

1.2. LASER SOURCES

1.2.1. Ruby Laser

Ruby is formed when a small amount of Cr_2O_3 is dissolved in sapphire, Al_2O_3. The pink color is due to absorption by Cr^{3+} ions in the broad bands that provide pumping for the red laser transitions. Energy levels for $Cr^{3+}:Al_2O_3$ are shown in Figure 1.1. Pumping occurs by absorption in the broad ${}^4A_2 \rightarrow {}^4T_2$, ${}^4T_1$ bands. This is followed by an efficient radiationless deactivation to the ${}^2E$ state. This state is split into two sublevels by the trigonal field distortion at the site of the Cr^{3+} ion. Laser emission occurs from the lower of these, $\bar{E}$, to the ground state. The wavelength of this line, the $\bar{E} \rightarrow {}^4A_2$ transition, is 694.3 nm

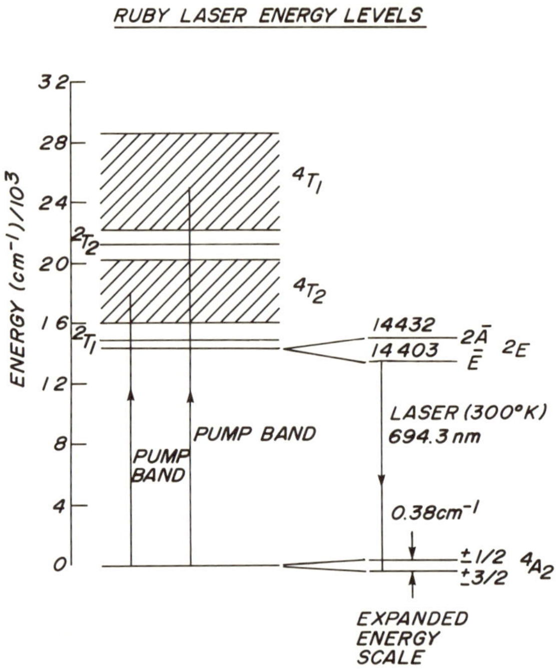

Figure 1.1. Energy levels for Cr^{3+} in ruby.

at room temperature. The ground-state zero-field splitting of 0.38 cm^{-1} is usually contained within the laser linewidth at 300 K.

By using frequency discriminating devices in the laser cavity, the laser can be operated on the $2\bar{A} \rightarrow {}^4A_2$ transition at 692.9 nm at room temperature. Both the R_1 694.3-nm and R_2 692.9-nm lines can be tuned over a ~50-cm^{-1} range by varying the temperature between 77 and 500 K. Widths of the R_1 and R_2 lines also are a function of temperature and become large at temperatures exceeding 100 K.

Typical Cr^{3+} concentrations in ruby are 0.01–0.5 at. %. Since the $Cr^{3+}:Al_2O_3$ system is a three-level laser, population inversion is obtained only when about 50% of Cr^{3+} ions are excited to the ^2E state. This implies a relatively high threshold for oscillation and sensitivity to cavity losses. A summary of optical properties for ruby laser material is given in Table 1.1.

Ruby lasers can be operated in either pulsed or continuous wave (CW) modes. The low overall efficiency (~0.1%) makes CW operation uneconomical when compared to other CW solid-state lasers. When operated in a pulsed mode without Q-switching, the output of the ruby laser consists of repetitive spikes each of high peak power and ~1 μs duration. Spiking occurs because stimulated emission rapidly depletes the population in the ^2E state while the

Table 1.1. Properties of Ruby Laser Material at 300 K[a]

Cr^{3+} concentration	1.6×10^{19} cm^{-3}
Wavelength	694.3 nm
Stimulated emission cross section	2.5×10^{-20} cm^2
Inversion for 1% gain per cm length	4.0×10^{17} cm^{-3}
Stored energy for 1% gain per cm length	2.3 J cm^{-3}
Loss coefficient	~0.001 cm^{-1}
Fluorescent lifetime	3 ms
Fluorescent linewidth	0.53 nm
Rod length	2–25 cm
Rod diameter	1 cm

[a]After Weber (1979).

pumping rate to this level is relatively small. Hence the output oscillates spontaneously between "on" and "off" conditions. The laser output commences ~0.5 ms after initiation of the pump pulse and continues for the duration of the pump pulse. This is typically several milliseconds. Maximum repetition rates are usually several ppm and are limited by heating of the laser rod.

Some output characteristics of the ruby laser are summarized in Table 1.2. Parameters are listed for each of the three pulsed modes of operation: long-pulsed, Q-switched, and mode-locked. No entry is given for a CW ruby laser

Table 1.2. Output Characteristics of Ruby Lasers

Pulse length: 1–3 ms	
Output energy	100 J/pulse (multimode) 1 J/pulse (TEM_{00})
Repetition rate	$\lesssim$1 Hz
Divergence	~5 mrad (multimode) ~1 mrad (TEM_{00})
Beam diameter	1 cm
Peak power in spike	10^6–10^8 W
Pulse length: ~10 ns	
Output energy	1–25 J/pulse (TEM_{00})
Repetition rate	$\lesssim$1 Hz
Divergence	0.3–0.5 mrad
Beam diameter	1 cm
Peak power	10^9 W
Pulse length: 20 ps	
Output energy	10 mJ/pulse (TEM_{00}) 40 mJ/pulse (multimode)
Repetition rate	4 ppm
Divergence	0.2 mrad
Beam diameter	0.2 cm
Peak power	10^8–10^9 W

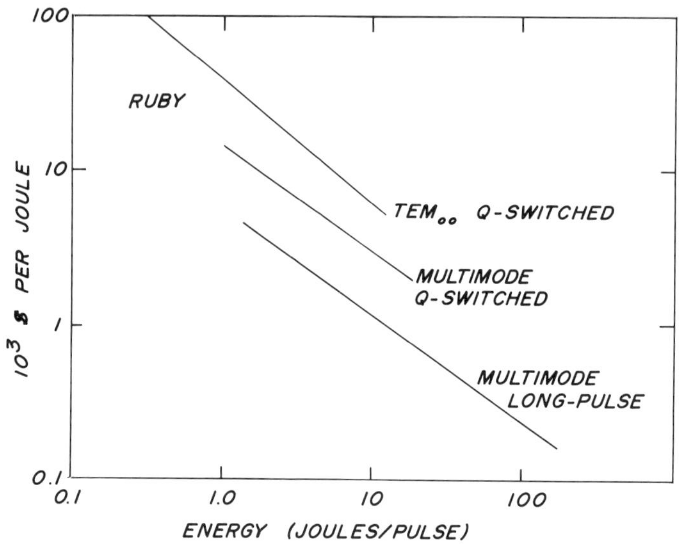

Figure 1.2. Cost per output joule for ruby lasers vs. output energy (1980 dollars).

since such a device does not appear to be commercially available. Pumping of the ruby laser is usually via helical or linear xenon flashlamps although Hg lamps are also used.

Figure 1.2 shows current (1980) costs per output Joule for long-pulse and Q-switched ruby lasers. These costs are representative of those quoted by manufacturers in the 1980 edition of the *Laser Focus Buyers Guide*. Some manufacturers offer prepacked frequency doubling to 347 nm on ruby laser systems.

1.2.2. Nd–YAG Laser

The Nd–YAG laser is the most widely utilized of all solid-state lasers. It is available in both pulsed and CW configurations. Many of the characteristics that make this laser attractive in terms of output power, pulse repetition rate, and pulse energy derive from the properties of the yttrium aluminum garnet (YAG) host. Large YAG crystals of high optical quality are readily available and the relatively large thermal conductivity together with four-level operation make high average output powers feasible. The relevant energy levels of Nd^{3+} in YAG are shown in Figure 1.3.

The complexity of the spectra of the rare earth ions in solids is due to the fact that possible $4f^n$ electronic configurations yield many electronic states with energies in the range $\lesssim 30{,}000\ cm^{-1}$. Spin–orbit interaction further splits these energy levels.

As the $4f$ electrons in the rare earth ions are shielded by electrons in outer shells, $f \rightarrow f$ transitions are sharp, even when these ions are in a solid-state environment. Transition energy is therefore not highly sensitive to the nature of the host crystal. However, static crystal field effects introduce some splitting of crystal field levels. As a result, the levels in the energy diagram of Figure 1.3 are shown as broad bands. In reality, each of these bands contains a number of discrete energy levels.

The laser line at 1064 nm is the transition in Nd–YAG with lowest pump threshold. It corresponds to a transition between crystal field split components of the ${}^4F_{3/2}$ and ${}^4I_{11/2}$ states. Other laser transitions are

$$
\begin{aligned}
{}^4F_{3/2} &\rightarrow {}^4I_{9/2} & \lambda &= 946 \text{ nm} \\
{}^4F_{3/2} &\rightarrow {}^4I_{13/2} & &= 1319 \\
{}^4F_{3/2} &\rightarrow {}^4I_{15/2} & &= 1833
\end{aligned}
$$

It is evident that these all correspond to transitions to excited states with energies that exceed thermal energies at 300 K. Hence, the Nd–YAG laser is a true four-level system. This implies that the threshold population inversion will be significantly smaller than that of the ruby laser which is a three-level system (compare Tables 1.3 and 1.1).

The relatively narrow linewidth of the 1064-nm line in Nd–YAG facilitates CW operation at room temperature. The result is that the CW Nd–YAG laser is second only to the CW CO_2 laser in terms of power generated. However, it does have a considerably lower overall efficiency (typically $<2\%$). Pumping of the Nd–YAG laser is usually with xenon of krypton arc lamps. Kryton lamps are well suited to Nd–YAG because their emission overlaps bands at $\sim$14,000–16,000 cm^{-1} of the Nd^{3+} ion (Figure 1.3).

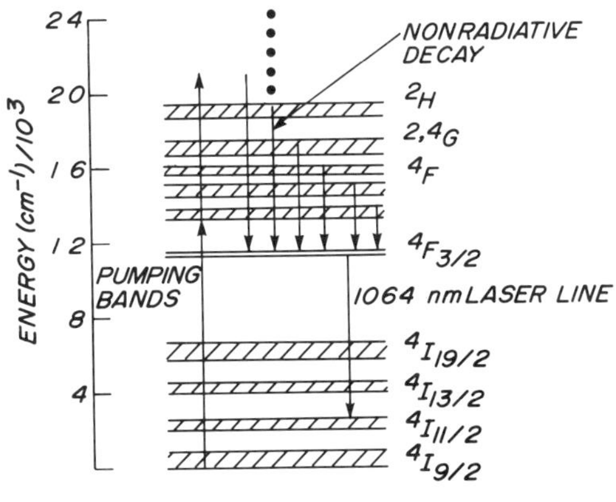

Figure 1.3. Energy levels in the Nd–YAG laser.

Table 1.3. Properties of Nd-YAG Laser Material at 300 K[a]

Nd^{3+} concentration	1.4×10^{20}
Wavelength	1064 nm
Stimulated emission cross section	$(5–8) \times 10^{-19}$ cm^2
Inversion for 1% gain per cm length	$(1.2–2) \times 10^{16}$ cm^{-3}
Stored energy for 1% gain per cm length	$(2–4) \times 10^{-3}$ J cm^{-3}
Loss coefficient	~0.002 cm^{-1}
Fluorescent lifetime	230 μs
Fluorescent linewidth	0.4 nm
Rod length	~15 cm
Rod diameter	~1 cm

[a]After Weber (1979).

The Nd–YAG laser can be operated in either the CW or pulsed mode. In the pulsed mode, long-pulse, Q-switched, and mode-locked operation are readily available. A summary of output characteristics in each of these modes is given in Table 1.4. Figures 1.4 and 1.5 show capital cost per watt and joule of output for Nd lasers.

1.2.3. Nd–Glass Laser

The energy levels of the Nd^{3+} ion shown in Figure 1.3 for Nd–YAG retain their identity in a variety of other solids including many glasses. The Nd–glass laser is a relatively efficient source of coherent radiation at or near 1.06 μm. Nd–glass laser material can be fabricated in a variety of sizes and shapes including rods with lengths of up to 2 m and slabs with diameters of ~10 cm. However, because of the low thermal conductivity of glass compared to that of YAG, cooling is a problem which tends to limit operation to a maximum repetition rate of ~2 Hz.

The electrostatic crystal field surrounding dopant ions in glasses exhibits more dispersion than in crystals. As a result, fluorescent linewidths tend to be somewhat broader than found in crystalline matrices. The large inhomogeneous linewidth of the laser line in Nd–glass lasers yields a wider spectral range for mode-locking. Thus, Nd–glass lasers are well suited to the generation of picosecond pulses. Broadening of other transitions, including those that couple pump light into the glass, yields improved pumping efficiency and enhanced energy storage. Nd–glass lasers are therefore often used when high pulse energies are required. Nd–glass laser systems are most often used in nuclear fusion experiments and in plasma diagnostics, but are not widely applied in industrial material processing.

Table 1.4. Output Characteristics of Nd–YAG Laser

CW operation:	
Average power	$\lesssim$1000 W (multimode)
	$\lesssim$20 W (TEM_{00})
Divergence	1–20 mrad
Beam diameter	0.1–1 cm
Pulse length: 0.1–20 ms	
Output energy	$\lesssim$500 J/pulse (multimode)
	5 J/pulse (TEM_{00})
Repetition rate	200 Hz
Divergence	10 mrad (multimode)
	3 mrad (TEM_{00})
Beam diameter	0.5–1 cm
Pulse length: 0.1–1 μs (repetitive switch)	
Output energy	1 mJ/pulse
Repetition rate	50–100 kHz
Average power	10–100 W
Peak power	10–50 kW
Pulse length: 10–20 ns	
Output energy	1 J/pulse (multimode)
	0.5 J/pulse (TEM_{00})
Repetition rate	10 Hz
Divergence	5 mrad (multimode)
	0.1 mrad (TEM_{00})
Beam diameter	0.5–1 cm
Pulse length: 30–200 ps	
Output energy	5 mJ/pulse (TEM_{00})
Repetition rate	10 Hz
Divergence	0.2–1 mrad
Beam diameter	0.2 cm
Peak power	10^8–10^9 W
Frequency-doubled:	
Output wavelength	532 nm
Output energy	0.4 J/pulse (multimode)
Output	0.4 J/pulse (TEM_{00})
Pulse length	15 ns
Repetition rate	10 Hz
Frequency-tripled:	
Output wavelength	355 nm
Output energy	0.1 J (TEM_{00})
Pulse length	15 ns
Repetition rate	10 Hz
Frequency-quadrupled:	
Output wavelength	266 nm
Output energy	0.1 J (TEM_{00})
Pulse length	15 ns
Repetition rate	10 Hz

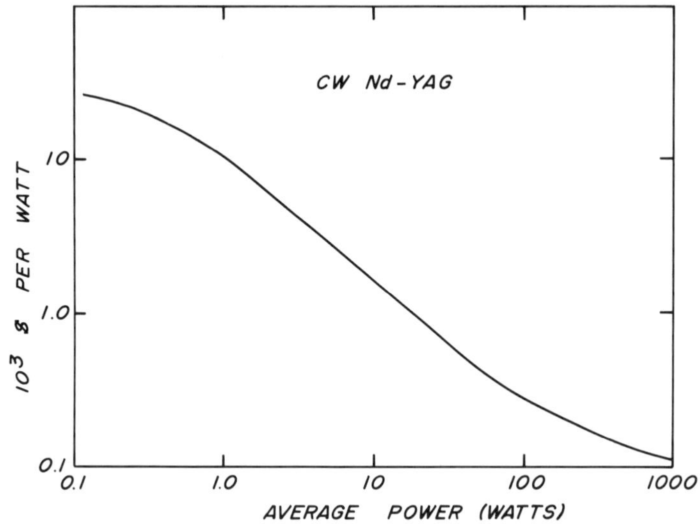

Figure 1.4. Cost per watt of output power in CW Nd–YAG laser vs. output power.

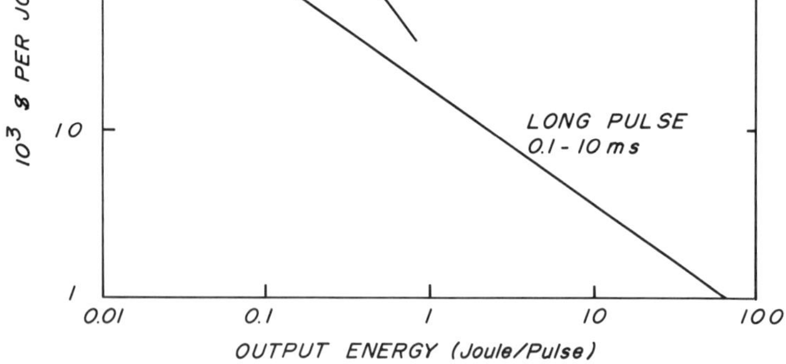

Figure 1.5. Cost per joule of output for Nd–YAG laser vs. output energy.

Table 1.5. Properties of Nd–Glass Laser Materials at 300 K[a]

Property	Silicate glass	Phosphate glass
Nd^{3+} concentration	2.8×10^{20} cm^{-3}	3.2×10^{20} cm^{-3}
Wavelength	1061 nm	1054 nm
Stimulated emission cross section	2.8×10^{-20} cm^2	3.9×10^{-20} cm^2
Inversion for 1% gain per cm length	3.6×10^{17} cm^{-3}	2.6×10^{17} cm^{-3}
Stored energy for 1% gain per cm length	0.07 J cm^{-3}	0.05 J cm^{-3}
Loss coefficient	~0.005 cm^{-1}	~0.001 cm^{-1}
Fluorescent lifetime	300 μs	290 μs
Fluorescent linewidth	26 nm	19 nm

[a]After Weber (1979).

A summary of the optical properties of Nd–glass lasers is given in Table 1.5 while output characteristics are listed in Table 1.6. It can be seen that the Nd–glass laser can be operated in long-pulse, *Q*-switched, or mode-locked configurations. With mode-locked operation and pulse selection, individual picosecond pulses can be switched out of the laser cavity. These can be amplified by passing through one or more amplifier stages. Peak pulse powers of 10^{12}–10^{13} W are obtained in such systems.

The flexibility inherent in glass fabrication has resulted in the construction of thin-film Nd–glass amplifiers and fiber lasers with lengths of several tens of meters. This offers interesting possible applications in integrated optics.

Table 1.6. Output Characteristics of Nd–Glass Lasers

Long pulse: 1–10 ns	
Output energy	1000 J/pulse (multimode)
Repetition rate	$\lesssim$1 ppm
Divergence	5–10 mrad
Beam diameter	0.5–1 cm
***Q*-switched:** 20 ns	
Output energy	100 J/pulse (multimode)
Repetition rate	$\lesssim$1 ppm
Divergence	0.5–5 mrad
Beam diameter	0.2–2 cm
Mode-locked: 10 ps	
Output energy	0.01–0.1 J/pulse (TEM_{00})
Repetition rate	$\lesssim$1 ppm
Divergence	0.2–0.3 mrad
Beam diameter	0.2–1 cm

1.2.4. Tunable Infrared Diode Lasers

Tunable coherent infrared radiation can be obtained from a variety of semiconductor diode lasers. The wavelength range covered by several important semiconductor laser materials is shown in Figure 1.6. The materials shown are all Pb salts although other compounds have been shown to emit in the regions shown [Melngailis and Mooradian (1975)].

The structure of a Pb-salt diode laser junction is shown in Figure 1.7. Emission occurs when a high current is passed through the junction formed between *p*- and *n*-type material. This emission is confined by the stripe geometry to the stripe region and thus appears as a beam from the end of the stripe. This results in operation in the fundamental mode for such geometry. Near-field emission patterns show an approximate similarity to the cross-sectional shape of the stripe. However, emission in the far-field shows a circular cross section. The beam divergence for such devices is often ~1 rad.

The composition in a compound such as $Pb_{1-x}Sn_xSe$ determines the width of the energy gap between the conduction and valence bands and hence the nominal wavelength range in which the laser operates. As an example, the bandgap of PbTe is 0.19 eV at 4 K. By incorporating Sn, the bandgap energy of $Pb_{1-x}Sn_xTe$ can be varied from 0.19 eV to ~0.04 eV. Since the bandgap

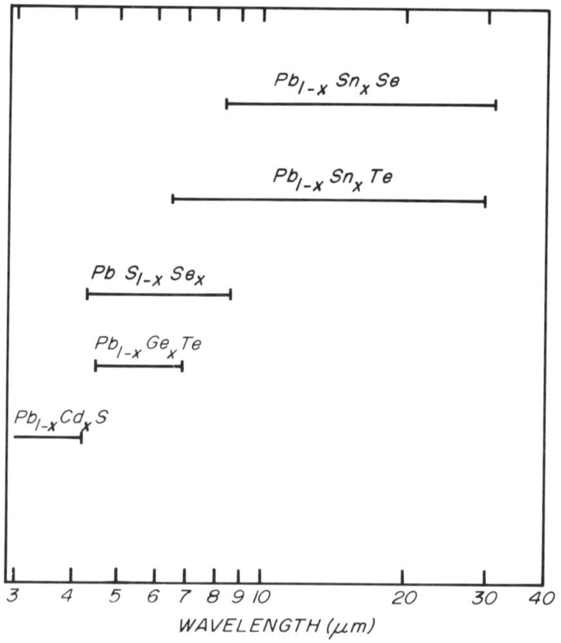

Figure 1.6. Tuning range for infrared diode lasers.

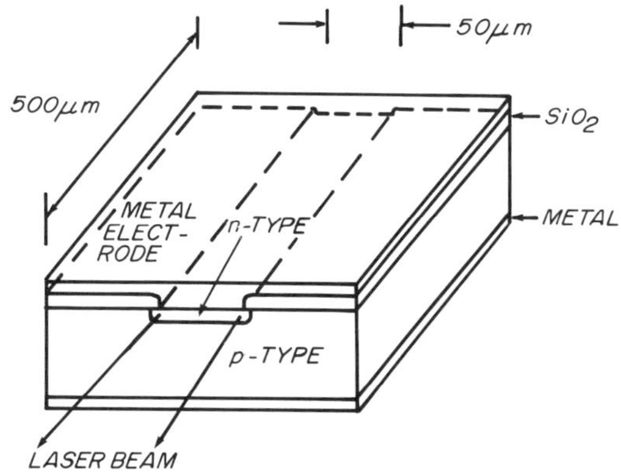

Figure 1.7. Construction of diode laser junction.

energy is determined primarily by composition in a particular device, only a relatively small tuning range is usually possible for each diode.

The output wavelength of these diodes can be tuned by varying any electrical or physical parameter that affects the energy of the bandgap. Common techniques involve tuning by variation of hydrostatic pressure and temperature or by the application of a magnetic field. The first two of these affect the bandgap energy directly, while magnetic field tuning splits energy levels in the valence and conduction bands.

The customary mode of operation of commercial tunable diode spectrometers involves temperature tuning by means of current variation. The diode is attached to a variable temperature cryostat that enables the bias temperature to be set at some fixed point between 4 and 77 K. When the temperature is stabilized at this point, a variation in diode current heats the junction and increases the bandgap energy. This causes the emission wavelength to decrease.

The tuning range of these devices is often covered discontinuously by a particular diode. This effect can be seen in Figure 1.8. The variation of laser output energy with injection current is found to be continuous only over small regions ($\sim$1–5 cm^{-1}) of the overall tuning range. These are separated by gaps caused by mode hopping. Hopping arises when the heating effect causes a change in refractive index altering the effective cavity length. In some cases, modes may overlap, permitting tuning over a somewhat wider spectral range, although this does not usually happen over an extended range and various gaps will occur over which given lasers will not emit.

Within the range of continuous tunability, the diode laser provides extremely high spectral resolution. Linewidths may be as small as 10^{-5}–10^{-6} cm^{-1}. Overall tuning ranges are 50–300 cm^{-1}.

The CW output power of these devices is typically several milliwatts when operated multimode although outputs of $\sim$30 mW have been reported. Opti-

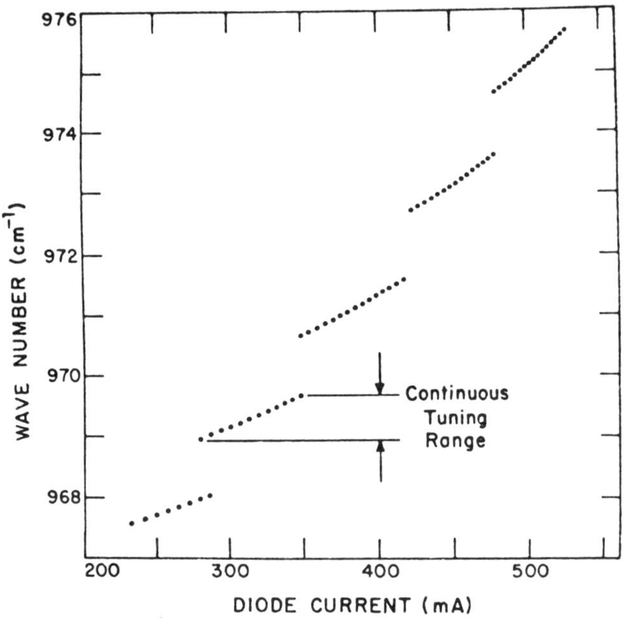

Figure 1.8. Fine tuning of modes from $Pb_xSn_{1-x}Te$ diode laser as a function of injection current. Tuning is continuous within a mode. (Melngailis and Mooradian, 1975, by permission of Addison-Wesley Publishing Co.)

cally coating the ends of the diode can suppress some of this mode structure as can operation near threshold. Under single-mode conditions, the CW output power is typically 0.1–1 mW. An example of the infrared spectroscopic capabilities of a diode laser spectrometer is shown in Figure 1.9. [Forrest (1980)]. This shows the absorption spectrum of a sample of cigarette smoke in the region near 1554 cm^{-1}.

1.2.5. Helium–Neon Laser

Laser emission is the He–Ne laser occurs from excited levels of neutral Ne atoms that have been populated by collision with excited He atoms. The relevant energy levels are shown in Figure 1.10.

The three dominant laser lines are at 632.8, 1152.3, and 3391.2 nm. The line at 1152 nm was the first observed from a neutral gas laser. Nowadays most He–Ne lasers are constructed to emit at 632.8 nm. The wavelength at which oscillation occurs can be controlled by variation of the coating applied to the laser mirrors or by application of an external inhomogeneous magnetic field. It

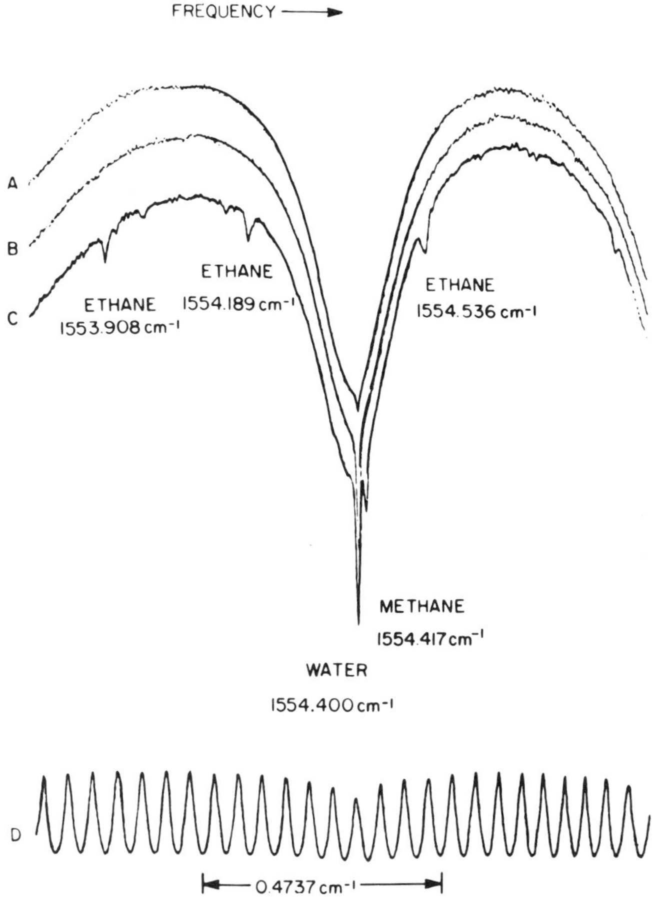

Figure 1.9. Transmission spectra in the 1553-cm^{-1} region. Samples of 50 cm^3 of either cigarette smoke (curve C) or room air containing 11.73 mm H_2O (curve B) were examined at a cell pressure of 12 torr. Lock-in amplifier gain was 25 mV with a 0.3-s time constant. Scan rate was 0.0077 cm^{-1} s^{-1}. The background (curve A) and etalon calibration spectra (curve D) are also shown [Forrest (1980)].

is particularly necessary to suppress oscillation on the 3391-nm transition if operation at 632.8 nm is desired since a high cavity intensity in the 3391-nm line tends to deplete the population in the upper state for the 632.8-nm line.

The He–Ne laser is usually operated in a mixture of He and Ne with proportions of 10:1 and at a total pressure of up to several torrs. Under these conditions, the spontaneous linewidth of the 632.8-nm line is limited by the

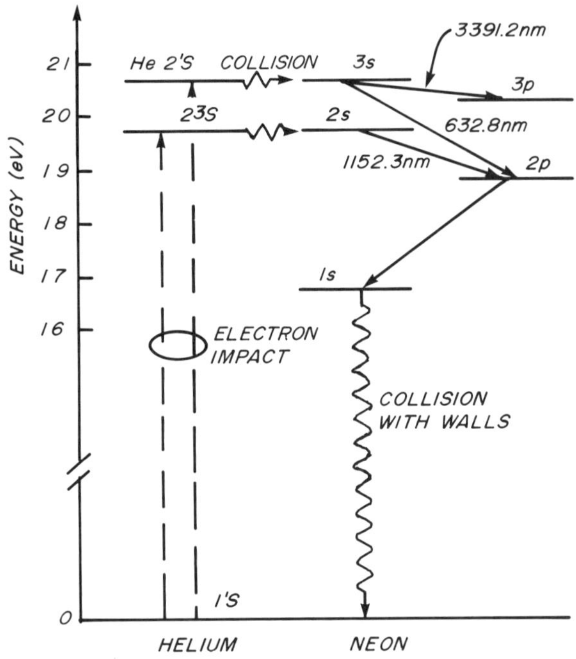

Figure 1.10. Outline of energy levels involved in the He–Ne laser. Although single states are shown for Ne, the levels actually contain many states.

Doppler width at ~300 K, the ambient temperature of the laser tube. A summary of operating parameters is given in Table 1.7.

Almost all He–Ne lasers that are commercially available are of sealed tube construction. These are produced on an assembly line basis using automated gas filling and processing procedures. Early models often had limited

Table 1.7. Operating Parameters for the Conventional He–Ne Laser[a]

Wavelength	632.8 nm
Efficiency	0.1%
Output power	50 mW m^{-1}
Power in discharge	50 W m^{-1}
Field strength × tube diameter	90 V mm cm^{-1}
Saturation parameter	0.3 W mm^{-2}
Current/tube diameter	6 mA mm^{-1}
Pressure × tube diameter	5 torr mm
Doppler width	1.5 GHz

[a]After Goldsborough (1972).

Table 1.8. Output Characteristics of Commercial He–Ne Lasers

Output power	1 mW	50 mW
Beam diameter	0.05–0.08 cm	0.2 cm
Beam divergence	1 mrad	0.5 mrad
Cost (\$/mW)	150–200	~200
Features	Various industrial packages for measurement, inspection surveying, etc.	Other wavelengths available

lifetime. However, improvements in cathode structure and composition and attention to gettering has increased the lifetime of a typical 1-mW He–Ne laser to in excess of 10,000 h. Over this period the laser power may drift by 10–12%. Short term power fluctuations are typically a few percent. Table 1.8 lists the output characteristics of some typical commercial He–Ne lasers.

1.2.6. Argon and Krypton Ion Lasers

These lasers provide the highest CW laser output power currently available in the visible and near ultraviolet regions of the spectrum. Powers of 10–20 W in the 500-nm region and 1–2 W near 350 nm are commonly available from commercial devices. However, these powers are achieved only with the expenditure of considerable amounts of electrical energy since the efficiency of inert gas ion lasers is rarely more than 0.1%. Thus a 10-W all lines Ar^+ laser might consume ~10 kW of line power. The dissipation of this heat places stringent constraints on the design of inert gas ion discharge tubes [Davis and King (1975)].

A partial energy level diagram for some of the transitions involved in the Ar^+ laser is shown in Figure 1.11. Emission of the laser lines between 454.5 and 528.7 nm occurs when the $4p^2$ and $4p^4$ levels are populated by electron excitation from the ground state of the Ar^+ ion or from excited metastable Ar^+ states. Population is also transferred by radiative cascade from excited Ar^+ states. One notes that each 3-eV quantum emitted in the laser line is the result of ionizing a neutral Ar atom and then exciting the Ar^+ ion to the states near 20 eV. Thus the overall quantum efficiency is small and equals $3/36 \sim 0.08$. This should be compared to ~0.4 for the CO_2 laser.

A list of the wavelengths of the strongest lines emitted from Ar^+ and Kr^+ lasers is given in Table 1.9. Wavelengths have been taken from the review by Davis and King (1975) and are quoted in air. Lines other than those listed in the table can also be obtained, however, these tend to be weak. Operation of the Kr^+ laser in the 860-nm region is possible. Some operating parameters for the CW Ar^+ laser are given in Table 1.10. These would be characteristic of a device with relatively modest output power. Laboratory and prototype Ar^+ las-

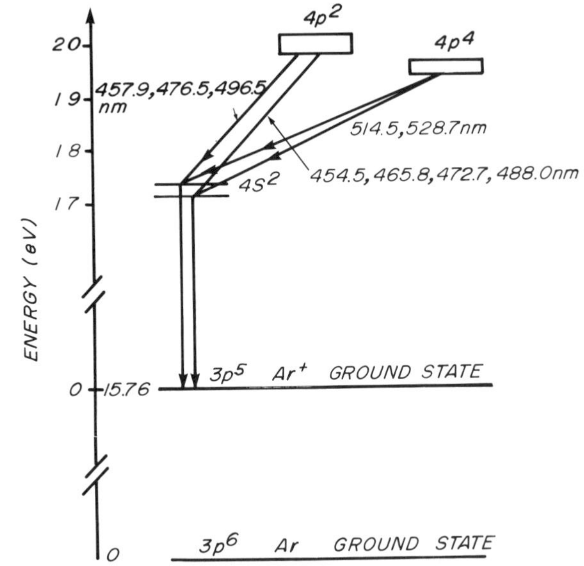

Figure 1.11. Energy levels involved in some transitions of the Ar^+ laser.

ers would have substantially different values of these parameters. For example, CW powers of ~500 W have been obtained from large-bore, low-pressure tubes. In such tubes, the power per unit length parameter may be ~100 W m^{-1}. Much higher currents are also used and the current density may be ~ 10^3 A cm^{-2}. Efficiencies are also increased.

Table 1.9. Wavelengths of Strongest Emission Lines for Ar^+ and Kr^+ Lasers[a]

Argon	
351.119 ± 0.004 nm (air)	476.488 ± 0.004
363.786 ± 0.004	487.986 ± 0.004
454.504 ± 0.01	496.509 ± 0.002
457.936 ± 0.016	501.717 ± 0.002
465.795 ± 0.002	514.533 ± 0.002
Krypton	
350.739 ± 0.005 nm (air)	482.518 ± 0.006
356.420 ± 0.006	568.192 ± 0.004
406.736 ± 0.006	647.10 ± 0.05
413.138 ± 0.006	676.457 ± 0.01
461.917 ± 0.01	752.55 ± 0.01
468.045 ± 0.006	793.14
476.244 ± 0.006	799.30 ± 0.05

[a]After Davis and King (1975).

Table 1.10. Operating Parameters for CW Ar^+ Laser[a]

Parameter	514.5 nm	488.0 nm
Output power/unit length	5 W m^{-1}	5 W m^{-1}
Discharge power input	250 W cm^{-1}	250 W cm^{-1}
Efficiency	0.02%	0.02%
Discharge electric field × tube diameter	10 V mm cm^{-1}	10 V mm cm^{-1}
Saturation parameter	7 W mm^{-1}	2 W mm^{-2}
Current/tube diameter	25 A mm^{-1}	25 A mm^{-1}
Pressure × tube diameter	0.25 torr mm	0.25 torr mm
Lorentzian linewidth	500–600 MHz	400–500 MHz
RMS noise 1 s–1 h	0.2%	
1 s–10 h	0.5%	

[a]After Goldsborough (1972), with additions.

A summary of the output characteristics of available CW Ar^+ and Kr^+ lasers is given in Table 1.11.

CW ion lasers can be stabilized to operate in a single axial mode providing single-frequency operation. With a Doppler limited linewidth of ~5–10 GHz, a 1-m laser will have several axial modes within the line profile. Without a mechanism that selects one of these modes the laser wavelength shifts discontinuously as the output shifts from one mode to another. A variety of techniques have been developed to suppress this instability and yield single-frequency oscillation [Davis and King (1975)]. The frequency drift in single-frequency operation can be ~100 MHz/h.

The ~10-GHz linewidth of the transitions in ion lasers means that mode-locking will produce pulse widths of 100 ps. Mode-locking has been performed with both Ar^+ and Kr^+ lasers. Typically, one obtains a pulse repetition frequency of ~150 MHz with a pulse width of 100 ps and peak pulse power of ~1 kW.

Using an intracavity modulator, cavity dumping is possible with the CW

Table 1.11. Output Characteristics of Commercial Ar^+ and Kr^+ Lasers

Characteristic	Argon		Krypton
Output power	2 W	20 W	1 W
Beam diameter	0.1–0.2 cm	0.2 cm	0.16
Beam divergence	0.6 mrad	0.6 mrad	0.6
Cost ($/W)	4–5000	1–2000	20–30,000
Features available	*a*	*b*	*c*

[a]Air-cooled operation, compact design; [b]Automatic filling, UV lines available; [c]High power on red lines.

Ar^+ laser. This can yield narrow pulses with repetition rates of $\gtrsim 1$ MHz. The peak power of these pulses can be ~ 100 times the CW power while average power is comparable to the CW average power.

1.2.7. Helium–Cadmium Laser

Excitation in the Ne–Cd laser occurs via collisions with He metastables and He^+ ions. The processes are

$$\mathrm{He}^*(2s\,{}^3S_1) + \mathrm{Cd} \rightarrow \mathrm{He} + \mathrm{Cd}^+ \quad (5s^2\,{}^2D) + e^-$$
$$\mathrm{He}^+ + \mathrm{Cd} \rightarrow \mathrm{He} + \mathrm{Cd}^+ \quad (9s,\, 6f,\, 6g)$$

The first of these leads to pumping of the $5s^2\,{}^2D_{3/2}$ and ${}^2D_{5/2}$ levels that give rise to laser transitions at 325 and 441.6 nm, respectively. The laser transitions are

$$325\ \mathrm{nm} \qquad 4d^95s^2\,{}^2D_{3/2} \rightarrow 4d^{10}5s5p\,{}^2P^0_{1/2}$$
$$441.6\ \mathrm{nm} \qquad 4d^95s^2\,{}^2D_{5/2} \rightarrow 4d^{10}5s5p\,{}^2P^0_{3/2}$$

These energy levels are shown in Figure 1.12.

To obtain laser emission, a CW discharge is operated in an axial configuration with a He pressure of 4–6 torr and a Cd pressure of ~ 2 mtorr. Difficulties arise due to cataphoresis that tends to cause Cd vapor to be swept toward the cathode. This can result in the coating of one laser window with metallic Cd. A variety of tube designs have been developed to reduce this prob-

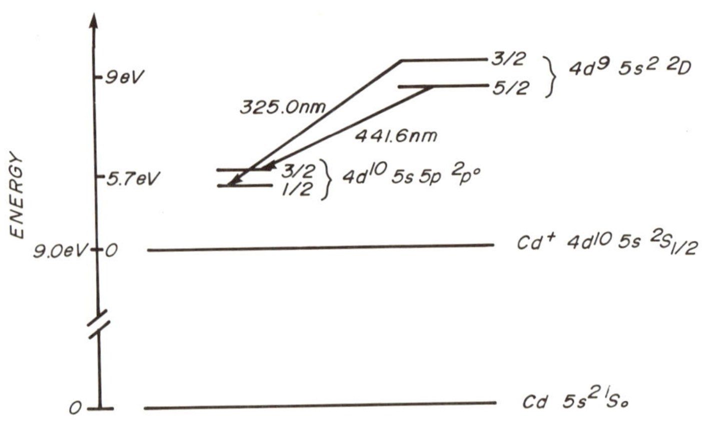

Figure 1.12. Energy levels in the He–Cd^+ laser.

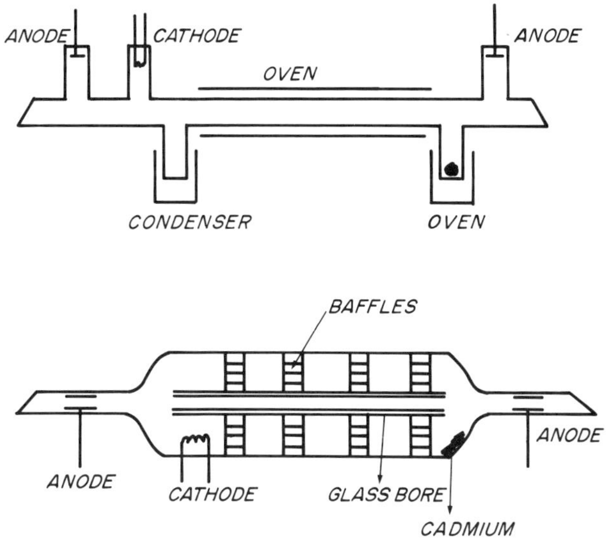

Figure 1.13. Two discharge tube configurations for the He–Cd$^+$ laser.

lem [Bridges (1979)]. Two of these are shown in Figure 1.13. In the first of these, a reservoir for Cd is provided at the cathode end and a second anode maintains a discharge that keeps the window at that end clear. The second configuration provides for continuous recirculation of Cd vapor. Vapor flow and depletion limits the lifetime of He–Cd$^+$ tubes. However, commercial He–Cd$^+$ laser tubes now have lifetimes in excess of 5000 h. Performance data for the He–Cd$^+$ laser is summarized below.

Output power	325 nm	1–8 mW TEM_{00}
		10 mW multimode
	441.6 nm	10–40 mW TEM_{00}
		60 mW multimode
Beam diameter		0.1 cm
Beam divergence		0.5–1.5 mrad

1.2.8. CO_2 Laser

One of the highest efficiencies for the conversion of electrical to optical energy is found in the CO_2 laser. This property, together with ease of operation at high power levels in both CW and pulsed configurations, has made the CO_2 laser widely applicable as a source in the laser processing of materials.

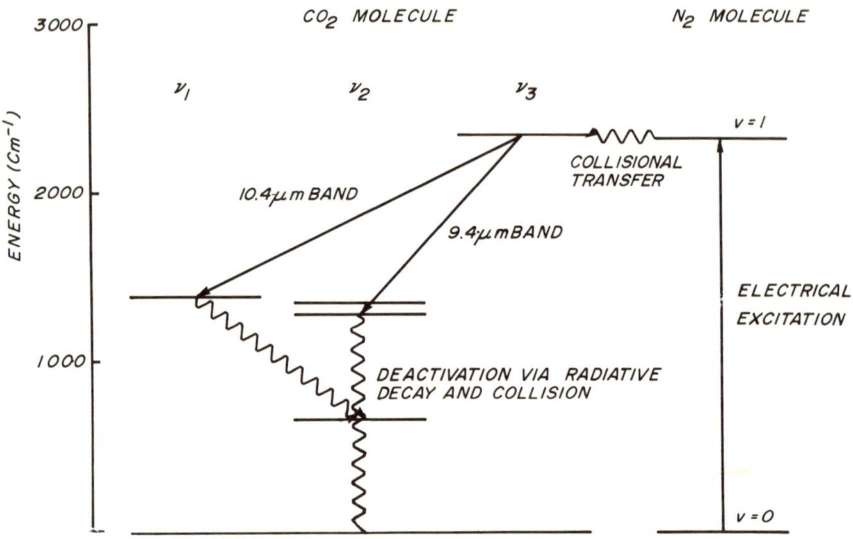

Figure 1.14. Energy levels for N_2 and CO_2 in the CO_2 laser.

Laser emission in the CO_2 laser derives from rotational lines in the 9.4- and 10.4-μm bands of the CO_2 molecule (Figure 1.14). Since many lines have similar gains under typical operating conditions, the CO_2 laser can be operated at a large number of wavelengths within the envelope of the 9.4- and 10.4-μm bands. These lines can be selected by a wavelength dispersive element (e.g., a grating) in the laser cavity. With no wavelength selection, operation occurs mainly on the $P(18)$, $P(20)$, and $P(22)$ lines of the 10.4-μm band. These lines occur near 10.6 μm; thus the output wavelength of the CO_2 laser is usually quoted as 10.6 μm.

Pumping in the CO_2 laser occurs directly by electron impact and by collision with vibrationally excited N_2. The CO_2 laser discharge is normally operated in a mixture of CO_2, N_2, and He gas at pressures of up to $\sim$1 atm. Relative concentrations of the three gas components must be chosen to optimize emission in a particular device.

Energy levels in the CO_2 laser involved in laser emission are shown in Figure 1.15. Emission occurs when the CO_2 molecule changes its vibrational state from ν_3 to either ν_1 or ν_2 accompanied by a change in rotational state. These processes are shown schematically in Figure 1.15. Frequencies and wavelengths of the corresponding emission lines are tabulated in Chang (1970), Freed *et al.* (1974), and Duley (1976).

The CO_2 laser can be operated in CW and a variety of pulsed modes. A description of CO_2 laser devices operating in these regions is given in the following sections.

1.2.8.1. CW CO_2 Laser. The CO_2 laser was first operated as a CW device, and it was only after a delay of several years that operation under true gain-switched pulsed conditions was obtained. The conventional CW CO_2 laser operates with a longitudinal electrical discharge through flowing gas in an axial tube configuration. All commercial CW CO_2 lasers with powers $\lesssim$800 W operate in this mode. Electrical pulsing can be used to obtain trains of millisecond or microsecond pulses with peak powers that are $\sim 10^2$ larger than the CW power. Electro-optical or acousto-optical intracavity modulation can be used with CW CO_2 lasers to produce trains of nanosecond pulses with peak power of $\sim$10 kW.

Operation under axial flow conditions at low pressure (10–50 torr) typically yields output powers of 50–100 W m^{-1} with the larger value being obtained with shorter tubes. The reduction in the power per unit length factor that occurs with an increase in tube length makes it unattractive to increase tube length beyond about 10 m. Most commercial CW lasers with powers in the 500–1000 W range use a folded tube configuration to achieve a relatively large tube length in a small volume. Some configurations are shown in Figure 1.16.

Higher CW laser powers can be obtained in axial flow configurations by an increase in gas pressure and gas flow rate. This results from a change to convective gas cooling from the diffusive cooling that ordinarily occurs in low-pressure, low-flow devices. An example of the advantages of fast axial flow can be seen in the 2-kW laser marketed by Control Laser. An output of up to 2 kW is obtained from $\sim$4 m of discharge tube. This represents a power per unit length ratio of 500 W m^{-1}.

The highest CW CO_2 laser powers are obtained from fast gas transport

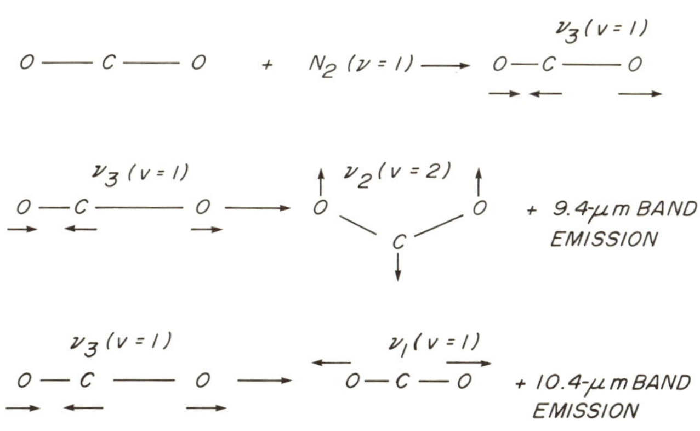

Figure 1.15. Schematic diagram of changes in CO_2 vibrational state during excitation and laser emission. ν_1, symmetric stretching mode; ν_2, bending mode; ν_3, antisymmetric stretching mode. Changes in rotational state also occur but are not shown.

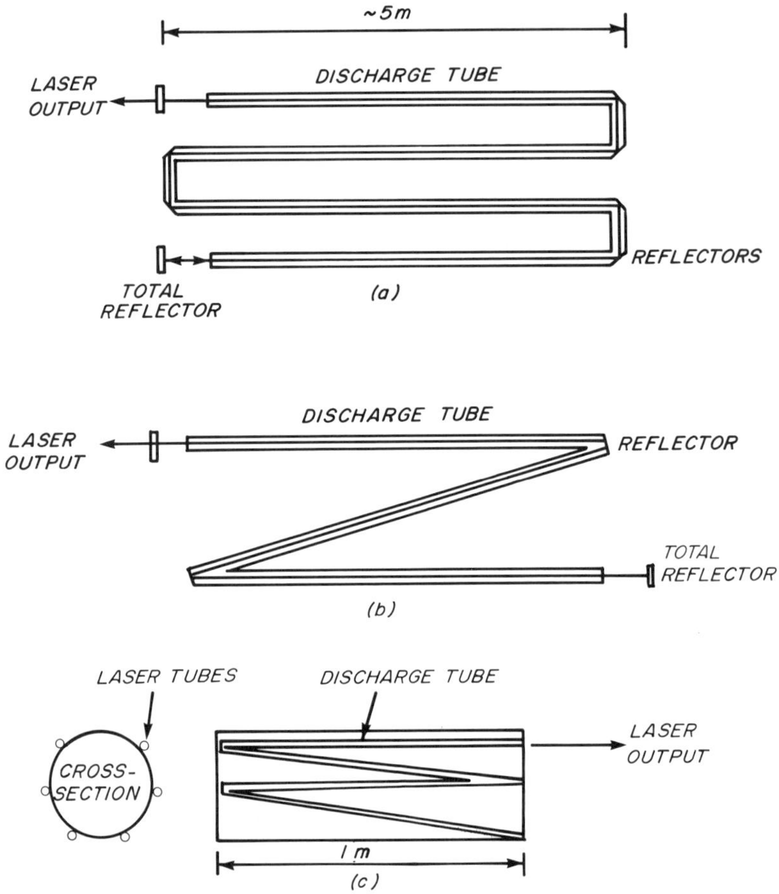

Figure 1.16. Folded-tube configurations for axial flow CW CO_2 lasers. (a) Coherent system: Each discharge tube has separate excitation and gas flow. (b) Photon system: Each section is separately excited. (c) Ferranti system: Folded-tube configuration around the exterior of a cylinder. Only a few of many ~1-m discharge tubes are shown.

and gas dynamic devices. In the gas transport laser, a blower circulates gas rapidly transverse to the long dimension of the laser cavity. The discharge can be operated either along the direction of the optical cavity or transverse to it. CW outputs of up to ~30 kW have been reported and a system with 15-kW output is available commercially (Avco Everett HPL).

The gas dynamic laser operates on the population inversion that can be produced under certain conditions downstream from a nozzle through which the gas has undergone a supersonic expansion. A description of the mechanism by which the population inversion is obtained has been described elsewhere [Gerry (1970), Hertzberg *et al.* (1972), Anderson (1976), and Duley (1976)]. These devices can produce in excess of 100 kW. However, they are not presently available as commercial lasers.

The smallest CO_2 lasers operate with the discharge confined to the interior of a waveguide. Bore diameters are typically 1 mm and the gas pressure is $\sim$100 torr. Since collisions with the walls of the waveguide is efficient in gas cooling under these conditions, high gain is possible and relatively large output powers can be obtained from small devices.

The output mode of waveguide lasers is determined primarily by the shape of the waveguide. With most devices the output beam is essentially TEM_{00}. High operating pressure and the output stability characteristic of waveguide operation facilitates frequency tuning over a wide range. Tunability over 1.2 GHz has been reported [Abrams (1974)].

Low-pressure CW CO_2 lasers can be operated in a sealed-off configuration if provision is made for the catalytic conversion and gettering of deleterious discharge products. CW output powers of up to 10 W are readily available from sealed-off CO_2 lasers. Lifetimes of 1 year are typical. Sealed-off waveguide CO_2 lasers are also commercially available. CW powers of $\sim$30 W have been obtained.

A summary of the output characteristics of commercially available CW $_2$ lasers is given in Table 1.12. These data are representative of that obtained

Table 1.12. Output Characteristics of Commercial CW CO_2 Lasers

	Conventional axial				
	Output power	Beam divergence	Beam diameter	Capital cost (per watt)	Available features
Conventional axial flow	10 W	1–5 mrad	0.5 cm	\$500–1000	Sealed operation, high stability, tunable
	10–100 W	1–3 mrad	0.5–1 cm	\$100–300	Gas recycling, pulsed operation, tunable
	100–1000 W	1–2 mrad	0.5–1 cm	$\sim$\$200	Gas recycling, pulsed operation
Fast axial flow	2 kW	1.2 mrad	1.3 cm	$\sim$\$100	Pulsing, gas recycling
Gas transport	6–15 kW	$\sim$1 mrad	7 cm	—	—
Waveguide	$\lesssim$30 W	10 mrad	0.1 cm	\$2000–10,000	Tuning range, line tunable, pulsing, sealed operation

from off the shelf devices. Laboratory or prototype devices may have much superior characteristics in certain categories.

1.2.8.2. TEA Laser. The transversely excited atmospheric pressure (TEA) CO_2 laser was first developed by Beaulieu (1970). In this device the electrical discharge is applied transverse to the optical axis. This permits operation at relatively high gas pressures (several atmospheres) with voltages of less than 100 kV. The voltage pulse applied transversely to the gas is of sufficiently short duration that ionization is produced during the pulse but the voltage is not sustained for long enough to precipitate gas breakdown and arcing. In this way, a large volume of electrically excited gas containing CO_2 molecules in the $v = 1$, ν_3 level can be produced.

TEA lasers exhibit the gain-switching or giant pulse phenomenon. Rapid pumping of the upper laser level ensures that an appreciable population inversion is present when the electromagnetic field intensity builds up to the point where laser output first occurs after initiation of the electrical pulse. Since the gain is then high, a rapid increase in output intensity follows. This generates an output pulse of 100–200 ns duration containing up to ~ 75% of the energy stored in the medium.

The gain-switched pulse equalizes populations in the upper and lower energy levels. Subsequent pumping via collision with excited N_2 molecules reestablishes a population inversion and leads to further emission of a low peak power laser pulse that can last several microseconds. The result of these two effects is the characteristic "double" pulse shape shown in Figure 1.17.

TEA lasers can generate output pulses with energies of up to $\sim 2 \times 10^3$ J. Repetition rates of 200 Hz are readily obtained with pulse energies of ~ 1 J. Operation at over 1 kHz has also been achieved. Indeed, one prototype TEA

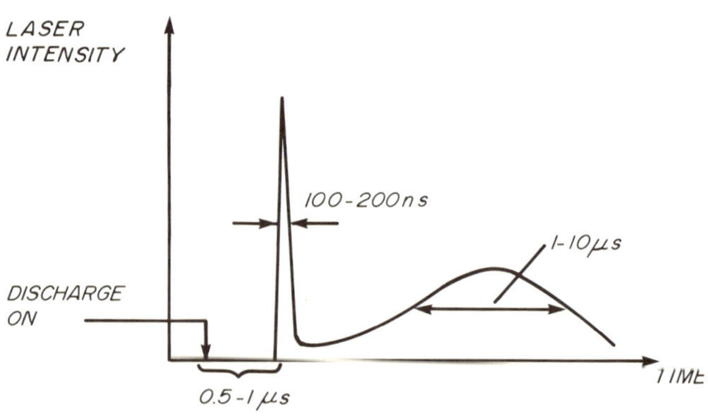

Figure 1.17. Schematic of TEA laser output. The initial gain-switched pulse may have many times the peak power of the subsequent broad pulse. However, both pulses can contain comparable energies.

Table 1.13. Output Characteristics of Commercial TEA-Type CO_2 Lasers

	Low energy	High energy	High average power
Output energy	1–3 J/pulse	1–2 kJ/pulse	0.5–1 J/pulse
Repetition rate	<10 Hz	0.05 Hz	300–600 Hz
Divergence	0.5 mrad	0.5 mrad	5 mrad
Beam diameter	2–3 cm	20 cm	2.5
Cost ($/joule)	6000	150	200[a]

[a]Dollars per watt of average power.

laser produces ~1.5 kW of average power at a repetition rate of 1–2 kHz. A summary of the output characteristics of some commercial TEA type CO_2 lasers is given in Table 1.13.

The TEA laser output can be mode-locked. Active, passive, and self-mode-locking have been reported [Duley (1976)]. Mode-locked pulses typically have durations of 1–2 ns, although operation at pressures at or above 1 atm can yield pulses with 100–200 ps duration.

Broadening of lines of the 9.4- and 10.4-μm bands at pressures approaching 10 atm forms a quasicontinuum between 9 and 11 μm. When operating under these conditions the TEA laser can be used as a source of tunable infrared radiation within this spectral range. Such devices can also be used as amplifiers for picosecond pulses.

1.2.8.3. Electron-Beam Controlled CO_2 Lasers. High pulse energies can be obtained from pulsed CO_2 laser devices in which the source of ionization and excitation are decoupled. This occurs in the electron-beam controlled laser where direct deposition of preionization is accomplished by a percursor pulse that sends a high current of ~100-keV energy electrons through the laser mixture. The subsequent deposition of pump energy can then be tailored in time and amplitude to maximize gain in the excited volume.

The result is that E-beam controlled CO_2 lasers generate large pulse energies over relatively long pulse lengths. Since such systems scale readily to large volumes, enormous pulse energies become available. Typical data for an E-beam controlled CO_2 device are:

Output pulse energy	10^3–10^4 J
Repetition rate	1 ppm
Pulse duration	2 μs
Beam diameter	35 cm
Beam divergence	0.3 mrad

1.2.9. Rare Gas Halide Lasers

The past five years have seen the rapid development of pulsed lasers based on transitions of rare gas halide (RGH) molecules. RGH lasers are one class of a group of lasers deriving from excimer emission by unstable molecules. In the excimer, laser stimulated emission occurs from a strongly bound excited state to a weakly bound ground state. Since the ground state of the excimer molecule may not be significantly populated under discharge conditions, a substantial population inversion can be established when molecules are formed in the excited state.

The relevant energy levels involved in the KrF RGH laser are shown in Figure 1.18. The minimum of the excited $^2\Sigma$ potential energy curve occurs at an energy of ~5 eV. Since the transition involves molecular states, fluorescent emission occurs over a range of wavelengths. This range can be as large as 2 nm.

The most important commercial RGH lasers are based on emission from the molecules ArF, KrF, KrCl, XeBr, XeCl, and XeF. Table 1.14 lists output wavelengths for these lasers together with performance parameters. The combination of high repetition rate and pulse energy results in average powers that are typically 10–20 W. The development of dye lasers compatible with RGH laser pumping offers the possibility of the extension of the tuning range of dye lasers into the 320–360 nm region or shorter wavelengths.

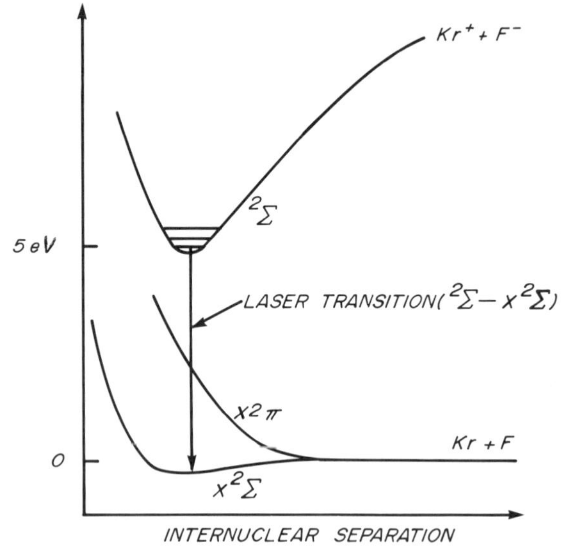

Figure 1.18. Energy levels in the KrF rare gas halide laser transition. The laser transition occurs at 248–249 nm and corresponds to emission from the bound $^2\Sigma$ state formed from $Kr^+ + F^-$ ions to the thermally unstable $X\,^2\Sigma$ state that dissociates to neutral ground-state atoms.

Table 1.14. Commercial Rare Gas Halide Lasers

	ArF	KrCl	KrF	XeBr	XeCl	XeF
Wavelength (nm)	193	222	248	282	308	351
Output energy (J/pulse)	0.2	0.05	0.25	0.01	0.08	0.08
Repetition rate (Hz)	100	100	150	100	150	100
Pulse width (ns)	10–20	5–15	10–20	7–20	4–20	10–20
Beam size			1 × 2.5 cm			
Beam divergence			2 × 5 mrad			

1.2.10. Dye Lasers

The availability of tunable coherent radiation in the wavelength range between ~0.2 and 1.0 μm has revolutionized spectroscopy and many areas of photochemistry. Tunability over this wavelength range is available because of the unique spectral properties of several groups of large organic dye molecules. These molecules have the property that they absorb efficiently over a wide spectral range and reemit this energy over a broad band at longer wavelengths.

Figure 1.19 shows a generalized energy level diagram for a typical dye

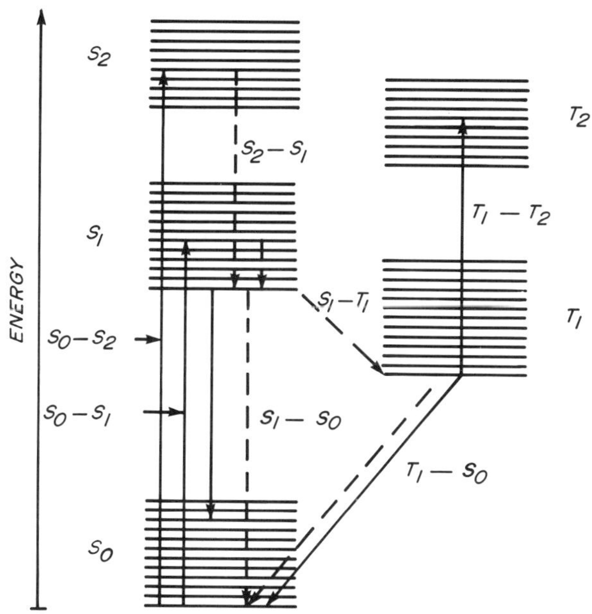

Figure 1.19. Energy levels in a dye molecule. The continuum of states in each electronic level results from overlapping vibrational and rotational states. The S_1 state is typically 15,000–30,000 cm^{-1} above the ground S_0 state. See Table 1.15 for a summary of individual excitation and deexcitation processes.

Table 1.15. Excitation and Deexcitation Processes in Dye Molecules

Transition	Type	Rate
	Radiative	
$S_0 \rightarrow S_1$	Singlet–singlet absorption	
$S_0 \rightarrow S_2$	Absorption to second singlet level	
$T_1 \rightarrow T_2$	Triplet–triplet absorption	
$S_1 \rightarrow S_0$	Fluorescence	$10^9\ s^{-1}$
$T_1 \rightarrow S_0$	Phosphorescence	$\lesssim 10^7\ s^{-1}$
	Nonradiative	
$S_1 \rightarrow S_0$	Internal conversion	
$S_2 \rightarrow S_1$	Internal conversion	$10^{11}\ s^{-1}$
$S_1^* \rightarrow S_1$	Internal conversion with S_1 state	$10^{11}\ s^{-1}$
$S_1 \rightarrow T_1$	Intersystem crossing	$\lesssim 10^7\ s^{-1}$
$T_1 \rightarrow S_0$	Intersystem crossing	

molecule which illustrates the processes involved in excitation and fluorescence. Identification of individual transitions is shown in Table 1.15.

Absorption of pumping radiation occurs via the $S_0 \rightarrow S_1$ and $S_0 \rightarrow S_2$ transitions. This excitation is efficiently transferred on time scales of $\sim 10^{-11}$ s to the lowest S_1 level via internal conversion. This can be followed by fluorescence to upper levels of the S_0 state with a lifetime $\sim 10^{-9}$ s or by transfer to the T_1 state via intersystem crossing. When the latter process occurs, population may accumulate in the T_1 state which is metastable and this can lead to absorption due to triplet–triplet transitions. This is usually undesirable as it represents a loss and lowers overall laser efficiency. To minimize this effect, the $T_1 \rightarrow S_1$ rate should be much smaller than the rate for the $S_1 \rightarrow S_0$ fluorescent transition. In addition, the T_1 lifetime should be as small as possible.

Fluorescence on the $S_1 \rightarrow S_0$ transition is shifted from the $S_0 \rightarrow S_1$

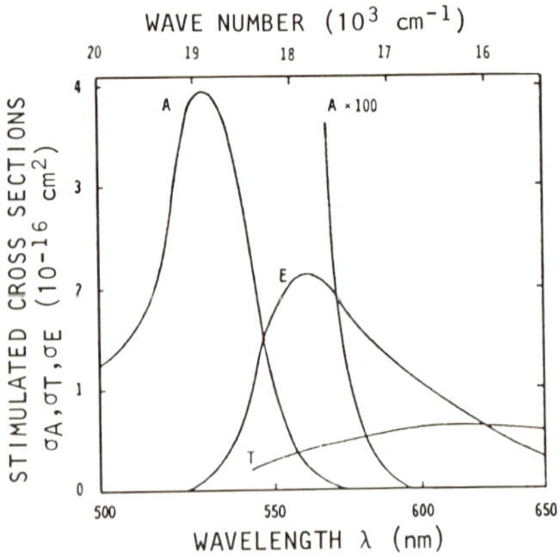

Figure 1.20. Singlet absorption, emission, and triplet absorption cross sections for rhodamine 6G [Dienes (1975)].

Table 1.16. Typical Pumping Regimes for Dye Lasers

Time scale	Pump source
~10 ns	Doubled Nd–YAG, N_2 laser, rare gas halide
~1 μs	Flashlamp
CW	Ar^+, Kr^+ lasers

absorption band and yields the broad band that can be used to obtain tunable laser emission. The operating range of a particular dye will depend on the spectral distribution of this emission in relation to the wavelength dependence of absorptive loss terms. These arise from triplet–triplet, $S_0^* \rightarrow S_1$ and $S_1 \rightarrow S_2$ transitions. The balance between these terms and their effect on the wavelength range of laser emission can be seen from Figure 1.20. This figure shows the wavelength dependence of the $S_0 \rightarrow S_1$, T_2, and $S_1 \rightarrow S_0$ cross sections for the dye rhodamine 6G. The normal operating range lies between 580 and 630 nm. The short-wavelength limit is determined by the tail of the $S_0 \rightarrow S_1$ absorption band which leads to absorption of the fluorescent emission. The long-wavelength limit arises from the decreasing efficiency for fluorescent emission for $\lambda > 630$ nm together with losses due to triplet–triplet absorption.

The tuning range for particular laser dyes depends on the time dependence of the pumping source. The three typical pumping regimes are given in Table 1.16. Triplet–triplet absorption is not usually a problem with 10-ns excitation since the singlet–triplet conversion is slow on a 10=ns time scale. On the other hand, appreciable triplet populations can build up with flashlamp and CW laser pumping. Thus the tuning range for a particular dye may be somewhat larger with short pulse excitation than with CW or quasi-CW (flashlamp pumped) excitation.

Only dyes with high fluorescent quantum efficiency are suitable as laser sources. Most common laser dyes belong to specific classes of chemical compounds. These compounds are listed in Table 1.17. Most laser dyes can be dissolved in alcohol solutions although some require exotic solvents for optimum efficiency. The wavelength of dye emission can often be varied substantially by changing the chemical properties of the solvent used. Indeed, emission wavelengths of many dyes can be changed significantly by a change in the acidity

Table 1.17. Chemical Compounds for Laser Dyes

Dye type	Fluorescent region
Coumarine, anthracene, oxazole	Violet, blue
Xanthene, acridine	Green, yellow, red
Polymethines	Red

Table 1.18. Properties of Laser Dyes[a]

Dye	Solvent	Tuning range (nm)	Pump source[b]
Butyl-PBD	Toluene	357–395	N_2
Coumarin 440	Ethanol	418–462	N_2
	Methanol	419–469	FL
	Ethylene glycol	427–477	Ar
Coumarin 450	Ethanol	427–473	N_2
	Methanol	427–488	FL
	Ethylene glycol	430–492	Ar
Coumarin 460	Ethanol	437–481	N_2
	Methanol	435–490	FL
	Ethylene glycol	446–506	Ar
Coumarin 580	Ethanol	451–498	N_2
	Methanol	456–531	FL
	Ethylene glycol	455–506	Ar
Coumarin 481F	Dioxane	480–522	N_2
	Ethanol	488–558	N_2
Coumarin 485F	Ethanol	490–563	N_2
Coumarin 500F	Ethanol	475–540	N_2
Coumarin 522F	Ethanol	500–572	FL
	MeOH/H_2O	515–570	FL
Coumarin 540	Methanol	516–562	FL
	Ethylene glycol	516–566	Ar
Coumarin 540AF	Dioxane	485–530	N_2
	Ethanol	518–570	N_2
Fluorol 555	Alcohol or water	530–608	FL
Rhodamine 560 (chloride)	Ethylene glycol	536–582	Ar
	Ethanol	535–620	Ar
Rhodamine 575	Alcohol	550–610	FL
Rhodamine 6G	Methanol	563–625	
	Alcohol	568–615	N_2
	Ethylene glycol	567–657	
Rhodamine B	Alcohol	595–645	N_2
	Ethanol	598–647	FL
	Ethylene glycol	608–682	Ar
Kiton Red 620	Ethanol	580–630	FL
	Methanol	600–660	FL
	Ethylene glycol	610–670	Ar
Rhodamine 640	Ethylene glycol	618–703	Ar
Cresyl violet	Alcohol + Rh6G	650–705	N_2
	Methanol	631–705	FL
	Ethylene glycol	650–696	Ar
	Methanol–water	620–670	Nd–YAG
Nile blue	Methanol	689–750	FL
	Ethylene glycol	692–782	Kr
Carbazine 720	Methanol	690–740	FL
	Aq. Ammonyx	680–760	Nd–YAG
Oxazine 720	Methanol	690–740	FL
Oxazine 725	Methanol/Rh6G	705–745	FL
	Ethylene glycol	690–780	Kr
	Ethanol/Rh6G	700–750	N_2

[a]Courtesy Exciton Corporation; [b]FL, flashlamp pumped.

of the solvent. A summary of data on laser dyes is given in Table 1.18. Tuning curves obtained on pumping with representative sources are shown in Figures 1.21–1.23.

1.2.10.1. N_2 Pumped Dye Lasers. The output of an N_2 laser is at 337.1 nm and pulse lengths are typically 5–10 ns. Output energies of 5 mJ per pulse are typical as are repetition rates of 50–100 Hz. The output beam is usually of rectangular cross section, making it ideal for transverse pumping of a dye in a dye cell as shown in Figure 1.24(a). This figure shows the elements of a basic N_2 pumped dye laser system. With an intracavity etalon, the emission linewidth from such a device would be typically 0.001 nm. A summary of N_2 pumped dye laser performance is given in Table 1.19. Figure 1.24(b) shows the simple N_2 pumped dye laser configuration developed by Littman and Metcalf (1978). This uses a grating at grazing incidence to obtain high spectral resolution (~0.001 nm) without the need for a telescope or etalon. When an intracavity etalon is used for line narrowing, the output can be scanned over a limited range by pressure tuning. This technique varies the refractive index in the etalon in order to change wavelength.

High peak power pulses are best obtained from oscillator–amplifier combinations rather than by increasing pump power in the oscillator stage. Since the dye laser pulse is typically of ~5-ns duration, the operation of oscillator–

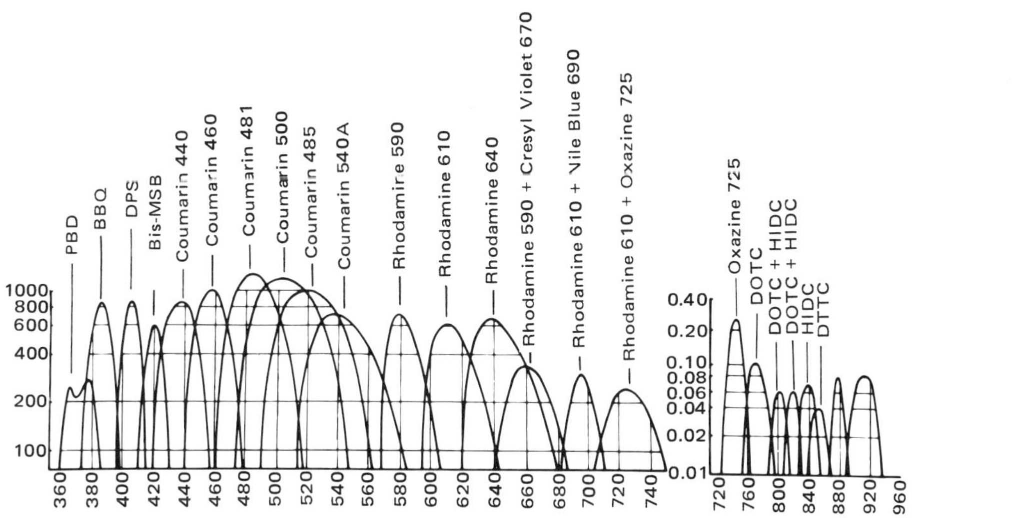

Figure 1.21. Tuning ranges for various laser dyes using N_2 pumped laser excitation. (Courtesy Exciton Corp.)

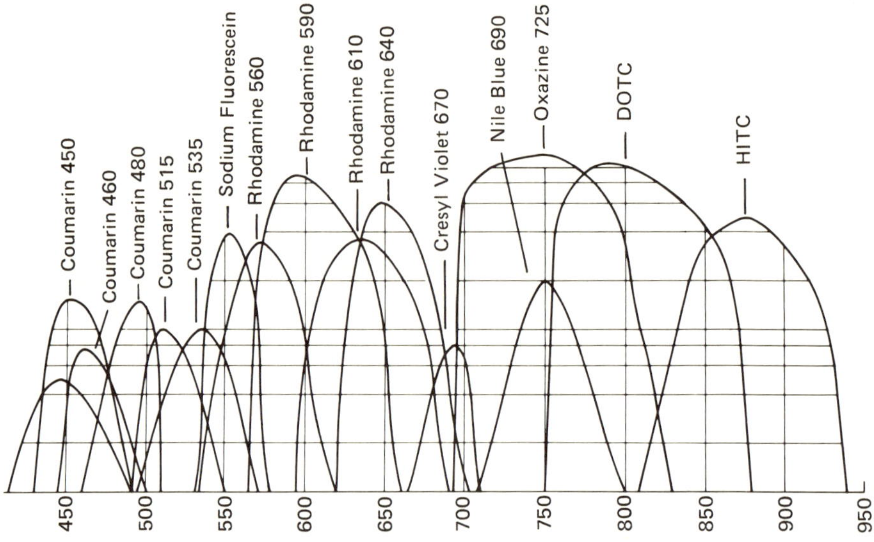

Figure 1.22. Tuning range for laser dyes using CW Ar^+ or Kr^+ laser excitation. (Courtesy Exciton Corp.)

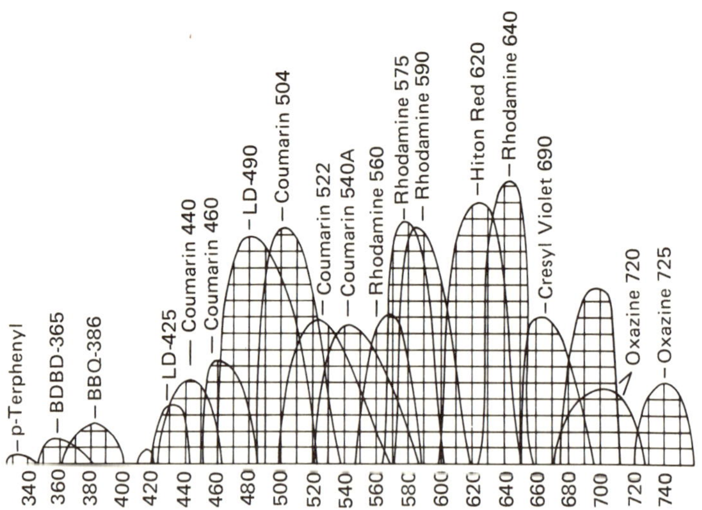

Figure 1.23. Tuning range for laser dyes in coaxial flashlamp pumped dye laser. (Courtesy Exciton Corp.)

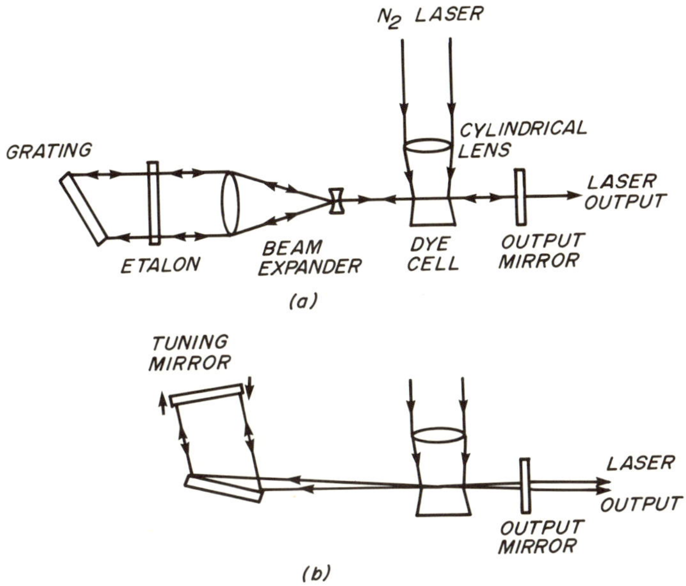

Figure 1.24. Two simple N_2 pumped dye laser configurations: (a) Hänsch design; (b) design of Littman and Metcalf (1978).

amplifier stages with different laser sources would require accurate synchronization. The common way of avoiding this problem is to use a beam splitter to select part of the N_2 laser beam to be used for pumping of amplifier stages. This ensures synchronization of amplifier gain with the oscillator pulse.

1.2.10.2. Nd–YAG Pumped Dye Lasers. Nd–YAG lasers are attractive as dye laser pumps because of high pulse energy and high repetition rates. Furthermore, the pump wavelength can be changed from 532 nm to 355 nm by frequency tripling the 1.06-μm fundamental. This is necessary when pump-

Table 1.19. Output Characteristics of N_2 Pumped Dye Lasers

Peak power	$\sim 10^5$ W
Repetition rate	50–100 Hz
Linewidth	0.01 nm
	0.001 nm (with etalon)
Pulse length	6 ns
Beam divergence	$\sim$10 mrad
Tunable range	350–1000 nm
	217–300 nm (doubled)

Table 1.20. Output Characteristics of Nd–YAG Pumped Dye Lasers

Output energy[a]	$\lesssim$250 mJ/pulse
Peak power[a]	25×10^6 W
Repetition rate	10–20 Hz
Linewidth	0.01–0.001 nm
Pulse length	~10 ns
Tuning range	200–1000 nm (with doubling)

[a]With rhodamine 6G and 532-nm pump wavelength.

ing blue dues but is also sometimes advantageous in improving the output of other dyes.

A summary of current performance data on YAG pumped dye lasers is given in Table 1.20.

1.2.10.3. Flashlamp Pumped Dye Lasers. Flashlamp pumped dye lasers satisfy the need in certain applications for large pulse energies. Commercial flashlamp pumped dye laser systems are capable of generating ~1–3 J per pulse at the peak of the rhodamine 6G emission curve. Repetition rates can be up to 200 Hz but this is obtained only at much reduced pulse energy. Nevertheless, the flashlamp pumped dye laser provides a useful alternative to N_2 or YAG pumped technology in applications such as isotope separation and photochemistry.

A variety of flashlamp–dye–cell geometries have been used in laboratory dye lasers [for a review see Peterson (1979)]. Commercial systems based on

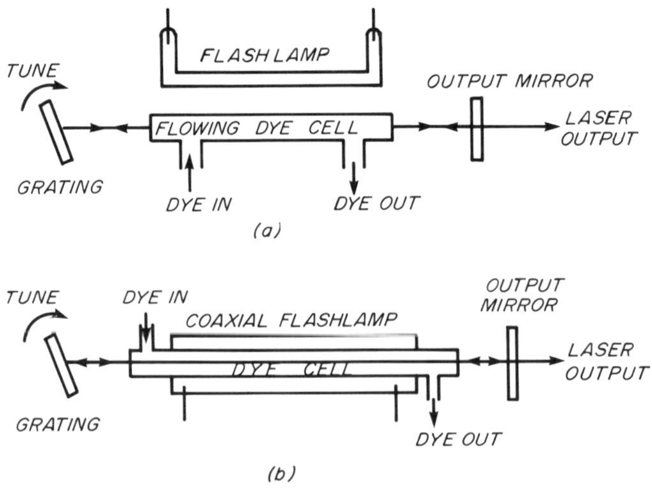

Figure 1.25. (a) Linear flashlamp pumped dye laser; (b) coaxially pumped system.

Table 1.21. Performance Data on Flashlamp Pumped Dye Laser Systems

	Coaxially pumped	Linearly pumped
Output energy	200–400 mJ/pulse	500 mJ/pulse
Peak power	1 MW	5×10^5 W
Pulse length	200–600 ns	0.5–1.5 μs
Repetition rate	1–10 Hz	10 Hz
Linewidth[a]	0.5 nm	0.05–0.1 nm

[a]To 0.001 nm with intracavity etalon.

pumping with linear and coaxial flashlamps are available. These two geometries are shown in Figure 1.25.

A summary of typical performance data on flashlamp pumped dye laser systems is given in Table 1.21. These data are representative of those obtained with commercial dye lasers.

1.2.10.4. CW Dye Lasers. CW dye lasers are pumped by the tightly focused output of Ar or Kr ion lasers. Tight focusing is required to achieve the $\sim$1 MW cm^{-2} needed to obtain population inversion in a dye on a CW basis. Focal diameters are customarily 10–30 μm. An intensity of this magnitude leads to rapid heating within the focal volume. This would rapidly decrease laser efficiency as a result of the population of higher lying S_0 states and increased $S_0 \rightarrow S_1$ absorption. Furthermore, triplet state population rapidly builds up with CW excitation.

These factors require that the excited dye volume be rapidly changed on a continuous basis. In practice, this is accomplished by exciting a small volume of a liquid dye jet in air. This configuration is shown in Figure 1.26. It is the basis of all commercial CW dye lasers. The jet-stream configuration ensures

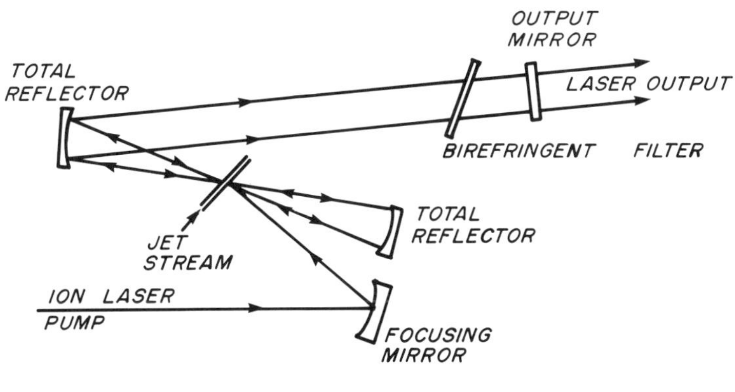

Figure 1.26. Jet stream dye laser. Tuning is via rotation of the birefringent filter.

that heated dye is rapidly swept out of the pump region. Furthermore, operation in air increases the dissolved oxygen content in the flowing dye. Since oxygen is an efficient quencher for triplet state molecules, this aids in lowering the equilibrium triplet state population. Typically, CW output powers are 100–200 mW with the exception of the rhodamine 6G band where up to 3 W can be obtained. These lasers are usually pumped with up to 5 W of CW ion laser power. High pump powers lead to dye heating and saturation and are of little value in obtaining higher tunable output power.

While the visible lines of the Ar^+ laser are useful in pumping rhodamine 6G and rhodamine B, red dyes are more efficiently pumped with the longer-wavelength output of the Kr^+ laser. Blue dyes such as the coumarins can only be pumped efficiently with ultraviolet light. Both the Ar^+ and Kr^+ lasers have lines near 350 nm that are suitable for pumping these dyes although special quartz optics are required for optimum efficiency and tube life when operated under the conditions required to generate ~1–2 W of UV radiation.

Innovative ring laser technology has recently been applied to CW dye laser systems to achieve higher power output and increased efficiency. Single-frequencey operation with linewidths of ~0.1 GHz are obtained. Tuning ranges under single-frequency operation are 10–30 GHz. Table 1.22 lists the output characteristics of representative CW dye lasers.

1.2.11. Stimulated Raman Scattering

The availability of high peak power Nd–YAG pumped dye lasers has resulted in the use of Stimulated Raman Scattering (SRS) for the generation of tunable UV and IR radiation. Under strong pumping of a Raman-active medium, gain can be achieved at the wavelengths of Stokes and anti-Stokes satellites. This can be used to establish oscillation at these wavelengths.

H_2, D_2, CH_4, and other molecular gases have been used to obtain tunable UV outputs via SRS [Byer (1980)]. With H_2, oscillation occurs at multiples of the H_2 fundamental vibration 4155 cm^{-1}. The H_2 gas is contained in a cell typically 30–100 cm in length and at a pressure of 2–20 bars. Gain at the anti-Stokes frequencies depends on H_2 pressure [Wilke and Schmidt (1979)]. Table

Table 1.22. Output Parameters of CW Dye Laser Systems

	Low resolution	High resolution
Linewidth	0.05–0.1 nm	$\lesssim$100 GHz
Output power[a]	0.5 W	2.5 W
Tunable range	400–1000 nm	400–1000 nm, 10–30 GHz
Beam diameter	0.05 cm	0.05 cm
Beam divergence	1.5–2 mrad	1.5 mrad

[a]At peak of rhodamine 6G band.

Table 1.23. Generation of UV Radiation via SRS in H_2[a]

Wavelength (nm)	Output energy per pulse (mJ)	Optimum H_2 pressure (bars)	Anti-Stokes satellites
558	37	Pump	—
453	5	>11	1
381	5	>11	2
329	0.7	>11	3
289	0.3	~11	4
258	0.2	~11	5
233	0.17	~9	6
213	0.09	~6	7
195	0.04	~5	8

[a]From Wilke and Schmidt (1979).

1.23 lists the energy per pulse obtained at various anti-Stokes frequencies using H_2 gas and a pump wavelength of 558 nm. The pump pulse had a duration of ~9 ns. Anti-Stokes output beams all had annular shaped beams whose diameters increased with increasing frequency. Wilke and Schmidt also report on the generation of tunable IR emission using SRS. The simplicity of SRS and the fact that it is a passive technique requiring no active optical elements makes it likely that this method will soon develop into a widely used method of UV generation.

1.3. LASER RADIATION

The usefulness of lasers in materials processing and analysis derives from the unique properties of laser light—properties that offer distinct advantages over light obtained from conventional radiation sources. While not all the unique properties of laser radiation may be important in a particular application, it is always found that the advantage of laser processing in a specific instance are due to one or more of the following properties

Monochromaticity
Low beam divergence
Coherence
High brightness

These parameters influence the practical application of laser light by determining the extent to which laser radiation can be focused to a small spot and directed to remote locations. They also determine the spread of wavelengths emitted from a laser source, i.e., the laser linewidth. This is of fundamental importance in spectroscopic and photochemical applications.

1.3.1. Monochromaticity

Laser linewidth is a function of both the intrinsic width of the laser transition and the existence of a range of longitudinal cavity modes. These cavity modes arise because the wavelength λ_n of standing electromagnetic waves in the laser cavity is quantized with

$$\lambda_n = \frac{2L}{n}$$

where L is the length of cavity and n is an integer. The spectrum of these modes consists of a series of sharp resonances at frequencies $\nu_n = nc/2L$ where c is the speed of light. The frequency interval between these modes is $\Delta\nu = c/2L$. For a 1-m cavity this corresponds to $\Delta\nu = 0.15$ GHz.

Figure 1.27 shows the result of the convolution of longitudinal cavity modes with the fluorescent lineshape function in the case in which many cavity modes lie under the fluorescent profile. Oscillation occurs alternately on each of the cavity modes that overlap the fluorescent line. The result is that the laser output wavelength varies discontinuously over the range of the fluorescent pro-

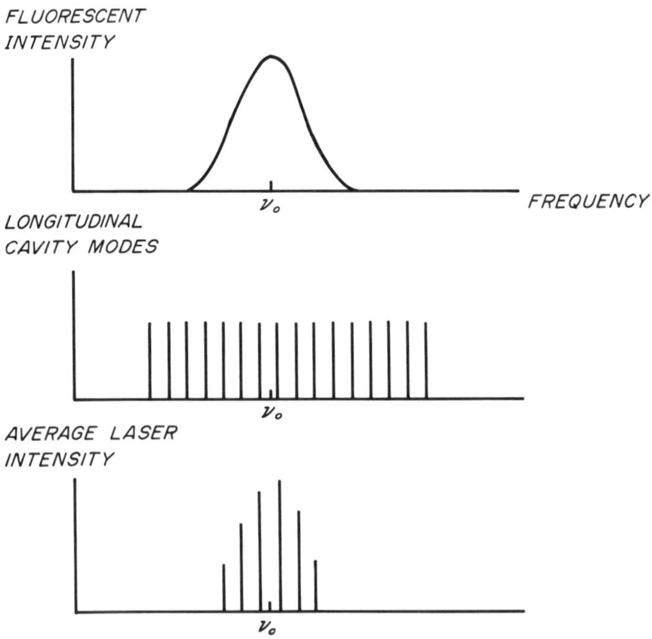

Figure 1.27. Relationship between the profile of the fluorescent line, the frequency spectrum of longitudinal cavity modes, and the time-averaged spectrum of laser output intensity. Note that oscillation need not occur at ν_0. Cavity modes in the wings of the fluorescent line may not oscillate due to low gain.

file. Operation on a single frequency can be obtained by including a frequency selective element (e.g., an etalon) within the cavity and tuning a resonant mode of this element to lie within the fluorescent line profile. Tunability of single-frequency lasers is accomplished by scanning the etalon frequency across the gain profile of the emission line.

The relationship between the frequency spacing of longitudinal cavity modes and the fluorescent linewidth of several common laser materials is shown in Figure 1.28. The line shown corresponds to the solution $\Delta\nu = c/2L$.

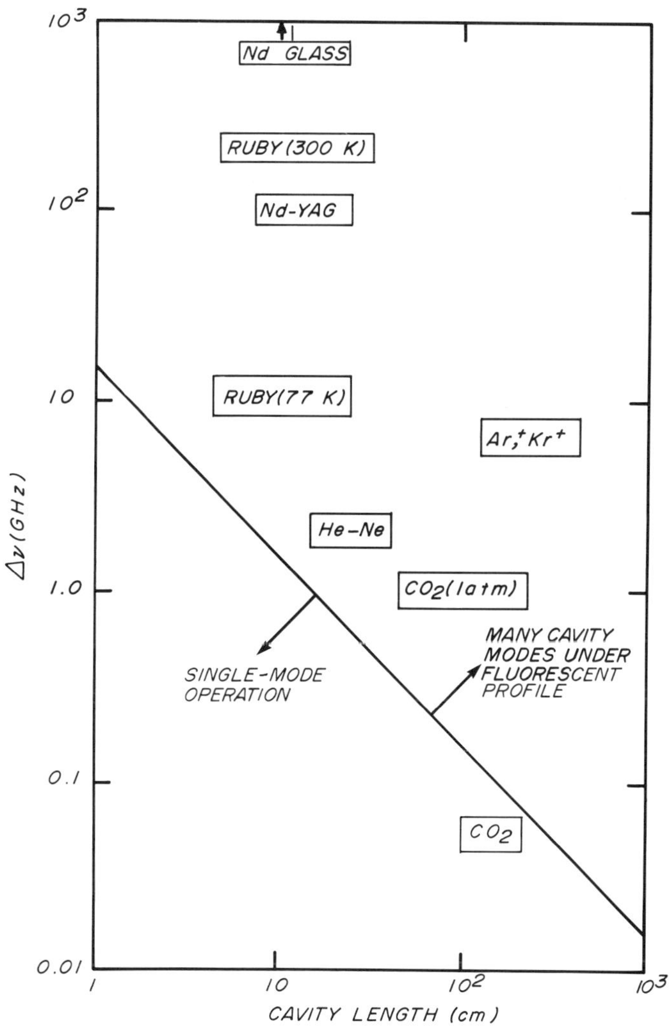

Figure 1.28. Frequency spacing of longitudinal cavity modes vs. fluorescent linewidth $\Delta\nu$ for various laser sources.

It is evident that with the exception of short, low-pressure, CW CO_2 lasers, all lasers will exhibit multifrequency operation as many cavity modes are contained under the laser gain profile. The result is that the effective linewidth of the laser output will be comparable to that of the fluorescent line and may be considerably larger than the width of an individual cavity mode.

Frequency stabilization to ensure operation at a single frequency can be accomplished by close control over a combination of mechanical and electrical factors that influence laser operation. Since a primary constraint on stable operation involves maintenance of a unique cavity length, close attention must be paid to mechanical stability and vibration isolation in the design of the laser head. Drifting can be corrected by attaching one cavity mirror to a piezoelectric element that provides control over the cavity length. When this is incorporated in a feedback loop which involves monitoring laser power, the cavity length can be continually adjusted to ensure operation at the peak of the gain profile. This is aided in practice by the existence of a small dip in the gain profile at ν_0, the center of the fluorescent profile. This dip which was first predicted theoretically by Lamb, has a width that is $\sim(\tau_{\text{rad}})^{-1}$, where τ_{rad} is the spontaneous lifetime of upper laser level. This is typically 10^7–10^9 Hz for visible lasers and 10^3–10^4 Hz for infrared lasers. The Lamb dip arises due to saturation and is extremely useful in stabilizing laser output as small changes from the central frequency minimum can be easily detected electrically and used to drive the output mirror to compensate for this change. In stabilized lasers at visible wavelengths, the frequency jitter of the emitted laser line can be limited to $\lesssim 10$ MHz. This represents a stability of $\gtrsim 1:10^8$.

1.3.2. Beam Shape

The spatial profile of a laser beam is determined by the geometry of the laser cavity. The shape of the laser cavity in a direction transverse to the optical axis provides boundary conditions on the wave equation that determines which configurations of electromagnetic field will be allowed in the cavity. When this cross section is symmetrical, as in cylindrically or rectangularly shaped resonators, the spatial profile of the permitted transverse electromagnetic modes is particularly simple.

The notation for these transverse electromagnetic modes is taken from the theory of waveguides. Particular modes are labeled TEM_{mn} where m and n are the number of nodes in two orthogonal directions. Both cylindrical and rectangular geometries have the TEM_{00} mode as the transverse mode of highest symmetry. Other higher-order modes are labeled TEM_{01}, TEM_{11}, TEM_{12}, and so on. Some of these modes for cylindrically shaped resonator cavities are shown in Figure 1.29.

Many "homemade" lasers often oscillate on one or more high-order modes (i.e., modes with $m, n \neq 0$). In order to inhibit operation on high-order modes

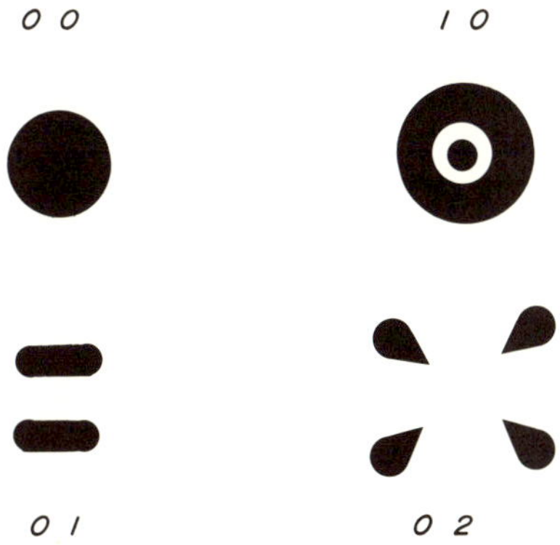

Figure 1.29. Some low-order TEM_{mn} cavity modes for cylindrical symmetry. The dark areas represent areas of enhanced output although the distribution of intensity within these regions is not uniform.

the diameter of the laser cavity can be reduced until operation on the TEM_{00} or fundamental mode is produced. The critical diameter for the initiation of operation on the TEM_{00} mode is a function of the tube length.

Operation in the fundamental mode provides true diffraction limited beam divergence. Higher mode outputs are accompanied by larger beam divergence. Operation in high-order modes may also be accompanied by instabilities in laser output as oscillation shifts from one transverse mode to another with similar gain.

Efficient extraction of power from a laser medium requires that the mode volume within the resonator fill the active medium as closely as possible. This is always the case with long thin tubes where the TEM_{00} mode is the dominant mode of oscillation. In volume discharges, however, such as those found in electron-beam or TEA-type devices, a large number of high-order modes can oscillate and only a relatively small fraction of available laser power is extracted in the fundamental mode. As a result, the beam divergence will be large. This problem can be eliminated by the use of an unstable resonator configuration [Siegman (1971)].

The output of small TEA-type CO_2, N_2, and rare gas halide lasers often has a rectangular cross section. In these cases the distribution of output intensity is determined primarily by the distribution of gain within the laser volume. The output from the tunable diode laser is primarily rectangular in cross section in the near-field but becomes quasi-Gaussian in the far-field (compare Section 1.2.4). This can be seen in Figure 1.30 [Ralston et al. (1973)].

Fundamental mode operation implies a specific distribution of laser intensity across the output beam. This distribution is

$$I(r) = I_0 \exp\left[\frac{-2r^2}{w^2}\right] \tag{1.1}$$

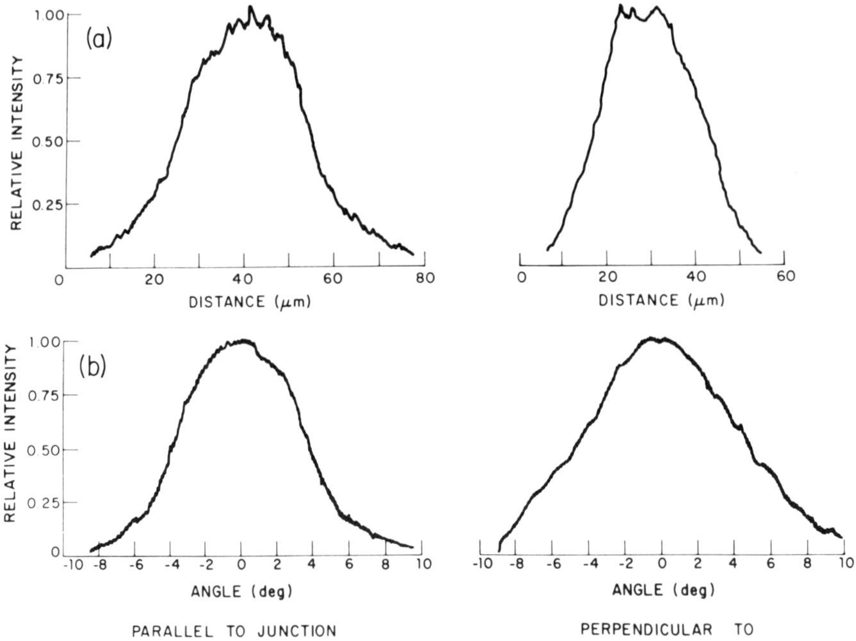

Figure 1.30. Output beam intensity distribution in (a) near-field and (b) far-field from Pb–Sn–Te laser [Ralston *et al.* (1973) by permission IEEE].

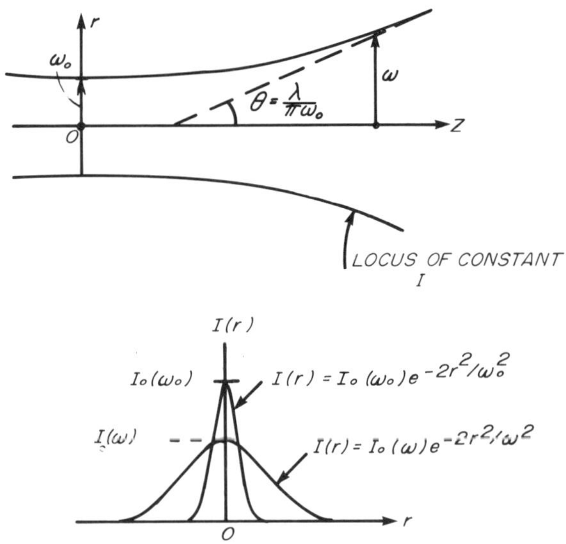

Figure 1.31. Intensity distribution in Gaussian beam (a) versus distance z; (b) cross section through beam at $z = 0$ ($w = w_0$) and $|z| > 0$ ($w > w_0$). θ is the beam divergence.

where w is the radius of the beam at the radial distance for which $I = I_0 e^{-2}$, r is a radial coordinate, and I_0 is the intensity at $r = 0$ (Figure 1.31.). The minimum diameter of this beam, $2w_0$, usually occurs within the laser cavity. The beam subsequently spreads but the Gaussian intensity distribution [Eq. (1.1)] is maintained at all distances z. This is true even when the beam has been focused.

1.3.3. Beam Divergence

The definition of beam divergence is shown in Figure 1.31. This definition is valid for all beam shapes, however, the relation between θ and w_0 is valid only for Gaussian beams. Typical θ values for lasers operating in the visible and infrared are 0.1–10 mrad. Semiconductor diode lasers can have θ exceeding 100 mrad.

While θ is directly related to the minimum beam radius, w_0, which is not generally known with any certainty, it can be estimated approximately from the output beam diameter at the exit window of the laser. As an example, the beam diameter of the Ar ion laser is typically 2 mm. With $\lambda = 0.5\ \mu$m, θ becomes 0.16 mrad. Measured values (Section 1.2.6) are about a factor of 3 larger than this value.

The importance of beam divergence can be seen through a comparison of the intensity at some distance z of conventional and laser light sources. For the same total power, P, one has

$$I(\text{conventional}) = \frac{P}{4\pi z^2}$$

$$I(\text{laser}) \simeq \frac{P}{\pi w^2} = \frac{P}{\pi z^2 \theta^2}$$

Then

$$\frac{I(\text{laser})}{I(\text{conventional})} = \frac{4}{\theta^2}$$

Since $\theta \sim 10^{-3}$ rad, the enhancement due to the small beam divergence of the laser source is on the order of 10^6. Note that this enhancement is independent of distance as long as one is in the far field. In addition, since $\theta \propto \lambda$ the enhancement is largest at short wavelengths.

The beam divergence can be changed with a suitable optical system. Two common configurations are shown in Figure 1.32. The diverging lens system offers a more compact design and is more suitable for high intensity beams since it does not bring the beam to a tight focus where air breakdown could occur.

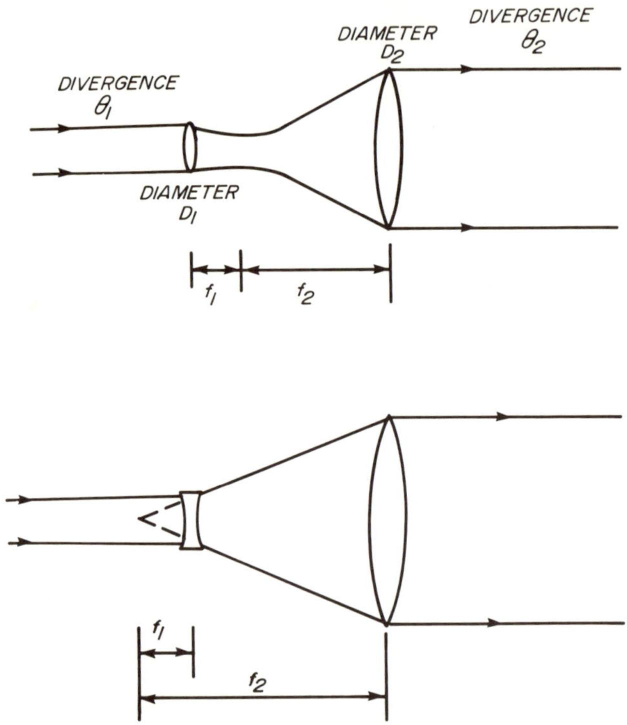

Figure 1.32. Beam collimating optical systems: (a) converging system; (b) diverging system.

When the *f*/numbers of the two lenses are matched, so that the beam fills both lenses, the beam divergence in the incident and transmitted beams are related as follows

$$\theta_2 = \frac{\theta_1 D_1}{D_2}$$

where D_1 and D_2 are the lens diameter. Thus a 10:1 beam expansion reduces the beam divergence by the same amount.

1.3.4. Brightness

Radiance or *brightness* is defined as power per unit area per unit solid angle. For a circular beam of radius A, having divergence θ and containing a total power P, the brightness, B, is

$$B = \frac{P}{\pi A^2 \theta^2}$$

Under diffraction limited conditions, $\theta = \lambda/\pi A$ and thus

$$B = \frac{P}{\lambda^2} \text{ W cm}^{-2} \text{ sr}^{-1}$$

The spectral radiance is defined as $B_\nu = B/\Delta\nu$, where $\Delta\nu$ is the laser linewidth in Hz. A summary of B and B_ν for several laser sources as well as for a 1-kW arc lamp is given in Table 1.24.

1.3.5. Focusing of Laser Radiation

The dominant parameter in most applications of lasers to materials processing is the laser intensity. If we consider a laser beam propagating through an optical system, the intensity will vary along the optical path, increasing in regions of beam convergence and decreasing in regions where the beam diverges. While the beam intensity varies, the beam power stays constant in the absence of absorption or other losses. Changes in intensity are the result of changes in the spatial extent of the propagating beam. These changes arise in response to diffraction and focusing. A prediction of the effects of laser radiation requires knowledge of the spatial extent of the laser beam as it propagates through the optical system. In particular, the distribution of laser intensity at or near the focal point is usually required.

A simple calculation of the distribution of intensity within the focal region is possible only when the propagating beam has a Gaussian profile. As we have seen, this is the beam profile obtained from lasers operating in the fundamental TEM_{00} mode. Figure 1.33 shows the focusing of a propagating Gaussian beam by a simple lens and the intensity distribution at three points along the optical

Table 1.24. Brightness, B, and Spectral Radiance, B_ν, for Various Laser Sources and a 1-kW Arc Lamp

Source	B^a ($\text{W cm}^{-2}\text{ sr}^{-1}$)	B_ν^b ($\text{W cm}^{-2}\text{ sr}^{-1}\text{ Hz}^{-1}$)
1-kW arc lamp	10^3	10^{-10}
10-W Ar^+	4×10^9	1
100-W CW CO_2	10^8	10
1-J Nd–glass 10-ns pulse	10^{16}	10^3

[a]Assuming diffraction limited performance; [b]Using fluorescent linewidth. B_ν for single-frequency lasers will be considerably larger.

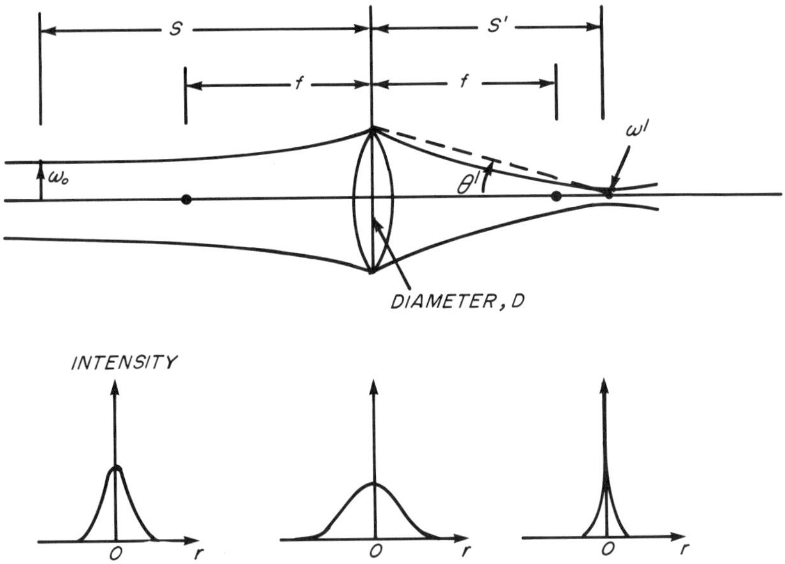

Figure 1.33. Focusing a Gaussian beam with a simple converging lens.

axis. Since the initial beam has small divergence, the focus is close to the focal point f on the right side of the lens ($s' \sim f$).

The theory of diffraction by a circular aperture shows that $w' = \lambda/\pi\theta'$, where λ is wavelength and θ' is the angle subtended by the lens at the focus. w' is the Gaussian beam radius at this focus. Since the beam fills the lens one can take $\theta' \simeq D/2f$, where D is the lens diameter. Then

$$w' = \frac{2\lambda f}{\pi D}$$

This is the radius of the diffraction limited spot at the lens focus. Since $f/D =$ f/number

$$w' = \frac{2\lambda}{\pi}\,(f/\text{number}).$$

As f/numbers substantially smaller than unity are not easily obtained in practice, the minimum spot radius obtained under ideal progagation and focusing conditions is $w' \sim \lambda$. The minimum spot size is therefore approximately one wavelength. This limit can be approached with TEM_{00} lasers operating in the visible and near infrared. However, in order to achieve diffraction limited performance other optical system aberrations must be minimized.

A simple estimate of the radius of the minimum spot size obtained with perfect focusing can be found from the expression

$$r_s = f\theta$$

where f is the lens focal length and θ is the beam divergence. Examples of r_s calculated for common laser types are given in Table 1.25. These values are calculated using the beam divergence typical of common commercial lasers and do not represent "state-of-the-art" performance.

Another important parameter for a focused beam is the depth of focus. This can be defined in a variety of ways since it is intended that this parameter simply give some estimate of the distance over which the lens–image distance can be varied without a substantial increase in spot size. A useful estimate can be obtained from

$$\Delta = \pm \frac{r_s^2}{\lambda}$$

where r_s is the spot radius and the $\pm$ sign appears since the depth of focus applies for both motion toward and away from the lens. Under conditions in which one has $r_s \sim \lambda$, the depth of focus $\Delta \sim \lambda$. This is the basis for the simple rule that depth of focus $\sim$ spot size. Of course when $r_s > \lambda$, the depth of focus is substantially larger than the spot size.

As an example of the information that can be obtained from a simple calculation, let us consider the focusing of the output of a 10-W Ar^+ laser with a lens with $f = 2$ cm, and $D = 1$ cm. We consider two cases. The first assumes that the lens is used directly to focus the beam. The second assumes that a 10$\times$ beam expander is used prior to the final focusing lens. We take the laser beam diameter to be 1 mm and the beam divergence to be 0.5 mrad.

Table 1.25. Calculated r_s for Different Laser Types

Laser	Beam divergence (mrad)	r_s[a] (cm)
He–Ne	0.5	10^{-3}
Ar^+, Kr^+	0.5	10^{-3}
Ruby	1	2×10^{-3}
CO_2	2	4×10^{-3}
Nd–YAG	3	6×10^{-3}
Nd–glass	5	10^{-2}

[a]Assuming $f = 2$ cm.

1.3.5.1. Simple Lens. Since $D \gg$ beam diameter we use $r_s = f\theta$, where θ is the beam divergence, to estimate the spot size. Then $r_s = 1.0 \times 10^{-3}$ cm $= 10\ \mu$m. The peak intensity, I_0, in the beam focus is related to beam power, P, by

$$
\begin{aligned}
P &= 2\pi \int_0^\infty I(r) r\, dr \\
&= 2\pi I_0 \int_0^\infty e^{-2r^2/r_s^2} r\, dr \\
&= \frac{\pi I_0 r_s^2}{2}
\end{aligned}
$$

Thus $I_0 = 2P/\pi r_s^2 = 6.4$ MW cm^{-2} The estimated depth of focus is $\Delta = \pm 10^{-2}$ cm $= \pm 100\ \mu$m.

1.3.5.2. Beam Expander Plus a Simple Lens. When the beam is expanded by a factor of 10 before entering the focusing lens, the beam divergence is reduced by a similar factor, i.e., $\theta = 0.05$ mrad. Then $r_s = f\theta = 10^{-4}$ cm and $I_0 = 6.4 \times 10^8$ W cm^{-2}. The depth of focus is $\pm 10^{-4}$ cm.

This simple calculation illustrates the advantage of expanding the beam to fill the aperture of the focusing lens. Such an expansion reduces the beam divergence and hence r_s. This in turn increases the intensity in the focal region. Note, however, that an increase in I_0 is obtained only with an attendant decrease in the depth of focus. In some applications, an effective limit on spot size may be provided by the need to have a large depth of focus so that beam expansion may not be that useful.

While the focused beam diameter can be calculated under given conditions, the absence of aberration-free optics usually implies that this calculation will yield only an extreme lower limit to the spot size that can be obtained from the given system. Thus measurements of beam diameter are usually required. Several methods have been devised to measure this quantity. All require that the beam profile be Gaussian for straightforward analysis. A compilation of available methods has been published by Sliney and Marshall (1979).

A particularly simple result is obtained if a knife-edge is scanned through the beam as shown in Figure 1.34. The transmitted beam contains power

$$
P(a) = \frac{P}{2} \operatorname{erfc}\left(\frac{a\sqrt{2}}{w}\right)
$$

where $a = x$ coordinate of the edge and erfc is the complementary error function. This measurement can be performed on a CW basis with a chopper [Skinner and Whitcher (1972)] and photodetector. The output of the detector can be recorded on an oscilloscope where the duration and shape of the pulses can be used to calculate w, the Gaussian beam radius.

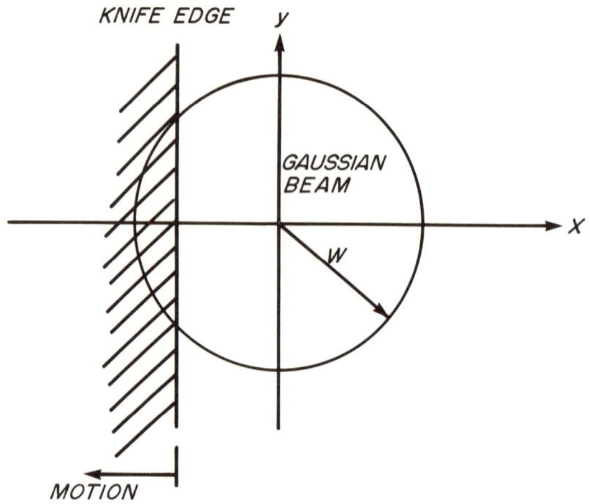

Figure 1.34. Method for determining Gaussian beam radius by scanning knife edge.

1.3.6. Coherence

Laser radiation exhibits both spatial and temporal coherence. Spatial coherence can be illustrated by examining the phase relationship between two points in space at identical times. If the electromagnetic waves at the two points are zero phase difference at time $t = 0$ and this zero difference is retained after some time interval, t, then perfect spatial coherence exists between the wavefront at the two points.

When the electromagnetic field at the same point in space has the same phase at times t and $t + \tau$ then one has perfect temporal coherence at the point. When the phase relationship is maintained for some finite time τ_0, τ_0 is referred to as the coherence time. The coherence time is $1/\Delta\nu$ where $\Delta\nu$ is the linewidth in Hz for CW lasers and is approximately equal to the length for pulsed lasers.

Temporal and spatial coherence are independent variables: one can have one type of coherence without necessarily having the other. Furthermore, complete coherence is never obtained so that one always has degrees of spatial or temporal coherence.

A useful parameter is the coherence length, L_c, which is related to the coherence time, τ_0, by

$$L_c = c\tau_0$$

This length is a rough estimate of the path difference than can exist between two arms of a Michelson-type interferometer and still result in interference. Alternatively, if a beam of light is broken into two separate beams that are

sent along different paths, L_c is the maximum difference in length of these paths that will allow the two beams to interfere when brought back together.

Single-frequency CW lasers have a high degree of temporal coherence and large coherence lengths. For example, a frequency stabilized He–Ne laser with $\Delta\nu = 1$ MHz has $\tau_0 = 10^{-6}$ s and $L_c = 3 \times 10^4$ cm. A multimode He–Ne laser might have $\Delta\nu = 1.5$ GHz, $\tau_0 = 6.6 \times 10^{-10}$ s and $L_c = 20$ cm. A Q-switched Nd–YAG laser with a pulse length of 10 ns has $\tau_0 = 10^{-8}$ s and $L_c = 300$ cm.

Spatial coherence depends on the transverse output mode of the laser. With a CW laser operating in a single transverse mode, spatial coherence is almost complete. A pulsed laser operating multimode may have only limited spatial coherence.

1.4. LENS ABERRATIONS

Lens aberrations often provide the practical limit to the spot size obtained in focusing a laser beam. Five basic types of aberrations are encountered in focusing nonchromatic light.

1.4.1. Spherical Aberration

Light rays that pass through the outer part of a lens are focused at a different distance from those passing through near the optical axis (paraxial rays). This effect is shown in Figure 1.35. It can be minimized by optimization of lens shape but, when present, the amount of aberration varies as h^2, where h is the height of a light ray above the optical axis (Figure 1.35). Spherical aberration is largest for short-focal-length lenses of large aperture (large $f/$ number system). It results in a blurring of the focal spot and an increase in effective spot size.

1.4.2. Coma

Light rays from off-axis points are imaged as ringlike structures in the focal plane. The net effect is to produce a comet-shaped image (Figure 1.36).

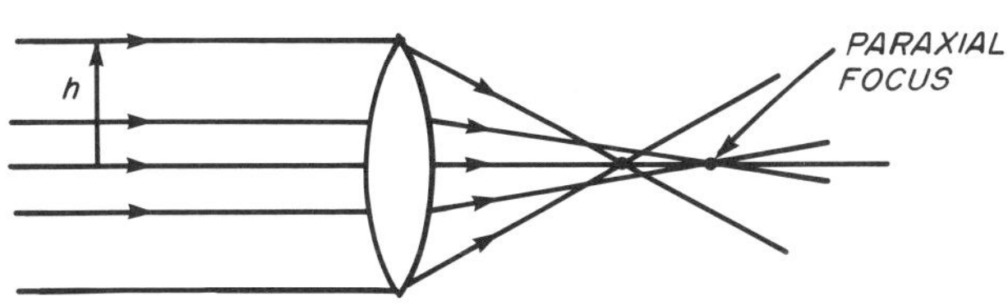

Figure 1.35. Spherical aberration in a simple lens.

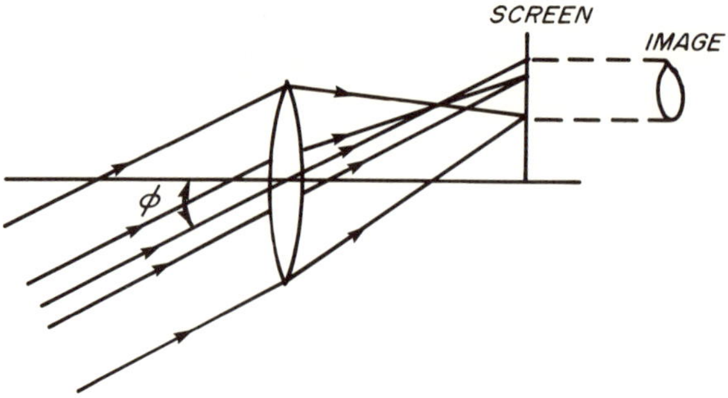

Figure 1.36. Example of coma in a single lens.

Coma can be eliminated by control over lens shape, although the conditions that minimize coma are not necessarily the same as those that minimize spherical aberration. The amount of coma scales as $\phi h^2 w$ where w is the image size.

1.4.3. Astigmatism

Blurring of an image is produced when off-axis rays, passing through a lens is a vertical plane, are focused at a different distance from those passing through in a horizontal plane. It may be compensated by the introduction of an additional lens. Astigmatism scales as the product $\phi^2 w^2$.

1.4.4. Field Curvature

The image is not formed in a flat plane but along a curved surface. Thus if the image is observed on a flat screen the edges may be blurred. Field curvature scales as $\phi^2 w^2$ and can be minimized by the introduction of stops.

1.4.5. Distortion

This is due to a variation in magnification across the image. Thin lenses are not usually subject to much distortion, but distortion can be *produced* by the incorporation of stops. Distortion scales as the product $\phi^3 w^3$.

1.5. WINDOW MATERIALS

Quartz, Pyrex, and other glasses are commonly used as laser window materials in the wavelength range between 350 and 1000 nm. Operation in the ultraviolet requires that windows be made of quartz, magnesium fluoride, or lithium fluoride. These materials are subject to discoloration due to radiation damage particularly after prolonged operation at high power levels as found, for example, in RGH lasers.

Quartz and glass can be used with Nd–YAG or Nd–glass lasers, however, these materials are not transparent farther out in the infrared. A variety of semiconductor and alkali halide materials are, however, transparent throughout the infrared and are useful as laser windows and focusing optics. The crystalline alkali halides NaCl and KCl are highly transparent near 10 μm and are favored as window materials even though they are also hygroscopic and are too soft to take high quality optical coatings. Semiconductor materials such as Ge, GaAs, ZnSe, and CdTe have higher resistance to chemical attack and can be easily coated but tend to be expensive.

Selection of a window or lens material for a particular application often involves a trade-off of several parameters. Some important considerations are the following:

1. *Optical absorption.* Materials can be compared via their absorption coefficients at the wavelength of interest. This is defined as

$$\alpha = -\frac{1}{x}\ln\left(\frac{I}{I_0}\right)$$

   where I_0 is incident intensity and I is transmitted intensity by a sample of thickness x. A small value of α is desirable.
2. *Thermal conductivity.* The thermal conductivity, K, is a measure of the capacity of the material to dissipate energy absorbed from the beam. A large value of K is desirable.
3. *Hardness and smoothness.* The hardness should be large to resist scratching and digging. Surfaces should be smooth to enable flat optical coatings to be applied.
4. *Chemical resistance.* Optics should have low solubility in water and resistance to other ambient chemicals, e.g., halogens in RGH lasers.

A combination of low absorption and high thermal conductivity is required when windows or lenses are used at high average power levels. Under these conditions, when other factors are equal, it is probably best to keep α as small as possible in order to minimize the initial absorption of power from the beam. Some relevant data for semiconducting and alkali halide materials that can be used for windows or lenses at 10.6 μm are summarized in Table 1.26. Also listed in this table [Lussier (1976)] is a figure of merit $= \sigma K/\alpha a E$, where σ is the rupture modulus, K the thermal conductivity, α the absorption coefficient, a the linear expansion coefficient, and E Young's modulus. The first entry is for perpendicular incidence while the second is for incidence at Brewster's angle. It is evident that GaAs and Ge have the highest figures of merit. Unfortunately, Ge suffers significantly from thermal runaway and so is not suitable for high power applications. However, it is useful for mirror substrates.

Table 1.26. Optical and Thermal Parameters for Several Semiconductors Used as Optical Elements at 10.6 μm[a]

	CdTe	ZnSe	GaAs	Ge	NaCl	KCl
Absorption coefficient (cm^{-1})	0.002	0.005	0.02	0.032	0.002	0.001
Thermal conductivity ($W\ cm^{-1}\ {}^\circ C^{-1}$)	0.06	0.18	0.48	0.59	0.07	0.07
Thermal expansion coefficient ($10^{-6}\ {}^\circ C^{-1}$)	5.9	8.5	5.7	5.7	44	36
Figure of merit	53	69	5300	1673	—	—
Figure of merit at Brewster's angle	151	179	17,214	6898	—	—
Resistance to thermal runaway	Very good	Very good	Very good	Poor	Very good	Very good
Cost	Moderate	Moderate	High	Low	Low	Low

[a]After Lussier (1976).

Polycrystalline GaAs material is often used in focusing lenses because of its ability to dissipate high power. ZnSe has the advantage that it transmits light in the visible spectral region—a useful alignment feature.

Optical components are usually coated to reduce reflection or to achieve a particular reflectivity. Common coating materials at 10 μm are Ge, Si, ThF_4 (thorium fluoride), and ZnS. Specifications of "off the shelf" commercial infrared components would customarily have the following tolerances:

Antireflection (AR) coatings
Reflectivity < 0.5% per surface
Absorption < 0.1% per surface
Partially reflective
Reflectivity 45–85% ± 3%
90–95% ± 1.5%
97–98% ± 1%
99% ± 0.5%
Absorption < 0.3% per surface

Lenses are usually AR coated on both sides to reduce losses due to reflection. Other optical components such as beam splitters and windows are coated to the desired reflectivity on one side and AR coated on the other. Unless the

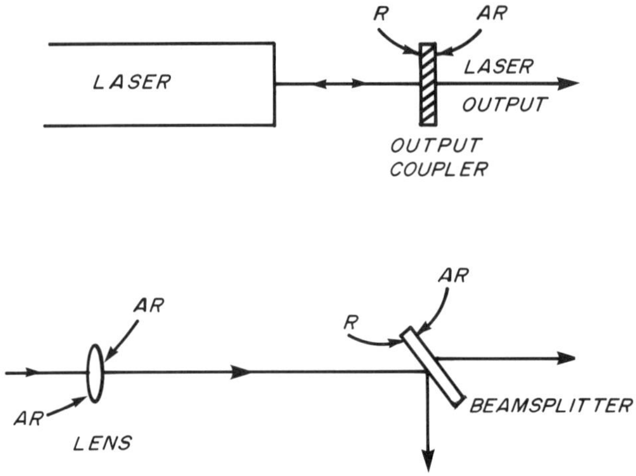

Figure 1.37. Use of coated optics in two common configurations: R, reflective surface; AR, antireflective surface.

desired components are standard items, coatings must be applied on a "special order" basis. This tends to increase the cost considerably, although the cost per part drops dramatically when several parts can be coated in the same batch. In many cases the cost of coatings may exceed the cost of the substrate. Figure 1.37 shows the use of coated optics in two common configurations.

1.6. MIRRORS AND POLARIZERS

With dielectric coating even a normally transparent material can be converted into a mirror with high reflectivity. This is accomplished by the application of several overlayers via vacuum evaporation. The thickness of each layer is controlled to yield an optical thickness (= refractive index × thickness) equal to a quarter wavelength at the wavelength desired. The required reflectivity can be achieved by increasing the number of quarter-wave layers. This technique can produce reflectivities in excess of 99.5%. Coating materials used in the ultraviolet include MgF_2, NaF, LaF_3, Al_2O_3, ZrO_2, ThO_2, HfO_2, and Y_2O_3.

Coatings used in the visible include SiO, SiO_2, sodium aluminum fluoride, TiO_2, and Al_2O_3. Infrared coatings have been given in the previous section.

High reflectivity mirrors in the infrared are usually produced on Si, Ge, or metallic substrates. Very high power applications often require the use of coated metallic mirrors. Typical substrates are Cu and Mo and these may be water cooled in order to minimize thermal distortion. Table 1.27 lists the reflectivities of several commercial metal mirrors.

Table 1.27. Reflectivities of Several Commercial Metal Mirrors[a]

Manufacturer	Material	Smoothness (nm)	Reflectivity 10.6 μm	5 μm	1.06 μm
Spawr	Cu	2.3	99.2	98.6	98.1
Spawr	Cu + ThF_4	2.3	99.1	98.5	98.0
Northrup	Cu + Ag + ThF_4	2.3	—	99.2	—
Spawr	Mo	1.8	98.3	—	—
Spawr	Mo	3.0	98.1	—	—
Typical	Au–glass	<2.5	~98.4	98.3	98.2
	Ag–glass	<2.5	~98.9	98.9	98.9

[a]After Spawr and Pierce (1975).

Operation of rare gas halide and other excimer lasers places a combination of severe constraints on mirror materials. Firstly, the number of transparent substrates and coating materials is limited as most materials absorb in the region below ~250 nm. Secondly, resistance to chemical corrosion particularly in the RGH lasers should be considered. However, commercial laser mirrors are available for the region below 200 nm. An example of the reflection and transmission of a reflector for the ArF laser is shown in Figure 1.38.

Gill and Newman (1979) have provided a useful summary of damage thresholds in UV coatings. Data were obtained at 266 and 355 nm using quadrupled and tripled Nd–YAG laser pulses of ~20-ns duration. Some of these data are given in Table 1.28. Damage thresholds at 266 nm for a single-layer

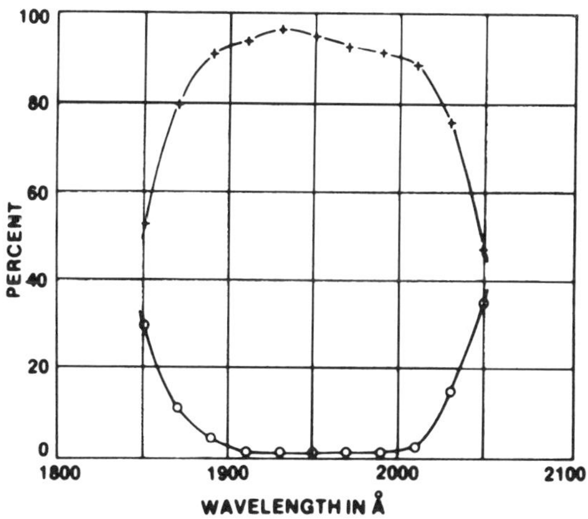

Figure 1.38. Reflectance and transmittance for a mirror designed to operate with the ArF laser at 193 nm. Lower curve (o) is transmission; upper curve (+) is reflectance. (Courtesy Acton Research Corp.)

Table 1.28. Damage Thresholds at 266 and 355 nm for Optical Coatings

		Wavelength			
		266 nm		355 nm	
Coating	Substrate	Fluence (J cm^{-2})	Intensity (MW cm^{-2})	Fluence (J cm^{-2})	Intensity (MW cm^{-2})
NaF	Suprasil 2	10.8	450	>38	>1390
ThO_2	Dynasil 1000	2.8	122	9.1	340
ZrO_2	Optosil 1	1.5	67	7.1	250
Al_2O_3	Suprasil 2	2.6	106	5.9	210
NaF/Al_2O_3 reflector	Suprasil 2	3.6	154	12.2	470
ThO_2/SiO_2 reflector	Dynasil 1000	1.3	71	4.9	180

[a]Pulse lengths were 22 ns at 266 nm and 27 ns at 355 nm [Gill and Newman (1979)].

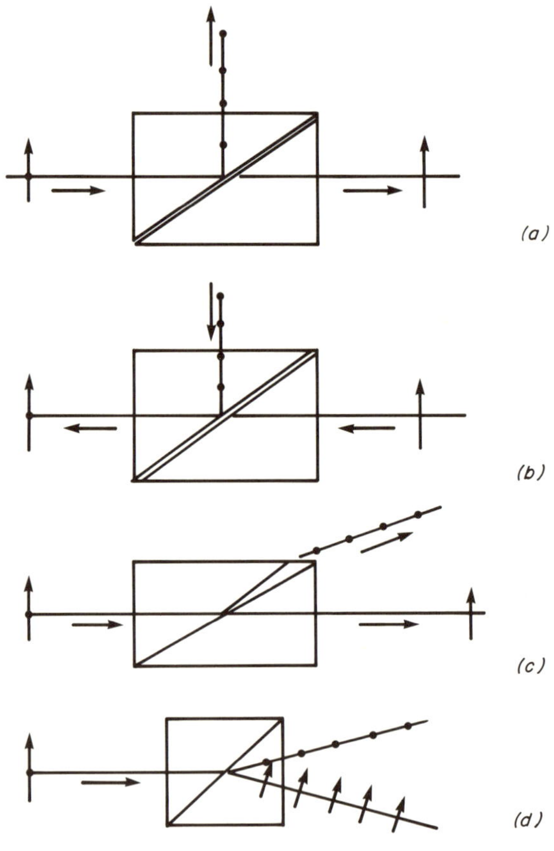

Figure 1.39. Some polarizing prism configurations: (a) Glan–laser; (b) Glan–laser beam combining; (c) Glan–Thompson; and (d) Wollaston.

coating was that of NaF while the best two-layer coating was the NaF/Al_2O_3 combination.

Polarizers

Polarizers are commonly used as intracavity elements and as beam splitters or beam combiners. Several standard prism polarizer configurations are shown in Figure 1.39. These prisms utilize the birefringence of crystals such as quartz, calcite, and magnesium fluoride to separate beams with different polarizations.

1.7. Q-SWITCHING

Q-switches introduce a loss into a resonant laser cavity which can be rapidly removed. If pumping of the active medium continues while the losses are large (high cavity Q) then a substantial amount of energy can be stored in the upper laser level. When the losses are removed (low Q) stimulated emission causes this energy to be rapidly dissipated through the output coupler. The result is that the stored energy is emitted in the form of a short pulse. The peak power in this Q-switched pulse can be substantial since significant energy is released in a relatively short time (typically 10–100 ns).

Q-switching requires that the cavity contain a component that can cause the cavity Q to change from a high value to essentially zero over a time scale that is short compared to the pumping time. Early studies of Q-switching used rotating mirrors or prisms to switch from a high-gain to a low-gain condition. However, this approach is rarely used at the present time with the availability of other active and passive Q-switching devices.

1.7.1. Acousto-Optical Q-Switches

When an acoustic wave is propagated in an elasto-optical solid, the presence of acoustic energy modifies the refractive index of the material. The amplitude of the refractive index variation depends on the intensity of the acoustic wave and hence a standing acoustic wave will generate a periodic refractive index variation in the solid. This may lead to scattering of an optical wave propagating in the crystal and can cause the wave to be deflected from its original direction.

The angle of deflection will depend on the frequency of the acoustic wave while the fraction of the optical beam that is deflected will depend on the amplitude of the applied acoustic field. Thus by varying the acoustic amplitude, an optical beam may be deviated from its original path over a short time scale.

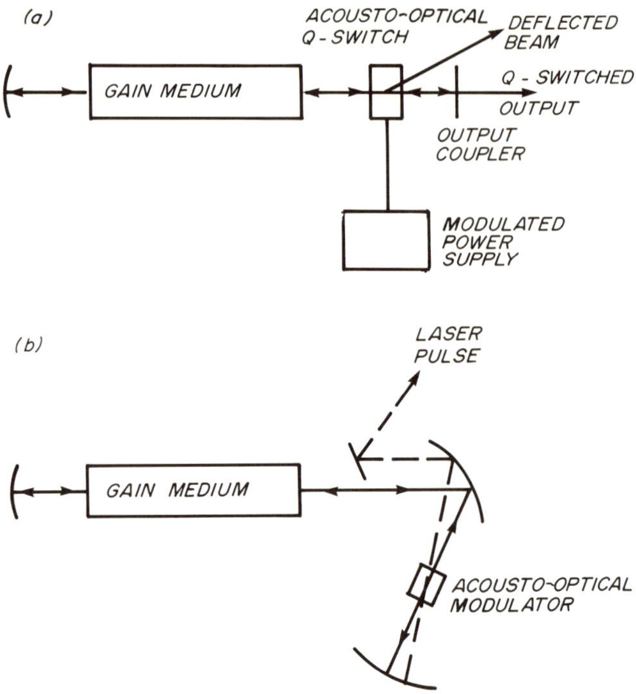

Figure 1.40. Use of acousto-optical element as (a) *Q*-switch and (b) modulator.

The use of an acousto-optical modulator as a *Q*-switch in a laser cavity is shown in Figure 1.40. This figure also shows how an acousto-optical deflector can be used for modulation or pulse selection.

Quartz is commonly used as an acousto-optical material at wavelengths in the visible and near infrared. Tellurium and GaAs can be used as modulators in the 5–10 μm region in the infrared. The performance of several com-

Table 1.29. Performance of Representative Commercial Acousto-Optical *Q*-Switches

Spectral range (μm)	Material	Rise time (μs)	Deflection efficiency (% at input power)	Aperture (mm)
0.33–1	SiO_2	0.007	15% at 0.1 W	0.1
1.06	Quartz	0.1	30% at 40 W	4
6–20	Te	1		6

Table 1.30. Performance of Electro-Optical *Q*-Switches

Material	Spectral range (μm)	Rise time (ns)	Maximum voltage (kV)	Peak intensity (MW/cm^2)	Aperture (mm)
KD*P[a]	0.25–1.2	0.45	9	10^3	8–15
KDP[b]	0.4–0.8	1	7	$(3–8) \times 10^2$	10
$LiNbO_3$	1–4	1	3–7	0.08	2–10
CdTe	2–4	1	5	50	3–10
GaAs	2–4	1	5	100	3–10

[a]KD*P = KD_2PO_4; [b]KDP = KH_2PO_4.

mercial acousto-optical *Q*-switches is summarized in Table 1.29. Typical rise times for these devices are ~100 ns mm^{-1}.

1.7.2. Electro-Optical Q-Switches

Application of an electric field to certain crystalline solids causes them to exhibit birefringence, i.e., different refractive indices for light polarized in two orthogonal directions. This effect results in a rotation of the plane of polarization for linearly polarized light propagating through a medium with birefringent properties. Thus, if such an element is placed in a laser cavity that normally has high gain only for light of one polarization, the gain can be decreased by applying a voltage to the element. When an electro-optical crystal is biased to provide low gain during pumping of the laser gain medium and then has its electric field reduced to zero, the gain is increased and the stored cavity energy is emitted in the form of a *Q*-switched pulse. Some performance data on commercial electro-optical *Q*-switches are given in Table 1.30.

Since the response time of electro-optical *Q*-switches does not depend on acoustic transit times as in acousto-optical beam shifters, electro-optical devices are inherently faster. Rise times of $\lesssim 1$ ns are easily obtained.

1.7.3. Passive Q-Switching

Passive *Q*-switching is provided by saturable absorbers that are incorporated in the laser cavity. These materials, which usually consist of a solution of organic dye molecules in a suitable solvent, normally absorb at the laser wavelength. However, as the electromagnetic field intensity builds up in the laser cavity the absorption coefficient at the laser frequency decreases due to depletion of ground-state molecules. When the material starts to become transparent or bleached, the electromagnetic field intensity builds up rapidly due to the initiation of feedback and *Q*-switched emission occurs. Since the bleaching of absorption is not controlled externally, the Q-switched pulse can be synchronized with other electrical signals.

The usual dyes used as saturable absorbers are phthalocynine in nitrobenzene or cryptocynanine in methanol for ruby lasers and Eastman 9740 and 9860 dyes for the Nd–YAG and Nd–glass lasers. Bleachable glass filters can also be used.

1.8. FREQUENCY CONVERSION

The electrical polarization, P, of a medium is related to the applied electric field, E, through the dielectric susceptibility

$$P = \chi E$$

χ is a tensor but we will consider that P and E are scalars for simplicity. P can be expanded in powers of E as follows:

$$P = \chi_1 E + \chi_2 E^2 + \chi_3 E^3 + \cdots$$

where $\chi_1 \gg \chi_2 \gg \chi_3$, and so on. If $E(t) = E_0 \cos \omega t$ then the first term yields a polarization that is of the form $\chi_1 E_0 \cos \omega t$, i.e., $P(t)$ oscillates at the same frequency as $E(t)$. When E_0 becomes large, the second term may be important. Then $P(t) \alpha \cos^2 \omega t = (1 + \cos 2\omega t)/2$ and $P(t)$ contains a term that oscillates at 2ω. If we suppose that instead of having E_0 strong enough to make the E^2 term large we have two individual fields $E_0(1) \cos (\omega_1 t)$ and $E_0(2) \cos (\omega_2 t)$, then $P(t) \alpha \cos (\omega_1 t) \cos (\omega_2 t) = [\cos (\omega_1 + \omega_2)t + \cos (\omega_1 - \omega_2)t]/2$ and sum and difference frequencies are generated in the medium.

In crystals without a center of symmetry $\chi_2 \neq 0$ and a strong beam of light at frequency ω will give rise to an oscillating polarization at 2ω. Under certain conditions, an output wave at 2ω can be generated in the material. In this case, the beam at 2ω is the frequency-doubled component of the fundamental at ω.

As the field strength is increased, higher terms in $P(t)$ will dominate and radiation will be generated at 3ω, 4ω, and so on. If beams at different frequencies are mixed in the crystal, sum and difference frequencies will be generated. Some nonlinear processes are as follows:

Frequency doubling	2ω
Sum frequency generation	$\omega_1 + \omega_2$
Difference frequency generation	$\omega_1 - \omega_2$
Frequency tripling	3ω
Frequency quadrupling	4ω

Table 1.31. Nonlinear Mixing to Generate UV and IR Radiation

Sources	Generated wavelengths	Input power	Output power	Crystal	Reference
Nd–YAG	354.7 nm	50 MW	6 MW	RDA	Zernike (1979)
	532.1 nm	50 MW	27 MW	RDA	Zernike (1979)
CW dye	285–315 nm	Intracavity	50 mW	ADA	Frolich *et al.* (1976)
Ar CW + Rh6G	2.3–4.6 μm	Intracavity	4μW	$LiIO_3$	Wellegehausen *et al.* (1976)
Q-switched Nd–YAG + RhB	1.25–1.6 μm	3.1 mW cm^{-2} + 4.7 kW intracavity	70 W	$LiIO_3$	Goldberg (1975)
N_2 pumped dyes 8–10 ns	5.5–18.3 μm	4.7, 5.7 kW	4 W	$AgGaS_2$	Seymour and Zernike (1976)
CO + CO_2	15.5–16.5 μm	1 kW, 200 kW	2.5 kW	$CdGeAs_2$	Piltch *et al.* (1975)

Frequency doubling and tripling can be used to convert a high power beam in the visible region to the ultraviolet. Sum frequency generation can produce tunable UV radiation when one of the input beams is tunable. Difference frequency generation can be used to create infrared radiation from the difference between two beams at visible wavelengths.

Table 1.31 lists some representative experiments in this area. This list is by no means complete but does serve to illustrate the range of nonlinear techniques available.

The output of a particular nonlinear crystal will depend on the angle between the incident beams and the crystal planes. Efficient conversion occurs only when the appropriate phase matching conditions are satisfied. The output wavelength at the converted frequency can therefore usually be tuned over a limited range by changing the angle of incidence. Further tuning can be accomplished by varying the temperature of the nonlinear crystal. Temperatures above or below room temperature may be required. An example of the tuning range obtained under these conditions is shown in Figure 1.41.

A schematic diagram of the optical system required to frequency triple the 1.06-μm output of a *Q*-switched Nd–YAG laser is shown in Figure 1.42. The frequency doubling crystal is placed in the YAG laser cavity so that the laser output contains both 1064- and 532-nm components. These are orthogonally polarized and so a quartz rotator is required to bring both beams into the same state of polarization. The third harmonic at 355 nm is generated in an external nonlinear crystal.

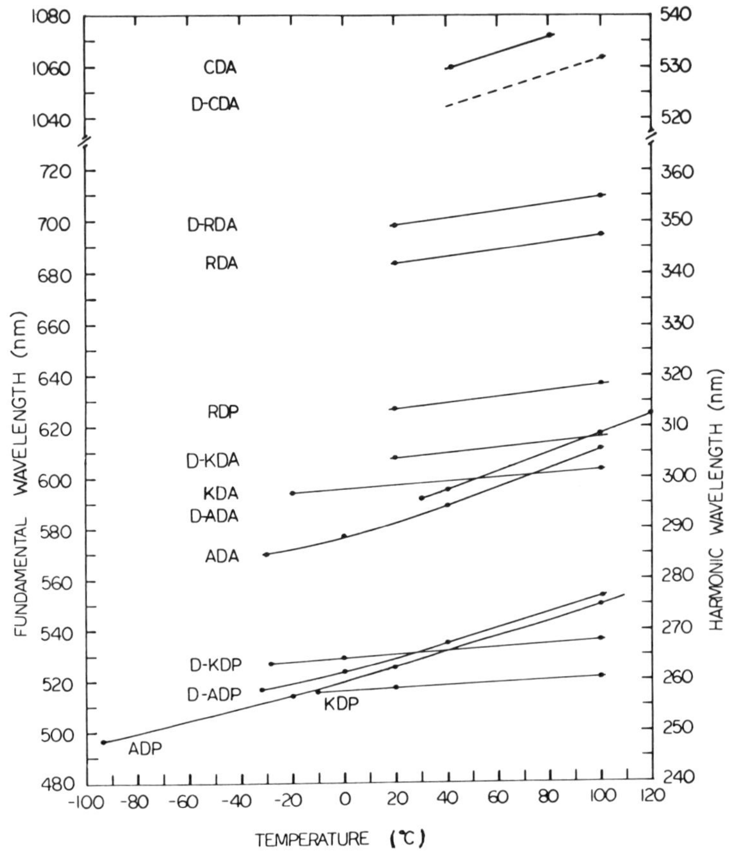

Figure 1.41. Frequency doubling range for 90° phase-matched crystals (courtesy Quantum Technology Inc.): CDA, cesium dihydrogen arsenate, D-CDA, cesium dideuterium arsenate; RDA, rubidium dihydrogen arsenate; D-RDA, RbD_2AsO_4; RDP, rubidium dihydrogen phosphate; KDA, potassium dihydrogen arsenate; D-KDA = KD_2AsO_4; D-KDA = KD_2PO_4; ADA, ammonium dihyrogen arsenate; KDP, dihydrogen arsenate; KDP, dihydrogen arsenate; KDP, KH_2PO_4; D-KDP, KD_2PO_4; ADP, ammonium dihydrogen phosphate.

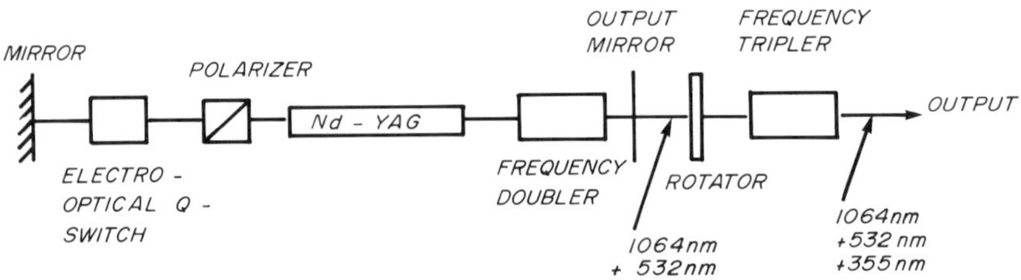

Figure 1.42. Example of frequency tripler for use with *Q*-switched Nd–YAG laser.

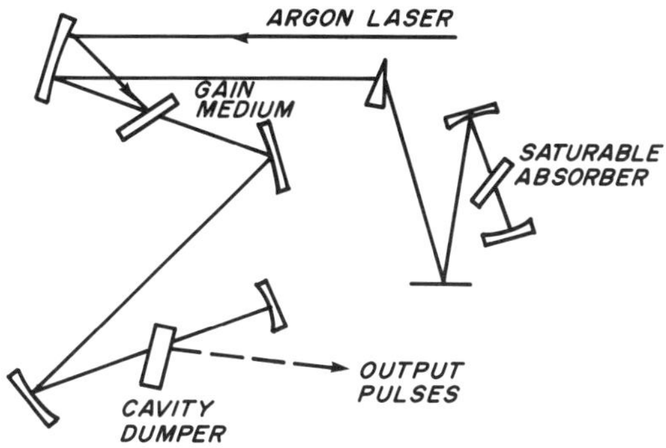

Figure 1.43. System for generation of subpicosecond pulses using CW dye laser. The acousto-optical cavity dumper provides pulses at $\sim 10^{10}$ Hz. Each pulse contains ~ 5 nJ and peak power is 5 kW [after Ippen and Shank (1979)].

1.9. MODE LOCKING

In a multilongitudinal mode, laser operation shifts from one mode to another and no distinct phase relationship exists between individual modes. Under certain conditions, however, several modes may oscillate together, maintaining a fixed phase relationship. This is the mode-locked condition.

Oscillating modes interfere to produce short light pulses separated in time by twice the time required for light to traverse the length of the cavity. This is $\Delta t = 2L/c$, where L is the cavity length and c is the speed of light.

The duration of these pulses is $\tau = 1/\Delta\nu$, where $\Delta\nu$ is the oscillating bandwidth of the laser line. For a gas this may only be 10^{-9} s. However, for a solid-state laser where $\Delta\nu \sim 10^{12}$–10^{13} Hz, $\tau \lesssim 1$ ps. In principle, with a dye laser where $\Delta\nu \sim 10^{14}$–10^{15} Hz, pulses as short as 10^{-14}–10^{-15} s could be generated via mode locking.

Although spontaneous mode locking occurs in some systems (e.g., TEA-type CO_2 lasers), mode-locking is usually induced by means of active or passive intracavity elements. Active mode-locking longitudinal modes will produce a time varying population difference between ground and excited states of the absorber with a component at the difference frequency between the modes. The decay time of the dye excited-state population must be less than the inverse of the mode separation frequency. Continuous trains of subpicosecond pulses have been produced in this way from a mode-locked CW dye laser [Ippen and Shank (1979)]. A system for cavity dumping individual mode-locked CW dye laser pulses is shown in Figure 1.43.

1.10. DETECTORS AND POWER METERS

Perhaps the most important parameter in laser processing is laser power. Measurement of laser power is straightforward but usually requires different instrumentation from that customarily used for measurements at low power levels. Conventional radiation detectors can be used, in fact, to measure laser outputs. However, the high sensitivity of these devices and their small power handling capability make them susceptible to damage at the power levels ordinarily obtained from lasers of even modest power. As a result, radiation detectors and power meters with higher power handling capability have been developed specifically for use with laser sources. These devices are commonly used in laboratory or industrial systems for monitoring of laser output characteristics. However, conventional radiation detectors are still used in many laser applications in analysis, particularly in remote sensing where power levels are low.

In this section we survey some representative detection systems for use with laser sources. Two separate types of detector are usually needed:

Power meter. Measures average power in a CW or pulsed beam without providing any information about the temporal behavior of the radiative source.

Radiation detectors. Measures the time-dependent intensity in a beam and can respond to rapid variations in intensity.

1.10.1. Power Meters

These devices are usually designed to dissipate high powers on a continuous basis. A common configuration involves the trapping of incident radiation in a blackened cone that is in intimate contact with a circulating water flow. Absorbed power can be related to the difference in temperature between input and output streams. An alternative design involves the use of thermistors to measure the temperature of the cone directly. These thermistors are in a bridge circuit whose imbalance when radiation is absorbed can be read out on a calibrated meter.

A variety of energy meters have been developed for use in the measurement of laser pulse energy. These devices relate the heating effect of a radiation pulse to the integrated energy in the pulse. Several commercial devices utilize the pyroelectric effect to measure pulse energies of up to several joules. These are commonly used to measure pulse energies in TEA-type CO_2, N_2, and RGH lasers. A detailed description of pyroelectric and other calorimeters has been given by Duley (1976). Table 1.32 gives a list of representative types of power and energy meters and summarizes their performance. Many of these instruments respond with similar efficiency and all wavelengths of interest. As such, they are true thermal detectors, i.e., detectors that respond only to the total energy absorbed irrespective of the spectral distribution of this radiation. This contrasts with the behavior of photon detectors which normally have high responsivity only over a limited spectral range.

Table 1.32. Performance of Commercial Energy and Power Meters

Spectral range	Minimum energy per pulse	Energy meters: Maximum energy per pulse	Rise time	Type
0.25–30 μm	0.1 J	800 J	30 s	Calorimetric
0.3–10.6 μm	0.005	1000	15 s	Calorimetric
1–100 μm	1	3000	20 s	Calorimetric
B.B.[a]	10^{-5}	0.15	5 ms	Pyroelectric
B.B.	5×10^{-2}	100	20 ms	Pyroelectric
B.B.	10^{-8}	1	50 ms	Pyroelectric
0.3–1.1 μm	10^{-3}	2	1 ns	Silicon
		Power meters		
Spectral range	Minimum power per pulse	Maximum power per pulse	Rise time	Type
0.3–10.6 μm	0.001 W	100 W	0.05 s	Calorimetric
1–100 μm	1	1000	20 s	Calorimetric
B.B.	10^{-6}	10	—	Pyroelectric
0.2–1.1 μm	10^{-12}	200	100 ns	Radiometer
8–14 μm	10^{-18}	10^{-4}	25 ns	Radiometer
0.2–1.1 μm	10^{-11}	10^{3}	100 ns	Silicon
0.4–1.1 μm	10^{-8}	0.1	5 ns	Silicon
B.B.	1	100	1 s	Thermopile
B.B.	10	1000	1 s	Thermopile

[a]B.B. = wide spectral response.

1.10.2. Radiation Detectors

These tend to be photon detectors, i.e., they respond to the absorption of individual photons. Each photon that is absorbed generates an electron–hole pair in the detector element. This charge is detected in an external circuit and can be related to the intensity of absorbed radiation. Since the response of these detectors does not depend on heat transfer, they tend to have superior response times to those obtained with thermal detectors. On the other hand, only certain photon energies are effective in creating electron–hole pairs in these materials, and so these devices tend to have a limited wavelength range over which operation is possible.

Several parameters can be defined that describe the operation of a radiation detector. These are as follows:

Responsivity. Ratio of output voltage or current to incident power. Units are A/W or V/W.

Noise equivalent power. The input radiative power that produces a signal voltage equal to the noise voltage in unit bandwidth. It is defined as

$$\text{NEP} = I_0 A V_N / V_S (\Delta f)^{1/2} \qquad \text{W Hz}^{-1/2}$$

where I_0 is the incident intensity, A the detector area, V_N the rms noise voltage, V_S the rms signal voltage, and Δf the bandwidth (Hz). A low NEP is desirable.

Detectivity. A measure of the sensitivity of a detector. The definition is

$$D^* = A^{1/2}/\text{NEP} \qquad \text{cm Hz}^{1/2}\ \text{W}^{-1}$$

Measured values of D^* must specify the wavelength of incident radiation, the chopping frequency in the incident beam, and the bandwidth over which measurements are made. The usual notation is $D^*(\lambda, \nu, \Delta f)$. A high value of D^* is desirable.

The three basic types of photon detectors are:

Photoemissive detectors. The photomultiplier tube is an example.

Photoconductive detectors. Absorption of radiation is accompanied by an increase in conductivity.

Photovoltaic/photodiode detectors. Absorption of radiation induces a change in current flow across a *p-n* junction.

A summary of values for D^* obtained at 0.5, 1.0, and 10 μm with different detector types is given in Table 1.33. Representative D^* values for the three regions are $D^*(10\ \mu\text{m}) = 2 \times 10^{10}$ cm $\text{Hz}^{1/2}\ \text{W}^{-1}$, $D^*(1\ \mu\text{m}) = 10^{11}$ cm $\text{Hz}^{1/2}\ \text{W}^{-1}$ and $D^*(0.5\ \mu\text{m}) = 4 \times 10^{11}$ cm $\text{Hz}^{1/2}\ \text{W}^{-1}$, typically at a repetition frequency of several hundred Hz.

If one had a detector operating at $\lambda = 1.06\ \mu$m with $D^* = 10^{11}$ cm $\text{Hz}^{1/2}\ \text{W}^{-1}$ and an active area $A = 1$ mm, then NEP $= 3 \times 10^{-12}$ W $\text{Hz}^{-1/2}$. Such a detector would yield a signal-to-noise ratio of unity for an incident power of 3×10^{-9} W in a bandwidth of 1 MHz.

The responsivity of photoconductive detectors is high with values of 10^4–10^6 V W^{-1} being fairly typical. Photovoltaic detectors have much lower responsivity. For example, a Si photodiode operating at 1 μm might have a responsivity of 0.5 A W^{-1}. Pyroelectric detectors have responsivities of $\sim 10^4$ V W^{-1} or $\sim 10^{-6}$ A W^{-1} although these values depend on detector response times. Large responsivity is obtained only with low-frequency operation. Response times for typical radiation detectors are given in Table 1.34.

A good example of a detector that has been designed exclusively for laser measurements is the photon-drag detector. Typical data for this detector when used at 10.6 μm is

Table 1.33. D^* for Various Detectors at 0.5, 1.0, and 10 μm

Detector	Temperature (K)	D^* 0.5 μm	D^* 1.0 μm	D^* 10 μm
HgCdTe	77			2×10^{10}
PbSnTe	77			4×10^{10}
PbS	200		2×10^{11}	
	300		6×10^{10}	
PbSe	200		10^{10}	
	300		2×10^{9}	
Ge:AuSb	77		3×10^{11}	
InAs	77		10^{11}	
	195		2×10^{10}	
InSb	77		3×10^{10}	
Ge	300	2×10^{10}	3×10^{10}	
GaAs	300	8×10^{10}		
Si	300	4×10^{11}	10^{12}	
		0.5	1.0	10.6
Thin-film[a] thermopile	300	←	$(1–5) \times 10^{8}$	→
Thermistor[a] bolometer	300	←	$(1–5) \times 10^{8}$	→
Pyroelectric[a]	300	←	$(6–8) \times 10^{8}$	→

[a]Thermal detector.

Table 1.34. Response Times for Typical Radiation Detectors

Detector type		Response time
Photoconductive	Ge	10 ns
	PbS	300 μs
	PbSe	1 μs
	HgCdTe	50–100 ns
Photovoltaic	Ge	100 ps
	InSb	2 ns
	InAs	5 ns
	HgCdTe	2 ns
	Si	100 ps
	Photon drag	<1 ns
Photoelectric		to <1 ns

NEP	600 W $Hz^{-1/2}$
Responsivity	200 μV W^{-1}
Response time	<1 ns

Despite the high NEP and low responsivity, the photon-drag detector is extremely useful in characterizing the output of high-power CW CO_2 and TEA-type CO_2 lasers. Its advantages include a fast response time and high-power handling capacity.

The choice of a radiation detector for a particular application often involves a trade-off of a variety of sometimes disparate factors. The usual gain–bandwidth considerations are dominant in some applications. In these it may be possible to sacrifice sensitivity for a wider bandwidth and improved response time or, alternatively, to sacrifice rise time for sensitivity.

Lowering the detector temperature and restricting the field of view increases D^* for most infrared detectors. The exception is the pyroelectric detector that operates only at room temperature. Cooling brings its own set of problems. However, the need for liquid cryogens has almost been eliminated these days with the availability of compact closed-cycle refrigerators that can cool to 4 K. Nevertheless, a cryostat is required and this may constitute a limitation to the portability of a detection system.

Many photoconductive and photovoltaic detectors are fairly fragile mechanically and can be easily damaged. They can also be damaged by exposure to CW radiative fluxes even at levels as small as $\sim$1 W cm^{-2}. As a result, care must be taken to ensure that in sampling high-power beams only a small fraction of the beam power is gathered by the device.

Many radiation detectors are now available in the form of multiplexed arrays. This is an advantage in beam profiling or in the optical readout of spectrometer signals. In the latter application an array of photodiodes placed at the focal plane of a monochromator or spectrometer provides simultaneous detection of several wavelengths.

Chapter 2

Materials Processing

2.1. ABSORPTION OF LASER RADIATION BY METALS

For an opaque solid, the fraction of incident radiation absorbed is

$$\epsilon = 1 - R_0$$

where ϵ is the emissivity and R_0 is the reflectivity at normal incidence. R_0 and ϵ can be calculated from measurements of optical constants or the complex refractive index. For a complex refractive index,

$$m = n - ik$$

the reflectivity at normal incidence is

$$R_0 = \frac{(n-1)^2 + k^2}{(n+1)^2 + k^2}$$

The emissivity is then

$$\epsilon = \frac{4n}{(n+1)^2 + k^2}$$

In general, n and k for metallic materials are functions of wavelength and temperature. The variation of n and k with wavelength and corresponding changes in ϵ for Ti at 300 K are shown in Figure 2.1. It is apparent that both n and k are relatively slowly varying functions of λ over the range $0.4 < \lambda < 1.0$ μm and ϵ is large in this range. At longer wavelengths, n and k both increase rapidly with λ and ϵ decreases to a small fraction of its value at shorter wavelengths. In the infrared $\epsilon \propto \lambda^{-1/2}$ at constant temperature. Since $\epsilon \propto$

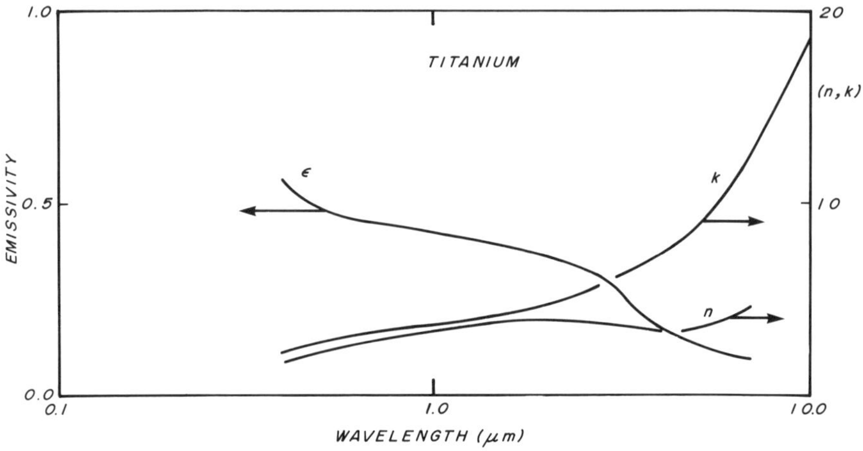

Figure 2.1. Wavelength dependence of ϵ, n, and k for Ti at 300 K.

$r^{1/2}$, where r is the electrical resistivity, while r increases with temperature, $\epsilon(\lambda)$ increases with temperature in the infrared [Duley (1976)].

The temperature dependence of $\epsilon(\lambda)$ for $\lambda \lesssim 1\ \mu m$ is more complex; however, the net change in ϵ is smaller than observed in the infrared. In the visible region of the spectrum ϵ often decreases slightly with increasing temperature.

Emissivities for metals at wavelengths characteristic of Ar^+, ruby, Nd–YAG, and CO_2 lasers are summarized in Table 2.1. The temperature dependence of ϵ for some metals at 1000 nm and 10.6 μm is shown in Figures 2.2 and 2.3, respectively. Data on ϵ(1000 nm) vs. T were obtained from the observations of Barnes (1966). That for $\epsilon(10.6\ \mu m)$ was calculated from the temperature-dependent emissivity from the expression given by Duley (1976).

$$\epsilon(10.6\ \mu m) = 11.2[R(1 + \gamma T)]^{1/2} - 62.9[R(1 + \gamma T)] + 174[R(1 + \gamma T)]^{3/2}$$

where R is the resistivity at 20°C and γ is the coefficient of resistivity change with temperature T°C.

Inspection of the data in Table 2.1 and Figures 2.1–2.3 shows that the absorption of laser light by metal surfaces at 20°C is almost an order of magnitude larger at visible wavelengths than at infrared wavelengths. However, values of ϵ given apply only for clean metal surfaces heated in vacuum. In most practical applications of laser heating this assumption will not be valid because of oxide formation or the presence of surface contamination. When this is the case, values of ϵ in the infrared can be increased substantially. Correspondingly large increases do not occur in ϵ measured at visible wavelengths. Thus, under practical conditions the difference between $\epsilon(10.6\ \mu m)$ and ϵ(visible) is unlikely to be as large as suggested by the data in Table 2.1.

Table 2.1. Values of Emissivity for Various Metals At Laser Wavelengths

Metal	Emissivity[a] Ar^+ (500 nm)	Ruby (700 nm)	Nd–YAG (1000 nm)	CO_2 (10 μm)
Aluminum	0.09	0.11	0.08	0.019
Copper	0.56	0.17	0.10	0.015
Gold	0.58	0.07	—	0.017
Iridium	0.36	0.30	0.22	—
Iron	0.68	0.64	—	0.035
Lead	0.38	0.35	0.16	0.045
Molybdenum	0.48	0.48	0.40	0.027
Nickel	0.40	0.32	0.26	0.03
Niobium	0.58	0.50	0.32	0.036
Platinum	0.21	0.15	0.11	0.036
Rhenium	0.47	0.44	0.28	—
Silver	0.05	0.04	0.04	0.014
Tantalum	0.65	0.50	0.18	0.044
Tin	0.20	0.18	0.19	0.034
Titanium	0.48	0.45	0.42	0.08
Tungsten	0.55	0.50	0.41	0.026
Zinc	—	—	0.16	0.027

[a]At 20°C.

Duley *et al.* (1979) have investigated the effect of oxidation on ϵ(10.6 μm) for several metals heated in air. This follows work by Wieting and de Rosa (1979) and Wieting and Schriempf (1976) on the absorptance of stainless steel and Ti–6A1–4V alloy at high temperatures in vacuum. Data obtained by Wieting and de Rosa on the absorptivity, ϵ, of type 304 stainless steel at 10.6

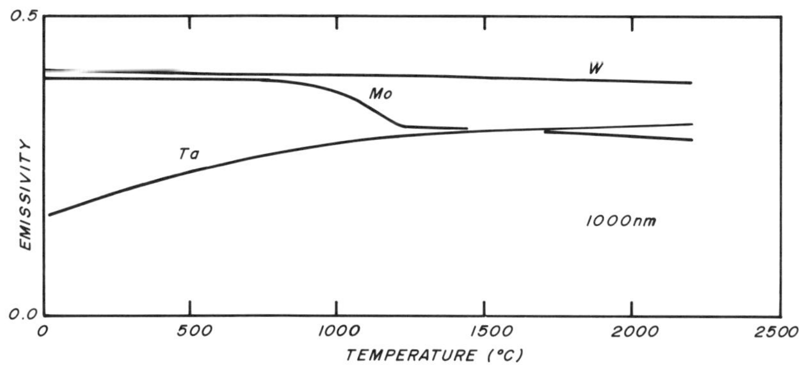

Figure 2.2. Temperature dependence of emissivity for several metals at a wavelength of 1000 nm.

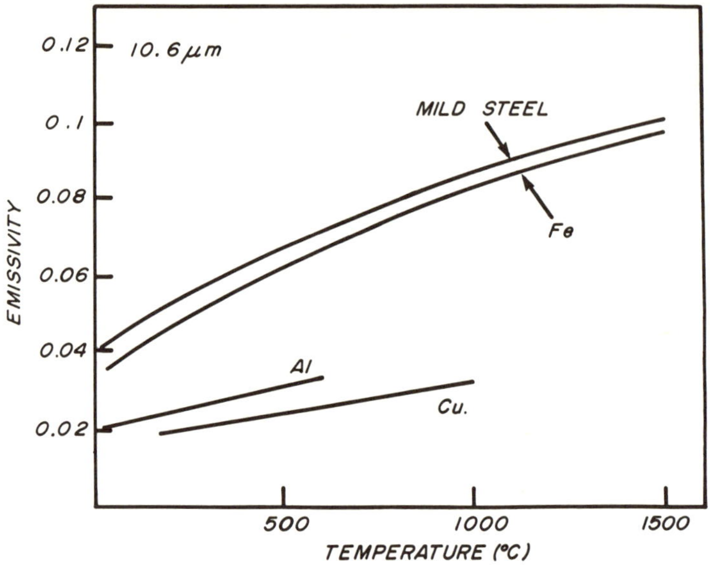

Figure 2.3. Temperature dependence of emissivity at 10.6 μm for several metals.

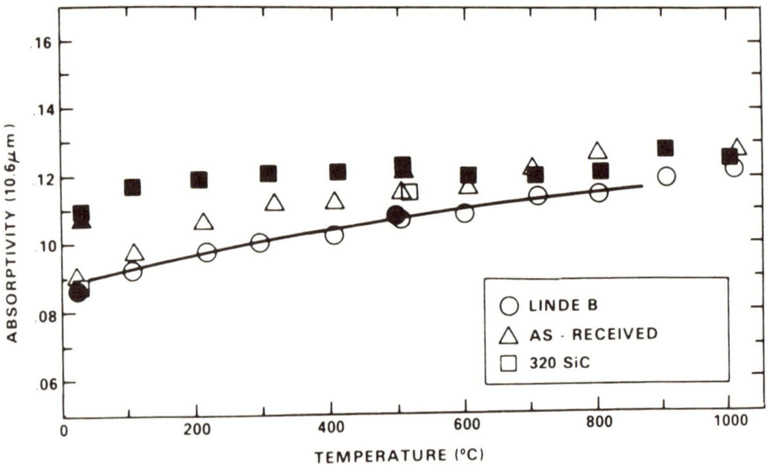

Figure 2.4. Absorptivity of 304 stainless steel at 10.6 μm as a function of temperature and surface condition. The black data points were obtained while increasing the furnace temperature from 25 to 1000°C; the other data points mark the return half of the temperature cycle. Each point represents an average of between two and four measurements [Weiting and deRosa (1969)].

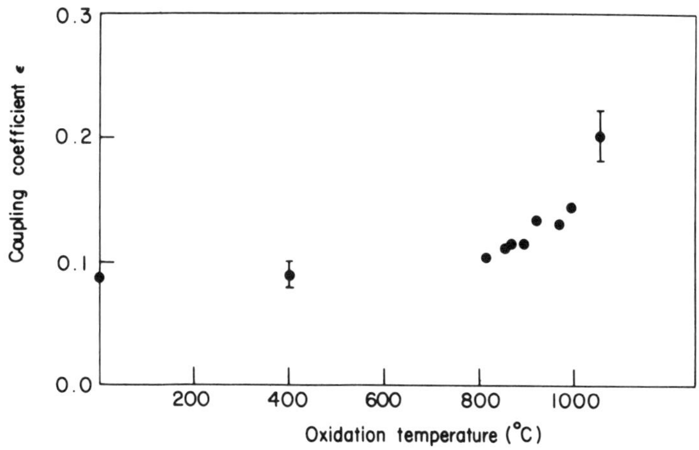

Figure 2.5. ϵ(10.6 μm) vs. T for 304 stainless steel oxidized for 1 min in air at indicated temperatures [Duley *et al.* (1979)].

μm are shown in Figure 2.4. Figure 2.5 shows the effect of a 1-min oxidation in air on ϵ(10.6 μm). It is evident that ϵ(10.6 μm) increases substantially when surface oxide is present. Since the growth of an oxide layer is time-dependent at a particular temperature, ϵ is also a function of time. This behavior can be seen in Figures 2.6 and 2.7 for Mo and Ta, respectively.

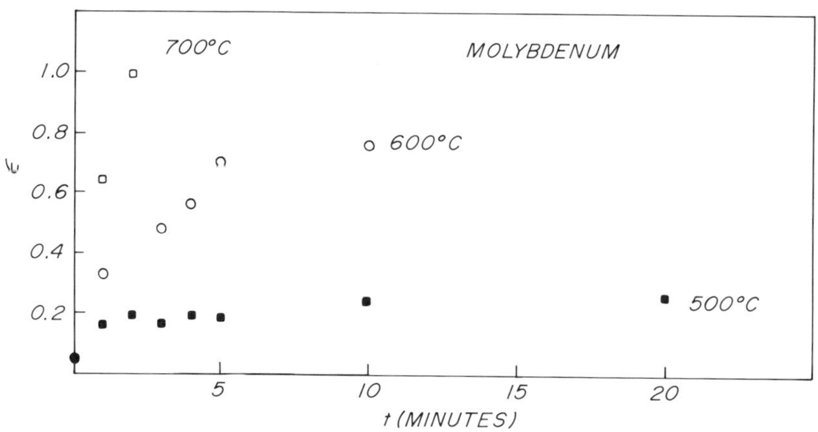

Figure 2.6. Variation of ϵ(10.6 μm) for Mo with oxidation time at various temperatures. Oxidation is in air.

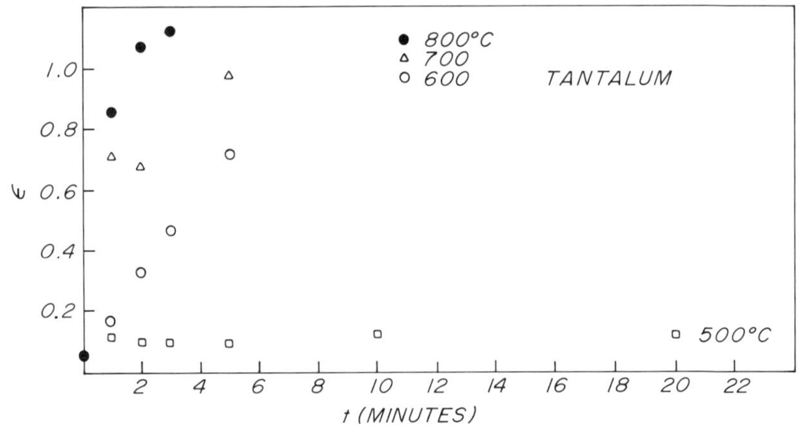

Figure 2.7. Variation of $\epsilon(10.6\ \mu m)$ for Ta.

These data were obtained under low-power irradiation with a CW CO_2 laser. A variety of dynamic effects can change ϵ at high irradiation levels. These effects have been extensively investigated [Arzuov *et al.* (1975), Marcus *et al.* (1976), Ready (1976), Maher and Hall (1978), Arzuov *et al.* (1979), Stürmer and von Allmen (1978), Robin (1978), Roessler and Gregson (1978), and McKay and Schriempf (1979)]. These experiments show that a substantial increase in the coupling coefficient for laser radiation at the target occurs when the intensity is sufficient to initiate a breakdown plasma in the focal region. This plasma absorbs some fraction of the incident laser radiation and transfers the resulting energy to the target via hydrodynamic expansion. While the energy coupled from the laser to the target can be increased by this mechanism, it is deposited over a larger area than the normal beam focus. As such it may inhibit thermal processes (drilling, material removal) at the focal point. An example of the increase in coupling coefficient that occurs with increasing intensity is shown in Figure 2.8. Note that the thermal coupling coefficient is equivalent to ϵ. These data were obtained with a 0.5 μs pulse from a TEA-type CO_2 laser. The increase in ϵ at an energy density of $\sim$10 J cm^{-2} was found to accompany the onset of surface damage at the focus.

The time evolution of target shielding via plasma formation has been followed in detail by Stürmer and von Allmen (1978). They identify three separate absorption regimes when a high-intensity pulse of long duration is incident on a metallic target in air or other atmosphere. These are:

1. Strong reflection from target
2. Absorption by plasma and target shielding
3. Dissipation of plasma and enhanced coupling to target

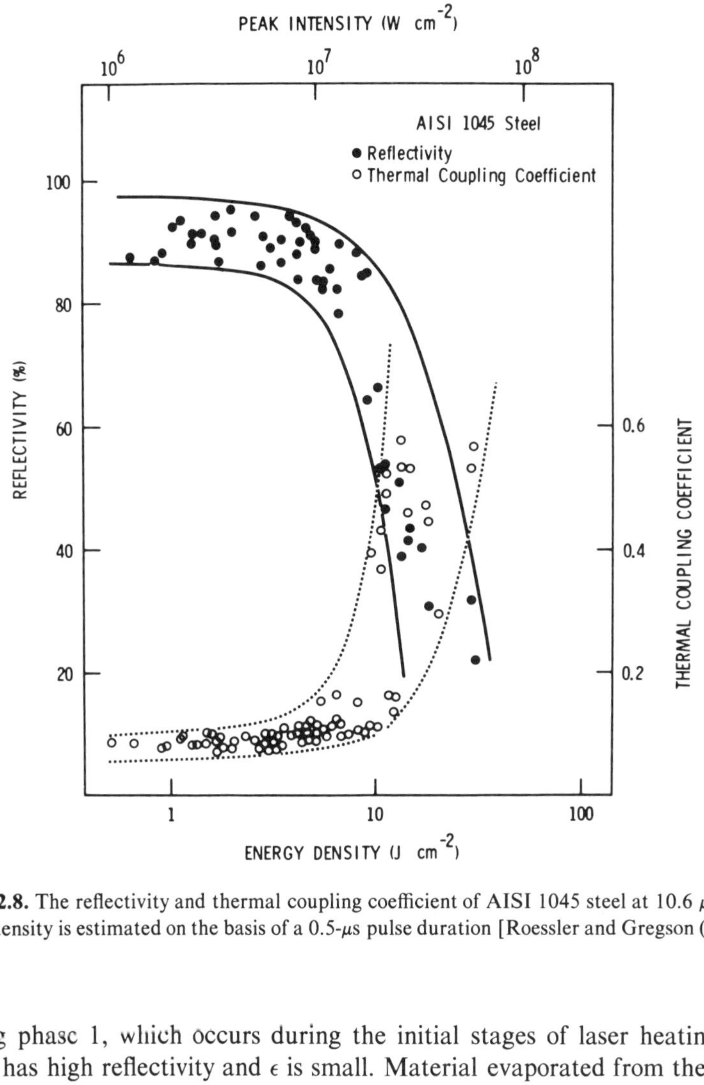

Figure 2.8. The reflectivity and thermal coupling coefficient of AISI 1045 steel at 10.6 μm. The peak intensity is estimated on the basis of a 0.5-μs pulse duration [Roessler and Gregson (1978)].

During phase 1, which occurs during the initial stages of laser heating, the target has high reflectivity and ϵ is small. Material evaporated from the focus may seed the gas in front of the target initiating plasma breakdown and lead to the formation of a laser supported detonation (LSD) wave. This wave absorbs practically all the incident laser radiation (phase 2) and shields the target. It dissipates by moving away from the target toward the focusing lens. This results in a reduction of plasma density followed by a decrease in opacity. The surface is then exposed to the last part of the laser pulse (phase 3) which couples efficiently to the damaged target. Subsequent LSD wave ignition is suppressed by the low gas density left in front of the target following dissipation of the initial LSD wave.

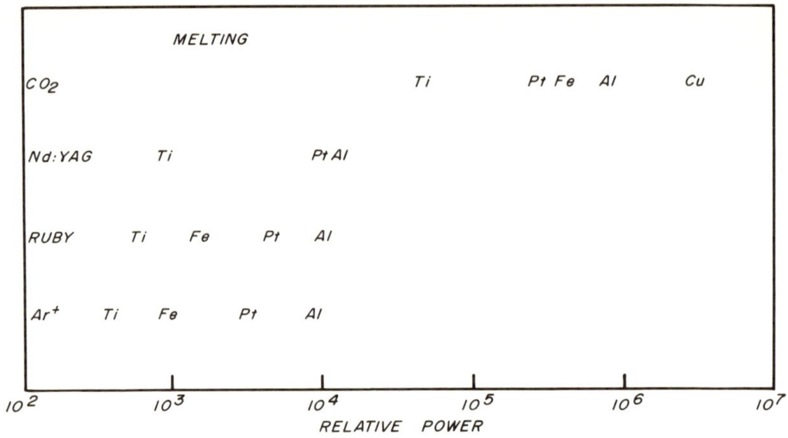

Figure 2.9. Relative power to produce surface melting with various laser sources.

While ϵ may be important in the initial stages of the heating of metallic targets with laser radiation, it is unimportant in many practical laser heating applications. The importance of ϵ is diminished when material removal has proceeded to the point where a cavity or keyhole has formed in the workpiece. In this case, the cavity acts as a blackbody absorber with ϵ effectively equal to unity. It has been shown [Shewell (1977)] that control over the fitting of parts to be joined can also be effective in increasing ϵ. Under conditions in which laser radiation is absorbed in a keyhole, Steen and Eboo (1979) have shown that plasma absorption occurs *within* the keyhole yielding $\epsilon = 1$.

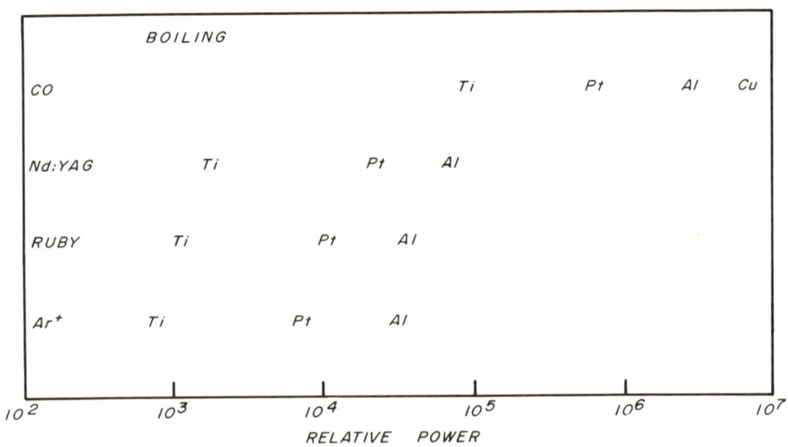

Figure 2.10. Relative power to produce boiling at surface with various laser sources.

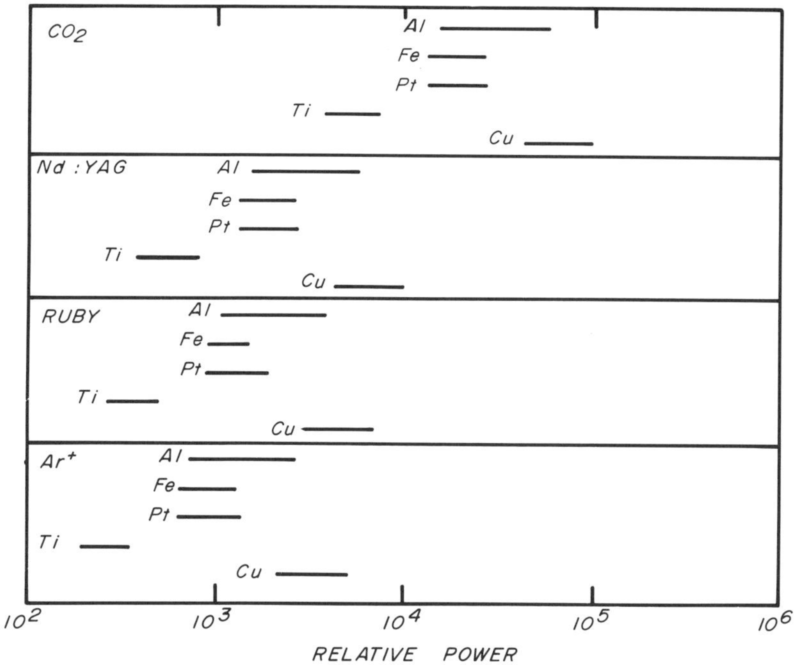

Figure 2.11. Relative power to reach temperatures between T_m and T_b assuming $\epsilon = 1$.

The data contained in Table 2.1 can be used to obtain an estimate of the relative merit of Ar^+, ruby, Nd–YAG, and CO_2 sources for laser heating. Heat transfer calculations show that the limiting temperature at the center of a Gaussian focal spot on a bulk target is

$$T = \frac{\epsilon I_0 d \pi^{1/2}}{K}$$

where I_0 is the peak laser intensity (W cm^{-2}), d the Gaussian beam radius, and K the thermal conductivity. With optimum focusing $d \propto \lambda$, where λ is the laser wavelength. Since $I_0 \propto P/\lambda^2$, where P is the laser power, one has

$$P \propto \frac{KT\lambda}{\epsilon}$$

If we assume that useful thermal effects are produced only when $T = T_m$ or T_b, where T_m is the melting temperature while T_b is the boiling temperature, then this expression can be used to obtain an estimate of the relative difficulty of machining with different laser sources. A comparison of this sort is shown in Figures 2.9 and 2.10 for the production of melting and boiling, respectively.

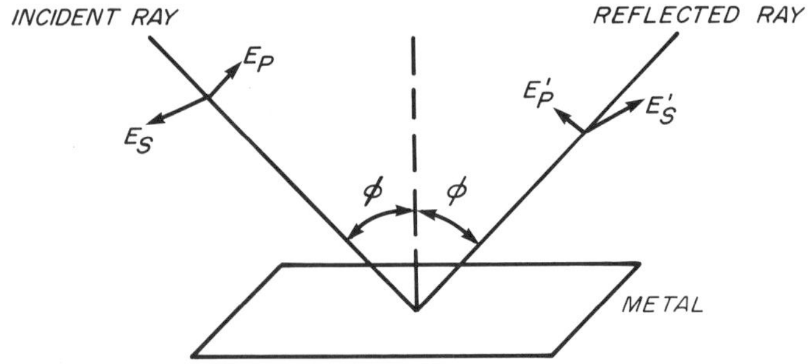

Figure 2.12. Incident and reflected waves at a metal surface.

The correlations shown in these two figures are, of course, highly approximate because they assume room temperature values for ϵ and K can be used to high temperatures. The uncertainty in these estimates can be reduced somewhat by taking $\epsilon = 1$. A comparison calculated on this basis is shown in Figure 2.11. If we compare powers required to melt or boil Ti from this figure we see that $P(CO_2)/P(Ar^+) \sim 20{:}1$. Taking $\epsilon < 1$ as shown in Figure 2.1, the corresponding ratio is ~ 120. Thus, while higher powers are required to initiate surface damage with infrared lasers than with visible lasers, when damage has occurred (i.e., $\epsilon \sim 1$), much of the effect of the increased spot size of infrared lasers has disappeared.

At anything other than normal incidence the reflection of laser radiation depends on polarization. The geometry at some general angle of incidence ϕ is shown in Figure 2.12. The reflectivity for the two polarization directions s and p, R_s and R_p, will in general be different. This means that the reflection coefficient for polarized laser light will be dependent on the orientation of the polarization vector relative to the metal surface.

An example of the angular dependence of R_s and R_p for Cu at 10.6 μm is shown in Figure 2.13. It can be seen that R_s is high for all angles. However, R_p becomes very small at close to grazing incidence. Thus ϵ for incident light polarized perpendicular to the metal surface is larger under these conditions and enhanced coupling occurs. This has some important consequences in laser machining where the efficiency of material removal depends on the relation between polarization direction and the direction of translation of the metal substrate. This effect will be discussed in more detail in Section 2.6.2.

2.2. ABSORPTION OF LASER RADIATION BY SEMICONDUCTORS AND INSULATORS

Absorption of light by insulating materials is a strong function of wavelength. In the infrared, absorption arises from vibrational modes of the crystal lattice or in organic solids by intermolecular vibrations. Absorption coefficients $\alpha \sim 10^2$–10^4 cm^{-1} are typical within these bands. In the visible, absorption may occur due to impurities (e.g., transition metal ions, crystal defect centers, etc.) or due to the tail of strong ultraviolet absorption bands. Absorption can also occur due to discrete electronic transitions in molecular crystals (e.g., many organic solids). Absorption coefficients are typically 10^3–10^6 cm^{-1} within absorption bands.

Figure 2.14 shows absorption coefficients, α, for several refractory materials in the visible and ultraviolet. α can be related to the transmission of a sheet of thickness t via

$$\frac{I}{I_0} \times 100 = \text{Transmission in percent}$$
$$= 100e^{-\alpha t}$$

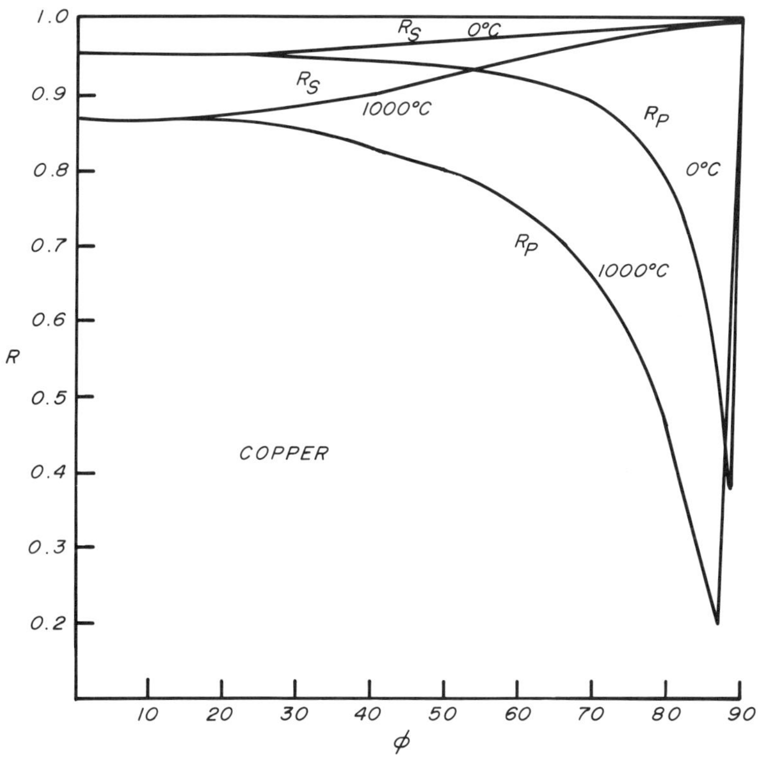

Figure 2.13. Angular dependence of reflectivity for s and p polarizations with 10.6-μm radiation incident on Cu at 20 and 1000°C.

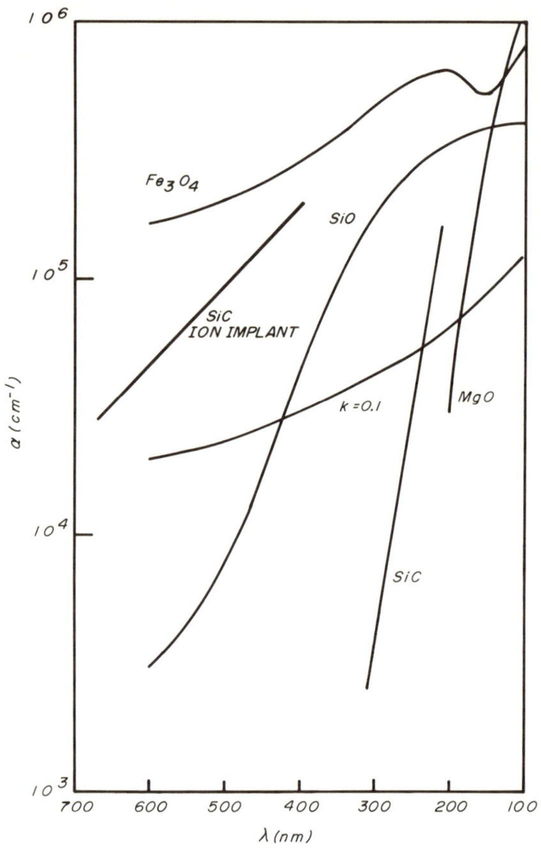

Figure 2.14. Absorption coefficient $\propto$ for several insulators at wavelengths between 100 and 700 nm. The curve for $k = 0.1$ corresponds to the absorption that would be produced by a material with an imaginary refractive index equal to this value.

or

$$\frac{I}{I_0} = e^{-\alpha t}$$

where I_0 is the incident intensity and I is the transmitted intensity. A useful measure of the thickness required for significant attenuation of incident radiation is given by

$$L = \alpha^{-1}$$

where L is the attenuation length. A strong absorber has $\alpha = 10^6\ \text{cm}^{-1}$ and $L = 10^{-6}$ cm while a relatively weak absorber has $\alpha = 10^1\ \text{cm}^{-1}$ and $L = 10^{-1}$ cm.

The relation between α and refractive index is given by

$$\alpha = \frac{4\pi k}{\lambda}$$

where k is the imaginary term in the complex refractive index

$$m = n - ik$$

while λ is the wavelength of incident light. In the visible region, nominally transparent materials will typically have $k \sim 10^{-5}$ or $\alpha \sim 10\ \text{cm}^{-1}$. The absorption due to $k = 0.1$ is shown in Figure 2.14.

Data on the transmission of several insulating materials in the infrared are shown in Figure 2.15. Further data can be found in the useful compendium of curves in *Laser Focus Buyers Guide*. Table 2.2 lists wavelength ranges over which a variety of insulating and semiconductor materials are nominally transparent in the infrared region. In the range over which these materials are transparent, α may be as small as $1\text{–}10\ \text{cm}^{-1}$.

Many semiconducting materials are opaque in the visible ($\lambda \lesssim 1\ \mu\text{m}$) and relatively transparent in the infrared. This is due to the influence of band-to-band absorption in the visible. In the infrared, absorption is primarily due to free-carrier absorption and transitions to impurity levels. This combination of factors means that most semiconductor annealing is done with Nd–YAG or other short-wavelength lasers (Chapter 3).

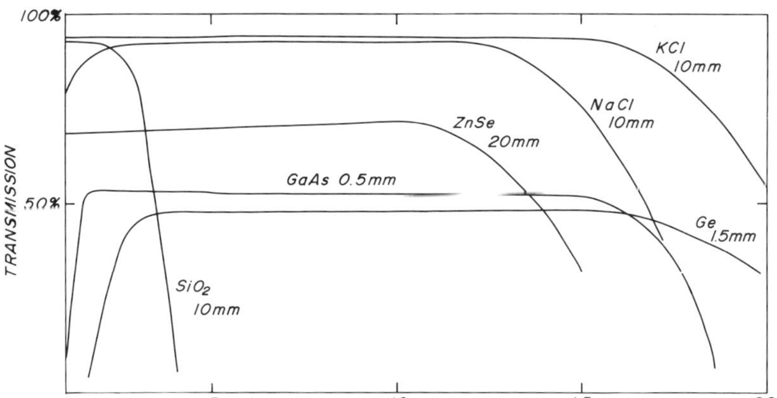

Figure 2.15. Infrared transmission of some insulators including losses due to reflection. Reflection losses can be large, especially for materials with high refractive index. For example, the normal incidence reflectivity of Ge at 10.6 μm is ~30%. Sample thicknesses are given [data from *Laser Focus Buyers Guide* (1981)].

Most materials with Si–O bonds are relatively transparent in the visible but absorb strongly at or near 10 μm. Thus, processing of quartz, glass, and silicate minerals is best accomplished with the CO_2 laser. Organic solids are strongly absorbing throughout the infrared but may be relatively transparent at shorter wavelengths (an example is polyethylene). Thus the CO_2 laser is ideal for the processing of these materials.

Unlike metals where radiation is absorbed in the skin depth at the surface, absorption in insulators and most semiconductors occurs over the attenuation length, L, which can be much larger than typical skin depths. In the infrared $L \gtrsim 10^{-4}$ cm and thus, in many instances, heating must be considered to be a volume effect. This is particularly valid for heating of thin films where L may exceed the film thickness.

Although reflection coefficients for polarized light depend on angle of incidence and direction of polarization as in metals, ϵ is almost unity during pro-

Table 2.2. Range of Transparency of Some Insulating and Semiconducting Materials in Infrared[a]

Material	Range between 10% cutoff points in transmission[b] (μm)
Al_2O_3	0.15–6.5
As_2S_3	0.6–13
BaF_2	0.14–15
CdSe	0.72–24
CdS	0.5–16
CdTe	0.3–30
CaF_2	0.13–12
CsBr	0.2–45
CuCl	0.4–19
Diamond (type IIa)	0.225–2.5, 6–100+
GaAs	1–15
Ge	1.8–23
InAs	3.8–7.0
PbS	3–7
MgO	0.25–8.5
Se	1–20
SiO_2 (fused)	0.2–4.5
Si	1.2–15
TiO_2	0.43–6.2
ZnSe	0.5–22
ZnS	0.54–10

[a]Compiled from data given in *Laser Focus Buyers Guide* (1980); [b]Typically for a thickness of 1–10 mm.

cessing and the effects seen in machining metals with polarized beams do not occur with insulators.

2.3. THERMAL CONSTANTS

The absorption of laser radiation by a solid is equivalent to a source of heat in or on the solid. The response of a material to this source can be calculated by solving the three-dimensional heat transfer equation

$$\rho C \frac{\partial T}{\partial t} = \frac{\partial}{\partial x}\left(K \frac{\partial T}{\partial n}\right) + \frac{\partial}{\partial y}\left(K \frac{\partial T}{\partial y}\right) + \frac{\partial}{\partial z}\left(K \frac{\partial T}{\partial z}\right) + A(x,y,z,t) \tag{2.1}$$

While the form of the solution to this equation, $T(x,y,z,t)$, is determined by the source distribution function $A(x,y,z,t)$ and the boundary conditions pertaining to the problem, the *amplitude* of the solution is controlled by the magnitude of the thermal constants, C and K. These constants are

K = thermal conductivity (W cm^{-1} °C^{-1})
C = heat capacity (J g^{-1} °C^{-1})

The role played by K can be seen by reference to Figure 2.16. This shows a block of material of cross-sectional area A and length L subjected to a heat flux at one end. This heat is conducted through the material to emerge at the other end of the block. In equilibrium, the heat input is equal to the heat output but the temperature within the block has risen in response to the transfer of heat. The distribution of temperature in equilibrium is shown in the lower part

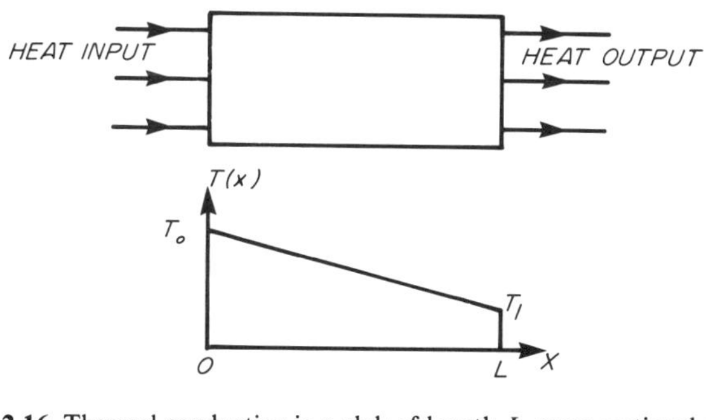

Figure 2.16. Thermal conduction in a slab of length L, cross-sectional area A.

of the figure. If dQ/dt is the power input then

$$\frac{dQ}{dt} = KA\,\frac{T_0 - T_1}{L} \tag{2.2}$$

Then the thermal conductivity, K, determines the temperature rise in equilibrium in response to dQ/dt. Since $T_0 - T_1 \propto K^{-1}$, a material with small K exhibits a higher temperature rise than that with large K in response to the same dQ/dt.

The fundamental nature of C can be understood by realizing that, until the equilibrium is achieved, the material stores heat in response to the externally applied heat source. The energy stored by a mass m of material that has undergone a temperature rise ΔT is

$$\Delta E = Cm\,\Delta T$$

If this incremental expression is integrated over the entire sample, then the excess heat stored in the material can be calculated.

The magnitudes of C and K determine the rate at which thermal equilibrium is achieved in response to the sudden application of the heat source. This can be seen most easily from the thermal diffusivity, κ, defined as

$$\kappa = \frac{K}{\rho c}$$

κ has the units $cm^2\ s^{-1}$ and ρ is the sample density (g cm^3). This distance x that the heat wave has penetrated the sample after time t can then be found approximately from

$$x = (Kt)^{1/2}$$

Upon creation of the heat source at $x = 0$ and $t = 0$ heat flows into the material with the heat wave traveling a distance x in time t. The material over the region $0 < x < x(t)$ undergoes a rise in temperature and stores heat according to its heat capacity. The material between $x = x(t)$ and $x = L$ is unheated at this time. As the heat source continues to supply energy at $x = 0$, the heating wave finally reaches $x = L$ and heat begins to leave the sample over this face. The temperature at $x = L$ will increase to the value required by the heat transfer equation

$$\frac{dQ}{dt} = \frac{KA(T_0 - T_1)}{L}$$

At this point, equilibrium is achieved and $(dQ/dt)_{x=0} = (dQ/dt)_{x=L}$

The processes outlined above occur in all materials heated by point or distributed energy sources, although an analytical description of $T(x,y,z,t)$ may not be obtainable because of complex boundary conditions involving heat transfer into and out of the sample.

A summary of thermal constants for a variety of solids is given in Table 2.3.

2.4. LASER DRILLING: HEAT TRANSFER

2.4.1. Heating without Change of Phase

The solution to the heat transfer equation can be expressed in analytical form only when the system possesses certian symmetrical boundary conditions. When this is not the case, the heat transfer equation can be solved numerically. Detailed analysis of almost any practical laser heating problem requires that a numerical approach be adopted. However, as considerable physical insight into laser heating mechanisms can often be obtained from approximate solutions expressed in analytical form, available analytical solutions are summarized in this section.

These solutions, together with relevant boundary conditions, are shown in Figures 2.17–2.29. While this does not represent a complete set of possible solutions, the examples shown have been chosen to be representative of boundary conditions most often encountered in laser drilling. Further examples and discussion can be found elsewhere [Carslaw and Jaeger (1959), Ready (1971), and Duley (1976)].

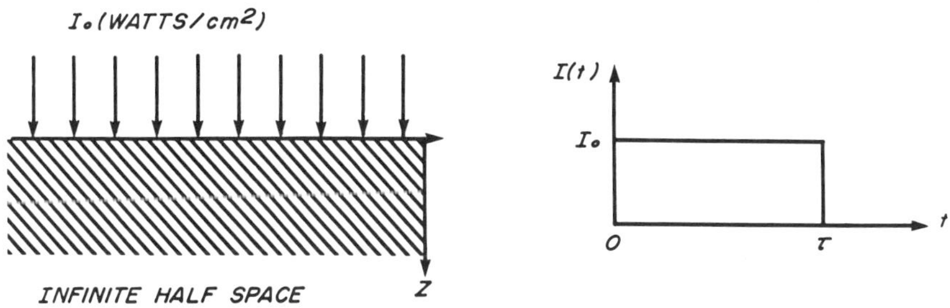

Figure 2.17. $0 < t < \tau$:

$$T(z,t) = \frac{2\epsilon I_0}{K}(\kappa t)^{1/2}\, ierfc\left[\frac{z}{2(\kappa t)^{1/2}}\right]$$

$$T(0,t) = \frac{2\epsilon I_0}{K}\left(\frac{\kappa t}{\pi}\right)^{1/2}$$

$t > \tau$:

$$T(z,t) = \frac{2I_0\kappa^{1/2}}{K}\left[t^{1/2}\, ierfc\frac{z}{2(\kappa t)^{1/2}} - (t - T)^{1/2}\, ierfc\frac{z}{2[\kappa(t - T)]^{1/2}}\right]$$

Table 2.3. Thermal Constants for Various Materials at a Range of Temperature: K, Thermal Diffusivity ($cm^2\ s^{-1}$); C, Heat Capacity ($J\ gm^{-1}\ {}^\circ C^{-1}$); κ, Thermal Conductivity ($W\ cm^{-1}\ {}^\circ C^{-1}$)

	0°C			300°C			700°C			1700°C			2700°C		
	K	C	κ	K	C	κ	K	C	κ	K	C	κ	K	C	κ
Aluminum	2.36	0.83	1.03	2.38	1.0	0.88	—	—	—	—	—	—	—	—	—
Chromium	0.95	0.5	0.27	0.85	0.49	0.24	0.65	0.58	0.16	—	—	—	—	—	—
Cobalt	1.04	—	—	0.75	—	—	—	—	—	—	—	—	—	—	—
Copper	4.01	0.38	1.19	3.88	0.42	1.04	3.56	0.47	0.85	1.82	—	—	1.80	—	—
Gold	3.18	0.14	1.22	3.09	0.14	1.19	2.78	0.15	0.93	1.20	—	—	1.25	—	—
Iron	0.84	0.42	0.28	0.62	0.54	0.15	0.33	0.98	—	0.43	—	—	0.46	—	—
Lead	0.36	0.13	0.24	0.33	0.14	0.21	0.2	—	—	—	—	—	—	—	—
Molybdenum	1.39	0.25	0.54	1.30	0.26	0.49	1.12	0.29	0.38	0.88	0.39	0.22	—	—	0.13
Nickel	0.94	0.44	0.2	0.72	0.52	0.17	0.72	0.56	0.14	—	—	—	—	—	—
Platinum	0.72	0.13	0.26	0.72	0.14	0.24	0.79	0.15	0.24	—	0.18	0.27	—	—	—
Rhodium	1.51	—	—	1.40	—	—	1.21	—	—	—	—	—	—	—	—
Silver	4.28	0.23	1.74	4.12	0.24	1.61	3.74	0.27	1.3	1.97	—	—	1.91	—	—
Tantalum	0.57	0.14	0.25	0.58	0.14	0.24	0.60	0.15	0.24	0.64	0.17	0.22	0.67	0.24	0.17
Tin	0.68	0.23	—	0.60	0.26	—	0.41	—	—	—	—	—	—	—	—
Titanium	0.22	0.52	0.097	0.20	0.58	0.08	0.21	0.74	0.06	—	0.7	—	—	—	—
Tungsten	1.82	0.13	0.70	1.49	0.14	0.56	1.20	0.15	0.41	—	0.18	0.30	—	0.20	0.23
Uranium	0.27	0.12	0.13	0.32	0.13	0.13	0.44	0.18	—	—	—	—	—	—	
Vanadium	0.31	0.47	0.11	0.33	0.51	0.11	0.39	0.62	0.10	0.51	0.83	0.10	—	—	—
Zinc	1.22	0.38	0.45	1.11	0.42	0.36	0.67	—	—	—	—	—	—	—	—
Zirconium	0.23	0.28	0.12	0.21	0.32	0.10	0.24	0.36	0.10	0.31	—	—	—	—	—
Armco iron	0.15	—	—	0.59	—	—	0.32	—	—	0.64	—	—	—	—	—
302 Stainless	0.13	—	—	0.16	—	—	0.24	—	—	—	—	—	—	—	—
303 Stainless	0.15	—	—	0.16	—	—	0.25	—	—	—	—	—	—	—	—
304 Stainless	0.17	—	—	0.19	—	—	0.26	—	—	—	—	—	—	—	—

Alumel	2.9	—	—	3.1	—	—	—	—	—	—	—	—	—	—	—
Brass	1.06	0.38	—	—	—	—	—	—	—	—	—	—	—	—	—
Inconel	0.15	0.41	—	—	—	—	—	—	—	—	—	—	—	—	—
Monel	0.21	0.42	—	—	—	—	—	—	—	—	—	—	—	—	—
Zircaloy	0.12	0.03	—	—	—	—	—	—	—	—	—	—	—	—	—
Berylium oxide	3.0	0.92	—	—	—	—	—	—	—	—	—	—	—	—	—
Nickel oxide	0.4	0.57	—	—	0.84	—	—	—	—	—	—	—	—	—	—
Thoria	0.15	0.23	0.07	0.06	0.26	0.02	0.03	0.29	0.01	—	—	—	—	—	—
Zirconia	0.02	0.46	0.007	0.02	0.54	0.006	0.02	0.63	0.005	—	—	—	—	—	—
Al_2O_3	0.4	1.00	1.0	0.2	1.07	0.05	0.08	1.22	0.02	0.06	1.52	0.01	—	2.1	—
Pyrolytic	22.3	0.64	—	11.3	1.17	—	5.3	1.67	—	2.5	2.1	—	—	—	—
Graphite[a]	0.11	—	—	0.05	—	—	0.03	—	—	0.01	—	—	—	—	—
MgO	0.53	0.70	0.16	0.27	1.12	0.07	0.10	1.25	0.02	0.09	—	0.02	—	—	—
SiO_2 (fused)	0.013	0.7	0.007	0.016	1.0	0.006	0.029	1.25	0.009	—	—	—	—	—	—
TiO_2	0.09	0.70	0.03	0.06	0.85	0.017	0.03	0.93	0.009	—	1.0	—	—	—	—
Basalt	0.02	—	—	—	—	—	—	—	—	—	—	—	—	—	—
Dolomite	—	—	—	0.012	—	—	—	—	—	—	—	—	—	—	—
Granite	0.015	—	—	—	—	—	—	—	—	—	—	—	—	—	—
Limestone	0.01	—	—	—	—	—	—	—	—	—	—	—	—	—	—
Steatite	0.03	—	—	—	—	—	—	—	—	—	—	—	—	—	—
Polyvinylchloride	0.0016	—	—	—	—	—	—	—	—	—	—	—	—	—	—
Polyethylene	0.004	—	—	—	—	—	—	—	—	—	—	—	—	—	—
Plexiglas	0.002	—	—	—	—	—	—	—	—	—	—	—	—	—	—
Hardwood	0.002	—	—	—	—	—	—	—	—	—	—	—	—	—	—
Pine	0.001	—	—	—	—	—	—	—	—	—	—	—	—	—	—
Cardboard	0.002	—	—	—	—	—	—	—	—	—	—	—	—	—	—
Wallboard	0.0005	—	—	—	—	—	—	—	—	—	—	—	—	—	—

We can illustrate the utility of these analytic solutions by means of an example—the heating of a stainless steel sheet with radiation from a pulsed CO_2 laser. The following values will be assumed:

$$K = 0.2 \text{ W cm}^{-1}\ {}^\circ\text{C}^{-1}, \quad \kappa = 0.06 \text{ cm}^2 \text{ s}^{-1}, \quad \epsilon = 0.1, \quad \text{and} \quad L = 0.1 \text{ cm}$$

We will estimate the temperature rise at the laser focus when $P = 10^3$ W (constant in time) and the beam is Gaussian with radius $d = 0.5$ cm. Since $d > L$, the temperature at the center of the laser focus at short times can be

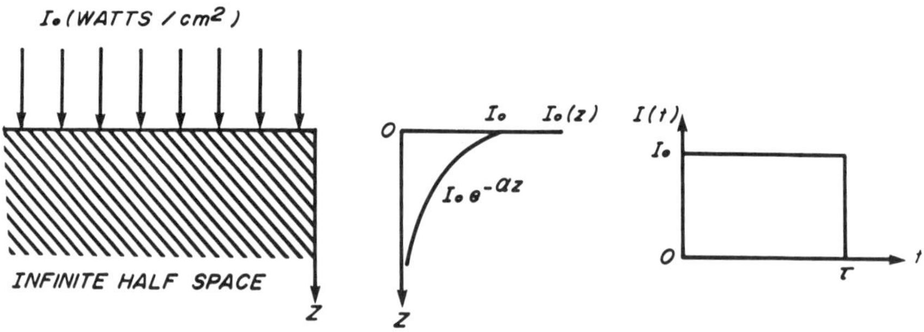

Figure 2.18. $t > 0$:

$$T(z,t) = \frac{2\epsilon I_0}{K}(\kappa t)^{1/2}\, ierfc\left[\frac{z}{2(\kappa t)^{1/2}}\right] - \frac{\epsilon I_0}{\alpha K} e^{-\alpha z} + \frac{\epsilon I_0}{2\alpha K}\exp[\alpha^2 \kappa t - \alpha z]\, erfc\left[\alpha(\kappa t)^{1/2} - \frac{z}{2(\kappa t)^{1/2}}\right] + \frac{\epsilon I_0}{2\alpha K}\exp[\alpha^2\kappa t + \alpha z]\, erfc\left[\alpha(\kappa t)^{1/2} + \frac{z}{2(\kappa t)^{1/2}}\right]$$

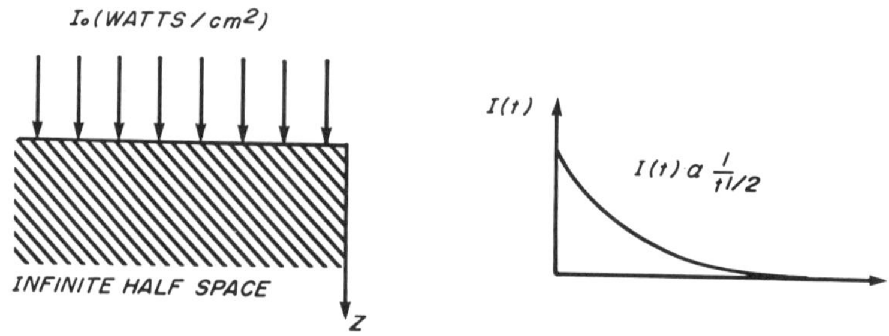

Figure 2.19. $t > 0$:

$$T(z,t) = \frac{I(t)\pi^{1/2}(\kappa t)^{1/2}}{K}\, erfc\left[\frac{z}{2(\kappa t)^{1/2}}\right]$$

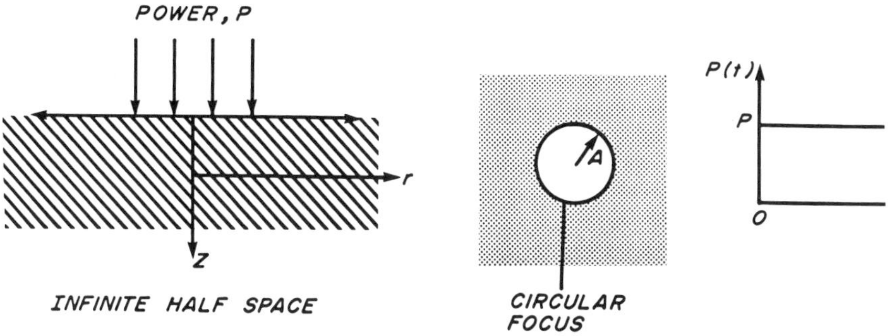

Figure 2.20. $t > 0$, $r = 0$:

$$T(0,z,t) = \frac{2P\epsilon(\kappa t)^{1/2}}{\pi A^2 K}\left[ierfc\left(\frac{z}{2(\kappa t)^{1/2}}\right) - ierfc\left(\frac{(z^2 + A^2)^{1/2}}{2(\kappa t)^{1/2}}\right)\right]$$

$$T(0,0,t) = \frac{2P\epsilon(\kappa t)^{1/2}}{\pi A^2 K}\left[\frac{1}{\pi^{1/2}} - ierfc\left(\frac{A}{2(\kappa t)^{1/2}}\right)\right]$$

$$T(0,0,\alpha) = \frac{P\epsilon}{\pi A K}$$

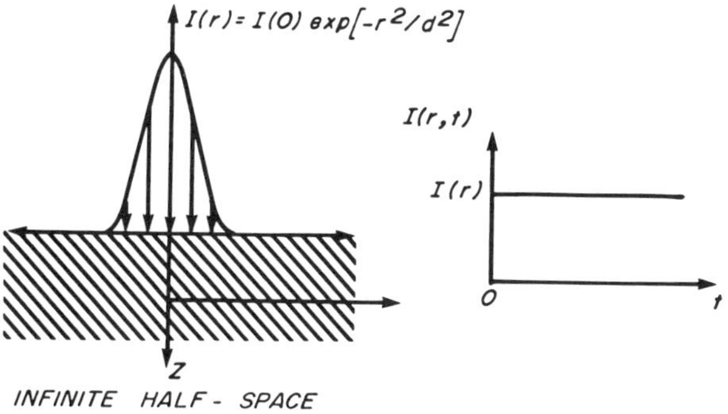

Figure 2.21. $t > 0$, $r = 0$, $z = 0$:

$$T(0,0,t) = \frac{\epsilon I(0)\,d}{K\pi^{1/2}}\tan^{-1}\left[\frac{4\kappa t}{d^2}\right]^{1/2}$$

$$T(0,0,\infty) = \frac{\epsilon I(0)\,d\pi^{1/2}}{2K}$$

calculated approximately by assuming that the sheet can be replaced by an infinite half-space. Then

$$T(0,t) = \frac{2\epsilon I_0}{K}\left(\frac{\kappa t}{\pi}\right)^{1/2} \tag{2.3}$$

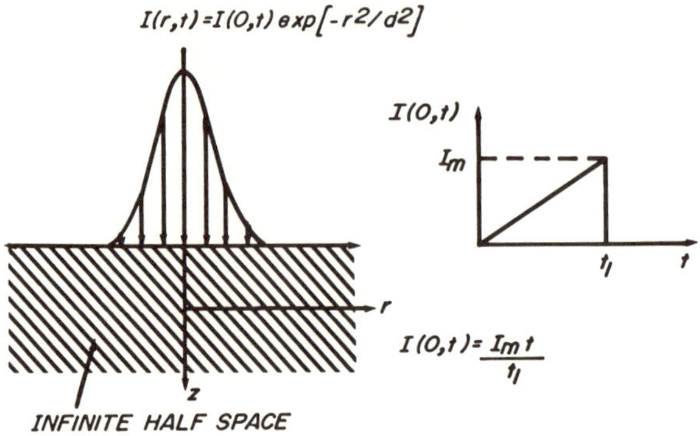

Figure 2.22. $t \leq t_1$:

$$T(0,0,t) = \frac{I_m d}{K\pi^{1/2}}\left(\frac{d^2}{4\kappa t_1}\right)\left[\left(\frac{4\kappa t}{d^2}+1\right)\tan^{-1}\left(\frac{2(\kappa t)^{1/2}}{d}\right)-\frac{2(\kappa t)^{1/2}}{d}\right]$$

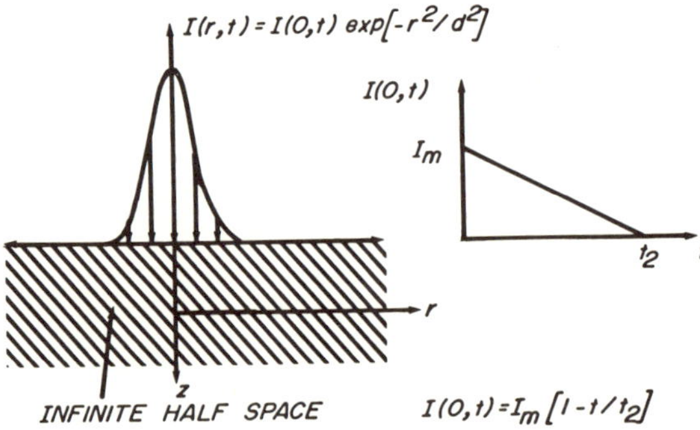

Figure 2.23. $0 \leq t \leq t_2$:

$$T(0,0,t) = \frac{I_m d}{K\pi^{1/2}}\left(\frac{d^2}{4\kappa t_2}\right)\left[\left(\frac{4\kappa t_2}{d^2}-\frac{4\kappa t}{d^2}-1\right)\tan^{-1}\left(\frac{2(\kappa t)^{1/2}}{d}\right)+\frac{2(\kappa t)^{1/2}}{d}\right]$$

with $I_0 = P/\pi d^2 = 1273 \text{ W cm}^{-2}$ the temperature at the center of the focal spot becomes $T(0,t) = 176t^{1/2}\ ^\circ\text{C}$.

If we use the exact solution for a Gaussian source but still assume that the sheet can be replaced by an infinite half-space then

$$\begin{aligned} T(0,0,\text{t}) &= \frac{\epsilon I_0 d}{K\pi^{1/2}}\tan^{-1}\left(\frac{4\kappa t}{d^2}\right)^{1/2} \\ &= 179.5\tan^{-1}(0.98t^{1/2}) \end{aligned} \tag{2.4}$$

Since $\tan^{-1}\theta \sim \theta$ for small θ, $\tan^{-1}(0.98t^{1/2}) \sim 0.98t^{1/2}$ and $T(0,0,t) \sim 176t^{1/2}$, as obtained above for the uniform surface source. This shows that as long as $d > (\kappa t)^{1/2}$ the temperature at the point at the center of the laser focus behaves as if the Gaussian source were replaced by a spatially uniform intensity distribution. As long as $d > (\kappa t)^{1/2}$ the point at the center of the focus has not been able to "sense" that the region $r > d$ is not heated.

When $t \to L^2/\kappa = 0.17$ s, the heating wave has reached the surface at $z = L$. If this surface is insulated no heat will be lost and the wave will be reflected back into the slab. The temperature within the slab then rises more rapidly than it would if the slab were in fact a semi-infinite half-space. The temperature at the center of the focal spot must then be calculated taking this boundary condition into account. The result is

$$\begin{aligned} T(0,0,t) &= \frac{\epsilon I_0 d^2}{4KL} \ln\left(\frac{4\kappa t}{d^2} + 1\right) \\ &= 398 \ln[0.96t + 1] \end{aligned} \tag{2.5}$$

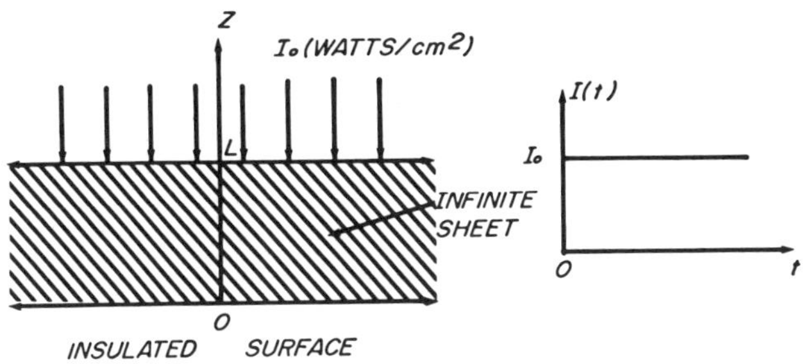

Figure 2.24. $t > 0$:

$$T(z,t) = \frac{\epsilon I_0 t}{\rho C L} + \frac{\epsilon I_0 L}{K}\left[\frac{3z^2 - L^2}{6L^2} - \frac{2}{\pi^2}\sum_{n=1}^{\infty}\frac{(-1)^n}{n^2}\exp\left(-\frac{\kappa n^2\pi^2 t^2}{L^2}\right)\cos\left(\frac{n\pi z}{L}\right)\right]$$

When the laser source has the time dependence

$$I(t) = I_0 t^{m/2}$$

where $m = -1, 0, 1, \ldots$, then

$$T(z,t) = \frac{2^{m+1} I(t)(\kappa t)^{1/2}\Gamma[(m/2) + 1]}{K} \times \sum_{n=0}^{\infty}\left[i^{m+1}\, erfc\, \frac{(2n+1)L - z}{2(\kappa t)^{1/2}} + i^{m+1}\, erfc\, \frac{(2n+1)L + z}{2(\kappa t)^{1/2}}\right]$$

where Γ is the Gamma function and

$$i^{m+1}\, erfc\, x = \int_x^{\infty} i^m\, erfc\, y\, dy$$

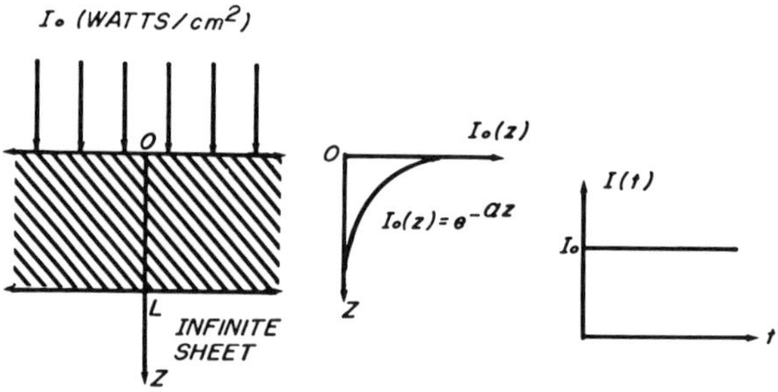

Figure 2.25. $t > 0$:

$$T(z,t) = \frac{\epsilon I_0}{2\kappa\alpha\rho C}\sum_{n=-\infty}^{\infty} A_n + \frac{\epsilon I_0}{\rho C}\left(\frac{t}{\kappa}\right)^{1/2}\sum_{n=-\infty}^{\infty} B_n - \frac{\epsilon I_0 e^{-\alpha z}}{\kappa\alpha\rho C}$$

where $A_n = \exp(\kappa\alpha^2 t \pm \alpha[z - 2nL])\left\{erfc\left[\alpha(\kappa t)^{1/2} \pm \frac{(z-2nL)}{(4\kappa t)^{1/2}}\right]\right.$

$$\left.- \, erfc\left[\alpha(\kappa t)^{1/2} \pm \frac{(z-2nL)}{(4\kappa t)^{1/2}} + \frac{L}{(4\kappa t)^{1/2}}\right]\right\}$$

$$B_n = 2ierfc\left[\frac{|z - 2nL|}{(4\kappa t)^{1/2}}\right] - \exp(-\alpha L)ierfc\left[\frac{|z-(2n \pm 1)L|}{(4\kappa t)^{1/2}}\right]$$

This solution together with that valid at small t is shown in Figure 2.30. Note that Eq. (2.5) is not valid at t such that $t < L^2/\kappa$ while the equation is not valid when $t > L^2/\kappa$. The region of overlap occurs when $t = 2L^2/\kappa = 0.33$ s, since it is only at this time that the front surface senses the presence of the rear surface.

When $t = L^2/\kappa$, the heating wave has just reached the back surface. We can consider that this corresponds to heating a right circular cylinder of area πd^2 and length L to a uniform temperature T. This would be

$$\begin{aligned} T(\equiv T(0,0,t)) &= \frac{\text{Energy input}}{\text{Cylinder volume} \times \text{heat capacity}} \\ &= \frac{10^3 \times 0.1t}{\pi d^2 L\rho C} \\ &= \frac{100t\kappa}{\pi d^2 LK} \\ &= 386\ t \end{aligned}$$

With $t = 0.16$ s, $T = 61.8\,^\circ$C. The exact value calculated from $T = 179.5 \tan^{-1}(0.98t^{1/2}) = 67.0\,^\circ$C.

To investigate the effect of tighter focusing we can redo this calculation with $d = 0.01$ cm and $P = 100$ W. In this case $d \ll L = 0.1$ cm and three

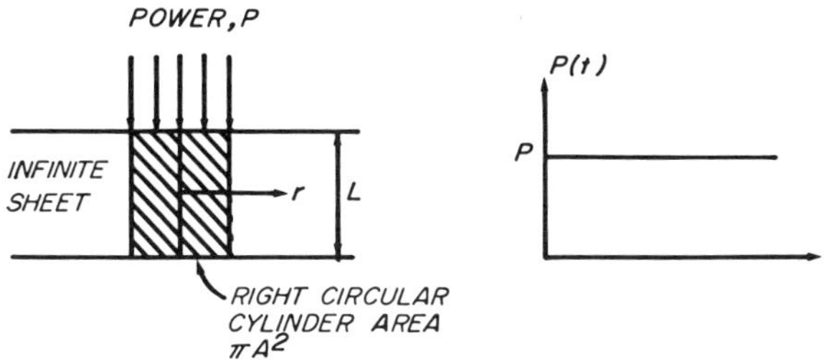

Figure 2.26. Cylindrical source function: $t = 0$, $A^2/\kappa t \ll 1$, $L < (\kappa t)^{1/2}$:

$$T(r,t) = \frac{P\epsilon}{4\pi KL}\left[\ln\left(\frac{4\kappa t}{1.78r^2}\right) + \frac{A^2}{2\kappa t}\ln\left(\frac{4\kappa t}{1.78r^2}\right) + \frac{1}{4\kappa t}\left(A^2 + r^2 - 2A^2\ln\frac{A}{r}\right)\right]$$

$A^2/\kappa t \gg 1$, $L < (\kappa t)^{1/2}$:

$$T(r,t) = \frac{P\epsilon}{\pi KLA^{1/2}}\left(\frac{\kappa t}{r}\right)^{1/2}\left[ierfc\left(\frac{r-A}{2(\kappa t)^{1/2}}\right) - \frac{(3r+A)(\kappa t)^{1/2}}{4Ar}\,i^2erfc\left(\frac{(r-A)}{2(\kappa t)^{1/2}}\right) + \cdots\right]$$

General case (no assumption of cylindrical source): $L < (kt)^{1/2}$:

$$T(0,t) = \frac{\epsilon P}{4\pi KL}\left[\frac{4\kappa t}{A^2}\left(1 - \exp\left[-\frac{A^2}{4\kappa t}\right]\right) - Ei\left(-\frac{A^2}{4\kappa t}\right)\right]$$

with

$$Ei\left(-\frac{A^2}{4t}\right) = \int_{-\infty}^{-A2/4\kappa t}(e^u/u)\,du$$

General L:
T depends on z as well as r. z measures from front surface into sheet.

$$T(0,z,t) = \frac{P\epsilon}{\pi A^2}\frac{(4\kappa t)^{1/2}}{K}\sum_{n=-\infty}^{\infty}\left[ierfc\frac{|z-2nL|}{(4\kappa t)^{1/2}} - ierfc\frac{[(z-2nL)^2 + A^2]^{1/2}}{(4\kappa t)^{1/2}}\right]$$

time scales can be identified. These are $t_1 < d^2/\kappa$, $d^2/\kappa < t_2 < L^2/\kappa$, and $t_3 > L^2/\kappa$. The solutions for $T(0,0,t)$ in the first two of these time regimes are

$$T(t_1) = \frac{2\epsilon I_0}{K}\left(\frac{\kappa t}{\pi}\right)^{1/2} \tag{2.6}$$

$$T(t_2) = \frac{\epsilon I_0 d}{K\pi^{1/2}}\tan^{-1}\left(\frac{4\kappa t}{d^2}\right)^{1/2} \tag{2.7}$$

Note that Eq. (2.7) cannot be used for $t > t_2$ since $d < L$.

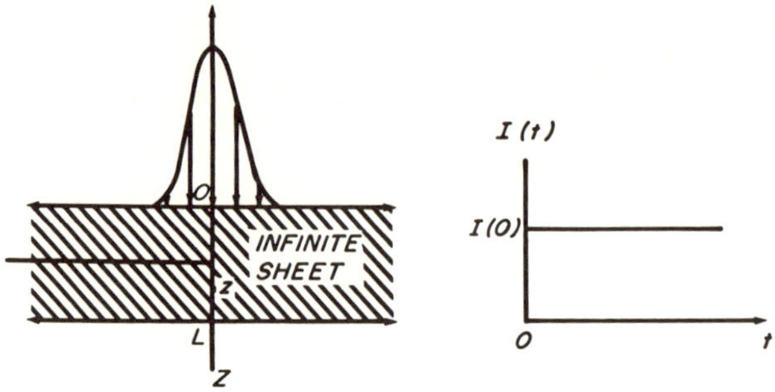

Figure 2.27. $t > 0$; $L < (\kappa t)^{1/2}$; $d > L$:

$$T(r,t) = \frac{\epsilon I(0)\, d^2}{4KL}\left[Ei\left(-\frac{r^2}{d^2}\right) - Ei\left(\frac{r^2}{d^2}\frac{1}{[(1 + 4\kappa t)/d^2]}\right)\right]$$

$$T(0,t) = \frac{\epsilon I(0)\, d^2}{4KL} \ln\left[1 + \frac{4\kappa t}{d^2}\right]$$

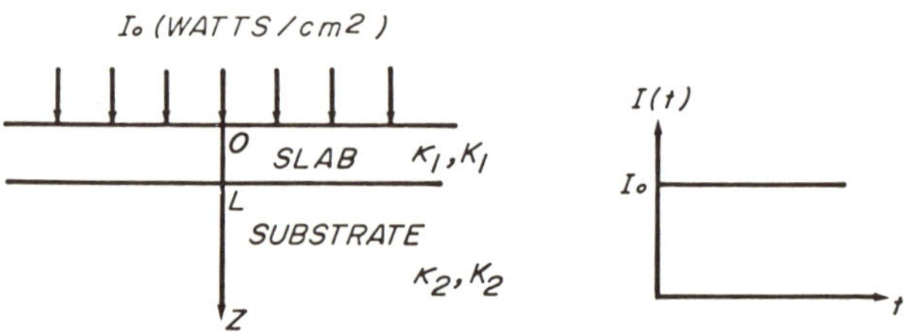

Figure 2.28. $t > 0$:

$$T_1(z,t) = \frac{\epsilon I_0}{K_1}[4\kappa_1 t]^{1/2} \sum_{n=-\infty}^{\infty} \zeta^{|n|}\, ierfc\left[\frac{|z - 2nL|}{(4\kappa_1 t)^{1/2}}\right]$$

where T_1 = temperature in slab and

$$\zeta = \frac{K_1(\kappa_2)^{1/2} - K_2(\kappa_1)^{1/2}}{K_1(\kappa_2)^{1/2} + K_2(\kappa_1)^{1/2}}$$

Defining $z' = z - L$

$$T_2(z',t) = \frac{2T_L}{(\Lambda + 1)} \sum_{n=0}^{\infty} \zeta^n[I_1(Z^*)]$$

where

$$T_L = \frac{2\epsilon I_0(\kappa_2 t)^{1/2}}{K_2 \pi^{1/2}}; \qquad \Lambda = \frac{K_1}{K_2}\left(\frac{\kappa_2}{\kappa_1}\right)^{1/2};$$

$$Z^* = \frac{z'}{(4\kappa_2 t)^{1/2}} + \frac{(2n + 1)L}{(4\kappa_2 t)^{1/2}}; \qquad \text{and } I_1(x) = \pi^{1/2}\, ierfc\,(x)$$

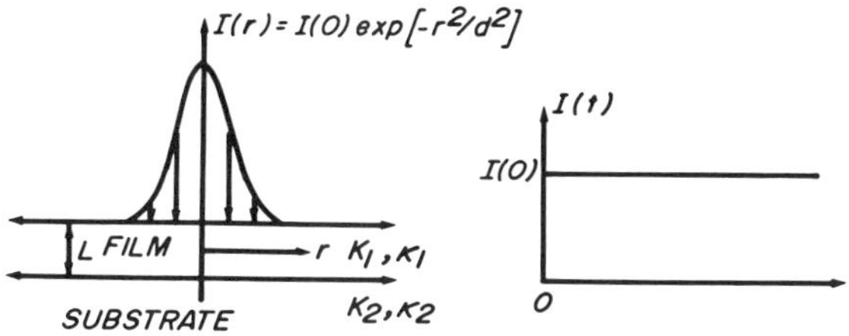

Figure 2.29. $t > 0$, $L < (\kappa_1 t)^{1/2}$, $t \leq d^2 | \kappa_1$:

$$T_1(r,t) = T_v \exp\left[-\left(\frac{r}{d}\right)^2\right]\left[\frac{(\pi\sigma)^{1/2}}{\beta} S\tau^{1/2} - \frac{\sigma\pi S}{2\beta^2} \ln\left(1 + \frac{2\beta\tau^{1/2}}{(\pi\sigma)^{1/2}}\right)\right]$$

where

$$\sigma = \frac{\kappa_2}{\kappa_1}, \beta = \frac{K_2}{K_1}, S = \frac{\epsilon I(0) L}{K_1 T_v}$$

and T_v = vaporization temperature of film

$$\begin{aligned} \text{Since } I_0 &= 3.18 \times 10^5 \text{ W cm}^{-2} \\ T(t_1) &= 4.4 \times 10^4 t^{1/2} \\ T(t_2) &= 902 \tan^{-1}(49 t^{1/2}) \end{aligned}$$

These temperatures are plotted in Figure 2.31. In this case, the temperature in the focus rises rapidly at short times due to the higher incident intensity. One can see that the infinite half-space constant-intensity solution follows the infinite half-space Gaussian beam solution until $t \sim 10^{-4}$ s. At this point, the finite area of the heated spot starts to affect the solution for the temperature at the center of the focus. At t_1, $T(t_2)$ is already significantly less than $T(t_1)$. The temperature continues to rise but less rapidly for $10^{-4} < t < 10^{-1}$ s. $T(t_2)$ approaches 1407°C as a limiting value. A detailed solution for the temperature when $t > t_2 = 0.17$ s shows that T continues to rise slowly. This increase occurs because of the presence of the surface at $L = 0.1$ cm. Note, however, that the effect of this surface is much less here than when $d > L$. When $d < L$, the temperature vs. time solution within the focus is well approximated by the Gaussian source infinite half-space solution at all but the longest times.

These two examples illustrate how approximate solutions can be used to estimate the temperature rise in a solid heated over a localized area by laser radiation. Further examples of this approach are given in the paper by Sparks (1976).

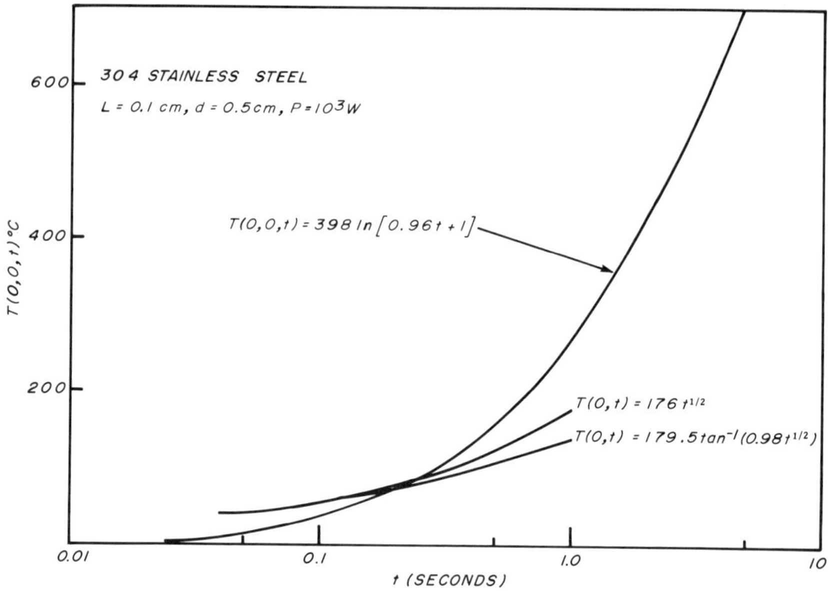

Figure 2.30. Solutions for temperature calculated from Eqs. (2.4) and (2.5).

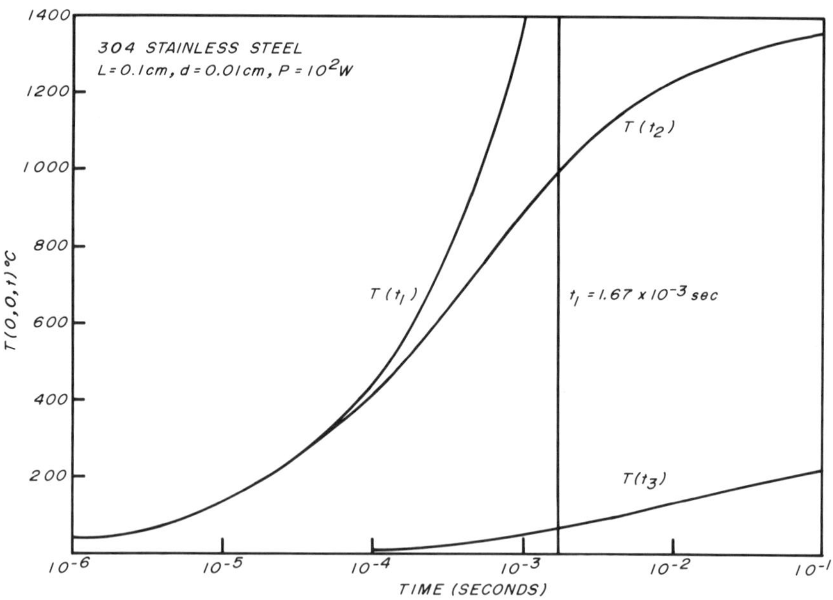

Figure 2.31. Temperature vs. time curves calculated from Eqs. (2.6) and (2.7).

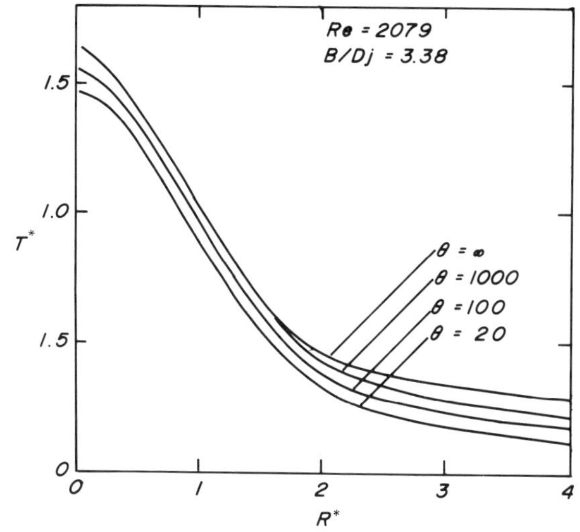

Figure 2.32. Thermal history of a 0.48-mm diameter spot subject to P/D_b = 10.4 W cm^{-1} [Steen (1977)].

As an example of the power of numerical techniques in the solution of the heat equation with complex boundary conditions, Figure 2.32 shows dimensionless temperature vs. distance profiles for points on the surface of an infinite half-space heated with a Gaussian beam [Steen (1977)]. Normalized units are

$$T^* = \frac{2K\pi Td}{\epsilon P}$$

$$R^* = \frac{r}{d}$$

$$\theta = \frac{4\kappa t}{d^2}$$

and it is assumed that the focal area is covered by gas flow from a nozzle with given Reynolds number, Re. The notation in this figure is D_b—beam diameter at ⅓ point in intensity distribution; B—jet-plate distance; and D_j—jet diameter. These curves are general and can be applied to any material simply by adopting specific values for thermal constants. Figure 2.33 shows a plot of T^* vs. R^* when $\theta \rightarrow \infty$ for various values of the heat loss parameter η where

$$\eta = \frac{[h_c + h_{\text{rad}}(\infty)]\ [T_0^* - T_a^*]d}{2K}$$

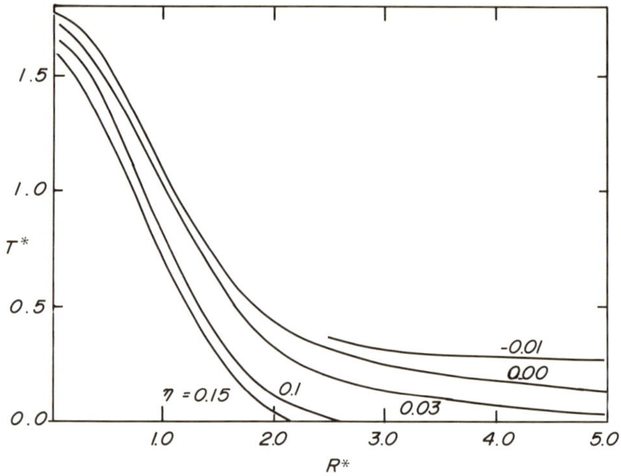

Figure 2.33. Dimensionless thermal profiles: T^* vs. R^* vs. η [Steen (1977)].

where h_c is the convective heat transfer coefficient h_{rad} (∞) the radiative heat transfer coefficient at $\theta = \infty$, T_0^* the reduced temperature at the center of the focal spot, and T_a^* the reduced ambient temperature. The clear role of convective losses in narrowing the temperature profile can be seen in this figure.

Steen used this numerical solution to the heat equation to investigate the effect of a variety of parameters on the temperature profile. This included the role played by jet Reynolds number, and jet-plane distance on T^* vs. R^*. The flexibility of the numerical approach is clearly evident in this study.

2.4.2. Heating with Change of Phase

Many applications of lasers in materials processing involve the initiation of phase changes. For example, drilling involves heating to vaporization while laser welding involves both melting and vaporization. In this section we briefly discuss the results of some theoretical models of relevance to laser drilling. More complete descriptions can be found in the reviews by Carslaw and Jaeger (1959), Ready (1971), and Duley (1976).

When the surface temperature reaches the melting temperature, T_m, a melt region is formed adjacent to the surface. In the absence of perturbations this melt will propagate into the substrate at a speed

$$v_m = \frac{\epsilon I_0}{\lambda_m + \rho C T_m} \exp\left(-\frac{v_m \Delta}{\kappa}\right) \tag{2.8}$$

where the geometry is shown in Figure 2.34 and λ_m is the latent heat of fusion (J km^3), ρ the density (g cm^{-3}), C the heat capacity (J g^{-1} °C^{-1}), Δ the melt thickness, and v_m is in cm s^{-1}.

When drilling metals it is often possible to replace the exponential term by its approximate value for a small argument. This implies that $v_m\Delta/2\kappa \ll 1$. Making this approximation

$$v_m^* = \frac{\epsilon I_0}{\lambda_m + \rho C T_m} \tag{2.9}$$

where v_m^* is now the maximum speed of the fusion front.

The time development of Δ, the width of the melt region, can be found from the conservation equation [Robin and Nordin (1975a,b)]

$$A\frac{d\Delta}{dt} = A v_m(\Delta) - \frac{dV_{\text{loss}}}{dt}$$

where dV_{loss}/dt is the volume of melt lost per second due to some external force, such as a forced gas flow, and A is the area of melt. This equation can be integrated to generate a time history of Δ. An example of the result of such a calculation for Al is shown in Figure 2.35. This calculation includes the effects of transverse air flows at Mach 1.5 and 3.0 on $\Delta(t)$. When the melt material is removed instantaneously, the speed of the fusion front is given by v_m^* and is independent of time. In this case

$$\Delta(t) = v_m^* t$$

This relation is also shown in Figure 2.35 for $\epsilon I_0 = 7.2 \times 10^3$ W cm^{-2}.

When the liquid melt is not removed as the melting wave propagates into the solid, the temperature of the melt rises in response to the continuation of absorption at the melt-vapor interface. If the incident radiation is of sufficient

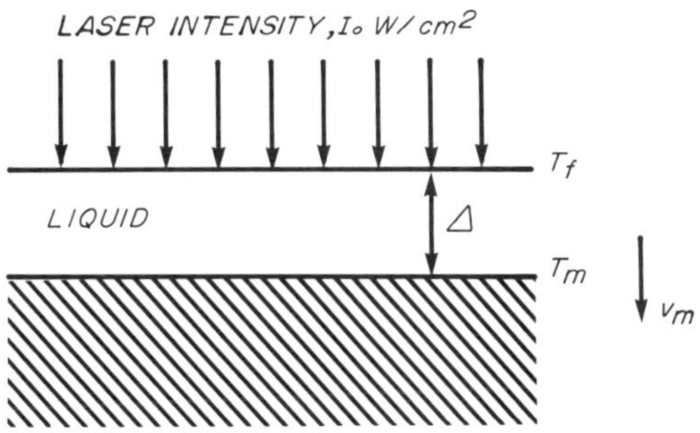

Figure 2.34. Geometry of melt region on surface of semi-infinite region heated uniformly over its surface.

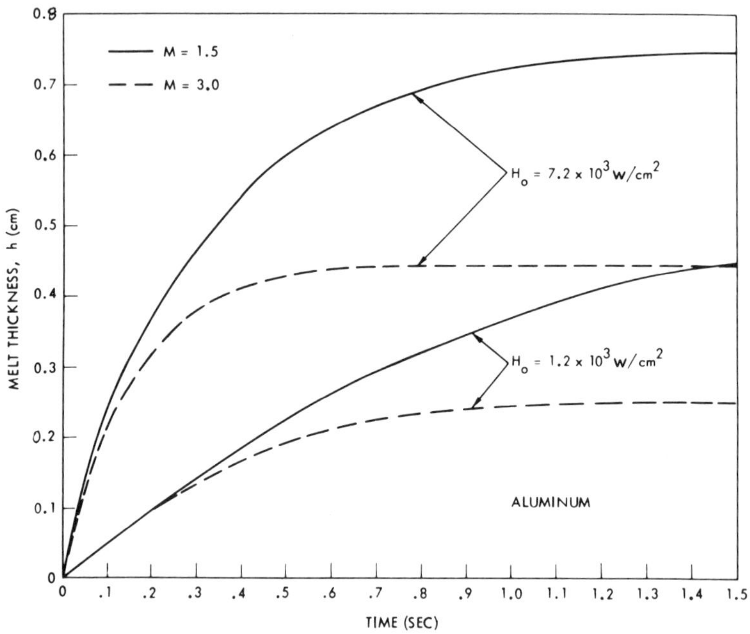

Figure 2.35. Melt-thickness time history for aluminum for gas flow velocities of $M = 1.5$ and 3.0, and $\epsilon I_0 = 1.2 \times 10^3$ and 7.2×10^3 W cm^{-2} [Robin and Nordin (1975)].

intensity then the temperature may rise to the boiling point, T_v, or higher. This is accompanied by the onset of a vaporization wave with a speed (when $T = T_v$).

$$v_v = \frac{\epsilon I_0}{\lambda_V + \rho C T_v}$$

where λ_V is the latent heat of vaporization and all quantities have the same units as in the expression for v_m. As I_0 increases, v_v increases until it approaches the speed of sound, v_s, in the material. When $v \to v_s$ the expression for v becomes

$$v = v_s \exp\left(-\lambda Z/\rho N_A k_B T_v\right) \tag{2.10}$$

where Z is the atomic number of material, N_A the Avogadro number, and k_B the Boltzmann constant. Since v is not dependent on I_0 in this regime, the speed of the vaporization rate saturates at high flux levels. For most metals this saturation occurs when $v \sim 10^5$–10^6 cm s^{-1} and $I_0 \gtrsim 10^8$ W cm^{-2}. Obviously vaporization rates of this magnitude can be sustained only for short periods of time, i.e., with pulsed laser excitation.

Drilling of metals with pulsed lasers usually occurs via a melting flushing mechanism in which material evaporating from the end of the hole flushes liquid out the sides of the hole. This may be accompanied by subsurface explosions that physically eject liquid droplets or even particulate matter [see the review by Uglov and Kokora (1977)]. Under such conditions modeling of temperature profiles and evaporation rates is difficult although Dabby and Paek (1972) have performed calculations which take some of these effects into account.

This problem has been analyzed by von Allmen (1976). When both liquid and vapor are being removed from a metal, the energy balance gives

$$I_0 = j_v L_v + j_e L_l \tag{2.11}$$

where j_v and j_l are the expulsion rates for vapor and liquid, respectively in g cm^{-2} s^{-1} and L_v and L_l are the specific absorbed energy (J g^{-1}) of vaporized and liquified material, respectively. When the tangential pressure at the tip of the drilled hole exceeds the surface tension of the liquid metal, evaporating material forces the liquid to flow out the end of the hole. The speed of this liquid is

$$u = (2p/\rho)^{1/2}$$

where p is the pressure of the evaporating surface and ρ is the density. j_l then becomes

$$j_l \simeq \left[\frac{2\kappa}{r}\ln\left(\frac{T}{T_m}\right)\right]^{1/2} p_s^{1/4}\rho^{3/4}$$

where r is the beam radius, T the surface temperature at the tip of the hole, and p_s the saturation pressure. The corresponding expression for j_v is

$$j_v = \beta p_s(T)\left(\frac{m}{2\pi k_B T}\right)^{1/2}$$

where $\beta \sim 0.8$.

Von Allmen defines a drilling efficiency $E = v/I_0$ where v is the drilling speed and

$$v = \frac{1}{\rho(j_v + j_l)}$$

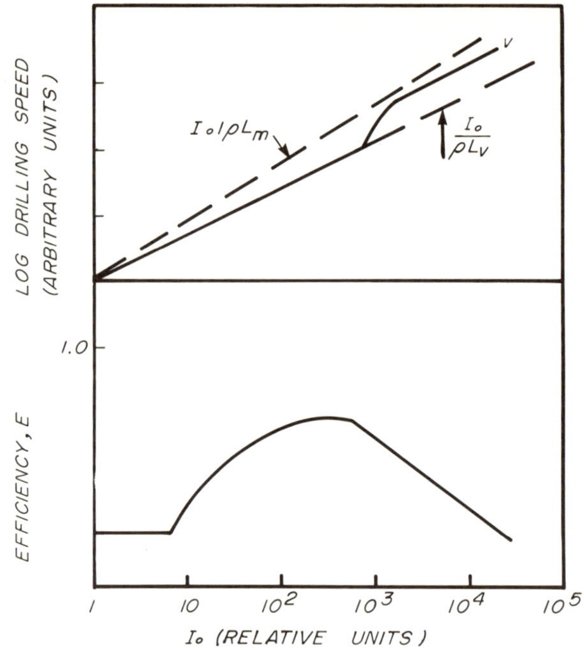

Figure 2.36. Drilling efficiency vs. laser intensity.

Calculations using the expressions for j_v and j_l given above show that E is constant at a relatively low value until I_0 reaches a threshold value at which point E increases. This threshold is achieved at the intensity where p exceeds the value required to eject liquid. Since the ejection of liquid also removes heat from the system it represents a power loss. Further increase in I_0 in this regime causes v to increase rapidly from $\sim I_0/L_v$ to $\sim I_0/L_m$. As I_0 increases further, v saturates, and E decreases because most material is now being removed as liquid. This behavior is shown schematically in Figure 2.36.

Duley (1976) has described thermal interactions in the drilling of holes in bulk nonconducting solids with a CO_2 laser. The speed of the vaporization wave is

$$v = \frac{\eta}{\zeta^{1/2}}\left(\frac{\zeta}{T}\right)^{1/2} \exp\left(\frac{-\zeta}{T}\right) \tag{2.12}$$

where

$$\eta = \frac{\alpha}{\rho}\left(\frac{\overline{m}}{2k_B}\right)^{1/2} \exp(A)$$

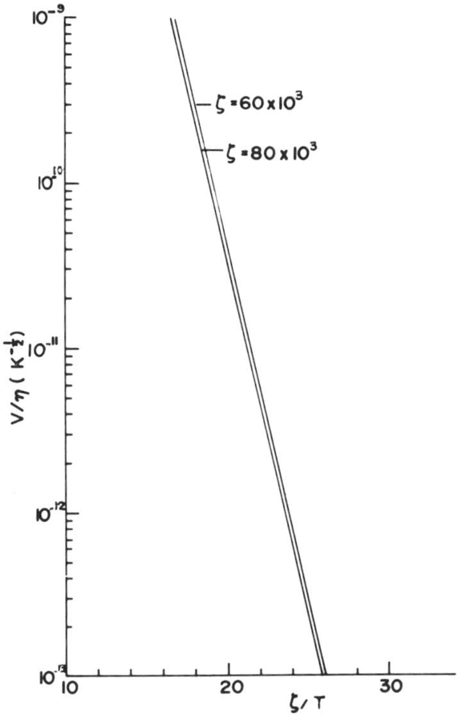

Figure 2.37. Plot of reduced velocity v/η vs. reduced temperature ζ/T for $\zeta = 60 \times 10^3$ and 80×10^3.

and A and ζ are evaporation constants with $\alpha = 1.013 \times 10^5$ N m^{-2} for p in N m^{-2}. $\overline{m}$ is the average mass of evaporated molecule, k_B the Boltzmann constant, and ρ the density (kg m^{-3}), while T is the temperature in K. Figure 2.37 shows a plot of v/η vs. ζ/T calculated from these expressions. Data on thermal and evaporation constants for a variety of refractory materials are given in Table 2.4.

Table 2.4. Evaporation and Thermodynamic Constants (Handbook Values) for Several Refractory Oxides

Oxide	A	m (amu)	η (m K$^{1/2}$ s^{-1})	ζ ($\times 10^{-4}$ K)	λ ($\times 10^{-10}$ J m^3)	c ($\times 10^{-3}$ J kg^{-1} °C^{-1})
MgO	18.3	30	1.05×10^8	6.0	4.41	1.34
CaO	18.1	47	1.17×10^8	6.45	3.14	1.00
Al_2O_2	18.9	27.5	1.69×10^8	7.35	2.34	1.46
ZrO_2	20.6	76	1.01×10^9	8.56	3.52	0.60
TiO_2	21.6	60	3.43×10^9	6.91	3.00	0.92
ThO_2	21.9	250	4.02×10^9	8.66	2.56	0.42
SiO_2	19.4	40	5.9×10^8	5.9	1.83	1.34

Table 2.5. Heat Transfer Terms at Two Different Radial Distances in a 330 μm Diameter Hole in Al_2O_3 Ceramic[a]

	$r = 150\ \mu m$ $I(r) = 32.1\ MW\ cm^{-2}$	$r = 25\ \mu m$ $I(r) = 136\ MW\ cm^{-2}$
Radial conduction	6.0 MW cm^{-2} (18.7%)	0.3 MW cm^{-2} (0.2%)
Axial conduction	0.7 (2.2)	1.0 (0.8)
Heating of material	25.4 (79.1)	134.7 (99.0)

[a]After Wagner (1974).

Wagner (1974) has investigated the physical processes occurring in the laser drilling of ceramics in some detail and has discussed their relation to the resulting hole shape. Wagner concludes that the resulting hole shape with its characteristic tapered profile is due to a melting–flushing mechanism. Since the thermal conductivity is small for ceramics, little peripheral heating occurs during drilling. As a result, the heat input at a particular point within the laser focus is primarily due to incident radiation and does not contain a term involving heat transfer from other points in the target. The hole shape is therefore closely related to the intensity distribution within the beam. A quantitative estimate of heat transfer terms in the drilling of Al_2O_3 ceramic with a YAG laser are given in Table 2.5 [Wagner (1974)]. The bottom line in this table indicates the fraction of the heat flux that is utilized in evaporation.

2.4.3. Experimental

One of the first industrial applications of lasers involved the drilling of fine holes in diamond dies [Epperson *et al.* (1966)]. The characteristics of the ruby laser that made it attractive in this application have subsequently proved valuable in a wide variety of other drilling operations. Some of these characteristics are summarized in Table 2.6.

The laser has gained acceptance as a means by which fine holes can be drilled under conditions that are difficult or impossible using conventional methods. It therefore supplements, but does not supplant, conventional technology. There is no advantage in using a laser to drill 1-mm holes in 3-mm Cu sheet; there is, however, an advantage in using laser heating to drill 0.02-mm holes in 0.02-mm sheet. Similarly, electrical perforation may be acceptable in the preparation of a random array of holes in paper sheet, however, laser perforation may be preferable if precise control over spatial location and hole size is needed.

In deciding whether or not laser technology is practical for a particular drilling application, the following questions should be asked:

1. Can the required hole be drilled with a conventional system?
2. If so, how is the laser competitive; cost per part, quality, reliability, throughput?
3. Can it be done with available laser technology?

Many potential applications of laser technology in drilling (or indeed elsewhere in materials processing) show, under close analysis, that the advantages of laser processing may be marginal when compared with conventional technology. It should be noted that the most successful laser applications have occurred in areas in which conventional technology is deficient.

Some of the considerations involved in the development of a laser-based drilling system in the presence of competitive conventional technology are well illustrated by the study of von Osenbruggen and Botden (1975) on the machining of diamond. Electron-beam, laser, and electric discharge machining were compared.

In the drilling of dies, electron-beam machining was found to be relatively inefficient. Typical drilling times were 8–27 min for ~ 1-mm thick dies. Laser drilling suffers from the fact that diamond is relatively transparent at the wavelength of pulsed solid-state lasers. Drilling can be accomplished in ~ 30 s. Hole finish was comparable to that obtained with electron-beam machining. A modification of the conventional electrical discharge method was used to drill holes in ~ 3 min.

The precision obtainable and hole finish were significantly superior to that obtained with laser drilling. It was concluded that both laser and electric discharge drilling had advantages over electron-beam drilling, but that these two techniques were competitive with each other. Properties of these methods are summarized in Table 2.7.

Every known material can be drilled with one or more types of laser. Reviews of experimental work done in laser drilling are numerous. A few are those by Cohen and Epperson (1968a,b), Gagliano *et al.* (1969), Ready (1971), Cohen (1972), Gagliano and Zaleckas (1972), Shkarofsky (1975), and Duley (1976). In view of this reference material, the present discussion will be

Table 2.6. Useful Characteristics of Laser Sources in Drilling Materials

Ability to focus to a fine spot
Contactless machining
Variable wavelength, hence source can be tailored to workpiece
Adjustable pulse duration providing close control over hole depth
Small heat affected zone
Good spatial control over hole location
Compatibility with computer control

Table 2.7. Properties of Different Methods For Drilling Diamond

	Time to drill (min)	Finish (μm)	Efficiency
Electron beam	8–27	10–20	<10%
Laser	0.5	20–30	~5%
Electrical discharge	3	~ 0.5	Low

limited to a general description of results obtained and will emphasize new developments in this field.

Figure 2.38 shows a cross-sectional view of a hole drilled in bulk stainless steel with a repetitively pulsed CO_2 laser. Notable properties are

Large aspect ratio
Tapering profile
Resolidified material on periphery of hole

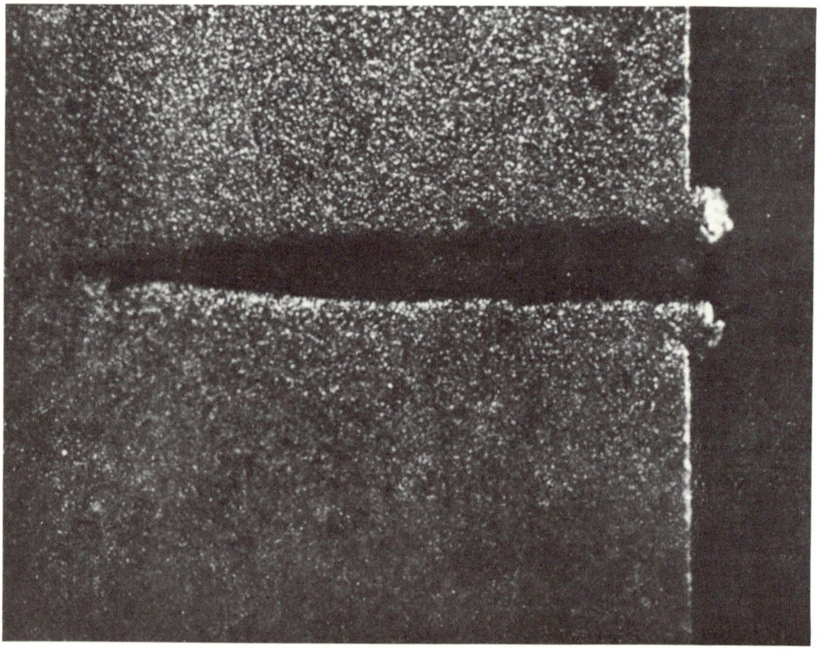

Figure 2.38. Cross section of hole drilled in stainless steel: Hole depth is 1.3 mm, TEA-type CO_2 laser [Hamilton and James (1976) by permission Institute of Physics].

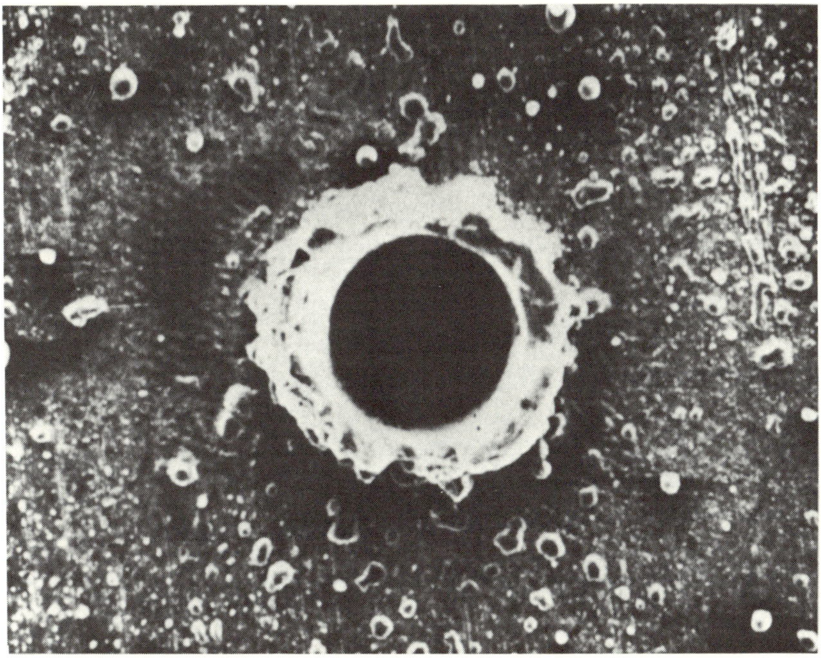

Figure 2.39. Electron micrograph of hole drilled through 0.6-mm steel with 100-mJ pulses from CO_2 laser and a 10-cm focal lens. Hole diameter is 70 μm [Hamilton and James (1976) by permission Institute of Physics].

These properties are characteristic of holes drilled in metals with a pulsed laser. The tapered profile and smooth internal surface of the hole as well as the presence of ejecta at the lip of the hole are characteristic of the flow of liquid from the tip of the hole along its sides to the entrance. This ejecta is shown clearly in Figure 2.39.

In drilling thin metal plates, the exit aperture of the hole may also be tapered outward to increase the exit hole diameter. This effect may arise from explosive evaporation of the remaining material as the depth of the hole approaches the sheet thickness.

Engel (1978) has discussed some different methods used to drill fine holes in thin metal sheets with the pulsed Nd–YAG laser. These are summarized in Table 2.8. The technique for producing small diameter holes involves the superposition of a series of pulses over the same focal area (percussion). With slightly larger holes, trepanning (rotation of target under laser focus) becomes the optimum technique. Larger diameter holes can be drilled by direct scanning of the beam around the hole circumference. Trepanned holes can be drilled with depth to widths ratios of $\sim$ 5:1, while percussion drilling gives

Table 2.8. Various Laser Drilling Techniques for Producing Fine Holes in Hastalloy-X[a]

Hole diameter (cm)	Method	Maximum pulse width (ms)s^{-1})	Size control
0.0025–0.03	Percussion drilling	1.2-ms vaporize	Mode control
0.03–0.045	Percussion drilling	1.2-ms vaporize	Defocus
0.045–~0.25	Trepanning	2.0-ms vaporize + melt	Mechanical adjustment
>0.25	Numerically controlled contour	2.0 ms	Numerically controlled program

[a]After Engel (1978).

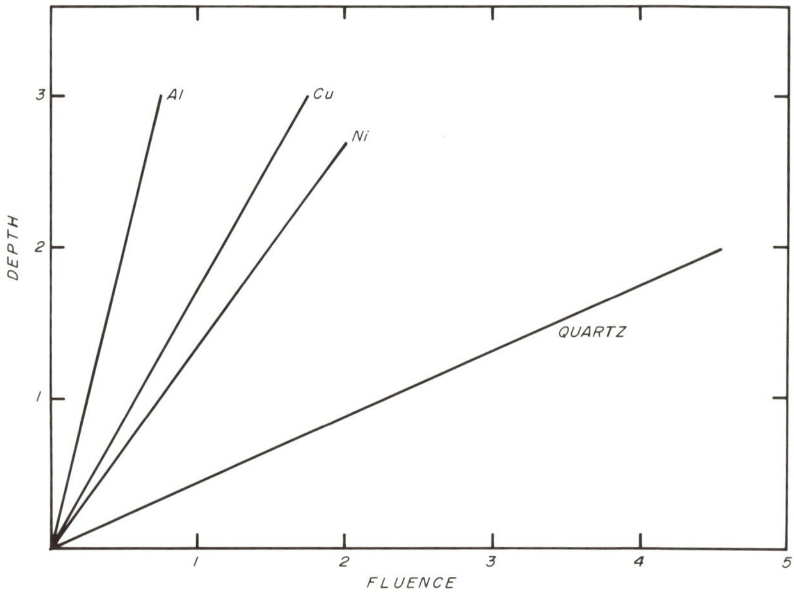

Figure 2.40. Hole depth vs. fluence for various materials drilled with laser pulses. Scales are as follows. Cu [von Allmen (1976)]: depth, units of 100 μm; fluence, units of 10^2 J cm^{-2}; YAG laser, 10 MW cm^{-2}, variable pulse length. Al, Ni [Cohen (1972)]: depth, units of 25×10^{-3} cm; fluence, units of 2.8×10^3 J cm^{-2}; ruby laser, 0.8-ms pulse length, variable power. Quartz [Duley (1976)]: depth: units of 1 mm; fluence, units of 6×10^2 J cm^{-2}; CW CO_2 laser 6×10^3 W cm^{-2} variable pulse length.

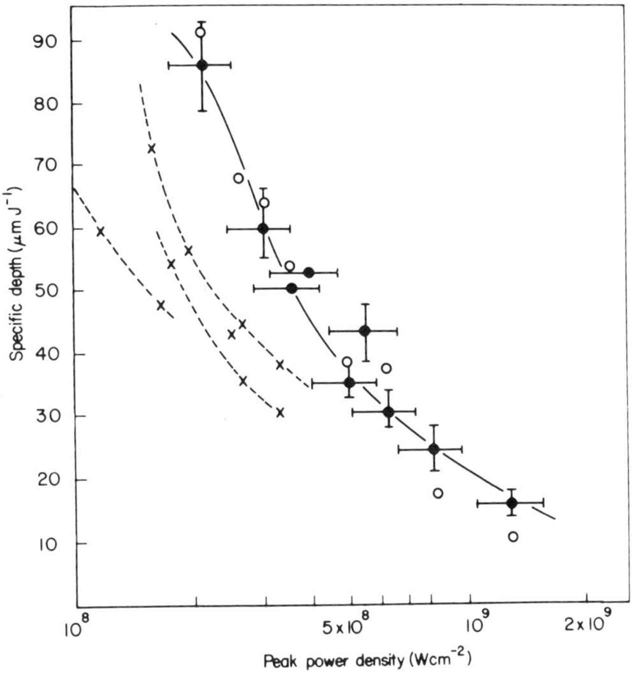

Figure 2.41. Variation of specific depth with peak power density for mild steel and aluminum where (○) Aluminum, peak power/average power ~25; (●) mild steel, peak power/average power ~25; (x) mild steel, peak power/average power ~10 [Hamilton and Pashby (1979)].

ratios of ~ 10:1. A commercially available computer-controlled laser hole punch has recently been described [*Laser Focus,* December (1978), p. 22].

In the absence of complicating factors, such as the initiation of laser supported detonation waves in front of the surface, hole depth increases linearly with the total amount of energy delivered during a pulse or series of pulses. This relation is usually valid until the hole depth approaches the depth of focus of the lens used. Some data supporting this observation are shown in Figure 2.40. The scales are different for each set of data shown in this graph, however, a linear relation is maintained between hole depth and total energy delivered to target (J cm^{-2}) in all cases.

The effect of laser breakdown and the formation of laser supported detonation (LSD) waves in front of the focal spot on drilling efficiency have been discussed by Sturmer and von Allmen (1978) and Hamilton and Pashby (1979). The presence of a shielding plasma in front of the target results in a decrease in drilling yield defined as volume removed per joule. Figure 2.41

Table 2.9. Applications of Laser Drilling

Application	Laser	Reference
Perforation of cigarette paper	Chopped CW CO_2	*Laser Focus*, February (1979), p. 47
Preparing nuclear fuel rods	Nd–YAG	*Laser Focus*, November (1979), p. 104
Sampling of coated nuclear fuel particles	Nd–YAG	Allelein *et al.* (1979)
Corrosion monitoring in reactors	Ruby	Tozer (1976)
Drilling drug capsules for release of medication	CO_2	*Laser Focus*, July (1980), p. 27
0.0043 in. holes in automobile spark advance mechanism	Nd–YAG	*Laser Focus*, June (1977), p. 36
Holes in baby bottle nipples	Chopped CW CO_2	*Laser Focus*, June (1977), p. 40
Matrix of holes in extensible rubber matrix	Pulsed CO_2	Longfellow (1970)

shows some data on specific depth vs. peak laser intensity for steel and aluminum obtained by Hamilton and Pashby (1979). This was obtained using multiple pulses from a TEA-type CO_2 laser with pulse length $\sim 40\ \mu s$. The decrease in specific depth at high intensity was due to absorption of part of each incident pulse in the breakdown plasma. Robin (1978) has developed a model which describes this phenomenon.

Enhancement in drilling efficiency by the superposition of pulsed and CW radiation has been discussed by Fox (1975), Robin and Nordin (1976), Towle *et al.* (1979), and Steverding *et al.* (1979). The initial experiments of Fox showed that the time required to drill thin metal plates could be reduced by up to 50% by superimposing a short pulse from a Nd–YAG laser on the spot heated by a CW CO_2 laser. It was postulated that the impulse imparted to the target by material ejected under pulsed irradiation was sufficient to dislodge material that had been melted by the CW laser. This hypothesis was supported by the observation that much cleaner penetration was obtained with CW + pulse excitation. Subsequent analysis has demonstrated [Towle *et al.* (1979) and Steverding *et al.* (1979)] that penetration can occur when CW heating proceeds to temperatures near but not at the melting point. A reduction in drilling time occurs because the material does not have to accumulate the latent heat of fusion. The effect of the superimposed laser pulse is to mechanically shear the heated plug of material, removing it from the target. This is possible because the strength of metals decreases substantially at temperatures just below the melting point.

Similar effects may be possible with superimposed gas jets [Robin and Nordin (1975a)] or ultrasonic agitation [Mori and Kumehara (1976)]. The effect of a gravitational field has also been discussed [Robin and Nordin (1975b)].

Two major laser applications are based on the drilling of matrices of fine holes. The first involves drilling of ceramic substrates for circuit boards [Longfellow (1971) and Nakada and Giles (1971)]. The second consists of the production of arrays of holes in thin metallic layers on a variety of substrates for image recording and data storage. Further information on this application will be found in Section 2.7.4.

Many other ingenious uses have been found for lasers in drilling. Some of these are proprietary and are only briefly described in the technical literature. A list of some applications that have been reported is listed in Table 2.9. This list is by no means comprehensive.

2.5. WELDING

2.5.1. Heat Transfer—Penetration Welding

Prediction of thermal effects produced by a CW laser beam scanning the surface of an absorbing solid requires that the three-dimensional heat transfer equation be solved subject to the boundary conditions shown in Figure 2.42. The information required from this calculation is (i) the temperature distribution in the workpiece, (ii) depth of melting, and (iii) temperature distribution during cooling.

An approximate solution to this problem was obtained by Swift-Hook and Gick (1973) using the line source model of Carslaw and Jaeger (1959). Their

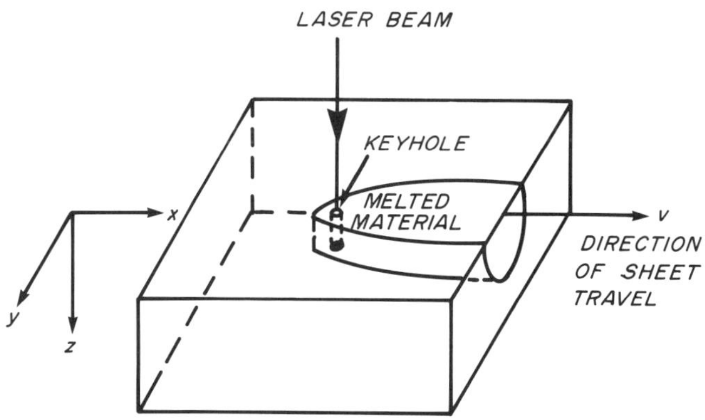

Figure 2.42. Geometry in penetration welding with focused laser source.

result for the maximum melt width, W, is

$$W = 2r[1 - K_0^2(Ur)/K_0'^2(Ur)]^{1/2} \tag{2.13}$$

where $U = v/2\kappa$, $r = (x^2 + y^2)^{1/2}$, K_0 is the modified Bessel function of the second kind [see Carslaw and Jaeger (1959), p. 48], and K_0' is the first derivative of K_0. This expression, when combined with

$$\frac{P}{aKT} = \frac{2\pi}{K_0(Ur)} \exp\left[\frac{UrK_0(Ur)}{K_0'(Ur)}\right]$$

to eliminate Ur, can be solved for the relation between P and W. In these expressions P is the absorbed laser power and a is the penetration depth. The result can be expressed analytically in both the high-speed (Ur large) and low-speed (Ur small) limits.

These are

$$Y_{\max} \sim 0.484M \qquad (Ur \text{ large}) \tag{2.14}$$

and

$$Y_{\max} \sim \exp[(1.50 - 2\pi)/M] \qquad (Ur \text{ small}) \tag{2.15}$$

where

$$Y_{\max} = \frac{vW}{K} \quad \text{and} \quad M = \frac{P}{aKT}$$

The relation between $Y_{\max}$ and M is shown graphically in Figure 2.43.

The validity of this approximate solution can be tested by comparison with experiments in which a complete set of welding parameters have been specified. The following data have been reported by Mazumder and Steen (1980) for mild steel welded with a CW CO_2 laser

$$\begin{aligned} P &= 1570 \text{ W} \\ v &= 3.35 \text{ cm s}^{-1} \\ a &= 0.13 \text{ cm} \\ W &= 0.11 \text{ cm} \end{aligned}$$

Thus taking $\kappa = 0.12 \text{ cm}^2 \text{ s}^{-1}$, $Y_{\max} = 3.35 \times 0.11/0.12 = 3.07$. The graphical solution $Y_{\max}$ vs. M in Figure 2.43 gives $M = 9$ with $Y_{\max} = 3.07$. Taking

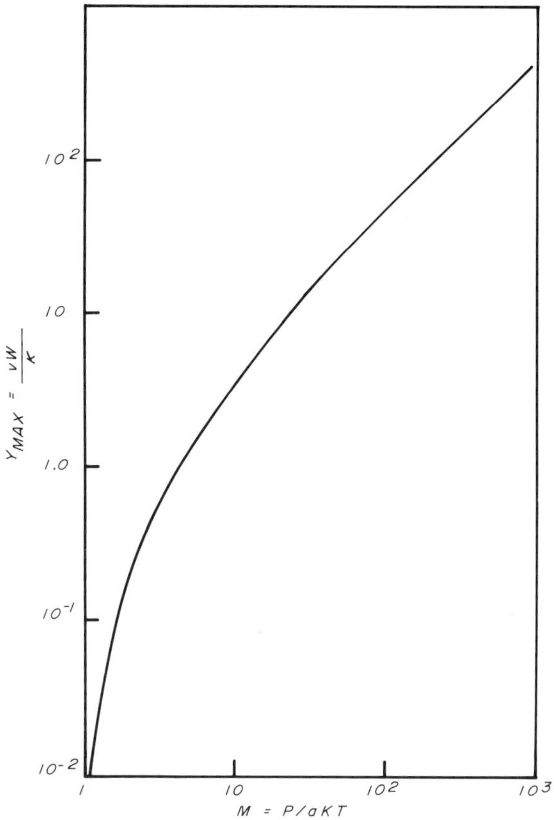

Figure 2.43. Y_{max} vs. M for penetration welding.

$K = 0.5$ W cm^{-1} °C^{-1}, $a = 0.13$ cm, and $T = 1600$°C gives

$$P(\text{calculated}) = 940 \text{ W}$$

Given the uncertainty in thermal constants together with the fraction of incident power absorbed and the fact that the point for Y_{max} vs. M in this case lies in the range where neither analytical solutions are valid, the agreement with observation is satisfactory.

While this simple expression is useful for an approximate estimate of welding parameters, further refinement of these predictions requires that the heat transfer equation be solved numerically. Solutions for a Gaussian beam shape have been reported by Cline and Anthony (1977) and Mazumder and

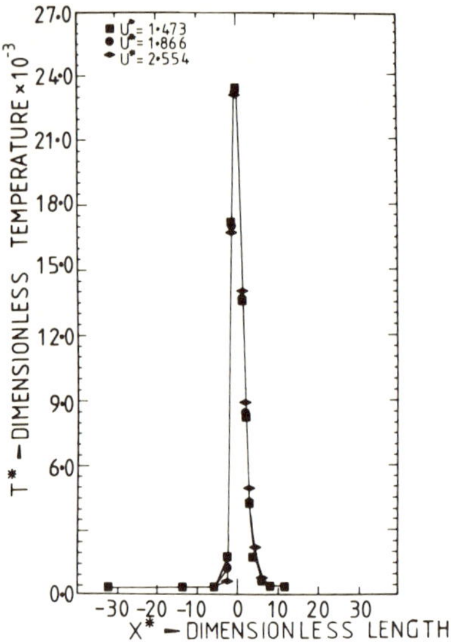

Figure 2.44. Center line T^* vs. X^* [Mazumder and Steen (1980)].

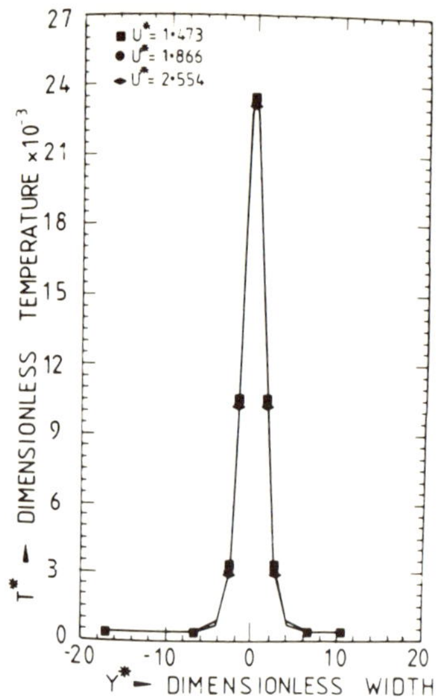

Figure 2.45. T^* vs. Y^* through the point of interaction [Mazumder and Steen (1980)].

Steen (1980). These calculations offer the possibility that studies of the effect of various parameters on heating and welding performance can be performed theoretically.

The model of Mazumder and Steen (1980) calculates temperature profiles in a sheet of finite thickness heated by a moving beam of Gaussian profile. The sheet is assumed to have finite width but infinite length in the direction of travel. Formation of a keyhole is assumed and convective effects due to a superimposed gas jet are taken into account. Absorption of incident laser radiation occurs in the plasma within the keyhole once this keyhole has formed. A Beer–Lambert law of the form $I = I_0 e^{-\alpha z}$ was assumed for the variation of intensity with distance z into the keyhole. Typical values of 700–800 m^{-1} were chosen as representative of α for mild steel. These values were inferred from the saturation of penetration depth at $\sim$ 2.7 mm at high laser powers for mild steel. Saturation is expected if the absorption of incident radiation is attenuated exponentially on entering the keyhole. Thus an increase in laser power above the value for saturation yields an increase in welding speed but no increase in penetration depth.

Defining the dimensionless variables

$$T^* = \frac{\Delta T \, K_g D_b}{\epsilon P}$$

$$Z^* = \frac{z}{D_b}$$

$$Y^* = \frac{y}{D_b}$$

$$X^* = \frac{x}{D_b}$$

where K_g is the thermal conductivity of gas, ΔT the temperature rise above ambient, and D_b the Gaussian beam diameter. Computer generated plots of T^* and Y^* are shown in Figures 2.44 and 2.45. They show the extent of laser heating for various scan speeds.

$$U^* = \frac{vD_b}{\kappa}$$

The corresponding plot of T^* vs. Z^* is shown in Figure 2.46. Data may be left in this general form or can be made specific for a particular solid. Figure 2.47 shows the variation in melt distribution in mild steel welded with a 1570-W laser beam as a function of D_b and α. Empirical data for such a weld are also shown.

Mazumder and Steen (1980) discuss the range of application of such calculations in metallurgy. Temperature profiles in surface hardening, glazing, and coatings are readily calculated using this method. Furthermore, the effect

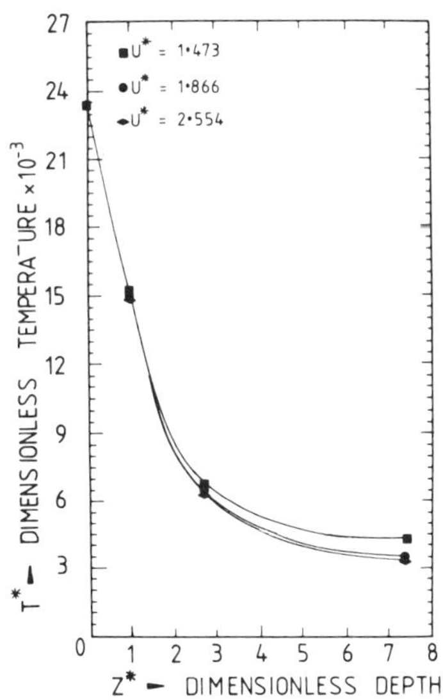

Figure 2.46. T^* vs. Z^* through the point of interaction [Mazumder and Steen (1980)].

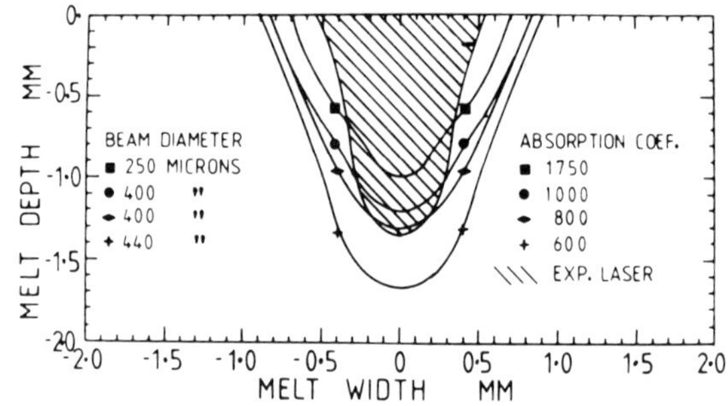

Figure 2.47. Variation in laser melt widths with beam diameter and absorption coefficient at a laser power of 1570 W and reflectivity of 0.8. Welding velocity is 33.5 mm s^{-1} [Mazumder and Steen (1980)].

of arc augmentation [Steen and Eboo (1979)] on welding speeds and thermal profiles can easily be incorporated into the model.

2.5.2. Heat Transfer—Conduction Welding

The analysis of thermal effects in this case involves the solution of the three-dimensional heat transfer equation when the heat input is limited to a surface source. Heat transfer from this source initiates a melting wave that propagates into the bulk material. Stable propagation of this wave will occur while the surface temperature is less than the boiling temperature. The melting depth can then be calculated from the speed of the melting wave and the time interval between T_m and T_v at the front surface. With a moving heat source, for example, as found in surface heat treating, the melting depth must be calculated from a dynamic model.

We first consider the example of a stationary heat source that uniformly heats the surface of an infinite half-space (Figure 2.19). Then if Z is the position of the melt surface below the heated surface

$$Z(t) = \frac{0.16}{\rho L_m} I_0(t - t_m) \tag{2.16}$$

where I_0 is the incident laser intensity (W cm^{-2}), L_m the latent heat of fusion (J g^{-1}), and t_m the time from start of laser pulse to melt front surface. Then, if it is assumed that the maximum depth of melting occurs when $t = t_v$, i.e., when the surface reaches the normal boiling point,

$$Z(t_v) = Z_{\max} = \frac{1.2K}{I_0}(T_v - T_m) \tag{2.10}$$

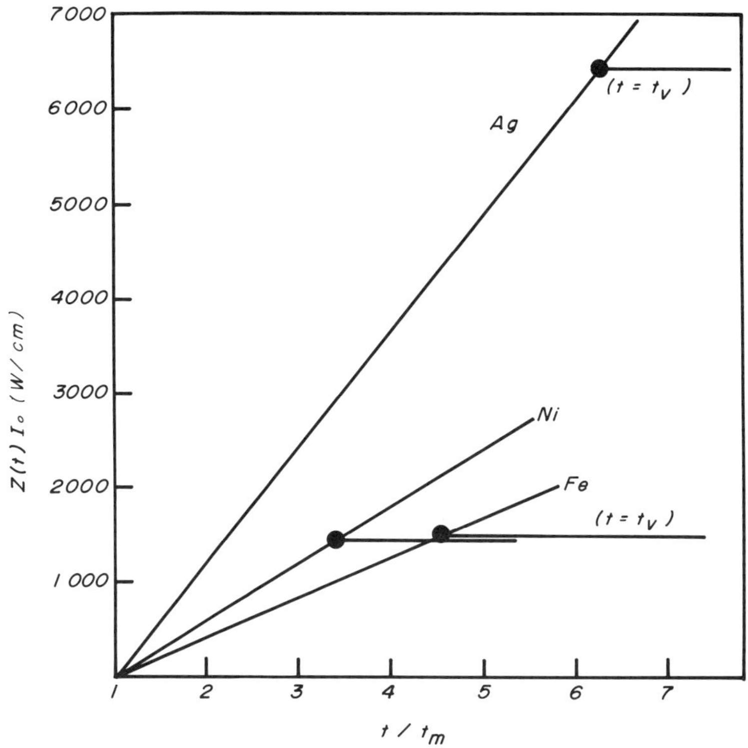

Figure 2.48. Melting depth vs. time for surface heating of an infinite half-space.

These expressions have been used to construct the plot of $Z(t)$ shown in Figure 2.48. The time, t_m, is given by

$$t_m = \frac{\pi K^2 T_m^2}{4\kappa I_0^2}$$

Numerical values for $Z(t)$ for the three metals chosen to illustrate this calculation are

$$Z(t)_{\mathrm{Fe}} = \frac{416}{I_0}\left(\frac{t}{t_m} - 1\right)$$

$$Z(t)_{\mathrm{Ag}} = \frac{1219}{I_0}\left(\frac{t}{t_m} - 1\right)$$

$$Z(t)_{\mathrm{Ni}} = \frac{596}{I_0}\left(\frac{t}{t_m} - 1\right)$$

where room temperature values for thermal constants have been used and $Z(t)$ is in cm. It can be seen that for a particular value of I_0, $Z(t)$ increases uniformly with time normalized to t_m. This behavior would continue indefinitely were it not for the limit imposed by taking $t = t_v$. This point occurs at $Z_{max}(\text{Fe}) = 1475/I_0$, $Z_{max}(\text{Ag}) = 6420/I_0$, and $Z_{max}(\text{Ni}) = 1440/I_0$. Thus for Fe and $I_0 = 10^5$ W cm^{-2}, $Z_{max} = 1.48 \times 10^{-2}$ cm.

Some general conclusions that can be reached from this simple analysis are as follows:

- **(a)** Z_{max} is inversely proportional to I_0.
- **(b)** Z_{max} increases as K increases.
- **(c)** Z_{max} is relatively independent of L but is sensitive to $T_v - T_m$.

A systematic parametric study of melting/solidification on the surface of metals using this simple one-dimensional heat transfer model has been carried out by Mehrabian *et al.* (1979). An example of this data on the solidification of Al, Ni, and Fe is shown in Figure 2.49. The parameters plotted are $(G_L v)_{avg}/I_0^2$ vs. S/Z_{max}, where G_L is the temperature gradient, v the speed of solid–liquid interface, and S the distance solidified. The product $(G_L v)_{avg}$ is the instantaneous average cooling rate in the liquid phase during solidification. Z_{max} here is slightly larger than that defined in Eq. (2.10). These data indicate that the average cooling rate scales as I_0^2.

Mehrabian *et al.* (1979) have performed similar calculations for heating with circular and Gaussian surface sources. They conclude that, with a circular surface source of radius A, the dimensionless temperature distribution in the substrate is independent of I_0 and A as long as $I_0 A$ is constant. With a Gaussian source, melt depths are ~2.3 times smaller than with a circular source when powers in the circular and Gaussian beams are identical and the same temperature is reached at the center of the focal spot.

We now consider the case of a moving surface heat source. The simplest solution is obtained when a point source is assumed. In this case the temperature distribution is (in equilibrium)

$$T(R) = \frac{\epsilon P}{4\pi K R} \exp\left[-\frac{v(R-x)}{2\kappa}\right] \tag{2.17}$$

where $R = (x^2 + y^2 + z^2)^{1/2}$ and v is the speed in x direction. The expression for the steady-state temperature due to a point source moving on the surface of a sheet of thickness l is [Carslaw and Jaeger (1959)]

$$T(R) = \frac{\epsilon P}{2\pi K l}\left\{K_0\left(\frac{vr}{2\kappa}\right) + 2\sum_{n=1}^{\infty} K_0\left[\frac{vr}{2\kappa}\left(1 + \frac{4\kappa^2 n^2 \pi^2}{v^2 l^2}\right)\right]\cos\left(\frac{n\pi z}{l}\right)\right\} e^{vx/2\kappa} \tag{2.18}$$

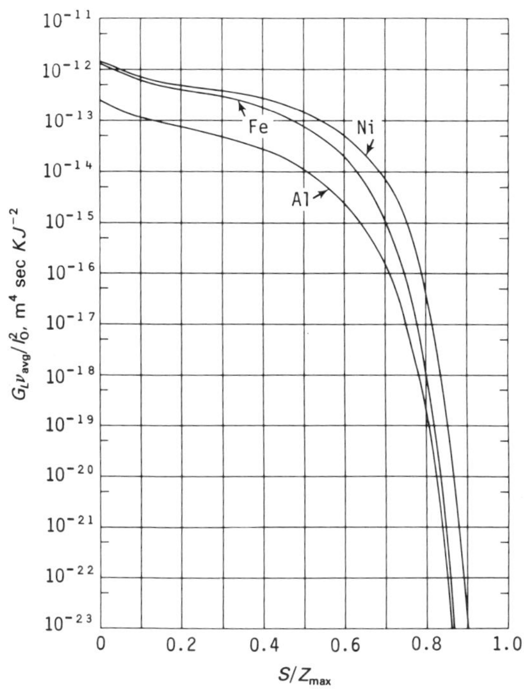

Figure 2.49. Ratio of instantaneous average cooling rate in liquid to I_0^2 vs. S/Z_{max} [after Mehrabian *et al.* (1979)].

where $R = (x^2 + y^2)^{1/2}$ and K_0 is the modified Bessel function of second kind of order zero.

Calculations for $T(R)$ with a moving Gaussian source have been performed by Cline and Anthony (1977) and Steen (1979). These calculations do not explicitly include latent heat effects and so are more relevant to heat treating than welding. However, they are still useful in estimating the penetration of the melting isotherm under conditions of a moving Gaussian surface source melting via conduction.

Figure 2.50 shows an example of the result of a calculation performed for 304 stainless steel by Cline and Anthony (1977). This shows how the penetration of the liquid zone depends on scan speed and absorbed laser power.

Figure 2.51, kindly supplied by Dr. W. M. Steen, shows temperature–time profiles during laser heating of En8 steel. Temperature profiles are shown for several depths. Under the conditions shown for the top set of curves, only depths $\lesssim$ 0.6 mm reach temperatures in excess of the eutectoid temperature during the laser scan.

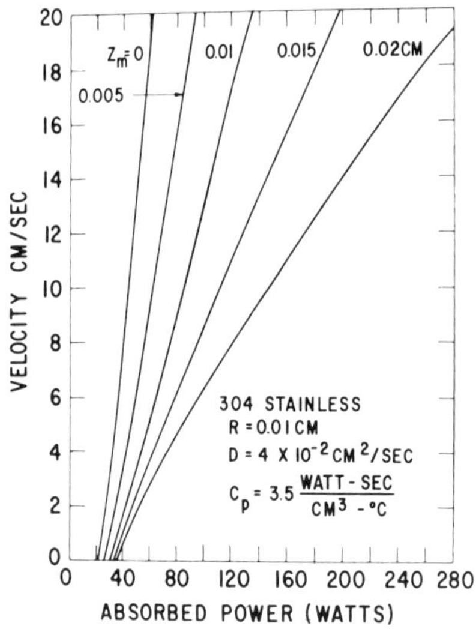

Figure 2.50. Calculated relationship between velocity and power absorbed for different depths of penetration of the liquid zone using constants for 304 stainless steel [Cline and Anthony (1977)].

2.5.3. *Welding with Multikilowatt Lasers*

While microwelding was one of the first industrial applications of high-power pulsed lasers, penetration welding has been one of the first industrial applications of multikilowatt CW lasers. The CW CO_2 laser with powers exceeding 1 kW has been shown to offer some important advantages over conventional penetration welding techniques. In terms of the ability to deliver a large power to a small area, intensities obtainable with the CW CO_2 laser can only be duplicated with electron-beam welding (EBW) devices. However, unlike EBW sources, laser welding can be carried out in air or other gases at atmospheric pressure. Furthermore, the beam can be transmitted through the atmosphere and can be directed to remote locations via lens–mirror systems.

The disadvantage of laser welding compared to EBW arises from the limited penetration depth possible with laser welding techniques. In practice, the penetration achieved with multikilowatt CO_2 lasers is limited to $\lesssim$ 2.5 cm at all but the highest laser powers while much larger depths are possible with EBW devices. This difference arises because the coupling of laser radiation to the workpiece occurs via plasma absorption in the keyhole. This absorption attenuates the beam intensity over a scale length $\sim$ 1 cm at the plasma densities obtained in welding metals. Hence little radiation penetrates the keyhole beyond this depth. As electrons are not subject to this absorption, enhanced penetration is possible. A quantitative comparison of laser and EBW performance has been reported by Ball and Banas (1974). A much more extensive comparison of laser and conventional welding processes has recently been published by Willgoss *et al.* (1979).

Several parametric studies of penetration welding with laser sources have been described [Locke and Hella (1974), Locke *et al.* (1972), Ball and Banas (1974), Mazumder (1977), Crafer (1976), and Willgoss *et al.* (1979)]. An analysis of these data suggests some general correlations between these parameters. It is found, for example, that the relation between penetration depth a, laser power P, and welding speed v can be expressed as $a = \beta P^{1/2} v^{-\gamma}$ where β and γ are constants that depend on the laser source/focusing system and the material welded. As an example, the data of Locke and Hella (1974) obtained in welding 304 stainless steel can be matched with $\beta = 1.46$ and $\gamma = 0.65$ when P is expressed in kilowatts, v in in. min^{-1}, and a in in. Taking $v = 30$ in. min^{-1}, and $P = 9$ kW one obtains $a = 0.48$ in. which is almost identical

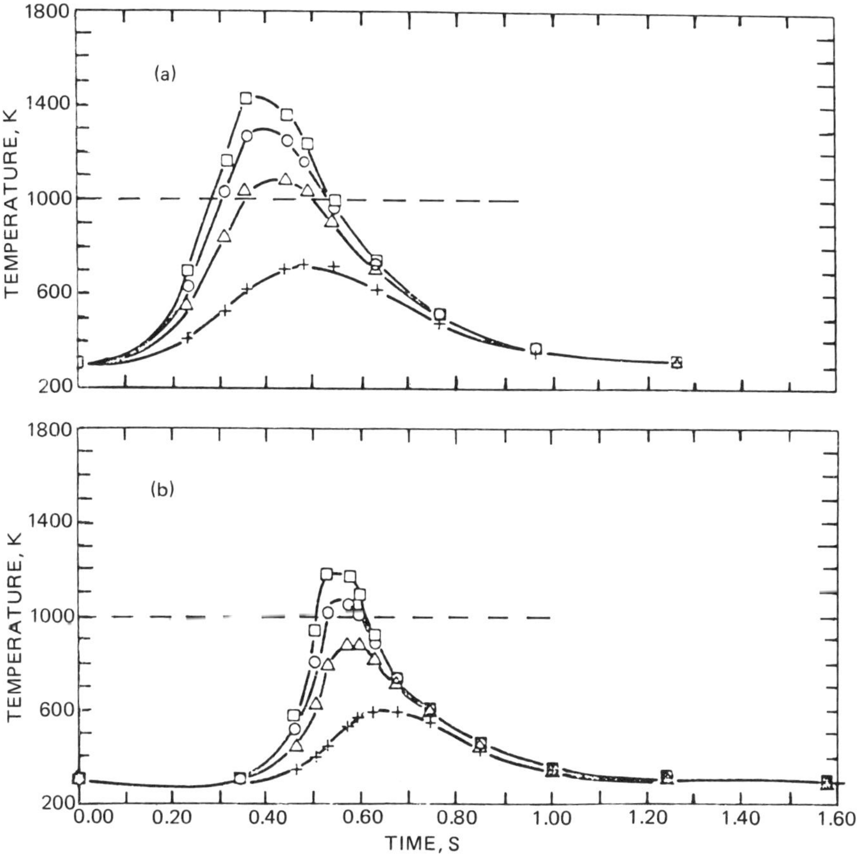

Figure 2.51. Theoretically predicted thermal cycle during laser heating of En8 steel. Power, 2 kW; beam radius, 3.0 mm; reflectivity, 0.4: (a) speed = 22.5 mm s^{-1}; (b) speed = 42.5 mm s^{-1}; - - -, eutectoid temperature. Depth in mm: □, 0.0; ○, 0.22; Δ, 0.6; +, 1.62 [Steen (1979)].

to the measured value. A similar fit to the data of Ball and Banas (1974) for welding 1010 rimmed steel is $a = 0.81\ P^{1/2}v^{-0.55}$ with the same units as above.

Plots of normalized weld speed vs. normalized laser power such as that derived from theory in Section 2.5.1 show that penetration welding utilizes laser power efficiently with overall efficiencies of 20–70% being typical. Greater efficiency tends to be obtained at high welding speeds, since at large v, less incident power is lost by conduction in the workpiece. Figure 2.52 summarizes some data obtained with the AVCO HPL-10 laser.

An example of laser welds produced in Ti–6Al–4V alloy at various welding speeds is shown in Figure 2.53. Similar data have been obtained by

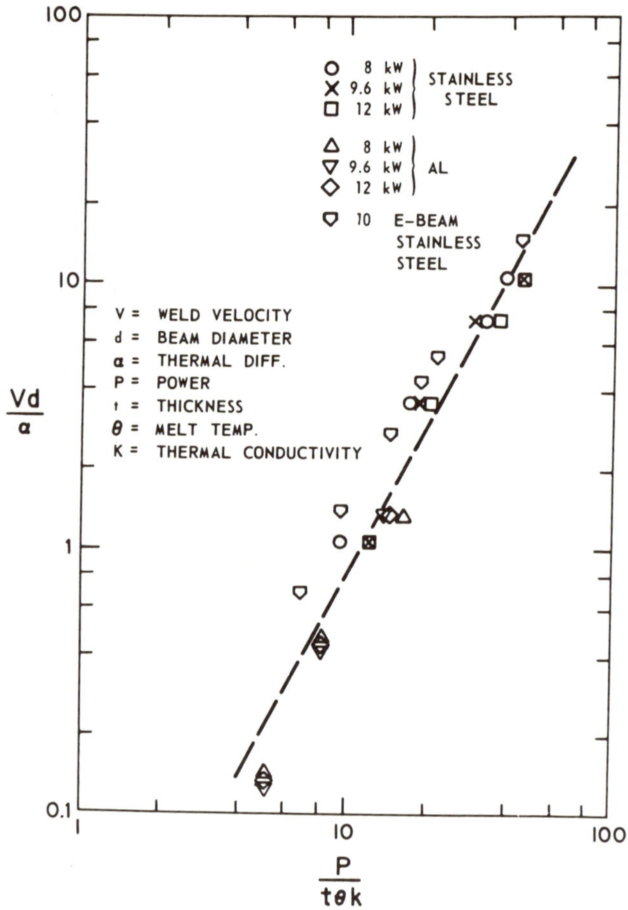

Figure 2.52. Normalized welding speed vs. normalized laser power for welds in stainless steel and Al made with AVCO HPL-10 laser. [Locke and and Hella (1974); copyright 1974 IEEE].

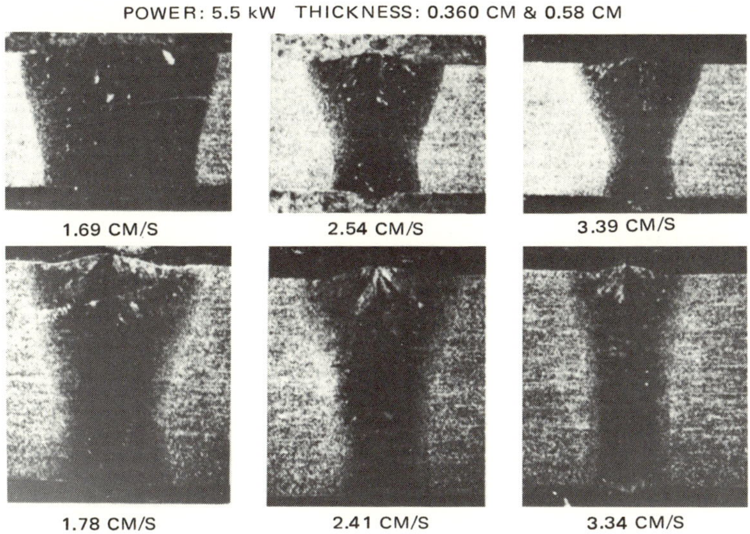

Figure 2.53. Weld profiles in Ti alloy at various weld speeds. A 5.5-kW laser was used [Breinan and Banas (1981)].

Mazumder and Steen (1977). Mazumder and Steen have reported that laser welding is the preferred method for Ti alloys. It was shown that laser welding produced no porosity or chemical contamination of the weld area in contrast to results obtained with EBW. The mechanical properties of the weldment were found to be almost identical to those observed in unwelded material. This is a general property of laser welds: the tensile strength of welded material is usually comparable to or even higher than that of the parent material. Figure 2.54 shows the variation in hardness across laser welded zones in 316 stainless steel and DUCOL W30, respectively [Willgoss *et al.* (1970)]. No significant change in hardness is observed in the stainless steel sample, however, the other alloy shows a dramatic increase within the weldment. This could be reduced by preheating to reduce cooling rates.

The morphology of resolidified melt regions on the surface of metal plates has been discussed by Anthony and Cline (1977), Copley *et al.* (1979), and Moore *et al.* (1979). Surface structures are shown to derive from a balance between liquid flow due to surface tension and gravitational effects. Longitudinal and cross-sectional views through a linear weldment are shown in Figure 2.55. Stopping and starting transients form accentuated depressions and elevations, respectively, of the melt region. The scale size for this structure has been shown by Anthony and Cline (1977) to depend on the thermal properties of the workpiece together with focusing conditions and scan speed. Typical val-

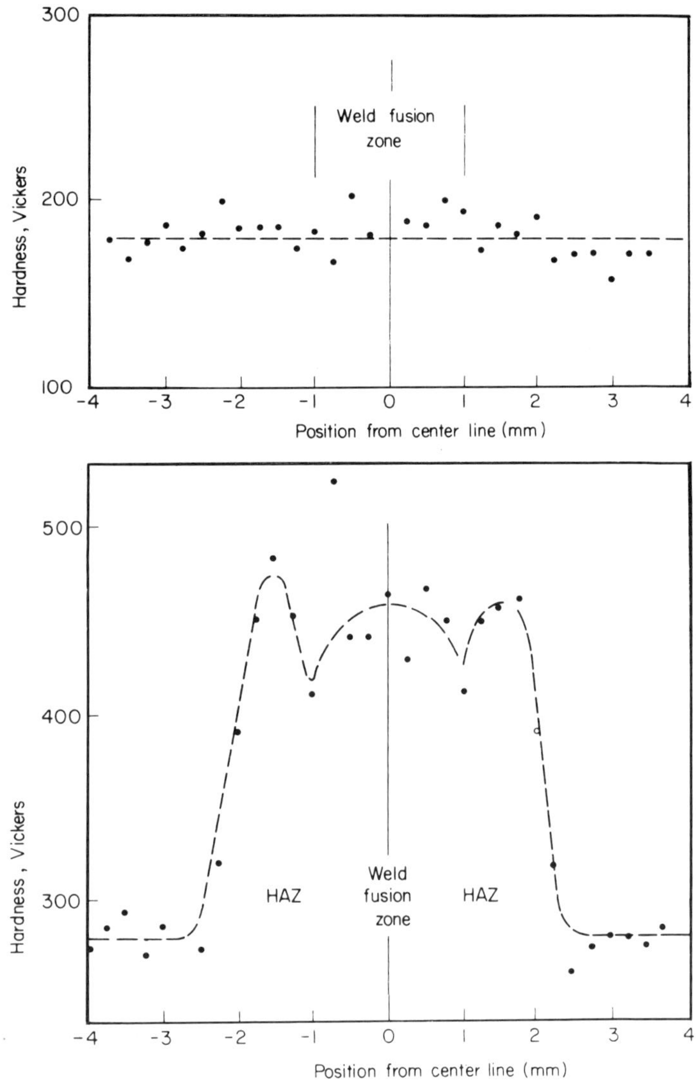

Figure 2.54. Hardness test on DUCOL W30 laser weld (no preheat): plate thickness, 6 mm; traverse, 16 mm s^{-1}; beam power, 5 kW [Willgoss *et al.* (1979)].

ues are ~1.3 mm for iron and ~0.8 mm for Ni. Surface rippling within the resolidified region can be suppressed by rapid scan rates.

Penetration depth and the shape of the weldment are also sensitive to the point at which the laser beam is focused relative to the surface of the workpiece. Locke and Hella (1974) have shown that defocusing by as little as 0.060

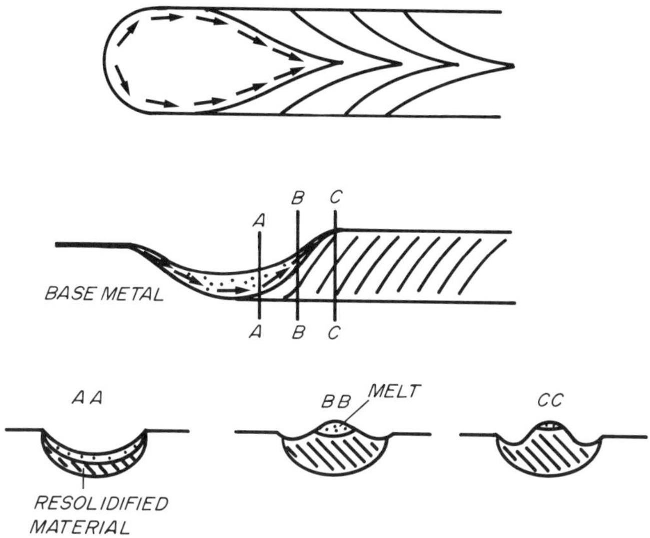

Figure 2.55. Morphology of resolidified regions in laser melted material. Arrows denote mass flow [after Moore *et al.* (1979)].

in. greatly influences both weld properties. It was found that the optimum location for focusing appeared to be slightly below the surface of the workpiece.

A useful addition to laser welding technology has been reported by Eboo *et al.* (1978), who show that the addition of an electrical arc greatly improves welding efficiency and enables welding to be carried out under conditions in which the laser itself is unable to weld. Apparatus for arc-augmented laser welding is shown schematically in Figure 2.56. The arc can be run on either side of the workpiece but best results are obtained with the arc and laser on opposite sides as shown in the figure. Enhancements of welding speed and penetration with arc augmented laser welding arise because the arc roots to the spot heated by the laser. Thus arc power is efficiently delivered to the region of the laser–metal interaction where it improves welding efficiency. Eboo *et al.* (1978) have shown that the combined effect of arc + laser is greater than the sum of the effects produced by each source acting independently of each other. This is most noticeable near threshold where welds are produced under conditions in which neither arc nor laser acting alone produces welds. A quantitative example of the effect of arc augmentation on welding speed is shown in Figure 2.57. Optimum augmentation occurs with the arc power comparable to absorbed laser power.

With the arc on the same side of the workpiece as the laser, electrical power is coupled to the target when the arc roots to the laser generated plasma. Thus electrical energy couples efficiently to the ionized material in the keyhole.

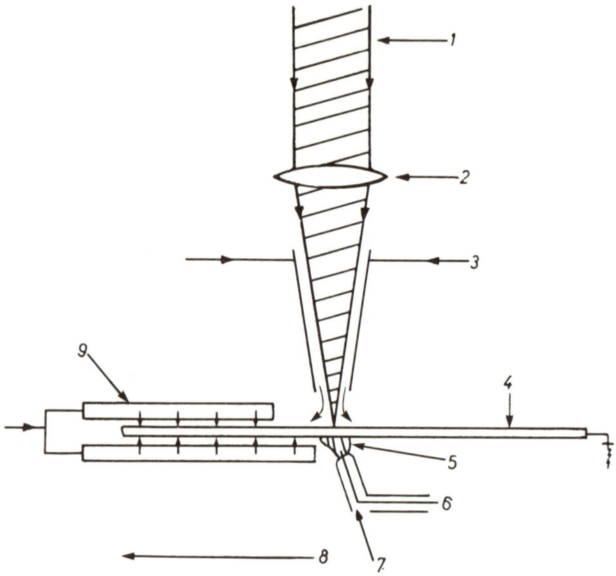

Figure 2.56. Schematic diagram of laser/arc welding system. 1, laser beam; 2, 75-mm KCl lens; 3, He shield; 4, $+ve$ workpiece; 5, arc plasma; 6, tungsten cathode; 7, argon; 8, welding direction; 9, argon shielding jets [Eboo *et al.* (1978)].

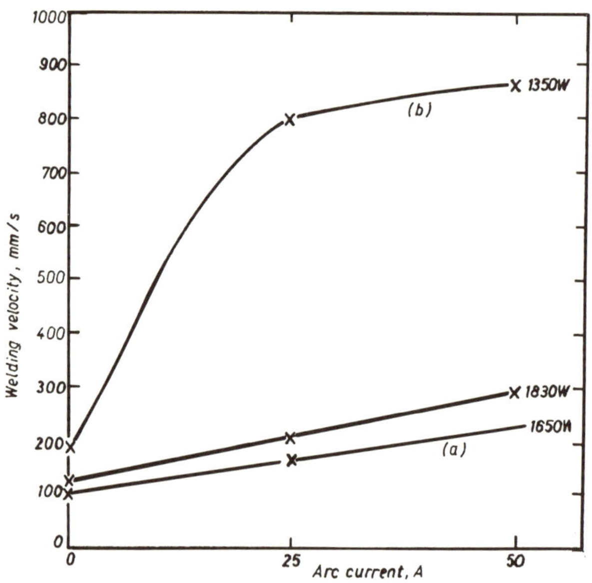

Figure 2.57. Graph of welding velocity against arc current: (a) -2 mm cp titanium with both laser and arc above sheet; (b) 0.2-mm tin plate with laser above and arc below [Eboo *et al.* (1978)].

Macrographs of weld sections on arc augmented laser welded mild steel are shown in Figures 2.58 and 2.59. In Figure 2.58 the arc roots to the keyhole as mentioned above. Figure 2.59 shows the weld produced with arc and laser on opposite sides of the steel sheet. Further discussion of the arc augmentation technique will be found in Section 2.6.2.

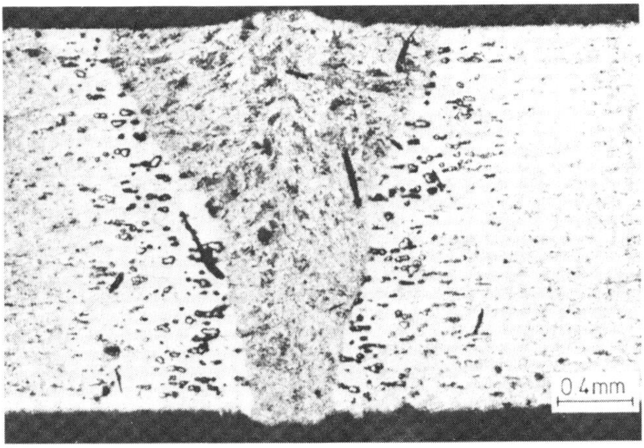

Figure 2.58. Macrograph of a section through a laser/arc weld with the arc on the same side as the laser. Laser power, 1.57 kW; arc power, 1.0 kW; welding speed, 33.5 mm s^{-1}; material, 2-mm thick mild steel. [Steen and Eboo (1979)].

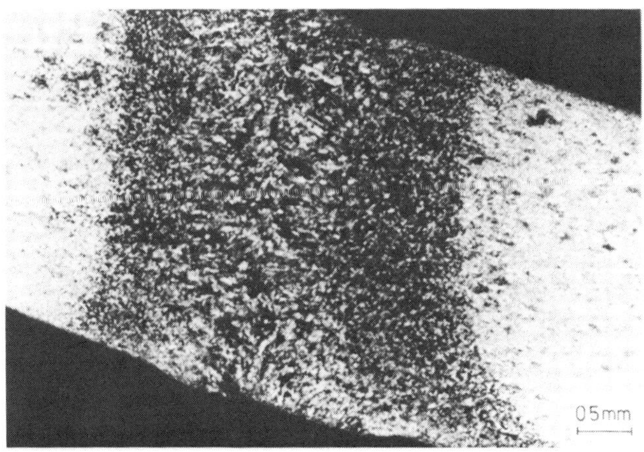

Figure 2.59. Macrograph of a section through laser/arc weld with the arc on the opposite side of the substrate to the laser. Laser power, 1.8 kW; arc power, 1.6 kW (from bottom surface); welding speed, 24.5 mm s^{-1}; material, 3-mm thick mild steel. [Steen and Eboo (1979)].

Recent developments in welding with multikilowatt lasers include the extension of measurements to the 100-kW level [*Laser Focus,* March (1977), p. 14]. At this power level welds are possible in 5-cm thick steel. Full penetration welds in 5-cm steel were made under single-pass conditions with a 90-kW laser at a rate exceeding 5 cm s^{-1}. The depth/width ratio of these welds was ~10. A more modest application involves the use of a 25-kW laser to weld 3.8-cm mild steel and 1.9-cm carbon steel at rates exceeding 1.2 cm s^{-1} [*Laser Focus,* December (1979), p. 26].

An established industrial application of multikilowatt laser welding occurs in the production of seam welds in lead–acid batteries [*Laser Focus,* July (1976), p. 30]. Limitations to acceptance of lasers in other industrial applications involves the relatively tight fit-up tolerances required for laser welding. This may be overcome in part by the use of arc augmented laser welding.

A prototype system for the laser welding of automobile underbodies has been described [*Laser Focus,* February (1974), p. 10]. This system used a 6-kW CO_2 laser to make seam welds along preprogrammed tracks at a rate of up to 450 in. min^{-1}.

The use of a multikilowatt CO_2 laser to weld arctic pipeline steel has been reported by Breinan *et al.* (1975). Full penetration welds in 1.3-cm thick X-80 steel could be made at a rate of 1 cm s^{-1} under single-pass conditions and 2.5 cm s^{-1} under double-pass conditions with laser power of 12 kW.

2.5.4. Welding with Low-Power Lasers

Much useful material processing can be done with lasers having powers in the subkilowatt range. In fact, most industrial applications of lasers involve the use of laser sources at this power level. Many of these applications involve material removal and are described in Section 2.6. Here some data related to welding with low-power CW and repetitively pulsed lasers will be discussed.

At powers below 1 kW, welding of metals occurs under conduction-limited as opposed to deep-penetration conditions. The result is that the utilization of laser radiation is less efficient than under conditions in which a keyhole is formed. Table 2.10 summarizes some data obtained on butt welding with a CW CO_2 laser under these conditions [Webster (1970)]. The laser power was 250 W and a gas nozzle with flowing Ar was used. A variety of other welding configurations (e.g., edge, lap fillet, and corner) were investigated by Webster.

The product, LvW, where L is the sheet thickness, v the weld speed, and W the weld width is relatively constant when data on the materials shown are compared. This product, which gives the volume of metal welded per second, varies between 0.9×10^{-3} cm^3 s^{-1} for Ni 600 and 3×10^{-3} cm^3 s^{-1} for Ti.

Calculating values for the variables $M = P/aKT$ and $Y_{max} = vW/\kappa$ for 302 stainless sheet with $a = 0.025$ cm, $W = 0.1$ cm, $v = 0.42$ cm s^{-1}, and P

Table 2.10. Welding Data Obtained with 250-W CW CO_2 Laser[a]

Material	Thickness (cm)	Speed (cm s^{-1})	Weld width (cm)
321 Stainless steel	0.013	3.8	0.046
	0.025	1.5	0.07
	0.04	0.46	0.076
302 Stainless steel	0.013	2.1	0.05
	0.02	1.3	0.05
	0.025	0.42	0.10
Inconel 600	0.01	6.4	0.025
	0.025	1.7	0.045
Nickel 200	0.013	1.5	0.045
Monel 400	0.025	0.64	0.06
Titanium	0.013	5.9	0.038
	0.025	2.1	0.056

[a]After Webster (1970).

= 250 W gives $M = 28$ and $Y_{max} = 0.84$ with $T = 1600\,^\circ$C and $K = 0.22$ W cm^{-1} °C. These values, when plotted on a graph such as that shown in Figure 2.43 of the previous section, indicate that the welding efficiency in this case is only 5–10%—almost an order of magnitude less than that attainable with penetration welding.

Figure 2.60 summarizes data obtained on penetration depth vs. welding speed obtained with a variety of laser sources under conduction-limited conditions. Welding with repetitively pulsed lasers involves translating the workpiece between pulses to produce an overlapping weld. The curve for welding 302 stainless sheet with a 1500–W CW CO_2 is from Engel (1976) and shows the effect of keyholing at low welding speeds with a concomitant increase in penetration depth.

Engel (1976) also defines some criteria for fit-up tolerances in laser welding of sheet stock. Figure 2.61 shows these relations for butt and lap welded joints. It should be noted that these tolerances limit some potential laser applications in welding thin sheet stock. This is particularly restricting in the butt welding of thin sheets where the fit-up would require the maintenance of ~ 10^{-3} cm clearances over distances of many centimeters. As a result, many laser welding applications have involved welding discrete components that are ordinarily manufactured with close mechanical tolerances. As an example, a pulsed Nd–YAG laser has been used to make closure welds in the cases of implantable heart pacemakers [*Laser Focus,* February (1978) p. 114]. Nath (1974) describes a fiber optic delivery system that permits hand held pulsed welding with the Nd–YAG laser.

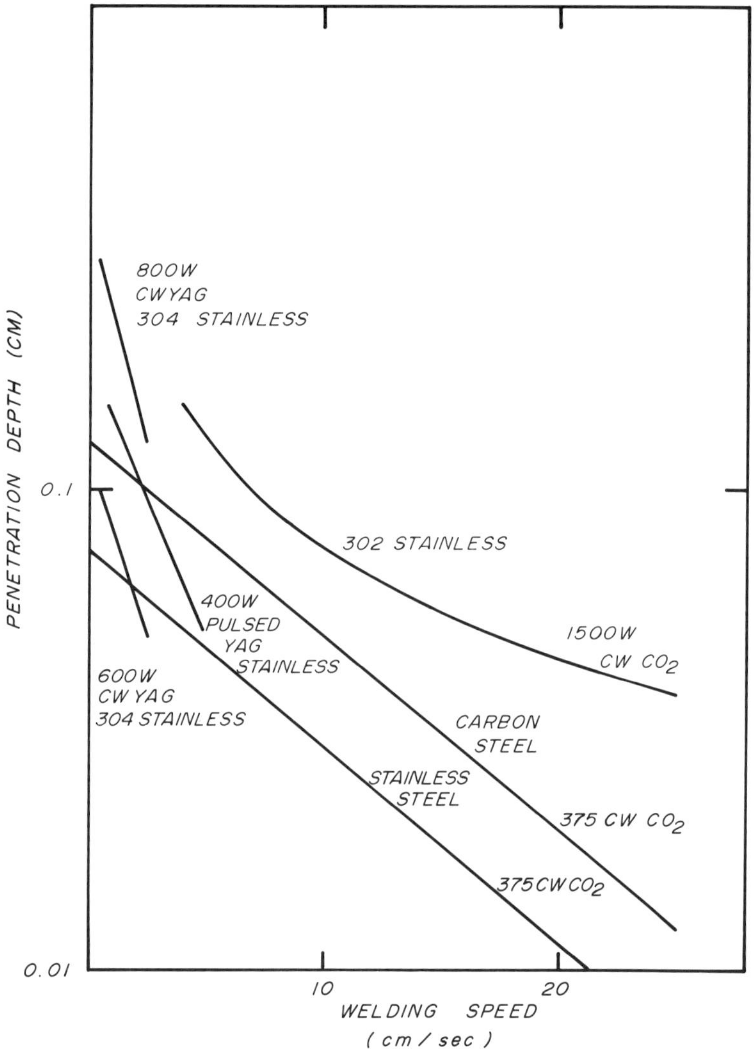

Figure 2.60. Penetration depth vs. welding speed for welding of steels under conduction-limited conditions.

2.5.5. *Laser Spot Welding*

Some of the earliest industrial applications of lasers occurred in the area of spot welding leads for electronic circuitry. Surveys of early work in this area have been published by Cohen and Epperson (1968a,b), Gagliano *et al.* (1969), Gagliano and Zalechas (1972), Ready (1971, 1978), and Duley (1976). The

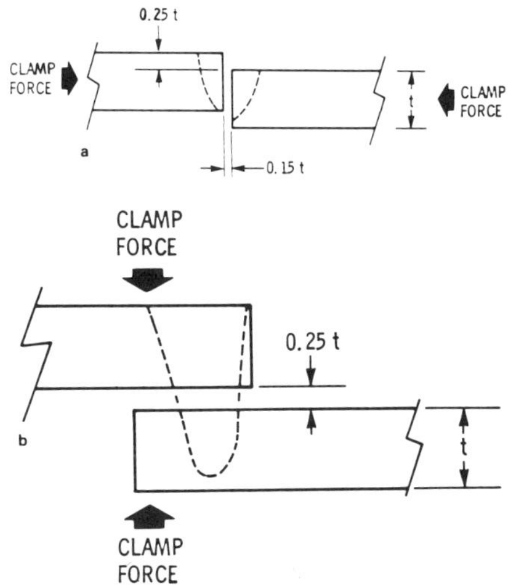

Figure 2.61. Clamping and tolerances for butt joint (a) and for lap joint (b). Dimensions are in units of material thickness [Engel (1976)].

technique is straightforward but some care must be taken in determining optimum pulse energy and focusing conditions for specific applications.

Most laser heating problems in this area are solved empirically since the requisite laser source is usually available and it simply becomes a question of determining the operating condition that yields an acceptable weld. The Nd–YAG laser is the laser of choice in most microwelding applications because of its high power and the fact that high repetition rates are easily obtained. For single-pulse operation or low repetition rate the ruby laser can also be used. It should be noted that *Q*-switched laser sources are not usually used for welding because of the low penetration depths possible with short pulses. Pulses of millisecond duration are optimum for most applications, however, since with $\kappa = 1\ \text{cm}^2\ \text{s}^{-1}$ the penetration depth in copper is $(\kappa t)^{1/2} = (1 \times 10^{-3})^{1/2} \simeq 0.03$ cm, which is typical of lead dimensions found in microcircuitry.

Small (and controllable) heating depths attainable with pulsed lasers represent one of the advantages of laser microwelding. Since these depths *are* small, welding can be carried out in close proximity to heat sensitive components. Moorehead (1971) has reported an interesting comparison of laser and conventional microwelding in the welding of support tabs to a Ti–6Al-4V cylinder. The cylinder wall to which tabs were attached was 0.05 cm thick. Details on the welding processes used together with the resulting depth of penetration and measured temperature rise in the cylinder near the weld are shown in Table 2.11. The small heat affected zone with laser welding is apparent from these data.

The advantages of laser spot welding are well illustrated by the system developed by Western Electric for laser welding of terminals on miniature

Table 2.11. Processing Parameters for Welding of Tabs to a Ti–6Al–4V Thin Wall Cylinder[a]

Welding process	Welding parameters	Peak temperature (°C)	Depth of penetration (cm)
Plasma arc	6 A, Ar plasma	970	0.04
Gas tungsten arc	10 A, 1-s weld time, 0.06-cm arc length	510	0.01
Electron beam	120 kV, 2.2 mA 0.5-s weld time	270	0.055
Laser	3.0 kV, 3.5 ms 1.9 J, 0.085-cm diameter spot	80	0.025

[a]After Moorhead (1971).

relays [*Laser Focus,* December (1978), p. 22]. This system uses a 8J Nd–YAG laser to produce 20 welds per second with 5-ms pulses. A microprocessor positions the relays and initiates firing of the laser. Advantages over conventional technology in this area involve a fourfold increase in speed and improved quality control.

2.6. CUTTING

2.6.1. Heat Transfer

Much of the theory that has been developed in Section 2.5.2 for welding with lasers is also applicable in general terms to laser cutting. Cutting occurs when a semicylindrical volume is driven through the workpiece in response to laser heating and translation of the workpiece. The analogy with penetration welding is evident with the exception that material removed at the leading edge of the "keyhole" is not replaced at the trailing edge. In practice, a gas jet is directed coaxially with the laser beam into the interaction region. The effect of a gas jet is to remove liquid from the focus, improving the quality of the cut. With a reactive gas such as O_2, exothermic reactions occur in the region of the laser focus releasing additional energy that assists cutting.

An approximate solution to the three-dimensional heat transfer equation can be obtained by using the point source model [Babenko and Tychinskii (1973) and Duley and Gonsalves (1974)]. In this approximation the equilibrium temperature in a moving sheet of thickness l at distance r from the origin is

$$T(r) = \frac{\epsilon P}{2\pi K l} \exp\left(\frac{vx}{2\kappa}\right) K_0\left(\frac{rv}{2\kappa}\right) \tag{2.19}$$

where v is the sheet speed and other symbols have the same meaning as in Section 2.5.

With a gas jet, additional heating/cooling terms must be added. These are P_1, the power transferred to point source due to chemical reactions, and P_2, the power loss due to convective cooling. Defining normalized variables $C_0 = \epsilon P/2\pi KlT_0$, $C_1 = P_1/2\pi KlT_0$, and $C_2 = P_2/2\pi KlT_0$ one obtains

$$C_0 + C_1 - C_2 = \frac{\epsilon^{-X}}{K_0(R)}$$

where $X = vx/2\kappa$, and $R = vr/2\kappa$, and T_0 is the kindling temperature. The term C_1 becomes

$$C_1 = \frac{2}{\pi}\frac{q}{CT_0}Y$$

where q is the net heat liberated from the chemical reaction (J cm^{-1}), C the heat capacity (J g^{-1}), and $Y = vy/2\kappa$. With $q/CT_0 = \psi$,

$$C_0 - C_2 = \frac{\epsilon^{-X}}{K_0(R)} - \frac{2}{\pi}\psi Y$$

Differentiating with respect to X gives

$$C_0 - C_2 = \frac{1}{K_0(R)_m}\exp\left[\frac{-R_m K_0(R_m)}{K_1(R)_m}\right] - \frac{2\psi Y_{\max}}{\pi}$$

where R_m is the value of R for $Y = Y_{\max}$. It can be shown that

$$Y_{\max} = R_m\left\{1 \quad \left[\frac{K_0(R_m)}{K_1(R_m)}\right]^2\right\}^{1/2} \qquad (2.20)$$

The value of ψ can be either positive or negative depending on whether or not the net effect of the gas jet (exclusive of convective cooling) is to produce heating or cooling. For example, if the jet acts to remove liquid metal q may be the latent heat of fusion, L_m, and $\psi \sim L_m/CT_m \sim -0.4$.

Analysis of experimental data [Duley (1976)] shows that cutting of stainless steel seems to be characterized by a value of ψ of about this magnitude. For stainless steel one has an approximate empirical relation

$$C_0 - C_2 = 1.34\, Y_{\max} + 0.3$$

which is equivalent to taking $\psi \sim -0.2$ in a theoretical plot of $C_0 - C_2$ vs. Y_{max}.

2.6.2. Cutting Metals

Most cutting of metals has been carried out with the CW CO_2 laser. At first inspection this is somewhat surprising in view of the high reflectivity of metals in the infrared (Section 2.1). However, this is a problem only for pristine metal surfaces at room temperature, since we have seen ϵ tends to increase with increasing T and oxidation. Furthermore, the onset of surface damage increases ϵ to $\sim$1. Despite the fact that ϵ for metals is therefore not as small in practice as in theory, high average powers are required to cut metals. The high average powers required are most easily obtained from CW CO_2 laser sources.

Some representative data on the laser cutting of metals is summarized in Table 2.12.

Cutting rate usually scales linearly with laser power for constant sheet thickness. The dependence of cutting rate on sheet thickness at constant laser

Table 2.12. Metals

Metal	Thickness (cm)	Cutting rate ($cm\ s^{-1}$)	Laser power (W)	Reference
Mild steel	0.32	1.48	400	Clarke and Steen (1979)
	0.40	3.1	1 600	
	0.20	9.0	2 000	
	0.32	6.8	4 000	
Carbon steel	0.63	3.75	15 000	Locke *et al.* (1972)
302 Stainless steel	0.0075	10	200	Duley and Gonsalves (1974)
	0.03	3.75	200	
304 Stainless steel	0.47	2.1	20 000	Locke *et al.* (1972)
410 Stainless steel	0.15	2.1	250	Williamson (1976)
	0.28	0.42	250	
Titanium	0.017	10	240	Harry and Lunau (1972)
	0.63	5.8	3 000	Wick (1975)
	5.0	0.83	3 000	
	2.5	8.3	10 000	Locke and Hella (1974)
Titanium alloy–6Al–4V	0.125	12.5	210	Coherent Radiation Ltd. Bookle
	0.22	6.25	210	
	0.625	4.6	250	
	0.98	4.2	260	
Aluminum	0.1	10.4	3 000	Wick (1975)
	0.3	4.2	4 000	
	1.25	1.25	5 700	
	1.25	3.75	15 000	Locke *et al.* (1972)
Inconel 718	1.25	2.1	11 000	Locke and Hella (1974)

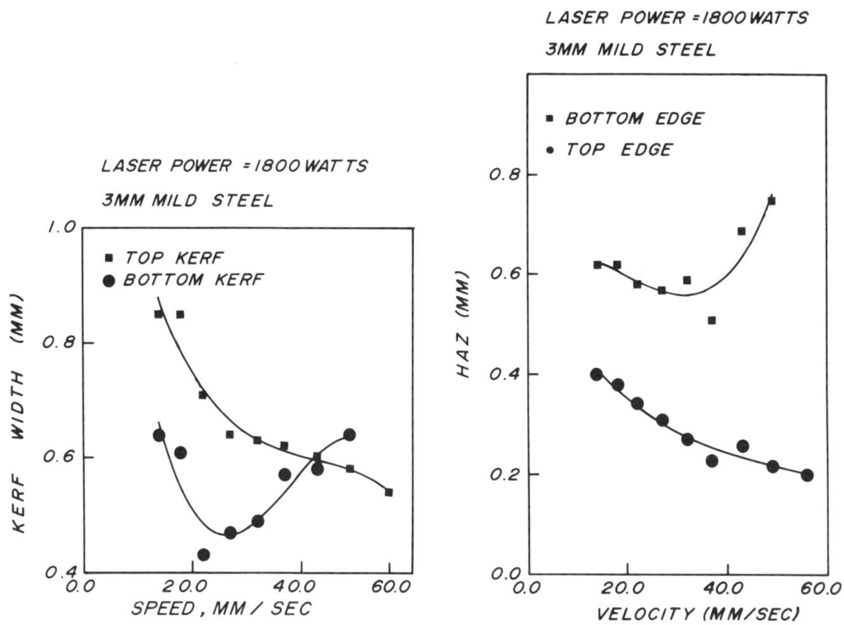

Figure 2.62. (a) Kerf width vs. cutting velocity for 3-mm mild steel and a laser power of 1800 W. (b) Heat affected zone (HAZ) vs. cutting velocity in 3-mm mild steel with a CW CO_2 laser power of 1800 W. The focal spot diameter was ~250 μm [Steen (1979)].

power is less well defined since the quality of the cut may become involved in subjective decisions as to the acceptability of a particular cut. Typically, however, one has $v \propto l^{-\beta}$ where $\beta \sim 1$.

The heat affected zone (HAZ) and kerf width are strongly dependent on cutting speed. Some data provided by Steen is shown in Figures 2.62(a) and (b). Both the HAZ and kerf width at the top surface decrease monotonically with increasing cut speed. Both parameters measured at the bottom surface exhibit minima. The value of v at which this occurs corresponds to optimum cut quality. The parameter P/v appears to characterize this effect. For mild steel the minimum HAZ (bottom surface) and optimum cut quality occurs with $P/v \sim 70$ J mm^{-1}.

Arc augmented laser cutting [Clarke and Steen (1979)] has been shown to result in substantial increases in cutting speed and penetration depth. Particularly good results are obtained when the laser and arc are on opposite sides of the sheet. The enhancement possible is shown for 3-mm mild steel in Figure 2.63.

This figure shows the cutting speed plotted vs. total incident power (laser + arc). The laser power is 1870 W so that powers in excess of this value are

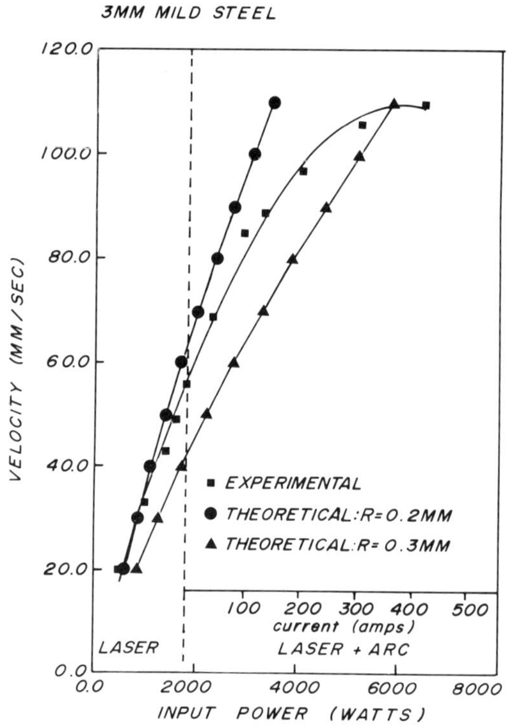

Figure 2.63. Cutting speed vs. total input power for laser + arc augmentation for 3-mm mild steel [Clarke and Steen (1979)].

contributed by the arc. It is evident that up to the point where arc power ~ laser power the cutting proceeds as if the laser power had been increased. Cut quality is maintained over this range. Above 4 kW total power the cutting speed saturates and there is a deterioration in cut quality. This is due to dross accumulation on the arc side of the cut.

Evidently, arc augmented laser cutting may lead to a significant increase in ultimate cutting speeds. The advantage is that this increase is obtained by the provision of relatively inexpensive electrical energy from an arc rather than by an increase in laser power or by the use of more than one laser.

As discussed in Section 2.1, the absorption of light at a metal surface is polarization dependent. Optimum coupling occurs when the polarization vector of incident laser light is perpendicular to the metal surface. This phenomenon means that cutting rates should be dependent on the relative orientation of **E**, the polarization vector and **v**, the cutting velocity. Furthermore, since $\epsilon_{\perp}$ is a strong function of angle and reaches a maximum value for angles of incidence close to, but not identically equal to, 90°, the balance of absorption and cutting with linearly polarized laser beams can lead to sloping of the sides of a cut. A schematic representation of this effect is shown in Figure 2.64. The effect is most noticeable when cutting along circular or other curved paths.

2.6.3. Cutting Nonmetals

Many practical applications of lasers in industry involve the use of CW CO_2 lasers to evaporate nonmetallic materials Most industrial lasers with powers $\lesssim$ 500 W are designed with these applications in mind. An example is the Coherent Everlase® System that can simultaneously provide CW laser powers of ~150 W to up to four stations. Alternative configurations of 2 × 150 W or 1 × 600 W are available simply by connecting plasma tubes in series in the laser head.

Since the thermal conductivity of nonelectrically conducting materials is small, little absorbed laser energy is lost by conduction. As a result, the maximum cutting rate can be simply related to laser power as follows

$$P = QWlv \tag{2.21}$$

where Q is the energy required for evaporation (kJ cm^{-3}), W the kerf width, l the sheet thickness, and v the cutting speed (cm s^{-1}). Q may be as small as 0.4 kJ cm^{-3} for a soft wood and as high as 100 kJ cm^{-3} for some glasses.

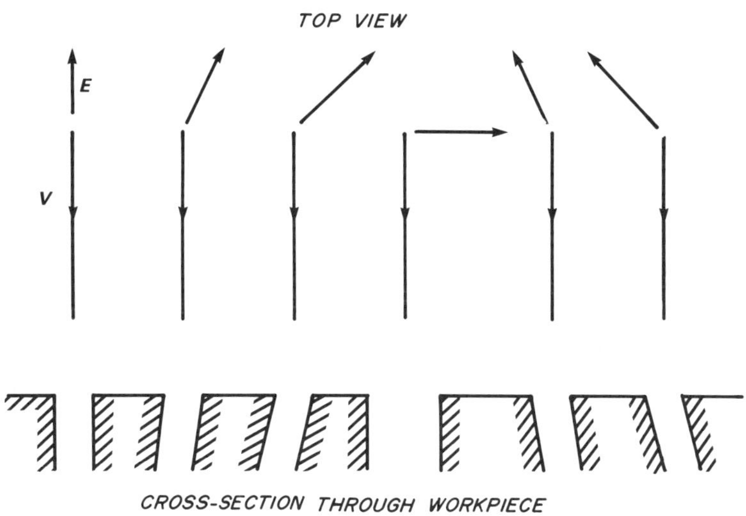

Figure 2.64. Effect of relative orientation of polarization vector and cutting direction on shape of cut.

Table 2.13 gives some representative data on cutting speeds for a variety of materials using the CW CO_2 laser as a heat source. Unfortunately, while there is much data on cutting rate vs. thickness and laser power, the kerf width is not often quoted. Since this quantity is required to obtain Q, it is difficult to use the data in Table 2.13 to make general predictions of P vs. v for given materials. Babenko and Tychinskii (1973) have published some values for Q/ρ, where ρ is the density for a variety of materials. These can be used to obtain approximate cutting rates. Some values for Q obtained from measurements of cutting rates with a multikilowatt laser [Locke *et al.* (1972)] are given in Table 2.14. Further discussion of laser cutting of wood, textiles, paper, and cardboard and a variety of nonmetallic materials can be found elsewhere [Duley (1976)].

Table 2.13. Laser Cutting of Nonmetallic Materials

Material	Thickness (cm)	Laser power (kW)	Cutting rate (cm s^{-1})	Reference
Bond paper	0.005	0.25	500	Miller and Osial (1969)
Newsprint	0.0075	0.25	265	
Plywood	0.5	0.225	2	Harry and Lunau (1972)
	2	0.225	0.46	
	2.5	8	2.5	Locke *et al.* (1972)
Corrugated board	0.45	3.9	175	Banas *et al.* (1971)
Borsical™ tape	0.02	3.9	50	Banas *et al.* (1971)
Sodalime glass	0.4	0.2	0.16	Lunau (1970)
	0.2	0.35	1.25	Harry and Lunau (1972)
	0.02	0.2	8.3	
Glass	0.94	20	2.5	Locke *et al.* (1972)
Quartz	0.2	0.25	0.16	Harry and Lunau (1972)
	0.31	0.5	1.2	Ready (1978)
Alumina	0.075	0.1	2.5	Harry and Lunau (1972)
Concrete	3.75	8	0.08	Locke *et al.* (1972)
Cellulose	0.025	0.375	2000	Photon sources
ABS plastic	0.025	0.375	1000	Photon Sources
Polycarbonate	0.025	0.375	100	Photon Sources
Polyethylene	0.025	0.375	30	Photon Sources
Expanded polystyrene	2	0.3	0.017	Harry and Lunau (1972)
Leather	0.3	0.225	5.1	Harry and Lunau (1972)
Simulated leather	0.15	0.35	28	Uglov *et al.* (1978)
Rubber	0.45	0.35	2.5	Uglov *et al.* (1978)
Limestone	3.75	3.5	0.2	Banas *et al.* (1971)

Table 2.14. Derived Values of $Q = P/Wlv$ from Measurements of Locke *et al.* (1972)

Material	Q (kJ cm^{-3})
Plywood	7.9
Plexiglass	7.9
Glass	78
Concrete	42
Boron epoxy composite	69
Fiberglass epoxy composite	36

One of the earliest and most successful industrial laser applications in this area involves the use of a CW CO_2 laser to cut slots in plywood sheet for the production of steel dies (Figure 2.65). Pattern information is supplied to a microprocessor that drives the x-y table and controls laser power. Laser powers of $\sim$300 W are typical. After slotting is produced by laser cutting, metal die blades are manually driven into the slots to form the final die. Advantages over conventional technology include much reduced slotting time and better tolerance in die dimensions.

A similar arrangement can be used to machine rubber for printing rolls or to engrave complex patterns in ceramic tubes [Cockayne and Gasson (1971)]. Since these objects have cylindrical symmetry they are placed on a lathe and the numerically controlled laser "writes" the required pattern as the beam is translated down the axis of the cylinder. Helical grooves have been machined in this way on the surface of vitreous silica and zirconia tubes.

Micromachining of grooves in absorbing glass has been shown to be useful in forming channel optical waveguides [Pavlopoulos and Crabtree (1974)]. Light from an Ar^+ laser operating at 488 nm was focused to a fine spot on the

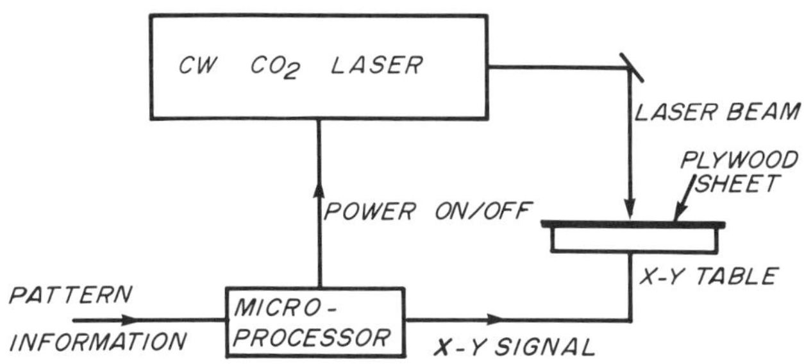

Figure 2.65. Numerically controlled system for cutting patterns in plywood sheet using CW CO_2 laser.

surface of a glass substrate while this was translated under the beam. Grooves 20–40 μm deep with widths of 2–8 μm were formed. Typical cutting rates were 100–1000 $\mu m\ s^{-1}$ at a laser power of 0.2–0.4 W. Chen *et al.* (1976) have shown that waveguides in glass can also be formed by laser-induced refractive index variation along a laser heated track.

A modification of the basic laser cutting technique involves the use of a laser to "saw." An example of laser sawing is found in techniques developed for the cutting of diamonds [*Laser Focus,* February (1977), p. 27]. In the sawing process, overlapping holes are drilled by moving the workpiece under the focused laser beam. The process, when applied to cutting diamond with a *Q*-switched Nd–YAG laser, was found to result in a five-fold decrease in cutting time for a 4.1-mm diameter stone. The kerf width was 5×10^{-3} cm. This technique is particularly useful in cutting stones with knots because the evaporation process is not affected by variations in mechanical hardness.

The differential absorption of CO_2 laser radiation by plastics and metals has been exploited in an application involving the removal of plastic overlayers from metal substrates [Slivinsky and Ogle (1977)]. Figure 2.66 shows a cross

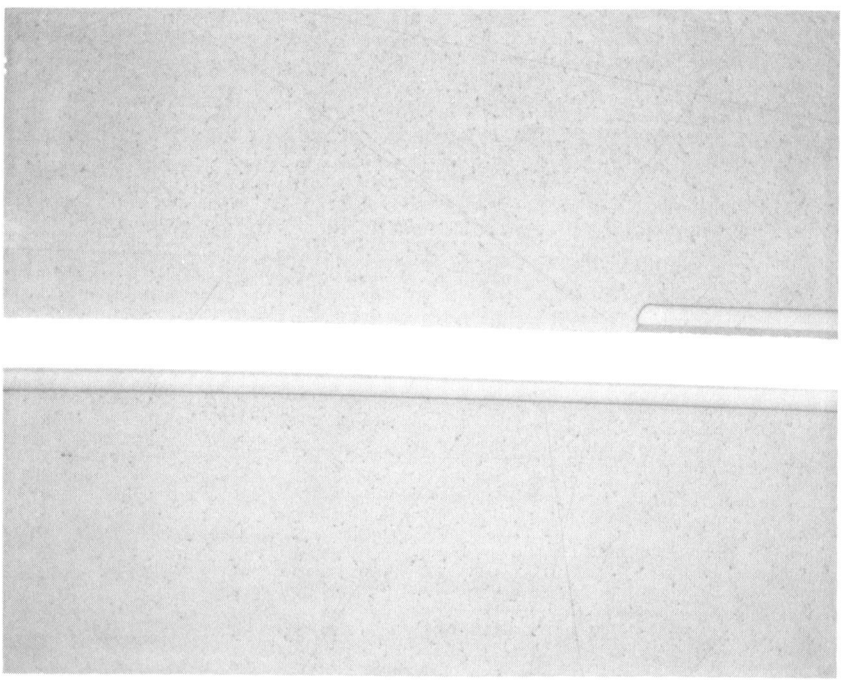

Figure 2.66. One side of a typical laser skiving result—50× magnification. Note clean plastic edge at cut and absence of copper removal [Slivinsky and Ogle (1977)].

section of a kapton-coated Cu sheet that has been skived with a beam from a CW CO_2 laser at an intensity of $\sim 7 \times 10^4$ W cm^{-2}. It is evident that the removal of plastic is accompanied by no noticeable substrate damage.

2.6.4. Scribing and Controlled Fracture

Both scribing and controlled fracture are alternative methods of separating nonmetallic materials such as ceramics, silicon, or glass. Both techniques utilize laser power to define a line along which the part will subsequently fail mechanically.

In scribing, the laser cuts a groove which extends only part way into the substrate. The substrate can subsequently be separated along this line by application of mechanical force. Alternatively, a pulsed laser may be used to perforate along the desired line.

Controlled fracturing occurs when laser heating is used to generate thermal stresses that lead directly to fracture along the requisite direction. Since the laser only provides a small fraction of the energy required to separate the material, controlled fracture requires only low laser power. However, it does suffer from the fact that the direction of fracture cannot always be maintained as desired and that complex shapes are not easily machined with this technique. Nevertheless, where applicable, controlled fracture leads to rapid separation rates. As an example, Lumley (1969) shows that 0.07-cm thick alumina ceramic can be parted at 0.5 cm s^{-1} with a CO_2 laser at the ~10-W power level.

Scribing is extensively used in the microelectronics industry for the production of silicon wafers. Since Si is more strongly absorbing at 1.06 μm than at 10.6 μm, the Nd–YAG laser is preferred in this application. However, scribing of alumina or other ceramic substrates can be efficiently carried out with the CW CO_2 laser.

Narrow shallow grooves have been scribed in glass via laser-assisted thermal etching [Daree and Kaiser (1978)]. This process relies on the fact that the rate of reaction of HF gas with a glass surface is strongly temperature dependent. By heating selected spots or linear regions with a tightly focused laser beam, the reaction can be enhanced in this region and material is removed. Daree and Kaiser report that the enhancement in reaction rate produced by laser heating is by a factor ~1000.

2.7. MICROMACHINING

The need for micromachining of thin film resistors and thin film on silicon circuitry has led to several important industrial applications of lasers. The advantages of laser processing in this area include

Small heat affected zones
Control over machining depth
Rapid throughput
Accomplishment of machining at any time during or after processing

Exploratory work in laser micromachining has been summarized by Cohen and Epperson (1968), Gagliano *et al.* (1969), and Ready (1971). Present applications of laser micromachining are many [Cohen (1972), Duley (1976), Ready (1978), and Shkarofsky (1975)].

2.7.1. Resistor Trimming

Laser trimming of thin and thick film resistors has become commonplace in industry. Trimming is accomplished by vaporizing part of the film or by thermally modifying part of the film to alter its resistance. When vaporization is used, material may be removed either as a series of holes [Figure 2.67(a)] or continuously in the form of lines [Figure 2.67(b)].

Laser trimming by evaporation is affected by post-trimming drift. This is a function of the heat-affected zone adjacent to the vaporized region. In this regard, while resistance adjustment using many fine holes offers the possibility of a sensitive approach to the final resistance value, post-trimming drift is largest with this method. Cohen (1972) has shown that the percentage drift in Ta_2N resistors is largest for small hole diameters. In addition, the net drift (which involves an increase in resistance) increases with the initial change in resistance induced by trimming. Thus, for optimum results, holes of large diameter or continuous cuts should be used. Production techniques involve the generation of resistors whose resistance is less than required in the final circuit. The aim is to produce trimmed resistors whose final values closely approach

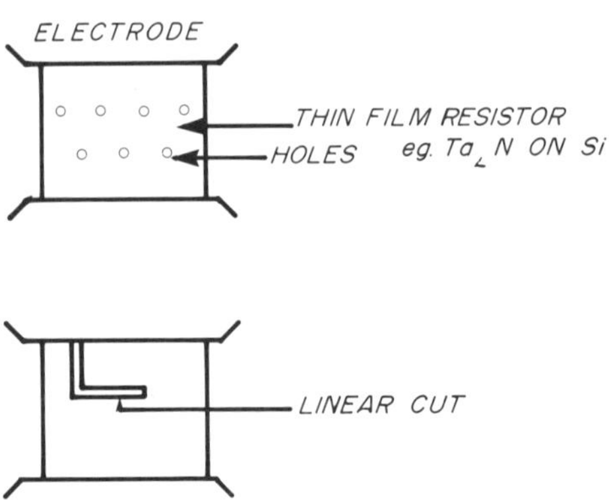

Figure 2.67. Resistor trimming by (a) removal of discrete areas and (b) by cutting lines.

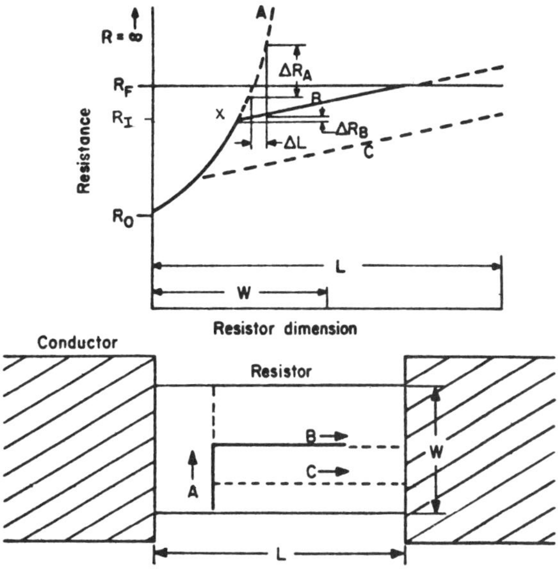

Figure 2.68. Benefits of the laser "L-cut" technique. If the resistor shown is trimmed to its final value, R_F, by cutting straight across along line A, two undesirable effects are noted. The resistance change, ΔR, per length of cut is relatively large, making accurate adjustment difficult. The hottest spot will occur at the narrowest remaining portion of the resistor. By approaching R_F along A, then completing the trim along B, a slope of much smaller ΔR per unit length is followed. This aids accurate adjustment and retains a greater minimum width, minimizing chances of a hot spot. If the combination A–C is chosen, such that the end of the resistor arrives before R_F is reached, the resistor must be rejected [Thompson (1970)].

desired resistances when post-trimming drift is taken into account. Final tolerances may be closer than 0.1%.

The approach to equilibrium in trimming of a thin film resistor by introducing linear cuts is shown in Figure 2.68 [Thompson (1970)]. If the desired final resistance value is R_F this is best approached by means of two cuts, A and B. Cut A across the film takes the resistance from R_0 to R_I. The final approach to R_F is made via cut B giving increased sensitivity since the slope $\Delta R/\Delta L$ is smaller along the path XB, than along AX. If the paths A and B can be kept relatively small then the chance of occurrence of a "hot spot" in the finished resistor is minimized.

The Q-switched Nd–YAG laser is the laser of choice in trimming thin film resisistors. Typical operating parameters are

Pulse length	200 ns
Peak power	1–10 kW
Repetition rate	1 kHz

Repetitive Q-switching at high rates permits the generation of linear cuts by the overlapping of individual pulses. This is accomplished by scanning the beam at the rate of spot radius × repetition rate. For example, for a 20-μm diameter spot size and a repetition rate of 1 kHz, the scan speed is

$$10 \times 10^{-6} \times 10^{3} = 10 \times 10^{-3}\ \text{m s}^{-1} = 1\ \text{cm s}^{-1}$$

The penetration of heat into the substrate can be roughly estimated from

$$\Delta x = (\kappa t)^{1/2}$$

where Δx is the penetration, κ the thermal diffusivity, and t the duration of laser pulse. When $\kappa = 0.1\ \text{cm}^2\ \text{s}^{-1}$ and $t = 200 \times 10^{-9}$ s, $\Delta x = 1.4 \times 10^{-4}$ cm. Thus Q-switched lasers are ideal for the removal of limited amounts of material on thermally sensitive substrates.

Constraints on laser sources for trimming thick film resistors are slightly different from those with thin film devices. These have been reviewed by Oakes (1978). The greater volume of material to be removed requires higher energies per pulse and longer pulse lengths. This means that the pulsed CO_2 laser may be more suited than the Nd–YAG laser for some applications in this area.

Comparison of CO_2 and Nd–YAG sources must also take into account the effect of sample reflectivity. CO_2 laser radiation is strongly absorbed by many nonmetallic materials and is reflected from metals. At the Nd–YAG wavelength (1.06 μm), many nonmetallic materials are relatively transparent. The reflectivity of metals at 1.06 μm is also smaller than at 10.6 μm so that Nd–YAG radiation is more strongly absorbed in metals.

Some practical problems associated with laser machining of thin and thick film resistors have been described by Headley (1973). These problems relate to factors that can lead to drift and instability following trimming. Possible problem areas are shown in Figure 2.69. Insufficient laser power during trimming results in detritus in the cut. This may consist of islands of unevaporated material or random debris ejected from the sides of the cut. Thermal shock can lead to crack formation with subsequent instability during operation. Headley identifies the following common causes of drift and instability:

- Final current paths are too narrow leading to overheating
- Trimming too close to a termination
- Long kerfs in the direction of current flow
- Machining that crowds current density into zones subjected to shock from processing

Thick film resistors are particularly susceptible to residual stresses and microcrack formation. Some studies of this effect have examined the relationship

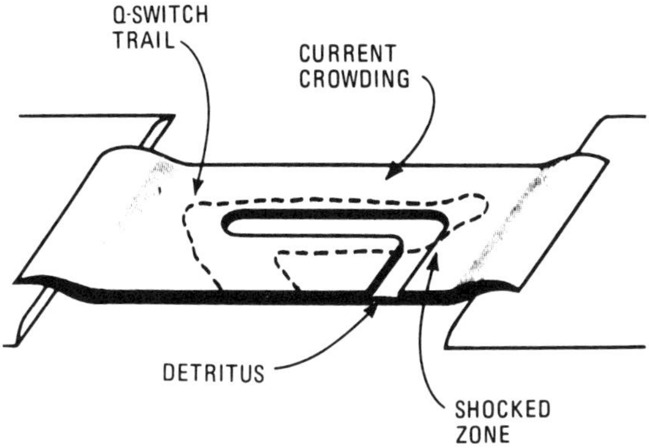

Figure 2.69. Possible problems in laser trimming of resistors. Detritus in the kerf, microcracks in the region surrounding the kerf, current crowding in the shocked zone (due to poor geometrical design), and *Q*-switched trail are the major potential sources of trouble in a laser trimming operation. [Thompson (1970)].

between stability and machining conditions [Bube (1975), Kummer and Taitl (1977), and Wood (1977)]. Differential expansion of resistor and substrate has been identified as a major problem [Kummer and Taitl (1977)].

The trimming of Cr–SiO cermet resistors without vaporization has been described by Braun and Breuer (1969) and by Berg and Lood (1968). Cermet films of 300-Å thickness were deposited by flash evaporation on oxidized Si wafers on which monolithic integrated circuit processing had been completed. After formation of the resistors, pulses from an Ar^+ laser were used to anneal the Cr–SiO film to reduce the final resistances to desired values. Adjustments to 0.01% were reported.

A significant development in laser trimming technology involves the trimming of elements in active circuits. A computer controlled laser can be combined with electrical measurements on the circuit to yield optimized circuit performance under dynamic operating conditions. A system for functional trimming has been described by Faber (1974). Some interactions between trimming operations and circuit response have been discussed by Mohr *et al.* (1977). The integration of resistive and capacitive functions eliminating the need for discrete resistors has also been reported [*Laser Focus,* March (1977), p. 43].

2.7.2. Machining of Conductor Patterns

The repetitively *Q*-switched Nd–YAG laser is ideally suited to removal of portions of metallic films on insulating substrates as required, for example, in the generation of conductor patterns. Radiation from the Nd–YAG laser is strongly absorbed by most metals while substrates such as sapphire or silica are transparent at 1.06 μm. Thus, careful control over operating conditions permits the removal of a metallic overlayer with little damage to the substrate.

Figure 2.70 shows a block diagram of the system developed by Weick (1972) for machining of conducting links in Au–Ti films on Al_2O_3 substrates. The system permits machining by modulated sequential scans instead of by point-to-point motion. Substrates to be machined are mounted on a rotating drum. Spatial and temporal information about the substrate are read from a code plate with a He–Ne laser. Processing time for 48 substrates mounted on each drum is 3.6 h with machining time of 17 s cm^{-2} of substrate. The laser had an average output power in the TEM_{00} mode of 4.9 W and was pulsed at 25 kHz with a 250-ns pulse width. The peak power per pulse was 785 W. Machining of 25–30 μm lines or gaps was possible with this system with resolutions between 14 and 25 μ. A typical pattern is shown in Figure 2.71.

Micromachining of conductor patterns has also been accomplished with pulses from a He–Ne laser operating at 1.14 μm [Brisbane and Jackson (1968)]. The substrate was moved under the beam on an x-y table. Possible methods of machining conductor patterns without beam or substrate translation have been discussed by Moran (1971) and Engel *et al.* (1974). These are based on optical processing of the laser output with a holographic lens or modulated zone plate.

The effect of interference on the machining of thin metal films on dielectric substrates has been investigated by North (1977) and by Kestenbaum (1979). Light reflected from dielectric layers on Si substrates was found to interfere with incident radiation so that the net intensity at the surface exhibited maxima and minima depending on the thickness of the dielectric layer. This variation was accompanied by a variation in the completeness with which

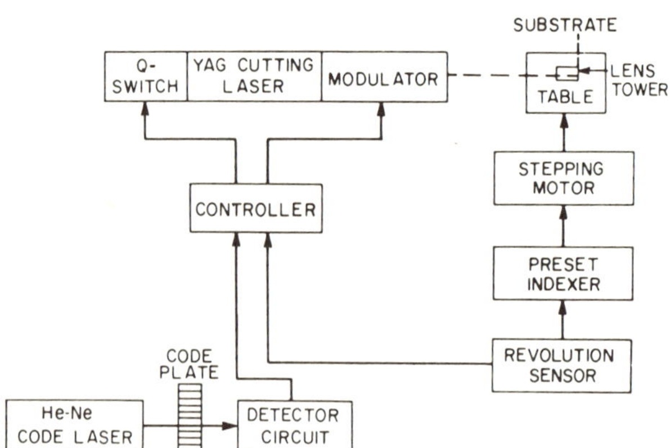

Figure 2.70. Nd–YAG micromachining system [Weick (1972)].

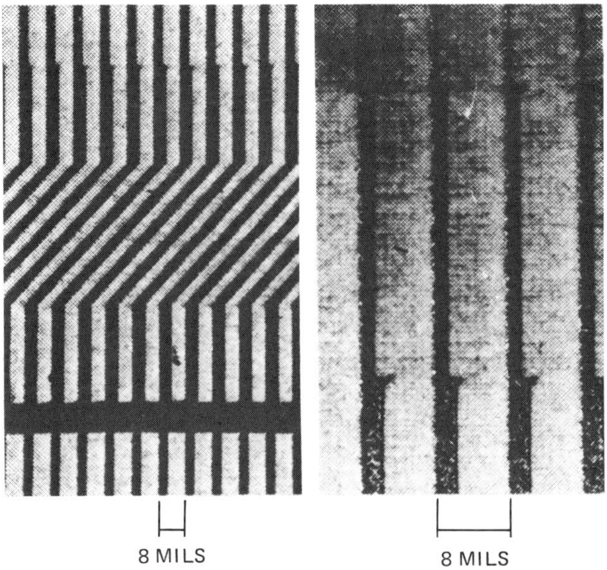

Figure 2.71. Laser machining of Au–Ti films on Al_2O_3 [Weick (1972)].

metallic overlayers could be removed with laser pulses of given energy. To quantify this effect, North (1977) measured the residual resistance of metallic lines evaporated from dielectrics of various thickness. This resistance varied by up to seven orders of magnitudes as the thickness of the underlying dielectric varied between 0.1 and 1.0 μm.

An alternative to micromachining of metallic films to produce conducting paths may involve direct photodeposition of conductor paths along paths defined by laser scanning. Preliminary experiments in this area have been conducted by Deutsch *et al.* (1979). Trimethyl aluminum and dimethyl cadmium were dissociated by 257.2-nm radiation from a frequency-doubled Ar^+ laser or by 10-ns pulses from an ArF laser at 193 nm. The photolytic reaction with these compounds results in the liberation of Al or Cd that coats the substrate in the vicinity of the beam focus. A typical deposition rate of 13 Å s^{-1} was obtained with 1 W cm^{-2} of Ar^+ radiation incident on the target. It was shown that CW excitation leads to metal deposition via photolysis. The higher temperature reached during pulsed deposition may initiate a pyrolytic reaction that enhances deposition. Features with 2-μm resolution were reported.

Further data on the direct deposition of metal layers by laser decomposition are found in the articles by Martin *et al.* (1968), Kim *et al.* (1971), and Inoue *et al.* (1973). Little *et al.* (1971) have used a laser to *inhibit* the deposition of thin metallic films along linear paths defined by the maxima of a standing-wave interference pattern. Regular arrays of submicron diameter holes have been produced in thin photoresistant films by an interferometric technique developed by Tsang and Wang (1975).

2.7.3. Fabrication of Gap Capacitors

Micromachining of metal films on dielectric substrates can be used to generate gap capacitors [Cohen *et al.* (1968)]. Figure 2.72 shows the structure of such a device. Machining of gaps as small as 6 μm gives capacitances that can exceed 200 pF cm^{-2} for composite metal films on sapphire. Leakage currents of 10^{-9} A at 40 V can be achieved if the capacitors are clear of machining debris.

2.7.4. Image Recording

Some early experiments in laser micromachining [Carlson *et al.* (1966)] demonstrated that drilling of closely spaced holes of small diameter in thin

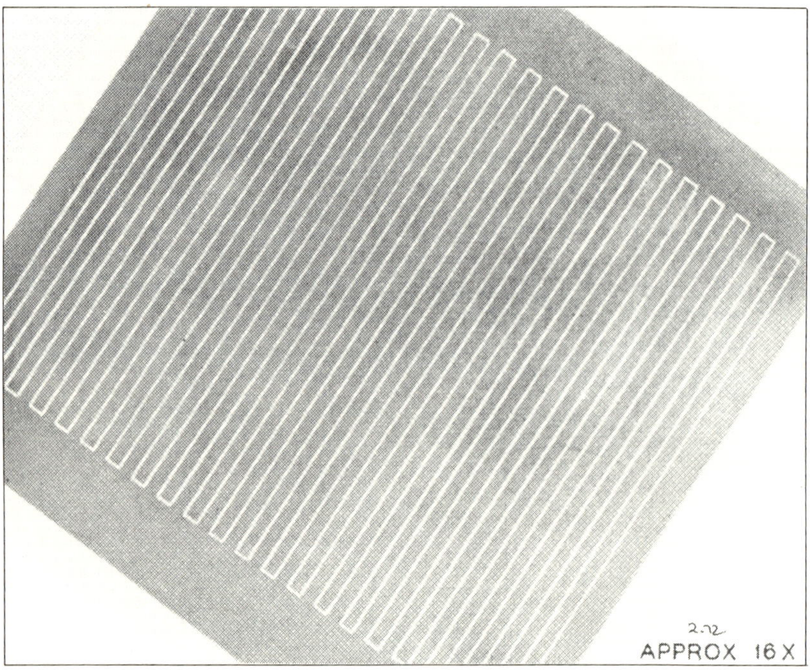

Figure 2.72. Serpentine gap capacitor on sapphire substrate. Gap width, 0.5 mil; separation between legs, 5.0 mils [Cohen *et al.* (1968)].

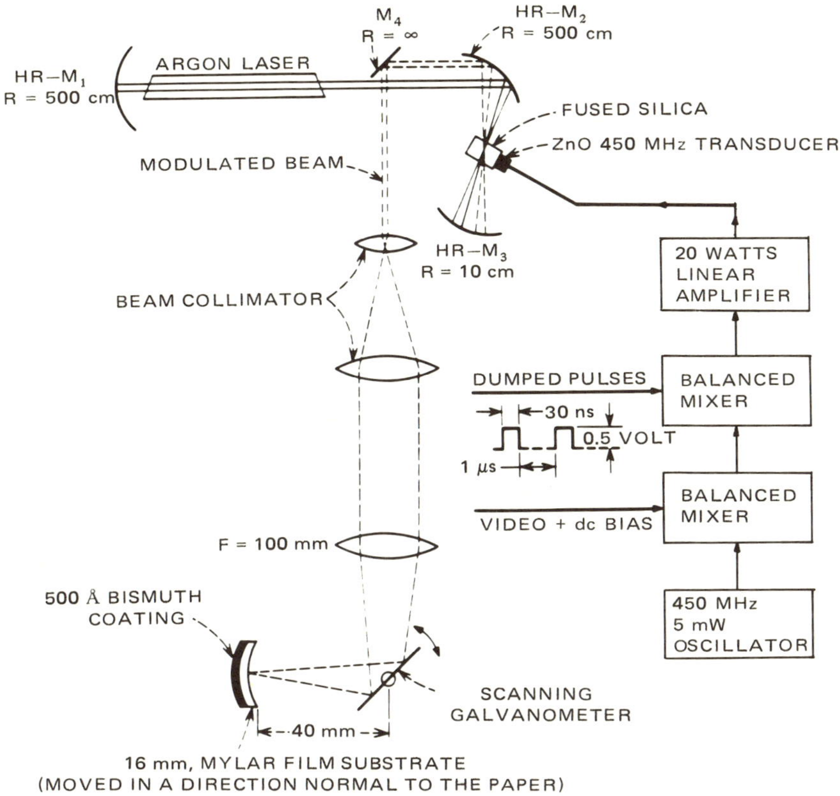

Figure 2.73. System set-up for recording video information by laser machining [Maydan (1971)].

metal films on transparent substrates could be used for image recording. Subsequent studies have resulted in great improvements in image quality and have led to an extension of this method to computer memory systems.

A system for the direct recording of visual images on metal-coated Mylar films is shown in Figure 2.73 [Maydan (1972)]. The recording medium consisted of 600-Å Bi films on Mylar. Images were recorded by focusing the output of an acousto-optical intracavity modulated Ar^+ laser on the film to create micron sized holes. By varying the intensity of the incident pulses, holes of various diameter could be drilled. This flexibility gives the system the capability of generating intermediate densities. The resulting gray scale was shown to possess 8–10 shades. The recording rate was 10^6 spots/second while a resolution of 175 lines/inch was reported.

A similar approach can be used for the storage of information on disks or tape. One of the first reports on the use of lasers in this application [Gray

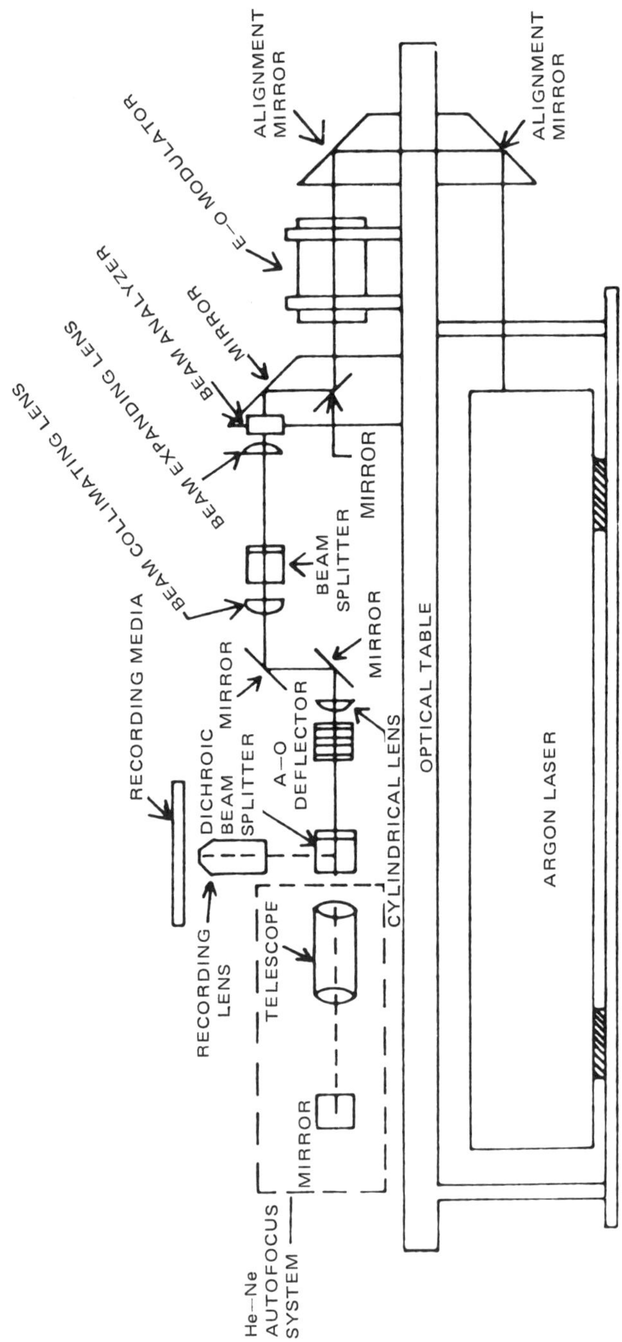

Figure 2.74. Optical schematic of the Slide Store document storage system. The argon beam is turned on and off by the electro-optical modulator, then scanned across the slide by the acousto-optical deflector. The slide is moved mechanically in the x direction while the beam scans in the y direction [*Laser Focus*, July (1980)].

(1972)] described the design and operation of a laser mass memory system with a gross storage capacity of 10^{12} bits. Data were stored on rhodium-coated polyester strips in the form of 1.5-μm holes machined by 90-ns pulses from an Ar^+ laser. The same laser is used to read data. With a 90-s WRITE time and the possibility of rapid READ after WRITE switching the accuracy of stored data is high. An error rate of 1 in 10^8 bytes read is reported.

Disk recording systems utilizing data storage by means of laser drilling in metallic or semiconductor films have also been reported [*Laser Focus,* May (1979), p. 67; March (1979), p. 28; Burns (1980)]. Both RCA and Philips use Te coatings as data storage media. A capacity of 10^{11} bits/disk was obtained.

Permanent storage of information can be achieved by laser recording on actinide films coated on glass substrates [Burns (1980)]. In a system developed by Omex, holes as small as 1 μm are drilled on 4×4 in.^{2} glass slides to yield a storage capacity of 2.9 gigabits/slide. Figure 2.74 shows a schematic diagram of the "slide Store" document storage system. Figure 2.75 is a photograph of bits recorded on a glass slide.

A study of recording media for video data has been reported by Terao *et al.* (1979). An As–Se–Te system was found to give optimum results and good hole definition.

Information storage via micromachining of alphanumeric data has been described by Evtikhiev *et al.* (1974). This method uses the focused image of a mask illuminated by a laser to thermally evaporate Ag, Fe_2O_3, or Si films on quartz substrates. A recording density of 10^6 characters/cm^2 was obtained with image sizes $\lesssim 10$ μm.

A variety of other laser-based methods for optical data storage are available [for a review see Chen and Zook (1975)]. These involve the utilization of

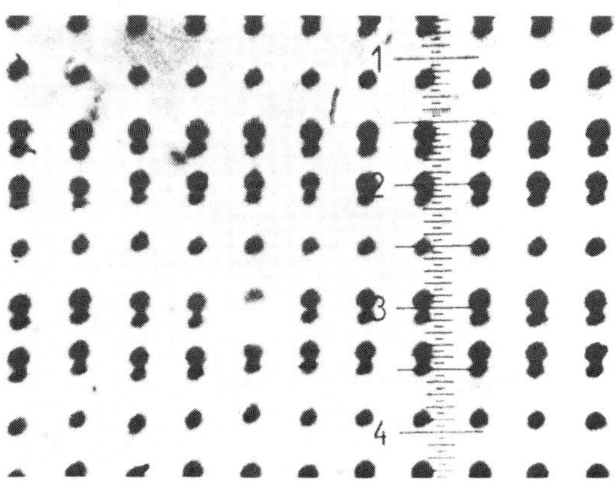

Figure 2.75. Bits recorded on a glass slide [*Laser Focus,* July (1980)].

laser heat or photon fluxes to induce physical or chemical changes in the recording medium. Many of these techniques involve erasable recording and are suited to temporary data storage. An excellent survey of available methods can be found in Ready (1978).

2.7.5. Laser Marking

The availability of high-power pulsed lasers with good beam quality and high repetition rates has opened up a novel laser application in the marking of individual pieces via laser evaporation. A schematic diagram of the equivalent of a laser "typewriter" is shown in Figure 2.76(a). Here information to be written on the part is typed in manually using a normal typewriter. The characters keyed manually are registered by rotation of the mask followed by triggering

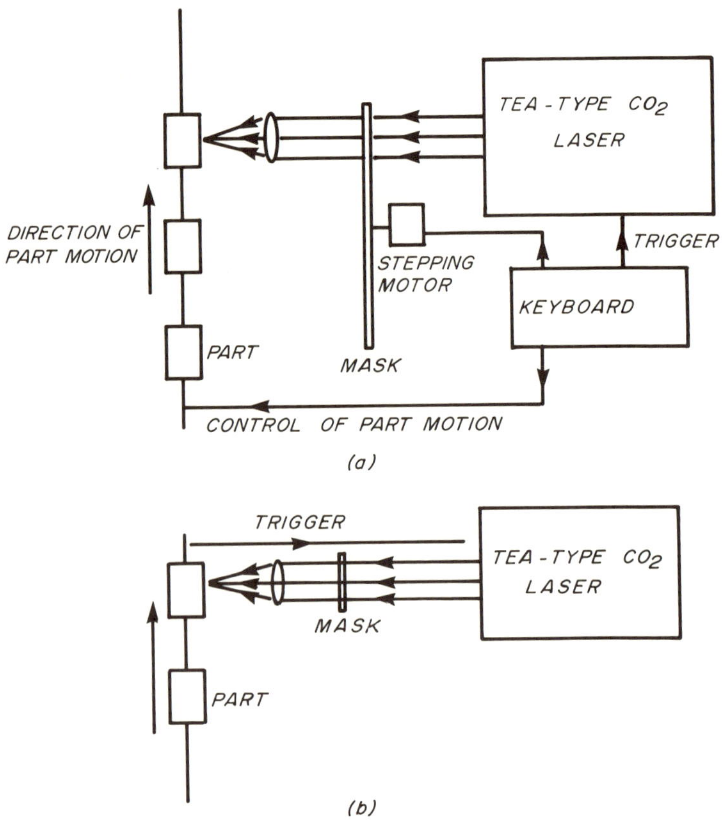

Figure 2.76. (a) Full alphanumeric capability and (b) repetitive marking of product information.

a pulse from the TEA-type CO_2 laser. This pulse reproduces the character on the part by evaporation of part of the surface.

Automatic registration of product information by laser marking is shown in Figure 2.76(b). Here the information (e.g., date code, product or batch number) is contained on the mask. Presentation of a part at the work station leads to the triggering of a pulse from the laser which burns the entire message written on the mask on the part using a single laser pulse. A commercial Laser Mark^{TM} system is manufactured by Lumonics Limited.

Since Si is not strongly absorbing at 10.6 μm the Nd–YAG laser is best used for marking on silicon substrates or wafers for microelectronics applications. The results of a pioneering study on this technique have been published by Holzinger *et al.* (1973). Both Nd–glass and ruby lasers were used to print numerical information on silicon wafers using the mask technique. Best results were obtained with the ruby laser although the irregular intensity profile within the beam required that a diffuser be used for optimum results.

An important development in the machining of complex patterns in parts via evaporation with CO_2 laser radiation has been reported by Campbell and Sweeney (1978). It was shown that the use of computer generated phase reflection holographic scanners simplifies the machining of complex patterns at moderate laser intensities. The motion of the holographic scanner in one dimension produces a two-dimensional pattern on the workpiece. This motion can be repeated to yield high production rates.

2.7.6. *Micromachining–Thermal Considerations*

The three basic operations used in micromachining with lasers are

1. Hole drilling.
2. Cutting with overlapping holes.
3. Cutting with scanning CW beams.

We have seen in previous sections that these operations are utilized in many important applications of lasers in the microelectronics industry. All involve material removal and, as such, are the result of thermal heating of a workpiece. In most cases the material to be removed is a thin metallic film. Removal of part of this metal film therefore involves several interactions:

1. Absorption of laser radiation by the film.
2. Absorption and reflection of laser radiation by underlying dielectric films or by the substrate.
3. Heat transfer from metal to substrate.

4. Heat transfer from substrate to metal.
5. Plasma coupling and/or explosive effect.

In this section, these effects are considered quantitatively and models are developed that describe the nature of the material removal process for thin metallic films on insulating substrates.

Several theoretical treatments of this problem have appeared in the literature. Of direct relevance are the articles by Maydan (1971), Brugger (1972), Paek and Kestenbaum (1973), and Stern (1973). The simplest approximation is to assume that the laser spot can be replaced by a uniform source as shown in Figure 2.28. In this case, the temperatures of the film (T_1) and the substrate (T_2) can be calculated from the equation for heat transfer

$$\frac{\partial^2 T_1}{\partial z^2} - \frac{1}{\kappa_1}\frac{\partial T_1}{\partial t} = -\frac{1}{K_1}\frac{I_0}{\delta}e^{-z/\delta}$$
$$\frac{\partial^2 T_2}{\partial z^2} - \frac{1}{\kappa_2}\frac{\partial T_1}{\partial t} = 0$$

where κ is the thermal diffusivity, K the thermal conductivity, δ the skin depth for laser radiation in metal film, I_0 the incident laser intensity (W cm^{-2}), and z the distance measured from metal surface (Figure 2.28). T_1 and T_2 are, of course, functions of z and t. The solution to these equations for Bi on Mylar is shown in Figure 2.77 calculated with $\delta = 140$ Å. At long times $T_1(x,t) \propto I_0 t^{1/2}$.

While the assumption of an extended source may be useful for approximate calculations, detailed predictions require that the spatial distribution of the focused beam on the target be taken into account. This distribution can be considered to be Gaussian when the laser output mode is TEM_{00}. In this case, the spatial distribution of intensity is

$$I(r) = I_0 \exp(-r^2/a^2)$$

where I_0 is the intensity at beam center, r the radial distance measured from position of maximum, and a the beam radius. A complete solution for the temperature rise in thin metallic films on substrates assuming this intensity distribution at the surface has been given by Paek and Kestenbaum (1973). The resulting solution is particularly simple in many cases for times less than 1 μs and may be used to estimate the minimum intensity required for vaporization. The graphical relation between normalized laser intensity and normalized time required to achieve vaporization is shown in Figure 2.78 where

$$F = \frac{\pi}{8.32}\frac{\kappa_2}{\kappa_1}\frac{K_1}{K_2^2}\frac{\alpha d}{T_v} I_0$$

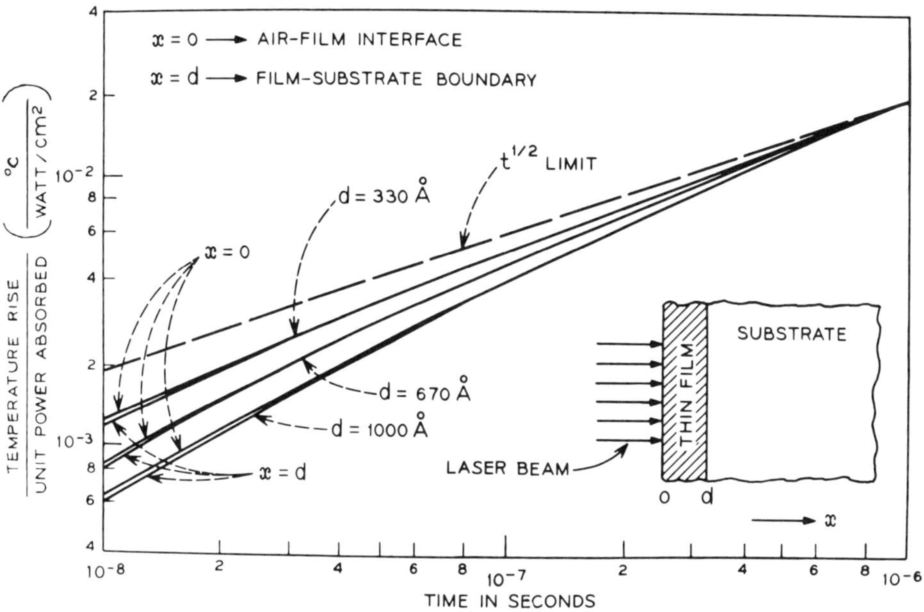

Figure 2.77. Temperature rise of a bismuth film on a Mylar substrate caused by front illumination [Maydan (1971)].

and

$$\tau = 2\frac{K_2}{K_1}\frac{\kappa_1}{d}\left(\frac{t}{\pi\kappa_2}\right)^{1/2}$$

with $\propto$ being the percentage absorption of incident intensity by film, and T_v the vaporization temperature. This result assumes that only the film absorbs radiation and that no temperature gradient is present in the film. Accurate predictions of critical vaporization intensity are obtained from this curve only while $\tau \lesssim 1$. With larger values of τ this solution becomes less accurate, but it is still useful for obtaining an order of magnitude estimate of critical intensity.

As an example of the way in which Figure 2.78 can be used to relate critical intensity I_0 to irradiation time and the thermal parameters of the system, suppose an estimate was required of the critical intensity for vaporization of a hole in a 1000-Å thick film of Au on glass with a pulse length $t = 50$ ns. Then one has $d = 10^{-5}$ cm, $K_1 = 2.95$ W cm^{-1} °C^{-1}, $\kappa_2 = 5.8 \times 10^{-3}$ cm^2 s^{-1}, $T_v = 2630$ °C. Assuming 3% absorption of laser light $\propto = 0.03$, $\tau = 1.54$, and $F = 1.7$ from Figure 2.78. Substituting in the expression for F one obtains I_0, the critical intensity for vaporization at center of focal spot, $= 8.8 \times 10^7$ W cm^{-2}.

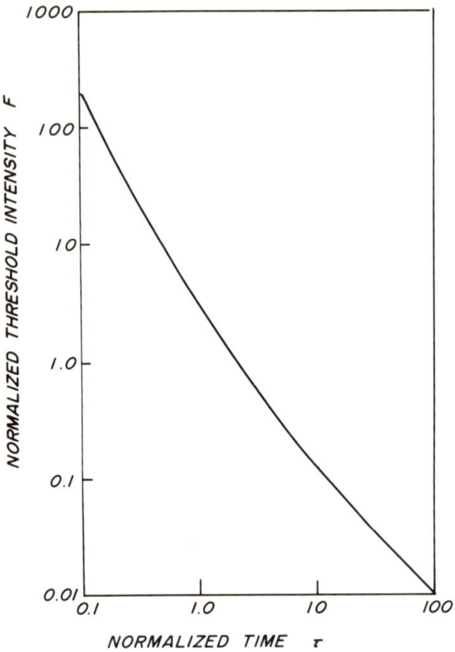

Figure 2.78. Threshold intensity vs. time for vaporization of a thin film on a substrate [after Paek and Kestenbaum (1973)].

Some further data for representative Bi–Mylar and Au–glass systems are given in Table 2.15.

Zaleckas and Koo (1977) have noted that the assumption of constant absorption for films of various thickness may not be strictly valid. Since the resistivity of a metal film depends on thickness in very thin films, $d \lesssim 1000$ Å, the absorption coefficient for laser radiation will also be a function of film thickness. As the resistivity increases with decreasing film thickness this results

Table 2.15. Calculated Critical Intensities for Drilling of Thin Metal Films on Transparent Substrates[a]

Metal substrate	Thickness (μm)	Absorption	Pulse length (ns)	Critical intensity[b] (W cm^{-2})
Bi–Mylar	0.1	50%	250	3.8×10^5
			50	1.2×10^6
Au–glass	0.1	3	250	3.1×10^7
			50	8.6×10^7
	1.0	—	250	2×10^9
			50	5×10^9

[a]Beam radius, $a = 10$ μm; [b] Exact calculation from Paek and Kestenbaum (1973).

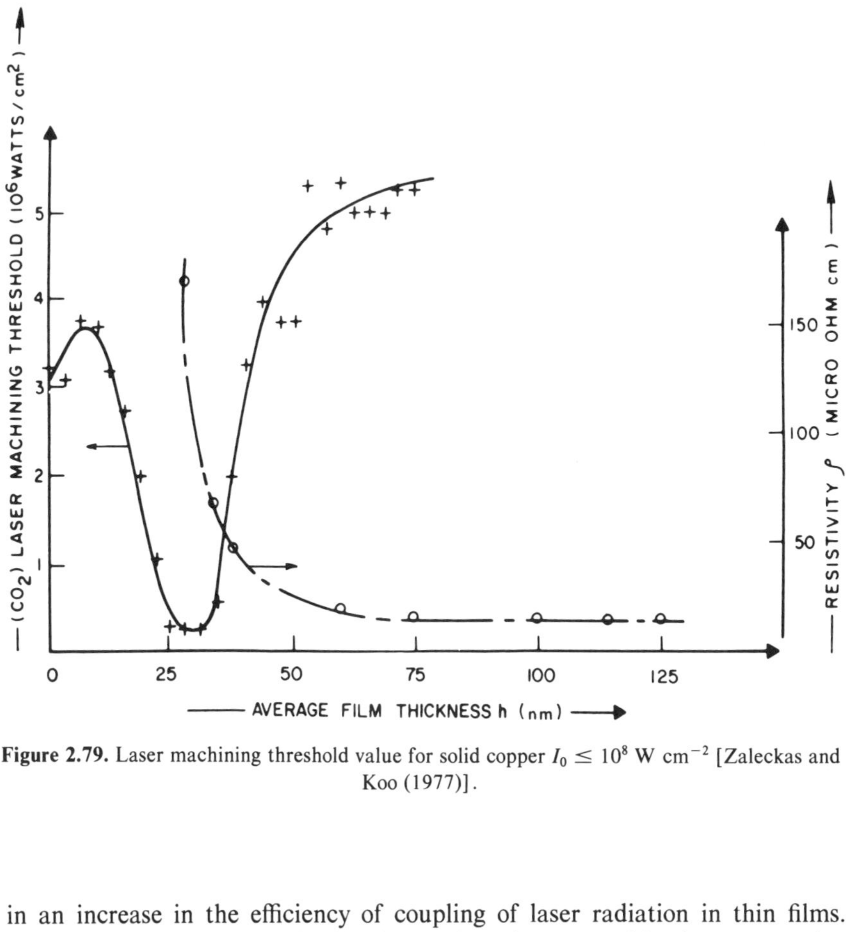

Figure 2.79. Laser machining threshold value for solid copper $I_0 \leq 10^8$ W cm^{-2} [Zaleckas and Koo (1977)].

in an increase in the efficiency of coupling of laser radiation in thin films. Radiation that penetrates the film heats the substrate and leads to gas evolution either as a result of desorption or evaporation. This gas on escaping from the substrate can rupture the film at the laser focus. Zaleckas and Koo show that these two effects combine to give a minimum in the laser intensity required for drilling in submicron metallic films. The result of an experimental verifi-

Lou (1977) has developed a method of modifying the optical properties of thin Bi films on Celanar substrates that improves the coupling of laser radiation to the film and results in enhanced machining efficiency. By depositing a layer of small particles between coarse grained Bi crystallites and the substrate, the efficiency, η, for machining can be enhanced. This efficiency is defined as

$$\eta = \frac{A}{d_b + Qd_s}$$

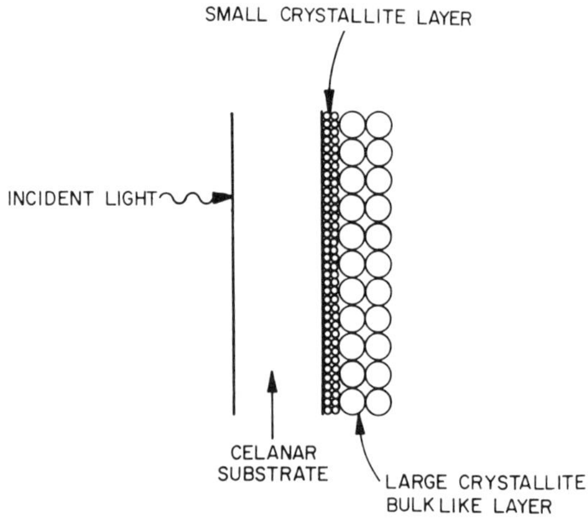

Figure 2.80. Structure of bilayer bismuth film [Lou (1977)].

where A is the film absorptance, d_b the thickness of coarse grained layer (Figure 2.80), d_s the thickness of the small grained layer, and Q the packing factor for the small grained layer. When laser light is incident from the Celanar substrate the small crystallite layer acts as an antireflection coating. This enhances absorption by the bulk layer and therefore improves machining efficiency. The effect of the small grained layer on reflectivity R, transmission T, and η and A are shown in Figure 2.81. It is evident that R is significantly reduced when $d_s \simeq 500$ Å. The corresponding increase in η is considerable with optimum coupling close to 90%. While this method places additional constraints on deposition conditions an improvement in film stability was reported.

Machining of thin metal films on insulating substrates with CW beams has been discussed by Seigman (1968, 1969). An analysis of heat transfer during CW micromachining is fully outlined elsewhere [Duley (1976)].

2.8. SURFACE HARDENING

Surface heat treating by induction or flame heating to produce a thin hardened layer is a widely used metallurgical technique. The past several years have seen the development of an alternative technology based on laser heating [Locke and Hella (1974), Hill *et al.* (1974), Engel (1976), and Breinan *et al.* (1976)].

Surface hardening in steels can be brought about by heating to a temperature at which the steel is austenitized, i.e., where carbon is present as a solid solution of iron carbide in γ-Fe. Subsequent rapid cooling results in a martensitic transformation in which the carbide is retained in solution in α-Fe. This phase is characterized by enhanced hardness.

The requisite rapid cooling rates and hardening depths can be easily obtained with laser heating using a rapidly scanned defocused laser beam. The incident intensity and scan rate are adjusted to provide heating to the depth required for hardening (typically <0.5 mm). Quenching occurs naturally on removal of the source of heat by conduction into the bulk metal. During heating the temperature at the surface may rise at a rate $\sim 10^4\,°\mathrm{C\ s^{-1}}$. Cooling occurs at $\sim 5 \times 10^3$–$10^4\,°\mathrm{C\ s^{-1}}$.

Typical intensities required to heat treat are $\sim 10^4$ W cm^2 and beams with cross-sectional areas of ~ 1 cm are desirable. Hence heat treating of extended areas is practical only with kilowatt CO_2 lasers. The scan speed is determined by the depth of hardening required and the beam area, but typical values are in the 1–10 cm s^{-1} range.

The high reflectivity of metals at 10.6 μm, together with the lack of surface damage that occurs in laser hardening, means that the metal surface is usually coated with an absorbing compound to improve coupling of incident radiation to the metal. Some materials that have been used include DAG col-

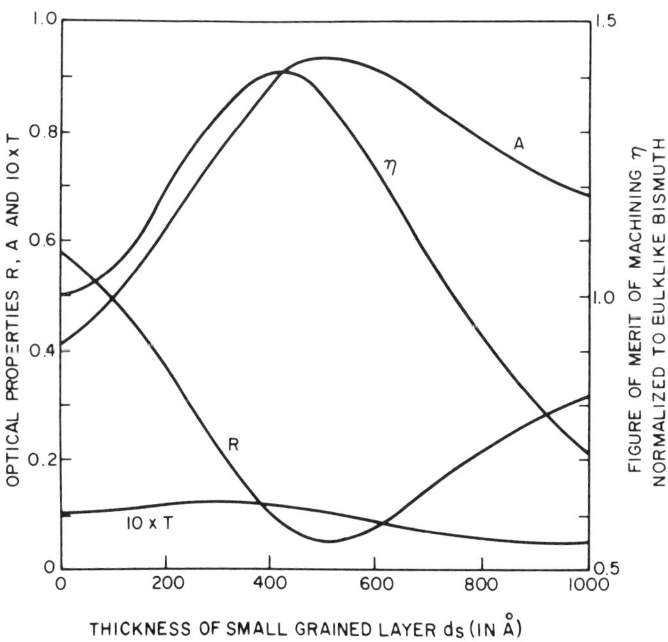

Figure 2.81. Optical properties and relative sensitivity of bilayer bismuth films [Lou (1977)].

loidal graphite, manganese phosphate, polycrystalline tungsten, and cuprous oxide.

Laser hardening has a variety of advantages over conventional technology. These have been summarized by Hill *et al.* (1974). They are as follows:

1. Low total heat input with resulting minimization of distortion.
2. Technique can be used on thin and small diameter workpieces.
3. Applicable to internal surfaces; the area heated is easily controlled.
4. Narrow hardened regions are readily obtained.
5. Hardening proceeds without contamination of environment.
6. Rapid temperature cycling minimizes oxidation.
7. Applicable to complex shapes, e.g., gear teeth.

A plot of the typical variation of microhardness with depth in EN8 steel hardened with a 2-kW laser is shown in Figure 2.82. It is evident that the Vickers microhardness of the base material (= 200) is increased to ~700 over a 0.5-mm surface layer.

Courtney and Steen (1978) have derived some constraints on the operating range for hardening of EN8 steel with laser powers of up to 2 kW. Their curves are shown in Figures 2.83 and 2.84. Figure 2.85 shows operating lines

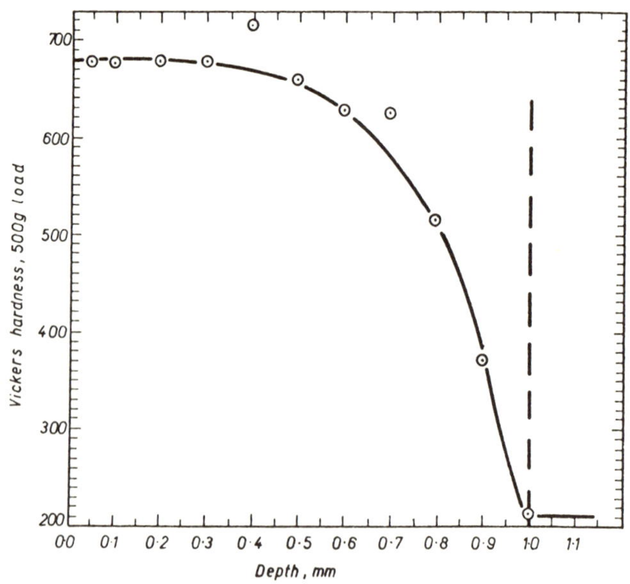

Figure 2.82. Microhardness vs. depth in EN8 steel hardened with a 2-kW CO_2 laser. Beam radius, 0.3 cm; speed, 0.8 cm s^{-1} [Courtney and Steen (1978)].

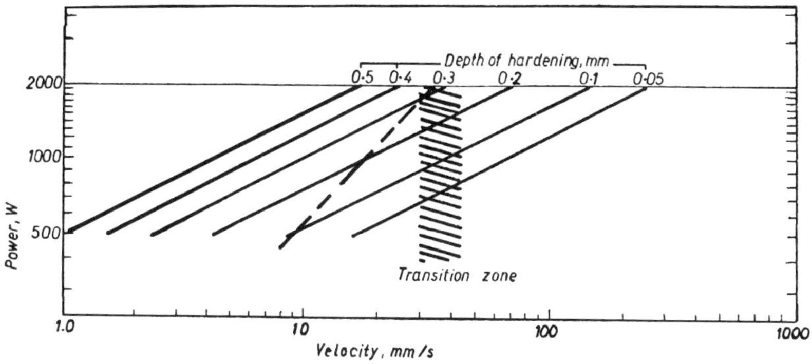

Figure 2.83. Operating chart for EN8 steel; constant beam diameter = 5.8 mm. Solid operating line for various depths of hardening; dashed line is surface melting line [Courtney and Steen (1978)].

relating scan speed to laser power for various hardening depths. The cross-hatched region in Figure 2.83 corresponds to that range where a significant change in surface microstructure was first observed as the scan speed increased. The dashed line defines the conditions that yield surface melting. Obviously hardening depths exceeding ~0.3 mm cannot be produced under these conditions without the initiation of surface melting.

Figure 2.84 shows a corresponding set of operating curves that relate scan speed and laser power to focused beam diameter for constant hardening depth.

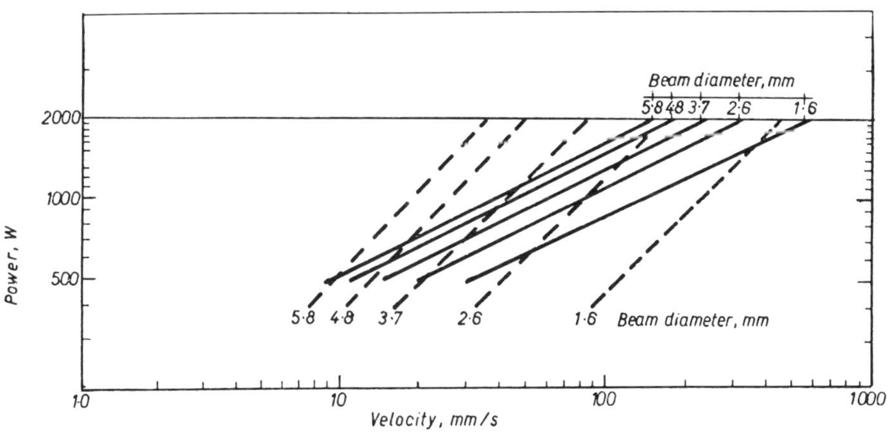

Figure 2.84. Operating chart for EN8 steel; constant hardening depth = 0.1 mm. Solid line is operating line for 0.1 mm hardening depth for various beam diameters; dashed line is surface melting line for each beam diameter. [Courtney and Steen (1978)].

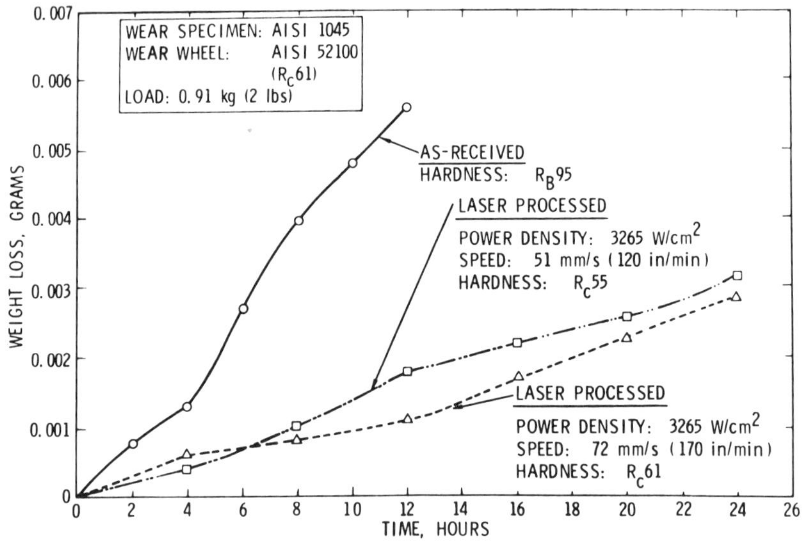

Figure 2.85. Weight loss measurements on treated and untreated AISI 1045 steel [Gnanamuthu *et al.* (1979)].

Individual melting curves are also shown as dashed lines. These show that the useful operating range for surface hardening is vastly increased as the laser spot size is increased.

A comprehensive study of laser transformation hardening of AISI 1045 steel by Gnanamuthu *et al.* (1979) has investigated residual stress, fatigue behavior, and the wear rate of laser hardened samples. Samples were transformation hardened with laser powers of up to 15 kW. Residual compressive stress was observed 0.25 mm below the surface of steel samples hardened with 8.2 $\times$ 10^3 W of laser power at the rate of 9.1 cm^2 s^{-1}. This was accompanied by an increase in medium fatigue life and a decrease in wear rate (Figure 2.85). The wear rate was decreased by a factor of $\sim$2.5 when compared to untreated steel.

Laser heat treating has found wide application in industry. General Motors has used laser hardening to process parts in its Saginaw steering gear division since 1973. Recent developments include the use of multikilowatt lasers to heat treat cylinder linings for diesel engines. Similar lasers have been used to harden camshafts [*Laser Focus,* November (1979), p. 44].

2.9. SURFACE MELTING, ALLOYING, AND CLADDING

At higher intensity than that used for hardening, the scanned laser beam can be used to induce melting of a thin surface layer on metals. The resulting melt is rapidly quenched by conduction into the bulk metal. The rate at which the temperature of this surface layer is reduced ($\sim 10^{6}\,^{\circ}\mathrm{C\,s^{-1}}$) can result in the formation of new metastable phases with useful properties.

If this procedure is carried out on a sample that has been previously coated with a suitable compound, laser heating can promote alloying with the substrate to form surface layers with novel composition.

Copley *et al.* (1979) have summarized the differences between these and other laser methods of influencing the composition of surface layers. Figure 2.86 shows the processes identified by Copley *et al.* in schematic form. Melt quenching occurs as noted above when the laser is used to melt a thin surface layer that subsequently cools rapidly by quenching. Surface alloying [Figure 2.86(b)] at the thin surface layer is melted into the substrate to yield an alloyed surface layer.

Cladding [diagram 2.86(c)] involves bonding of a thicker surface layer to the substrate using the scanned laser beam as a heat source. Part (d) of the figure shows a layer of inert particles being dispersed in the melt formed by laser heating.

Experimental studies of the microstructure of laser melted regions on the surface of M2 tool steel have been reported by Kim *et al.* (1979) and by Strutt *et al.* (1979). Anthony and Cline (1978) have used a low-power CW CO_2 laser to homogenize a heterogeneous composition distribution in the surface of sensitized type 304 stainless steel. This treatment resulted in the melting of a thin surface layer and subsequently decreased susceptibility to stress corrosion cracking.

The topography of laser melted Ni superalloy Udimet 700 has been discussed by Copley *et al.* (1979) and Narashimau *et al.* (1979). The structures observed could be understood on the basis of convective motion in the melt region. Greenwald *et al.* (1979) conclude that the regrowth occurring subsequent to laser melting occurs by epitaxial regrowth from the substrate. The microstructure of these regions shows that solidification proceeds via dendritic → cellular dendritic → planar front growth as the ratio G/R increases, where G is the temperature gradient and R is the solidification rate.

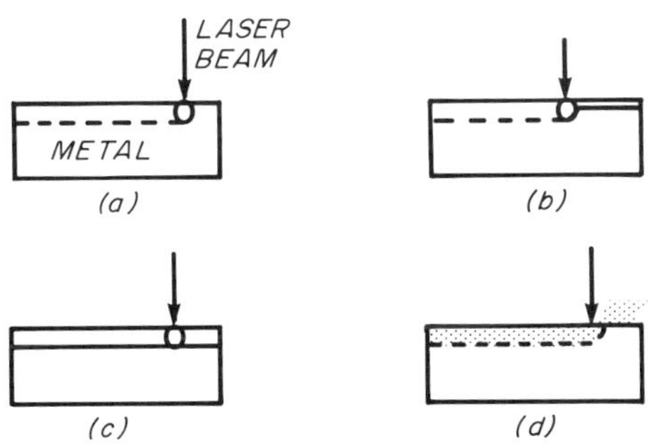

Figure 2.86. Surface treatment with lasers: (a) melt quenching, (b) alloying, (c) cladding, and (d) dispersing.

Weinman and Devault (1979) have studied laser-induced alloying of Cr and Ni on AISI 1018 carbon steel using the electron microprobe technique. They find that essentially complete homogenization of Cr occurs in the laser melt in the short time available. This appears to be due to mixing currents in the melt prior to quenching.

Some studies of melt quenching with pulsed laser heating have been reported by Elliott *et al.* (1973) and Weinman *et al.* (1978). Some results on the regrowth of Ni surfaces following melting with pulses from a *Q*-switched Nd–YAG laser have been reported by Buene *et al.* (1980).

Steen (1978) has shown that surfaces can be coated by depositing a thermally sensitive chemical on a surface heated by an imaged laser beam. The area of deposition then mimics the shape of the laser heated area. Steen identifies several areas where such a technique may be of particular value:

1. Coating for wear resistance, e.g., in machine tools and turbine blades.
2. Coating for corrosion resistance, e.g., in welds and electrodes.
3. To produce images, e.g., recording of part numbers.
4. In the initiation of electrical changes.
5. In the initiation of changes in magnetic properties.

Some practical applications of this technique in industry have been briefly discussed [*Laser Focus,* May (1979), p. 34; August (1979), p. 12].

Takei *et al.* (1980) have shown that thin metallic films can be formed on metallic substrates by laser melting of pressed powder layers. In their study, pulses from a Nd–YAG laser were used to melt Rh powder overlayers on permalloy substrates. Double pulsing was required, with the first pulse inducing melting while the second promoted adherence of the thin film to the substrate.

2.10. SURFACE CLEANING

Laser heating of metals is initiated by absorption of radiation within a thin surface layer. The thickness of this layer is of the order of the skin depth ($\simeq 10^{-6}$ cm). A rapid rise in temperature within this layer can lead to the release of adsorbed surface species. If the heating pulse is of short duration and of relatively low amplitude this desorption can occur without appreciable heating of the bulk material. In principle, therefore, laser-induced desorption offers an alternative to conventional methods of surface cleaning.

One of the first experimental investigations of this process was that of Bedair and Smith (1969). A *Q*-switched ruby laser was used to irradiate Ni and Si surfaces at intensities between 10 and 120 MW cm^{-2} under UHV conditions. It was found that both O_2 and S could be removed from Ni under a variety of irradiation conditions. However, this was accompanied by localized surface damage that could not be removed by annealing. It appears that this may limit possible applications of this technique in the surface science field. However, such damage does not appear to occur in Si [Bedair and Smith (1969) and Zehner *et al.* (1980)] suggesting that laser surface cleaning may be of value in the preparation of atomically clean Si samples. Zehner *et al.* (1980) note that laser cleaning could be used to shorten processing times for surface cleaning of Si from hours to seconds. Figure 2.87 shows an example of the Auger electron spectra of Si contaminated with C and O prior and follow-

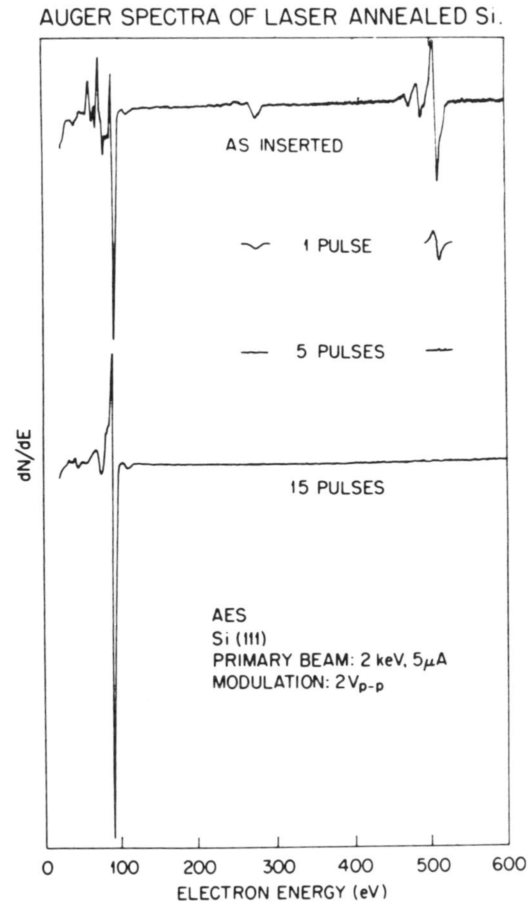

Figure 2.87. Auger electron spectra obtained from an uncleaned Si surface and after pulsed laser annealing at ~2.0 J cm^{-2} [Zehner *et al.* (1980)].

ing cleaning with pulses from a ruby laser with a fluence $\simeq 2$ J cm^{-2}. The C peak occurs at 272 eV while that of O is seen at 510 eV.

Schwirzke *et al.* (1974) found that the apparent rate of gas desorption increases dramatically under conditions of plasma production in front of the target surface. This enhancement in desorption rate, which can exceed a factor of 10^3, was shown to be due to impact of the laser plasma with the walls of the vacuum chamber. It was suggested that part of this effect may involve electric-field-induced desorption since the laser plasma creates a strong electric field.

A practical application of laser-induced desorption may involve the determination of impurities adsorbed on Tokamak walls [Schwirzke *et al.* (1978)].

Laser cleaning of stone statues has been reported by Asmus (1979). A normal mode ruby laser was used in a portable system to remove surface discoloration from stone artifacts in Venice.

2.11. CRYSTAL GROWTH

Heating with CW CO_2 laser radiation was first used to grow crystals by Eickhoff and Gürs (1969) and Cockayne and Gasson (1971). The process developed is identical to the floating zone technique with the addition of laser heat to the melt zone. Advantages of laser heating include the ability to heat in a variety of atmospheres and the reduction of thermal gradients that can generate residual strains in the crystals pulled. A summary of representative data on crystals grown with the laser heated floating zone technique is given in Table 2.16.

Nelson *et al.* (1970, 1972) have shown that small spheres of refractory solids can be formed by spinning a laser heated rod. For example, heating of

Table 2.16. Crystals Grown Via Laser Heated Floating Zone Technique

Compound	Details	Reference
Y_2O_3	Single crystal formed	Gasson and Cockayne (1970)
Al_2O_3	Single crystal, 5-mm diameter × ~3 cm long	Gasson and Cockayne (1970)
	3.5 mm h^{-1} draw rate	Eickhoff and Gürs (1969)
	Sapphire-clad ruby fibers	Burrus and Golden (1977)
$Mg_2Al_2O_4$	Single crystal	Gasson and Cockayne (1970)
Nd_2O_3	Polycrystalline	
GdO_3	Polycrystalline	
La_2O_3	Polycrystalline	
LaB_6	9.5 cm h^{-1} draw rate, 1-mm diameter 50 mm long	Takagi and Ishii (1977)

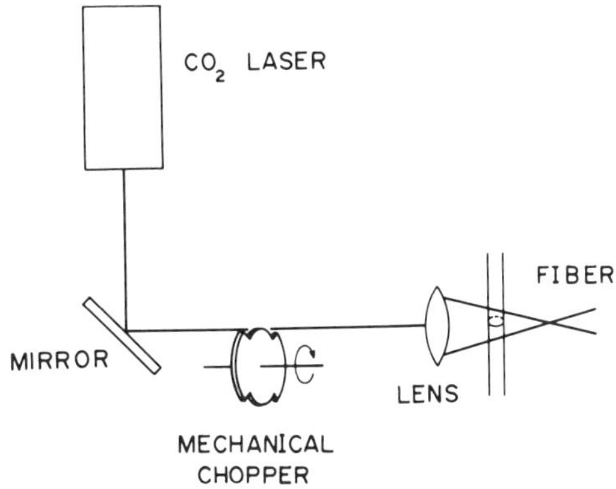

Figure 2.88. Splicing set-up using CW CO_2 laser radiation [Egashira and Kobayashi (1977a)].

high alumina tubing with a 250-W CW CO_2 laser generates Al_2O_3 particles with diameters in the 400–800 μm range. Further work in this area has been reported by Nelson *et al.* (1973) and Topol and Happe (1974).

Nelson and Richardson (1972) and Lydtin (1972) have shown that thin rods of pyrolytic graphite can be grown from an organic vapor at a laser heated spot. Growth occurs by the pyrolysis that proceeds selectively at the point heated by the laser.

2.12. OPTICAL-FIBER SPLICING

The first experiments on joining optical fibers using laser heating were reported by Fujita *et al.* (1976). Further work was discussed by Egashira and Kobayashi (1977a) who presented a theoretical model of the heating/splicing process [Egashira and Kobayashi (1977b)]. Radiation from a low-power CW CO_2 laser is chopped and focused on the junction between two fibers held in micromanipulators (Figure 2.88). The CO_2 beam is scanned across the join by means of a motor on the focusing lens. In these experiments the focal spot diameter was 1.0 mm at the fiber and the scan speed was 10 μm s^{-1}. A laser power of 2.3 W was applied to the fibers. A typical splice in a multimode step-index fiber is shown in Figure 2.89. Losses in these splices varied from 0.02 to 2.9 dB with 90% of splices having losses of less than 0.12 dB.

2.12.1. Optical Fiber—End Preparation

Optical communication systems involving fiber optic links require that high quality joints be made between optical fibers. A primary constraint is that the ends of the fibers to be joined be flat with surfaces perpendicular to the optical axis. A laer-based technique for forming endfaces with this characteristic has been described by Kinoshita and Egashira (1978). The process

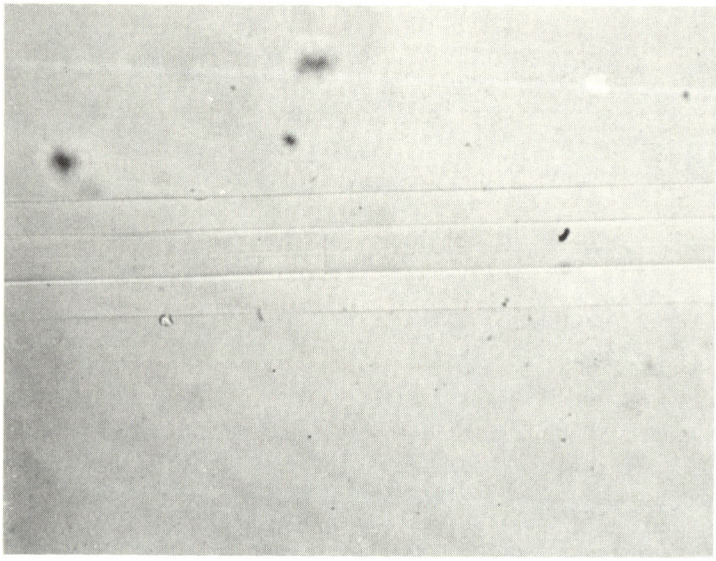

Figure 2.89. Typical splice [Egashira and Kobayashi (1977b).

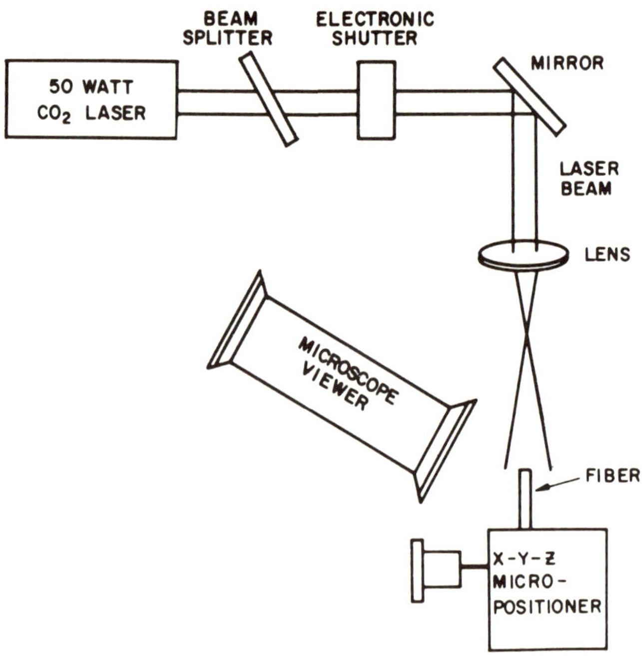

Figure 2.90. Schematic of optical set-up for laser system [Paek and Weaver (1975)].

involves introducing mechanical stress into a fiber by heating with a low-power CW CO_2 laser. The fiber can then be broken at the point of minimum strain. Since this method does not involve scoring, a cleaner break occurs.

It was found that best results were obtained with the attenuated beam from a CW CO_2 laser chopped at 100 Hz. This beam was scanned at a rate of 20–30 $\mu m\ s^{-1}$ across the fiber. Both step-index and graded-index fibers were fractured with this technique. Mirror endfaces were produced with average coupling losses of 0.35 dB without index-matching fluid. On splicing, losses were 0.05–0.06 dB.

Figure 2.90 shows the system developed by Paek and Weaver (1975) for producing spherical lenses on the end of optical fibers. When the end of the fiber is heated to its melting point, surface tension produces a hemispheric surface (Figure 2.91). This can be of good optical quality offering the possibility that fibers treated in this way could be used in imaging and machining. Hemispheric lenses were produced on Kimble R-6 glass and on fused quartz.

2.12.2. Optical Fiber—Drawing

Optical fibers suitable for optical transmission systems can be drawn from bulk materials if these are heated to the melting point. Conventional heat sources include resistance heaters, gas torches, and induction and plasma heat-

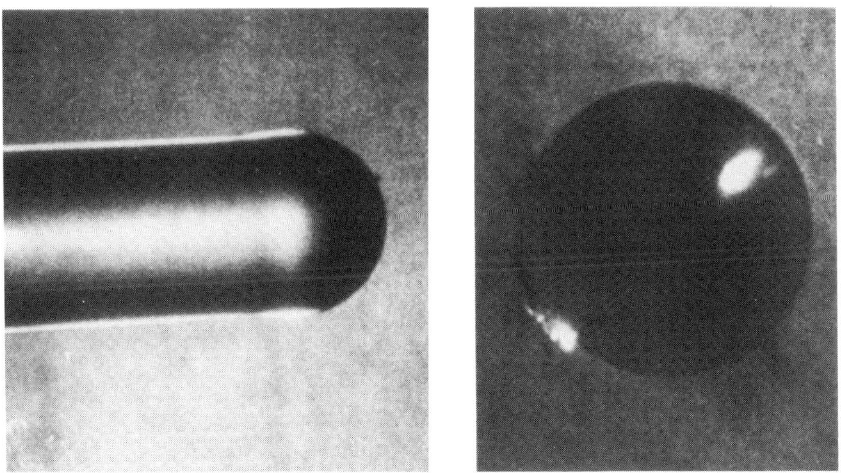

A B

Figure 2.91. Hemispherical lens formed at the end of Kimble R-6 glass fiber (250-μm diameter): (a) side view (100×); (b) front view (160×) [Paek and Weaver (1975)].

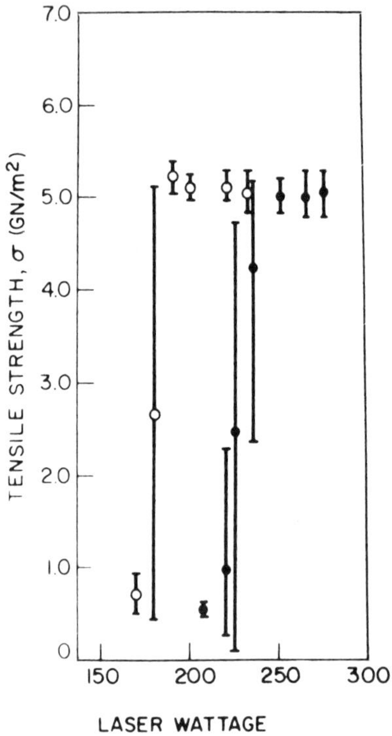

Figure 2.92. The tensile strength plotted as a function of wattage for 110-μm fibers drawn with a 250-W CO_2 laser from a 7-mm rod of fire-polished Suprasil-2: ○, 0.4 m/s; ●, 1.0 m/s [Schonhorn *et al.* (1976)].

ers. The pioneering studies of Jaeger (1976), Jaeger and Logan (1973), and Paek (1974) have shown that the CW CO_2 laser may also be useful in the drawing of low-loss cladded and uncladded fibers.

In the system developed by Paek (1974) fibers are drawn from the end of a feed rod heated by a 50-W CO_2 laser. To obtain uniform heating around the end of the rod the CW beam is focused with an ellipsoidal reflector. This results in a ring-shaped heating area at the end of the rod. The fiber is then drawn from the center of this ring and collected on a take-up reel. Fused quartz fibers with diameters between 10 and 25 μm have been drawn using this system. Diameter variation on 10-μm fibers was reported to be 5%. These fibers had losses of less than 100 dB km^{-1} at a wavelength of 6328 Å.

Since no contamination is produced by laser heating it is possible that low-loss fibers can be produced using this method [Jaeger (1976)]. Furthermore, the laser heat source can be easily incorporated into a feedback circuit to control fiber diameter.

The role of preparation conditions on the strength of laser-drawn silica fibers has been studied by Schonhorn *et al.* (1976, 1978). An example of their data is shown in Figure 2.92. This shows that the tensile strength exhibits a sharp increase with increasing laser power near 250 W. Their analysis indicates that at constant laser power the fiber tension scales as (drawing speed)^3. The ultimate strength of 110-μm coated fibers was seriously reduced when the fiber tension exceeded 9×10^8 dyn cm^{-2}.

2.13. LASER DEPOSITION OF THIN FILMS

The use of lasers to evpaorate refractory materials for the preparation of thin films was described as early as 1965 [Smith and Turner (1965)]. While interest in laser evaporation to produce thin films has continued [for a review see Duley (1976)], recent developments have seen the use of lasers for chemical vapor deposition, photodeposition, and electroplating. In addition, laser annealing has been used in the recrystallization of amorphous films produced by radio-frequency sputtering.

A summary of these methods is given in Table 2.17, together with references to representative studies in each area.

2.13.1. Evaporation

Ruby, CO_2, and Nd–YAG lasers have been used in the deposition of thin films by evaporation. Generally, the CO_2 laser is preferred for the evaporation of insulating or most semiconducting materials while ruby or Nd–YAG lasers are preferred for evaporating metals. CW lasers give better control over deposition rate and tend to yield evaporation conditions that maintain thermal equilibrium. The high evaporation rate produced with pulsed lasers leads to the

Table 2.17 Laser-Based Methods of Thin Film Production

Method	Material	Deposition rate (Å s^{-1})	Reference
Evaporation	SiO	150	Hass and Ramsey (1969)
	SiO_2	5.6	
	MgF_2	16–25	
	Cr, W, Ti, C	10^5–10^6	Schwarz and Tourtellotte (1969)
	ZnS	30	Cali *et al.* (1976)
	ZnSe	4–6	
	GaAs	5	
Electroplating	Au, Ni, Cu	5×10^4	von Gutfeld *et al.* (1979)
Chemical vapor deposition	Si	850	Baranauskas *et al.* (1980)
Photodeposition	Al, Cd, Sn, I	13	Deutsch *et al.* (1979)

ejection of particles and liquid droplets. While the overall effect is one of congruent evaporation, the composition of the resulting film has been shown to depend on geometrical factors and pulse intensity [Baldwin (1970, 1973) and Kliwer (1973)].

The deposition of thin films with pulsed laser heating is discontinuous since only discrete quantities of material can be evaporated with each pulse. Thus thick films require the superposition of several pulses and may, as a result, be of poor mechanical and optical quality. However, such effects do appear to depend on deposition conditions. For example, Cali *et al.* (1976) found that high quality films of a variety of refractory materials could be produced with a pulsed CO_2 laser having a peak power per pulse of 160–1600 W. The deposition rate under such conditions was <1 Å per pulse. It is likely that the relatively low peak power used in this experiment inhibited the formation of a liquid phase with subsequent ejection of liquid droplets. Ejection of particles by subsurface explosion would also be suppressed under such conditions.

Boyakov *et al.* (1978) have overcome the problem of substrate contamination between laser pulses by using a second laser for surface cleaning. In these experiments, a free running Nd–YAG laser was used to irradiate elemental solids with up to 120 J per pulse. Films of many elemental metals were produced with this technique. Electron diffraction showed that these films were polycrystalline.

2.13.2. Electroplating

Laser enhancement of electroplating efficiency has been described by von Gutfeld *et al.* (1979). The system used for experimental studies of this effect is shown schematically in Figure 2.93.

By focusing the output of a CW Ar^+ or Kr^+ laser on the front or back of a thin metal substrate on a glass support, enhancement rates of 10^3 could be produced at the focal spot. Substrate films of W, Mo, or Ni were used to electroplate Au, Ni, or Cu from solution. With the laser focused to a fine spot and the pulse length limited to 0.3 ms, 500 Å thick spots of ~4-μm diameter could be produced. Scanning of a CW beam at a rate of ~1 mm s^{-1} was found to generate stripes of electroplated metal with thicknesses in the range 2000–4000 Å and widths between 15 and 20 μm. The maximum growth rate observed in these experiments was ~5 μm s^{-1}. This was obtained in the generation of 5-μm thick 30–50 μm diameter spots using a 1-s laser pulse.

A preliminary model developed to describe the effect of laser radiation on electroplating efficiency attributes enhancement to the thermal gradient generated in the vicinity of the focal spot. This gradient introduces convective flow which decreases the thickness of the depletion layer in front of the cathode and increases the depleted ion concentration.

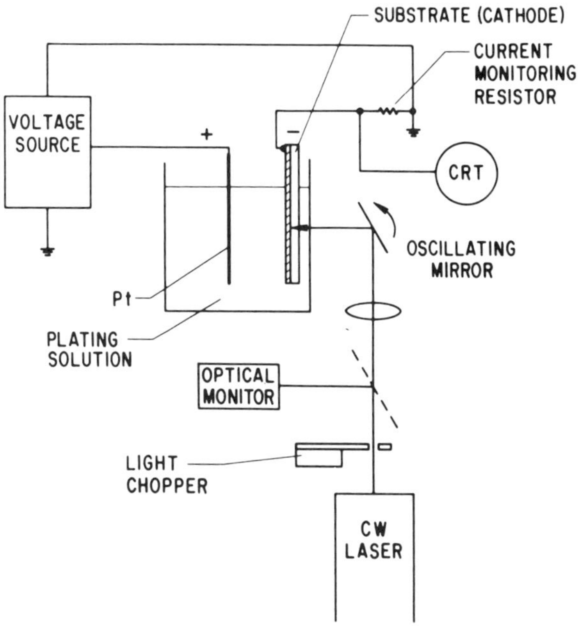

Figure 2.93. Schematic for laser plating enhancements in the region of optical absorption on the cathode. Here the laser is shown incident at a predeposited thin film optically transparent substrate interface [von Gutfeld *et al.* (1979)].

2.13.3. Chemical Vapor Deposition

Interest in chemical vapor deposition with lasers has centered on the production of granular Si films by pyrolysis of silane [Christensen and Lakin (1978), Hanabusa *et al.* (1979), and Baranauskas *et al.* (1980)]. Deposition of Si occurs when SiH_4 (or a compound such as $SiCl_4$) decomposes pyrolytically at the point heated by the laser. This makes it possible to generate Si patterns on the substrate without the need for subsequent masking and etching. A schematic diagram of apparatus for laser chemical vapor deposition (LCVD) is shown in Figure 2.94. The rate of deposition was found to increase rapidly with laser power at low laser power and then saturate at high laser power. With a focal spot diameter of ~3 mm Si growth was shown to proceed discontinuously at low laser power with the generation of separate nucleation centers. At slightly higher power these nuclei coalesced into islands to form a continuous film. At high power the intensity at the center of the focused beam is great enough to melt Si and nucleation is suppressed in this region. In this regime, Si growth from an annular ring toward the center was observed.

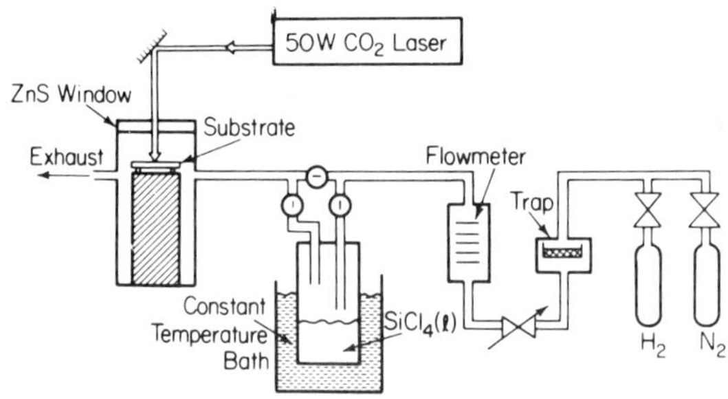

Figure 2.94. Schematic diagram of the LCVD apparatus. [Baranauseas *et al.* (1980)].

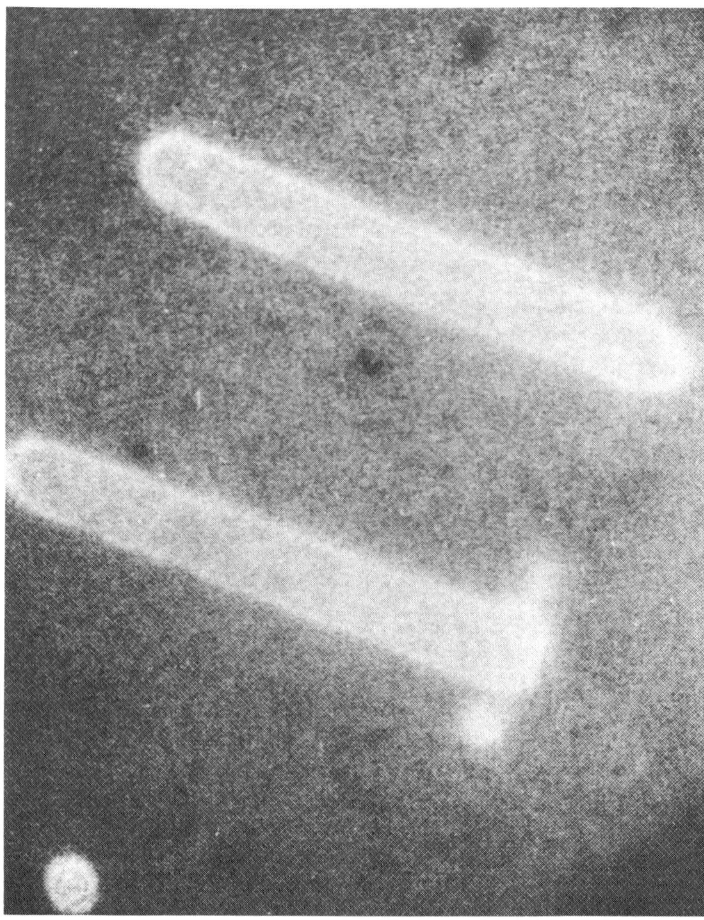

Figure 2.95. Two parallel lines obtained by tightly focusing 0.3 W of 10.33-μm radiation onto the substrate and translating it during illumination. The width of each line is approximately 50 μm [Christensen and Lakin (1978)].

Some experimental data on the spatial control of deposition have been reported by Christensen and Lakin (1978). With tight focusing, individual spots with 50-μm diameters were produced. Scanning by moving the substrate was used to produce linear depositions with $\sim$50 μm width (Figure 2.95).

2.13.4. Photodeposition and Photoetching

When the wavelength of a laser source is tuned to that of an electronic or vibrational transition in a gas phase molecule, the interaction between laser radiation and the molecule can result in photolytic decomposition. This mechanism is the basis for the photodeposition of metals by laser photolysis [Deutsch *et al.* (1979)]. In these experiments, the output from an Ar^+ laser was doubled to 257.2 mm and focused on a substrate in contact with $Al(CH_3)$ or $Cd(CH_3)_2$ gas. The subsequent photolytic reaction resulted in the deposition of metallic Al or Cd at the focal spot. An example of the deposit produced is shown in Figure 2.96. The fine scale structure in this photograph is due to Fresnel diffraction by an aperture in the beam. This structure has a characteristic size of $\sim$3 μm. Observation with a scanning electron microscope showed further structure on a scale of 0.2 μm. Thus the ultimate resolution of this method of pattern generation may be in the submicron range.

In a related experiment, Ehrlich *et al.* (1980) combined laser photodeposition with laser-induced substrate heating to produce stripe doping of *n*- or *p*-type InP with Cd. By varying the UV laser intensity it was found that the resultant dopant concentration could be varied over a wide range. Good quality ohmic contacts were generated using this novel technique.

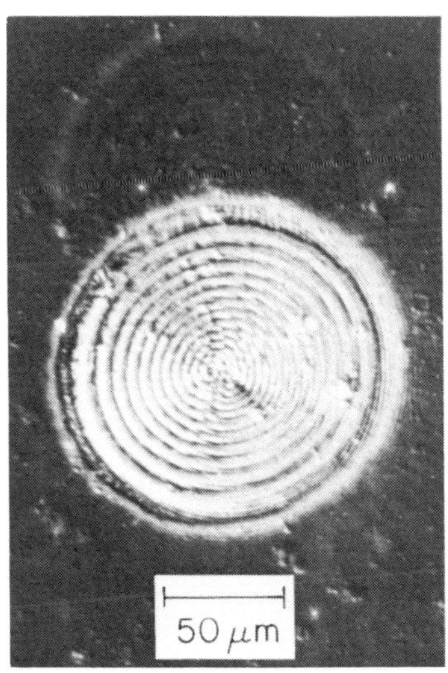

Figure 2.96. Cadmium deposit on SiO_2 substrate produced by 0.1-mW, 257.2-nm irradiation of a 3-torr DMCd: 700-torr He mix for 60 s [Deutsch *et al.* (1979)].

The process of deposition in this case occurs by UV laser-induced photolytic decomposition of the metal alkyl compound accompanied by solid-state diffusion in the substrate driven by heating with laser radiation that is not absorbed by the metal alkyl.

Laser photodeposition of reactive species has been used to chemically etch patterns on GaAs and InP [Ehrlich *et al.* (1980)] and Si and Ge [Sveshnikova *et al.* (1977) and Beterov *et al.* (1978)]. In this case the etching is enhanced when halogen atoms liberated by laser photolysis of a compound such as CH_3Br interact with localized regions of the substrate.

Chapter 3

Laser Processing of Semiconductors

3.1. INTRODUCTION

A wide variety of potential applications for lasers in the manufacture of semiconductors and semiconductor devices have been investigated in recent years. The advantages of laser processing in semiconductor preparation and utilization derive from the ability of laser sources to provide concentrated sources of heat in localized regions with extremely small areas. At the same time, beam scanning or stepping can allow scanning of large areas.

Some applications of lasers in this field are discussed in the following sections. While many of these are still at the laboratory stage of development, it is likely that several of these techniques will demonstrate engineering and economic feasibility and will soon find application in the manufacturing sector. Some laser processing techniques in semiconductor preparation are summarized in Table 3.1.

3.2. ANNEALING

One of the most dynamic areas of laser application over recent years has been that of laser annealing of ion-implantation damage in semiconductors. The implantation of ions of B, As, or P in a semiconductor such as Si leads to the formation of thin surface layers with *p*- or *n*-type characteristics. Since this implantation is accomplished by accelerating ions to keV energies before impact on the semiconductor surface, doping by this method is accompanied by damage to the host crystal. This damage takes the form of dislocations, stacking faults, dislocation loops, and a variety of point defects. At the doses customarily used for doping, this damage may be sufficient to drive the ion-

Table 3.1. Laser Applications in Semiconductor Processing and Device Manufacture

Annealing of ion-implantation damage
Recrystallization of amorphous layers
Laser-assisted doping
Ohmic contact formation
pn Junction formation
Silicide formation
Production of metal–semiconductor contacts in MOS structures
Machining of monolithic displays
Modification of resistance in Si resistors
Laser assisted oxidation to form SiO_2 overlayers
Gettering of wafers

implanted layer amorphous. Furthermore, many impurities do not end up in substitutional lattice sites and are not available for electrical conduction.

This damage can be partially removed by thermal annealing, i.e., by heating the entire semiconductor in a controlled atmosphere in a furnace. However, this rather drastic treatment often leads to contamination and degradation of electrical performance. In addition, bulk heating can lead to diffusion and recombination of dopants with precipitation in some instances.

Laser annealing of the ion-implanted layer has been shown to overcome many of these limitations [Antonenko *et al.* (1976), Shtyrkov *et al.* (1976), Kachurin *et al.* (1976), and Narayan *et al.* (1978a)]. In particular, since laser annealing heats only the damaged layer, no degradation of the substrate is produced. Heating with pulsed [Narayan *et al.* (1978a)] or CW [Gat and Gibbons (1978)] lasers yields essentially complete recrystallization of implanted layers with little residual damage and no diffusion of impurities. Dopant atoms are also observed to exclusively occupy substitutional lattice sites following laser annealing. Table 3.2 lists some semiconductors that have been subjected to laser annealing.

Although many of the drawbacks of thermal annealing are removed with laser annealing, others characteristic of the new process occur. For example, with pulsed laser annealing the semiconductor surface may become irregular and stresses may be introduced at the interface where melting terminated. With CW annealing some defects may remain.

Despite these limitations, it appears that laser annealing may soon find industrial application in the fabrication of semiconductor devices. Commercial laser annealing systems are now available [*Electro-Optical Systems Design,* December (1979), p. 29].

Table 3.3 [Kaplan *et al.* (1979)] summarizes the properties of some laser annealing methods. Other systems based on the CW CO_2 [Miyao *et al.* (1978)], CW Nd–YAG [Celler *et al.* (1979)], pulsed dye [Andrew and Lovato

Table 3.2. Representative Studies of Laser Annealing of Doped Semiconductors

Material	Laser	Reference
B:Si	*Q*-switched ruby	Narayan *et al.* (1978a)
As:Si	*Q*-switched ruby	White *et al.* (1979b)
	CW Ar ion	Auston *et al.* (1978)
P:Si	*Q*-switched ruby	White *et al.* (1978)
Ga:Si	*Q*-switched Nd–YAG	Bean *et al.* (1979)
Sb:Si	*Q*-switched ruby	White *et al.* (1979a)
Te:Si	*Q*-switched ruby	Foti *et al.* (1978a)
P:Ge	*Q*-switched ruby	Foti *et al.* (1978b)
In:$Pb_{0.8}Sn_{0.2}Te$	TEA CO_2	Bahir *et al.* (1979)
Bi:ZnTe	*Q*-switched ruby	Bontemps *et al.* (1980)
Si:GaAs		
N:GaAs	CW Ar ion	Makita *et al.* (1979)
Te:GaAs		
N:GaAsP	CW CO_2	Takai and Ryssel (1979)
Si:GaP	Mode-locked Nd–YAG	Murakami *et al.* (1979)
N:SiC	TEA CO_2	Makarov *et al.* (1979)

Table 3.3. Comparison of Laser Sources[a]

	CW Ar/Kr	Pulsed ruby	Pulsed YAG	Repetitively Q-S YAG
Wavelength (μm)	Visible	0.69	1.06/0.53	1.06/0.53
Power (W)	20	1	10/1	20/2
Spatial mode	TEM_{00}	multi	Multi/TEM_{00}	TEM_{00}
Pulse energy (J)	—	1–2	2/0.2	0.005
Spot area (mm)	0.025	1(local, in area of 10)	5	0.2
Operating flux, typical (J cm^{-2})	200	1	5/0.5	5/0.5
Overlap (%)	1 axis, 75	—	2 axis, 50	2 axis, 50
Throughput (cm^2 s^{-1})	0.1	0.01	0.5	1
Pulse duration (ns)	≈1	10–50	10–50	50–100
Heated depth (μm)	≈50	0.1–0.5	0.5–1	0.2–1
Stability	Very good	Poor	Fair	Very good
Controllability	Very good	Poor	Fair	Very good
Operating cost ($/h)	4.00	1.50	4.00	1.50
Operating cost (¢/cm^2)	1.1	4	0.2	0.04

[a]From Kaplan *et al.* (1979).

(1979)], and excimer [Yaron and Hess (1980)] laser have been reported. The highest laser power used in laser annealing seems to be 500 W. This was obtained from a CW Ar^+ laser [Klimenko *et al.* (1976)].

Annealing with pulsed lasers occurs when the ion-implanted layer is heated to the melting point. Recrystallization proceeds via liquid-phase epitaxial regrowth from the substrate with almost 100% of dopant atoms subsequently occupying substitutional sites [White *et al.* (1979a)]. This is accompanied by some redistribution of the dopant. The penetration of dopant atoms into the substrate is a strong function of laser power (Figure 3.1) [Celler *et al.* (1978)] or fluence [White *et al.* (1978)]. White *et al.* show that this redistribution can be understood in terms of diffusion in the melt prior to crystallization.

The relationship between ion dose and conditions for pulsed laser annealing has been investigated by several groups [Venkatesan *et al.* (1978), Muller *et al.* (1978), Bean *et al.* (1979), Wilcox (1980), and Prussin and von der Ohe (1980)]. Dopant levels influence annealing conditions by (in part) modification of the semiconductor absorption coefficient at the laser wavelength. This has an effect on the energy threshold for annealing [Bean *et al.* (1979)] and the time of formation and duration of the melt region [Venkatesan *et al.* (1978)]. At high dopant levels the energy threshold for annealing is reduced while the final impurity distribution can extend farther into the substrate. Miyao *et al.* (1978) conclude from a study of the laser annealing of P:Si at ion doses

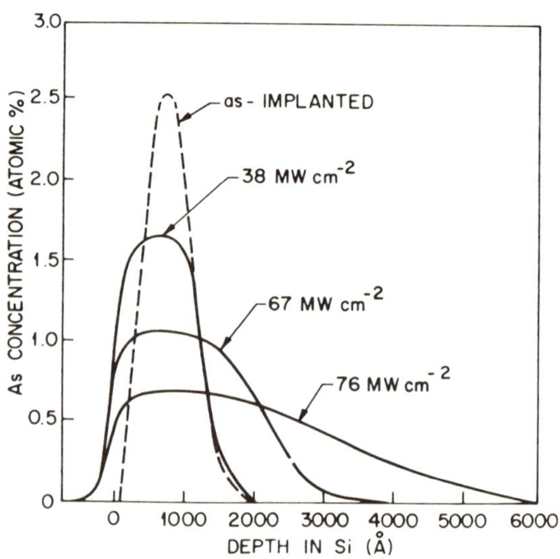

Figure 3.1. Depth distribution of As following Nd–YAG irradiation. Depth resolution of the backscattering measurement is ~ 200 Å [Celler *et al.* (1978)].

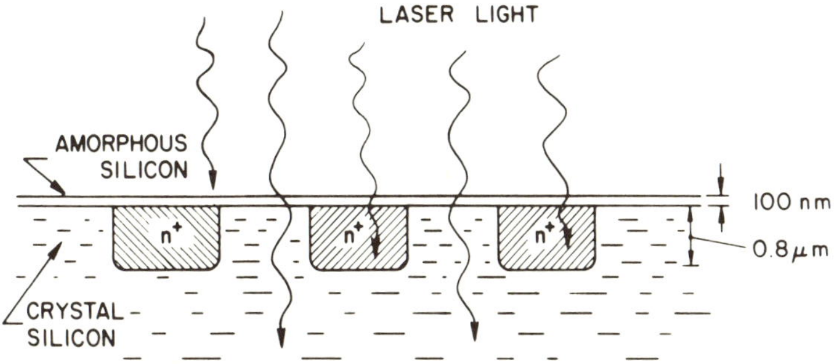

Figure 3.2. Arrangement used to generate fine linear patterns on doped Si [Fowler and Hodgson (1980)].

between 10^{14} and 10^{16} cm^{-2} that the degree of damage in the amorphous layer strongly influences the absorption coefficient and hence annealing behavior.

Since the wavelength of the Nd–YAG laser lies in the absorption edge of the fundamental band gap of Si, small changes in structure or increases in temperature can greatly influence the effective absorption coefficient. This effect has been noted by von Allmen *et al.* (1979) who show that under certain conditions increasing absorption can lead to thermal runaway over localized areas. They conclude that substrate preheating can be used to eliminate attendant inhomogeneities. Lietoila and Gibbons (1979) have performed a theoretical analysis of temperature profiles in Si heated with ruby or Nd–YAG lasers. They find that free-carrier absorption is important in *Q*-switched laser annealing.

Differential absorption in light and heavily doped regions for Nd–YAG laser radiation has been used by Fowler and Hodgson (1980) to generate fine linear patterns on doped Si. The configuration used is shown in Figure 3.2. A 1000-Å layer of amorphous Si covered the surface of crystalline Si previously prepared with lines of *n*-type material. On irradiation, light is absorbed more strongly by the *n*-type material causing the amorphous material above these regions to recrystallize. This may be advantageous in the preparation of small self-aligned surface structures.

Free-carrier absorption is a dominant mechanism for coupling of CO_2 laser radiation ($\lambda = 10.6$ μm) into Si semiconductors. Siregar *et al.* (1980) have studied the temperature dependence of the absorption coefficient, α, of Si at 10.6 μm as a function of temperature. They find an increase in α by a factor of 10 between 20 and 400°C. Details on laser annealing with CO_2 lasers can be found in Section 3.3.

The wavelength dependence of coupling of laser radiation to semiconductors is the basis for a novel dual-wavelength annealing system described by

Auston *et al.* (1979). This method involves the superposition of 1.06 and 0.53-μm pulses from a Nd–YAG laser. The weak pulse at 0.53 μm initiates surface melting. Since the melt in Si is metallic it strongly absorbs at 1.06 μm. Thus the 1.06-μm pulse efficiently couples into the substrate. By controlling the spatial distribution of the 0.53-μm pulse it may be possible to generate fine patterns on the semiconductor surface since the annealing pulse is effective only when preceded in time by the melting pulse.

Semiconductor devices are often constructed with an SiO_2 on Si structure. Experiments on the effect of SiO_2 layers on laser annealing have been reported by Tamura *et al.* (1979), Narayan (1980), and Okabayashi *et al.* (1980). It is found that the SiO_2 layer thickness plays a critical role in the extent to which annealing proceeds in the underlying substrate. Tamura *et al.* note that selective localized annealing could be possible when the SiO_2 layer thickness is adjusted to equal the laser wavelength.

A significant advantage of pulsed laser annealing is the nonequilibrium nature of melt thermodynamics. This offers flexibility in the preparation of annealed layers with a wider range of dopant concentrations than possible from thermal annealed samples. Several studies have shown that solid solubility limits for Sb, B, P, Ga, and As in Si can be greatly exceeded in pulsed laser annealed samples [White *et al.* (1978, 1979b), and Bean *et al.* (1979)] Golovchenko and Venkatesan (1978) have shown that the concentration of Te in annealed GaAs exceeds that from the maximum solubility limit by a factor $>$ 10.

The rapid heating and cooling that occurs during laser annealing with pulsed lasers results in the introduction of strain in the vicinity of the annealed spot. This effect has been studied by Larson *et al.* (1978) and Larson and Barhorst (1980). In B implanted Si, the depth distribution of strain was found to follow the distribution of B concentration. The nature of the lattice distortion is a function of the type of dopant and the substrate. B and P in Si yield a lattice contraction while Sb leads to expansion.

The sensitivity of the annealing process to laser intensity is often manifest in the occurrence of interference related structures in and near the annealed zone. Figure 3.3 shows the intensity dependence of these features on the surface of As:Si annealed with 110-ns pulses from a Nd–YAG laser [Leamy *et al.* (1978)]. At low intensity, a surface ripple attributed to the existence of polycrystalline Si regions is observed. With higher intensity pulses, the boundary of each spot shows the effect of rapid cooling. Interference fringes with spacings $\simeq$ 1 μm are observed at the perimeter of each spot. An analysis of these effects [Oron and Sørensen (1979)] shows that periodicities can be understood in terms of interference in the incident laser beam. Further studies of surface structures on Si irradiated with pulses from a ruby laser have been reported by Vitali *et al.* (1977).

Si on sapphire (SOS) technology is the basis for several important semi-

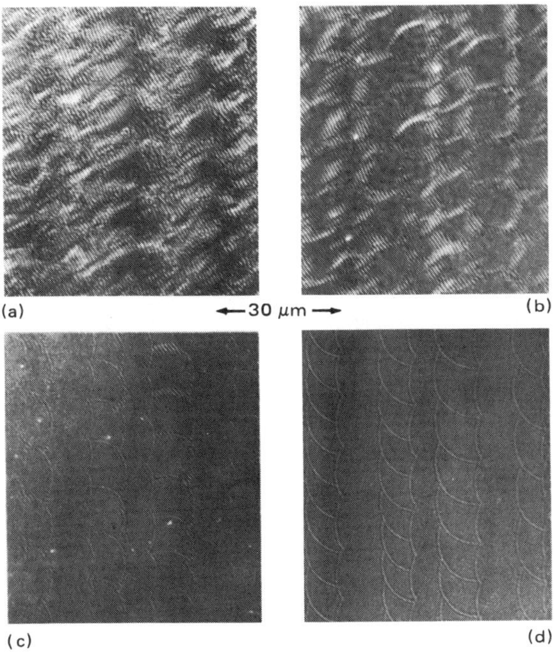

Figure 3.3. Normarski interference contrast micrographs of laser annealed silicon: (a) 38 MW cm^{-2}; (b) 49 MW cm^{-2}; (c) 57 MW cm^{-2}; and (d) 76 MW cm^{-2} [Leamy *et al.* (1978)].

conductor devices. The preparation of SOS wafers by conventional means leads to poor crystal quality particularly near the Si–Al_2O_3 interface where strains due to differential lattice contraction accumulate. These can be relieved in part by thermal annealing, however, the remaining defects degrade device performance.

Annealing with pulsed lasers has been used in attempts to alleviate this problem [Roulet *et al.* (1979), Lüthy *et al.* (1979), and Yaron and Hess (1980)]. It has been found that the strain level of Si films on SOS can be altered by laser annealing. Yaron and Hess speculate that it may eventually be possible to obtain essentially bulk Si behavior from Si films using this approach.

Strain relief in SOS layers has been shown to result from recrystallization of liquid Si seeded by the underlying Al_2O_3 substrate [Lüthy *et al.* (1979)]. Sai-Halasz *et al.* (1980) suggest that preheating of the Al_2O_3 substrate could introduce additional flexibility in the attainment of specific strain states via laser annealing.

In a related application, a Nd–YAG laser has been used to produce damage on the back of semiconductor wafers which acts to gather impurities

[Pearce and Zaleckas (1979)]. It was suggested that such treatment could be applied several times during the processing of wafers.

3.3. ANNEALING—CW LASERS

CW annealing of semiconductors occurs through solid-phase epitaxial regrowth. This is similar to the process occurring during thermal annealing in a furnace and does not involve melting. As such, CW annealing maintains dopant distributions and introduces little in the way of surface irregularities. On the other hand, since thermal annealing is an equilibrium process it does not offer the flexibility of pulsed annealing. Furthermore, processing times are lengthened over those characteristic of pulsed annealing.

A schematic diagram of a scannable CW annealing system as developed by Gat and Gibbons (1978) is shown in Figure 3.4. Typical operating conditions are as follows:

Laser power (multiline)	7 W
TEM_{00} spot diameter	38.5 μm
Intensity	6×10^5 W cm^{-2}
x scan speed	2.76 cm s^{-1}

Under these conditions the width of the annealed line was 22 μm. The y position was changed by $\sim$15 μm at the end of each scan to yield overlapping scans. Gat and Gibbons note that laser power must be stable for optimum results. Deviations of 8% were sufficient to give unsatisfactory annealing.

Figure 3.5 shows a comparison between As concentration profiles and Si subjected to a 30-min, 1000°C thermal anneal and that obtained after CW laser annealing [Gat *et al.* (1978a)]. This plot shows that the As distribution is essentially unaffected by CW laser annealing while thermal annealing leads to significant redistribution. In addition, a fairly sharp boundary is produced between recrystallized and amorphous zones. This boundary can be seen in the

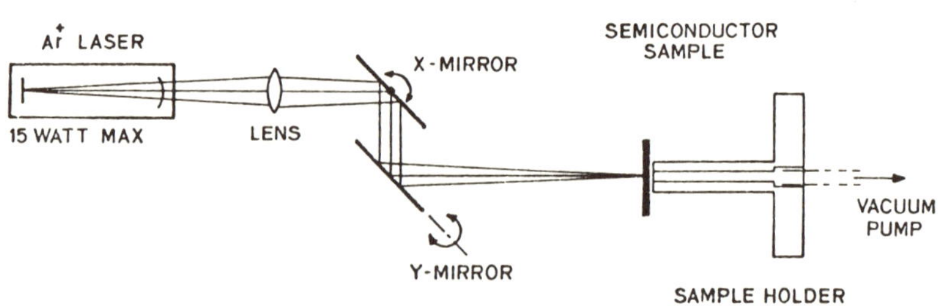

Figure 3.4. System for CW annealing with Ar$^+$ laser [Gat and Gibbons (1978)].

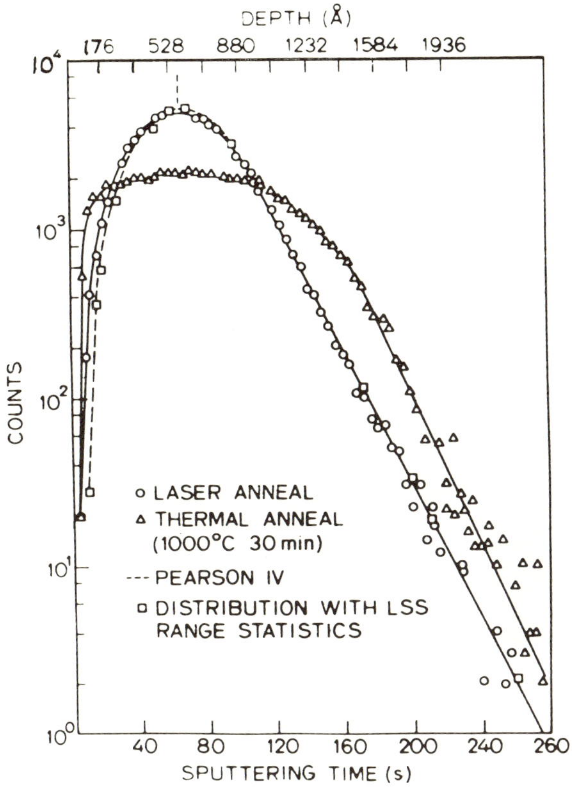

Figure 3.5. As concentration profile in As-implanted Si after laser anneal and thermal anneal [Gat *et al.* (1978a)].

electron micrograph of laser scanned B:Si shown in Figure 3.6 [Gat *et al.* (1978b)]. Gat *et al.* report that CW laser annealing provides complete recrystallization with 100% electrical activity. This is observed for both amorphous and nonamorphous surface implants on Si.

Two other general investigations of CW annealing of As implanted Si have been reported [Gat *et al.* (1979) and Auston *et al.* (1978)]. Both groups conclude that CW annealing is consistent with solid-phase epitaxial regrowth. Gat *et al.* show that the kinetics of this regrowth is compatible with thermal conduction through the sample together with a growth rate.

$$v = 1.55 \times 10^{14} \exp[-2.3/k_B T(r)]$$

where v is the growth rate in Å s^{-1}, k_B the Boltzmann constant (8.6×10^{-5} eV $K)^{-1}$, and $T(r)$ the temperature at radial distance r from the center of the focused beam.

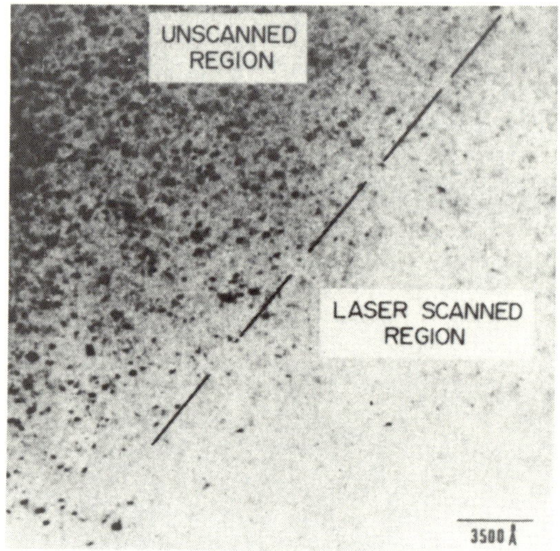

Figure 3.6. Transmission electron micrograph of B-implanted laser annealed sample at laser-scan boundary [Gat *et al.* (1978b)].

Since the band gap of Si is 1.1 eV, CO_2 laser radiation ($h\nu = 0.12$ eV) is not strongly absorbed. However, absorption does occur via free carriers and transitions to and from impurity levels. Celler *et al.* (1979) have studied the CW annealing of ion-implanted Si with CO_2 and CW Nd–YAG lasers. In view of the weak absorption at both laser wavelengths, it was found that preheating of substrates to 200–400°C was advantageous. Since the absorption coefficient is small, heating occurs over a larger volume than with Ar or Kr lasers. Electrical activity was smaller than that obtained with Ar or Kr annealing. CO_2 or Nd–YAG laser annealing therefore appears to offer a capability that is intermediate between that of CW Ar and thermal annealing.

CW annealing with a CO_2 laser has been used in the preparation of nitrogen isoelectronic traps in $GaAs_{0.6}P_{0.4}$ light emitting diodes [Takai and Ryssel (1979)]. The conversion efficiency of GaAs solar cells has been improved by annealing with CW ND–YAG laser [Fan *et al.* (1979)].

3.4. RECRYSTALLIZATION

While much effort has been devoted to the development of laser-annealing techniques for epitaxial regrowth of ion-implanted amorphous layers on semiconductors, similar methods can be used for the production of large grained semiconductor sheets. Initial studies in this area were briefly reported by Laff and Hutchins (1974). This work was extended by Fan and Ziegler (1975) using a scannable CW Nd–YAG laser to crystallize radio-frequency sputtered Si films on Al_2O_3 substrates. The resulting crystallite size was $\simeq 25\ \mu m$.

Gat *et al.* (1978) recrystallized 0.4-μm thick polycrystalline films of chemically vapor deposited Si that had been implanted with B using a CW Ar^+ system. A summary of their data is given in Table 3.4. It can be seen that laser annealing leads to a dramatic increase in grain size and raises carrier concentration to the limit provided by implantation ($\simeq 5 \times 10^{14}\ B^+\ cm^{-2}$). Carrier mobilities exceeded those obtained with thermal annealing.

Recrystallization of Ge films on Si substrates [Fan *et al.* (1980)] with a CW Nd–YAG laser focused to a slit image yields the structure shown in Figure 3.7. The spatial period of this structure was found to be a sensitive function of substrate temperature ranging from $\simeq 50\ \mu m$ at 20 °C to 250 μm at 475 °C. As shown in the inset of the figure, the crystal structure varies within each period from amorphous to mixed amorphous–crystalline to a region of larger crystallites. A model describing this behavior in some detail has been developed by Fan *et al.*

Irradiation of Ge films on Mo or graphite substrates heated to 400–500 °C over a localized region was found to initiate recrystallization over areas as large as $2.5 \times 2.5\ cm^2$. It is possible that this process may lead to a competitive method of forming single-crystal sheets of semiconductor material for use in the manufacture of solar cells.

3.5. SILICIDE FORMATION

Electrical contacts and conducting paths on silicon semiconductors can be made via metal silicides. Metal silicides are conventionally formed by first depositing a thin metal layer on the silicon substrate. The metal–silicon system

Table 3.4. Electrical Properties and Structure Comparison of Different Laser Anneal Condition on $5 \times 10^{14}\ B^+\ cm^{-2}$ Implanted Polysilicon[a]

Anneal condition	As-implanted	Thermal anneal	Laser anneal	Laser after thermal anneal
Grain size	300–600 A	900–1200 A	$\overline{25} \times \overline{2}\ \mu m$	$\overline{25} \times \overline{2}\ \mu m$
Sheet resistivity ($\Omega/\square$)	12.5 M	623	269	260
Sheet carrier concentration (cm^{-2})	Unmeasurable	$4.7 \times 10^{14}\ cm^2$	$5.2 \times 10^{14}\ cm^2$	$5.17 \times 10^{14}\ cm^2$
Mobility ($cm^2\ V^{-1}\ s^{-1}$)	Unmeasurable	24	44.5	46

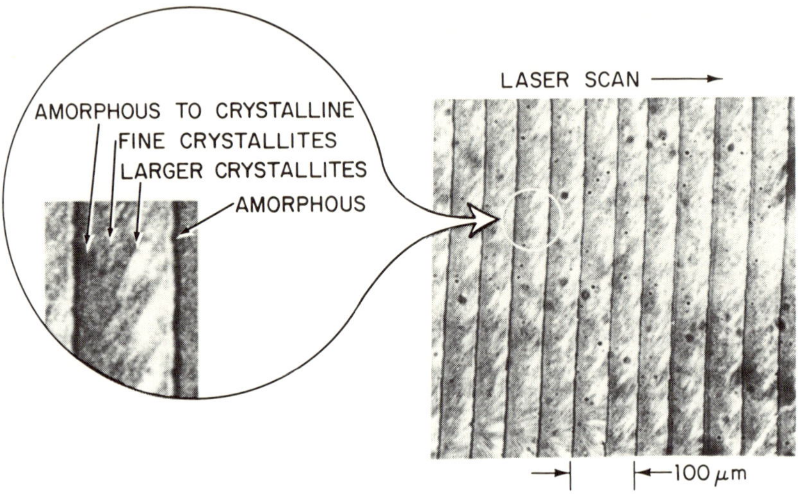

Figure 3.7. Optical transmission micrograph of a laser crystallized Ge film showing periodic structural features. Inset: expanded view illustrating four different microstructure regions [Fan *et al.* (1980)].

is then heated to several hundred °C in an inert atmosphere. Reaction between the metal and silicon generates the silicide compound.

Similar results are obtained with laser heating although in a shorter time. In addition, the laser heating process can yield higher annealing temperatures and can be carried out without heating adjacent structures.

A summary of experiments on metal–silicide formation is given in Table 3.5. With the exception of the work of Shibata *et al.* (1980), silicide formation has been initiated with pulses from a *Q*-switched Nd–YAG laser. A repetitively pulsed laser is used with beam scanning to cover areas greater than that of the focal spot.

With pulsed excitation, the rapid heating/cooling rates lead to the quenching in of several different compounds in the silicide film. von Allmen and Wittmer (1979) have found, for example, that irradiation of Pd on Si results in the formation of a silicide layer that is a mixture of PdSi, Pd_2Si, Pd_4Si, Pd_5Si, and pure Si. Poate *et al.* (1978) found that layers of PtSi consist of small Si spherules coated with Pt silicides. These structures can be attributed to a solidification and convection in the Pt·Si melt [Van Gurp *et al.* (1979)].

Figure 3.8 shows strips of PdSi formed by scanning with a CW Ar ion laser [Shibata *et al.* (1980)]. The darker center region of each scan corresponds to a distinct PdSi phase while lighter center regions are mixtures of Pd and Si without unique stoichiometry. The dark PdSi phase is evidently initiated

Table 3.5. Summary of Experiments on Laser-Assisted Silicide Formation

Silicide	Laser intensity	Reference
$Pt_{2.0}Si$–$PtSi_{10.4}$	18–26 MW cm^{-2}	Poate *et al.* (1978)
PtSi	20–40 MW cm^{-2}	Van Gurp *et al.* (1979)
Pt_2Si–$PtSi_2$	CW	Shibata *et al.* (1980)
PtSi	50–150 MW cm^{-2}	Wittmer and von Allmen (1979)
PdSi	40–90 MW cm^{-2}	von Allmen and Wittmer (1979)
PdSi–Pd_5Si	50–150 MW cm^{-2}	Wittmer and von Allmen (1979)
PdSi	20–40 MW cm^{-2}	Van Gurp *et al.* (1979)
PdSi	30 MW cm^{-2}	Poate *et al* (1978)
PdSi–Pd_2Si	CW	Shibata *et al.* (1980)
$NbSi_2$	CW	Shibata *et al.* (1980)
CoSi	20–40 MW cm^{-2}	Van Gurp *et al.* (1979)
NiSi	30 MW cm^{-2}	Poate *et al.* (1978)
MoSi	50 MW cm^{-2}	Poate *et al.* (1978)
MoSi	20–40 MW cm^{-2}	Van Gurp *et al.* (1979)
AuSi	—	von Allmen *et al.* (1980)

when the scanning beam encounters a surface defect such as a scratch, an edge, or a previously irradiated region. Shibata *et al.* concludes that silicide formation is a nucleation controlled reaction.

Laser-induced silicide formation may be accompanied by the creation of surface interference structures in the form of concentric rings [Wittmer and von Allmen (1979)]. It has been demonstrated [Affolter *et al.* (1980)] that these arise from scattering of incident laser radiation by small particles in front of the target. This results in a variation in the composition of metal silicides in the vicinity of maxima and minima in the interference pattern.

A study of the composition of mixed Au–Si solids heated by laser radia-

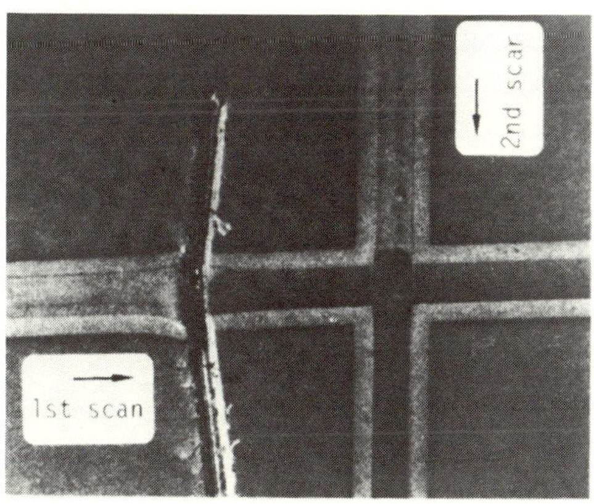

Figure 3.8. Nomarski contrast optical micrograph of a Pd/Si sample laser scanned from two different directions. The PdSi phase is the darker center region which forms when the scanning beam passes over the scratch (first scan) or the already existing PdSi phase (second scan) [Shibata *et al.* (1980)].

tion has been reported by von Allmen *et al.* (1980a). Samples with initial compositions between $AuSi_{10}$ and $Au_{10}Si$ were irradiated with pulses from Nd–YAG laser. Both 30-ns and 300-μs pulses were used. Structural differences where noted in the two cases, however, measurements of composition were compatible with the presence of a metastable Au silicide phase together with precipitation of either Au or Si. von Allmen *et al.* conclude that this method may be useful in the creation of metastable silicides with novel properties. Similar data have been obtained for Pt–Si and Pd–Si systems [von Allmen *et al.* (1980b)].

3.6. OHMIC CONTACTS AND JUNCTION FORMATION

Potential applications of lasers in the doping of semiconductors were first explored by Fairfield and Schwuttke (1968) and Harper and Cohen (1970). In laser-assisted doping, the semiconductor surface is first coated with a layer containing the dopant. This is then irradiated with one or more laser pulses. The resultant local heating causes the dopant to diffuse into the substrate forming a doped layer with electrical properties that differ from those of the substrate.

Ohmic contacts have been produced in Si by irradiating phosphorus-coated *n*-type Si or boron-coated *p*-type Si with Nd–YAG or CO_2 laser pulses

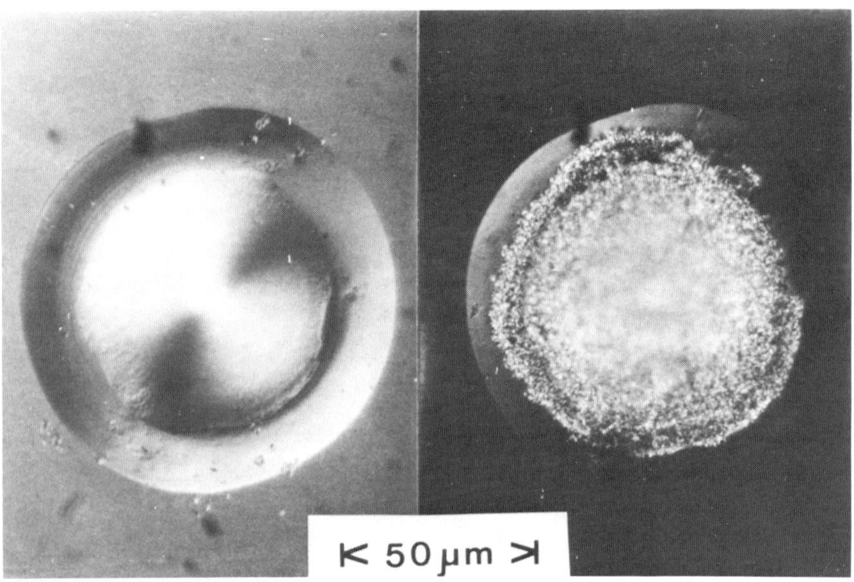

Figure 3.9. Micrograph of a laser doped ohmic zone on *n*-type Si after laser irradiation (left) and after galvanic deposition of copper (right). Pulse duration was 200 μs, light intensity 1 MW cm^{-2} [Affolter *et al.* (1978)].

Table 3.6. Electrical Parameters of Boron-Deposited Laser-Annealed Silicon[a]

Sample	Carrier concentration N_s (cm^{-2})	Sheet resistivity ρ_s (Ω/□)	Carrier mobility μ $cm^2\ V^{-1}\ s^{-1}$	Amount of boron deposited (nominal $\mu g\ cm^{-2}$)	Number of pulses
BD-10	2.10×10^{15}	80.0	40	<1	1
BD-11	8.62×10^{15}	23.8	31	5	1
BD-12	9.37×10^{15}	22.0	30	5	2
BD-13	6.10×10^{15}	32.0	33	4	1
BD-25[b]	3.13×10^{16}	8.2	24	10	1

[a]Narayan *et al.* (1978B)
[b]Preheat to 600°C

[Affolter *et al.* (1978)]. Irradiation of B-coated *n*-type Si or P-coated *p*-type Si generated np^+ and pn^+ rectifying junctions, respectively. A photograph of an ohmic contact on *n*-type Si is shown in Figure 3.9.

Large area *pn* junctions in Si have been produced using single 50-ns pulses from a *Q*-switched ruby laser [Narayan *et al.* (1978b)]. The electrical properties of these junctions were found to be comparable to those formed by conventional thermal diffusion techniques. A summary of the electrical properties of several laser-doped BiSi samples is given in Table 3.6.

In a related study, Young and Narayan (1978) found that the precipitates formed on the formation of *pn* junctions by thermal diffusion can be dissolved by laser heating. This leads to the activation of dopant atoms contained in the precipitates with an attendant improvement in electrical properties of the junction. It is also found that the final dopant concentration can exceed the solid solubility limit.

Other reports on the formation of ohmic contacts on semiconductors by laser irradiation are by Margalit *et al.* (1978, AuGe on GaAs ep:layers), Barnes *et al.* (1978, Te:GaAs), and Deutsch *et al.* (1980, Cd or Zn on InP). Deutsch *et al.* used pulses from an ArF laser ($\lambda = 193$ nm) to deposit ohmic contacts on InP by photolysis of $Cd(CH_3)_2$ or $Zn(CH_3)_2$. The deposition in this case is due to a combination of photolysis and surface heating. With 7-ns pulses, 10–100 pulses were utilized to obtain ohmic contacts. The contact resistance decreased rapidly after the first 10 pulses and then remained constant.

3.7. DEVICE FABRICATION

Laser annealing of semiconductors has been shown to offer many attractive alternatives to thermal annealing (see Section 3.2). Some recent studies have explored the relationship between this processing and device performance [Koyanagi *et al.* (1979) and Lee *et al.* (1979)].

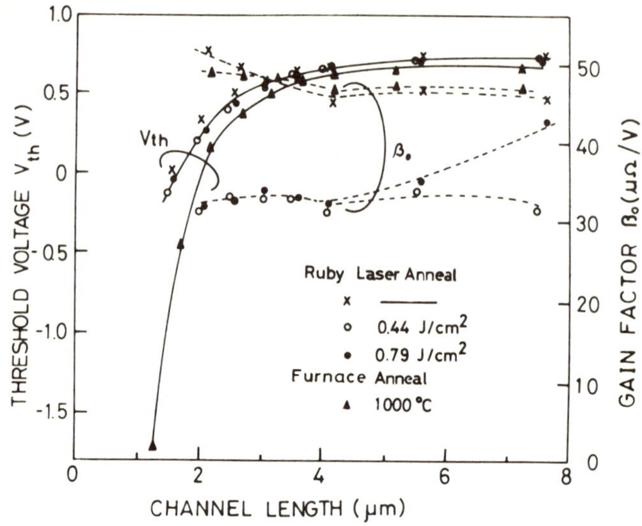

Figure 3.10. Threshold voltage vs. channel length and gain factor vs. channel length relations in laser annealed As^+ implanted source and drain structures in MOS FETs [Koyanagi *et al.* (1979)].

Koyanagi *et al.* fabricated short channel MOS FETs using laser annealed As^+-implanted source and drain structures. It was found that the lateral spread of both structures were smaller in laser annealed devices than in those fabricated by conventional means. The result was an improved threshold voltage vs. channel length curves (Figure 3.10). This was accompanied by smaller gain factors for laser annealed devices.

CW laser scanning of polycrystalline Si films has been shown to lead to an increase in grain size (Section 3.4). Lee *et al.* have shown that depletion-

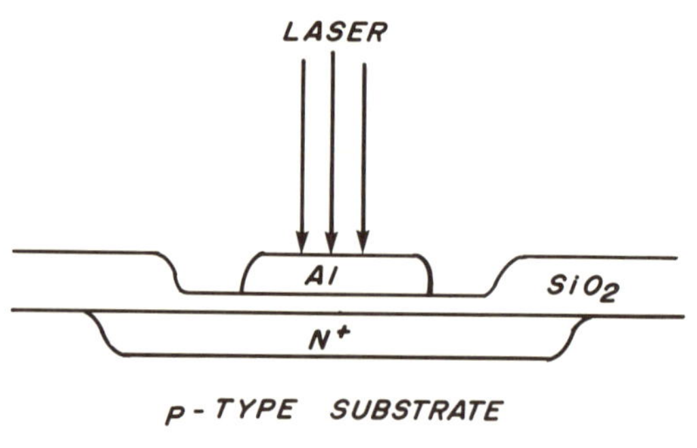

Figure. 3.11. Geometry of laser-induced electrical connection site on SiO_2:Si structure.

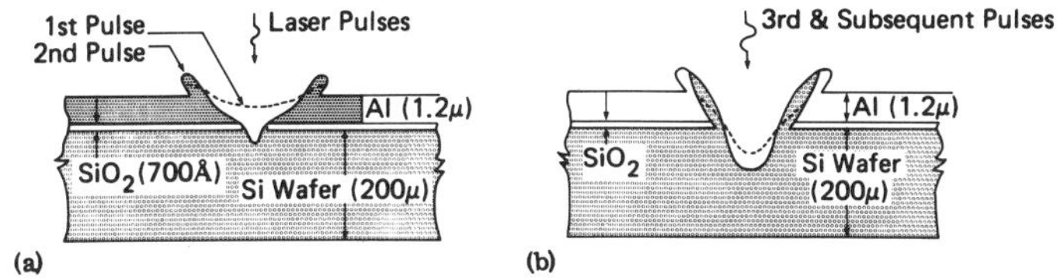

Figure 3.12. Model for the mechanism of M-S connection process induced by nanosecond laser pulses in MOS structures with thin SiO_2 layer [Platakis (1976)].

and enhancement-mode MOS FETs fabricated from laser annealed polycrystalline Si have similar electrical properties to those prepared on single-crystal Si. Almost total activity of implanted P in the depletion-mode device was noted.

These preliminary studies indicate that the improvement in electrical and chemical properties of semiconductor substrates obtained with laser annealing lead to significant improvements in devices manufactured with this material.

3.8. ELECTRICAL CONNECTIONS ON INTEGRATED CIRCUITS

A potential application of lasers in the electrical modification of integrated circuits has been described by Cook *et al.* (1975) and by Platakis (1976). This involves laser generated metal–semiconductor connections in MOS structures and the connection and disconnection of conductors on FET chips.

The geometry of a laser-induced electrical connection site is shown in Figure 3.11. A 1-μm thick Al line on SiO_2:Si is heated with two more more laser pulses. This causes Al to penetrate and make contact with the n^+ material on the substrate. The morphology of laser generated connections for both thin and thick SiO_2 layers has been thoroughly studied by Platakis (1976). Figure 3.12 shows the development of contacts in Al:SiO_2:Si layers as produced by pulses from an N_2 pumped dye laser. The final electrical contact to the Al layer occurs when molten Si flows from the Si wafer to coat the inside of the hole.

Potential applications of this process include repair and interconnection of integrated circuits after conventional processing as well as in the wiring of thin film memory/logic devices.

3.9. MONOLITHIC DISPLAYS

Dapkus *et al.* (1974) have reported on a novel method of producing optically and electrically isolated island devices from planar GaP *pn* junction material using micromachining with a *Q*-switched Nd–YAG laser. Numeric bars were isolated by cutting 20 μm $\times$ 74 μm grooves in GaP wafers. The resulting

Figure 3.13. Laser machined numeric in ON configuration [Dapkus *et al.* (1974)].

structure as seen in emission as shown in Figure 3.13. It is noted that a similar approach may be useful in the manufacture of devices such as GaAs lasers and in GaAs–GaAlAs laser optical waveguide fabrication.

Chapter 4

Chemical Processing

4.1. INTRODUCTION

The development of laser sources has revolutionized many aspects of photochemistry. The pace at which this revolution has progressed is remarkable: it is barely 12 years since the first laser photochemical experiments were carried out. The current literature extends to over a thousand articles.

Some areas, such as IR photochemistry and radiation-induced isotope separation, were almost unexplored prior to the advent of laser sources. Today they are the areas of most rapid development and are the source of much new information on the course of chemical reactions.

In this chapter these developments are described and work on laser processing of atomic and molecular systems is reviewed. While the mechanisms of the interaction between coherent radiation and chemical systems is discussed, our prime interest is in delineating present and future areas in which lasers will be applied in chemical processing. In keeping with the development of the field as a whole, emphasis has been placed on laser isotope separation and infrared photochemistry using high-power laser radiation.

4.2. SCHEMES FOR LASER ISOTOPE SEPARATION

In this section we summarize basic techniques that have been used for the separation of isotopes using laser excitation. Discussion of the details of these processes and their application to actual chemical systems will be given in subsequent sections.

1. *Laser-induced reaction.* For a molecule AB one has

$$\left.\begin{array}{l}\mathrm{AB} + h\nu \rightarrow \mathrm{AB}^* \\ \mathrm{AB}^* + \mathrm{C} \rightarrow \mathrm{Product}\end{array}\right\} \text{VIS, UV}$$

or
$$\left.\begin{array}{l} AB + h\nu(\mathrm{IR}) = AB^{v} \\ AB^{v} + C \rightarrow \text{Products} \end{array}\right\} \mathrm{IR}$$

For an atom

$$\begin{array}{l} A + h\nu \rightarrow A^{*} \\ A^{*} + B \rightarrow AB \end{array}$$

2. *Single photon predissociation.*

$$AB + h\nu \rightarrow A + B$$

3. *Two-photon dissociation.*

$$\left.\begin{array}{l} AB + h\nu_1 \rightarrow AB^{*} \\ AB^{*} + h\nu_2 \rightarrow A + B \end{array}\right\} \text{VIS–UV}$$

$$\left.\begin{array}{l} AB + h\nu(\mathrm{IR}) \rightarrow AB^{*} \\ AB^{*} + h\nu_2 \rightarrow A + B \end{array}\right] \text{IR–(VIS, UV).}$$

4. *Multiphoton dissociation.*

$$AB + nh\nu \rightarrow A + B$$

5. *Photoisomerization.*

$$\begin{array}{l} AB + h\nu \rightarrow AB^{*} \\ AB^{*} \qquad \rightarrow BA \end{array}$$

6. *Two-step photoionization.*

$$\left.\begin{array}{l} AB + h\nu_1 \rightarrow AB^{*} \\ AB^{*} + h\nu_2 \rightarrow AB^{+} + e^{-} \end{array}\right\} \text{VIS, UV}$$

$$\left.\begin{array}{l} AB + h\nu(\mathrm{IR}) \rightarrow AB^{v} \\ AB^{v} + h\nu_2 \rightarrow AB^{+} + e^{-} \end{array}\right\} \text{IR–VIS, UV}$$

$$\begin{array}{l} AB + h\nu_1 \rightarrow AB^{*} \\ AB^{*} + C^{*} \rightarrow AB^{+} + e^{-} + C \end{array}$$

For atomic species

$$\begin{array}{l} A + h\nu_1 \rightarrow A^{*} \\ A^{*} + h\nu_2 \rightarrow A^{+} + e^{-} \end{array}$$

7. *Selective Raman excitation.*

$$AB + h\nu_1 \rightarrow AB^{v} \qquad \text{(Stokes)}$$
$$AB^{v} + C \rightarrow \text{Products}$$

or
$$AB^{v} + h\nu_2 \rightarrow AB^{+} + e^{-}$$

or
$$AB^{v} + h\nu_2 \rightarrow A + B$$

8. *Deflection.*

$$A(\bar{v}_1) + h\nu \rightarrow A^*(\bar{v}_2)$$

where $\bar{v}_1$ is the velocity of incident atom and $\bar{v}_2$ is the new velocity.

The notation used in the above is as follows: AB^*, electronically excited molecule; AB^+, molecular cation; AB^v, vibrationally excited molecule.

4.3. THE ENRICHMENT FACTOR

We consider a sample that is a mixture of two species A and B. Let $[N_A]_0$ and $[N_B]_0$ be the concentration of A and B, respectively, in this mixture. After processing, new molecules containing the atoms or molecules A and B may be formed. The concentrations of these are $[N_{AR}]_f$ and $[N_{BR}]_f$, respectively. Then the enrichment factor (also called the coefficient of separation selectivity) is

$$\beta = \frac{[N_{AR}]_f}{[N_{BR}]_f} \frac{[N_B]_0}{[N_A]_0}$$

If $\beta = 1$ no separation has occurred while large separation efficiency corresponds to having $\beta \gg 1$. Although β depends on such factors as laser power and total energy delivered to the sample most values of β lie in the range $1 < \beta < 10^4$.

4.4. LASER-INDUCED REACTION

This process can lead to isotope separation if the reaction rate of an excited molecule or atom with a particular reactant exceeds the reaction rate with the ground-state atom or molecule. Selectivity arises because only the desired isotopic species is excited and therefore takes part in the subsequent photochemical reaction.

Excitation to a higher energy state does not automatically ensure that reaction will occur. Considering the laser-induced reaction of a molecule A, one requires that the rate

$$k(A^*,R)[A^*][R]$$

where $k(A^*,R)$ is the rate constant for reaction of A^* and R and $[A^*]$ and $[R]$ are the concentrations of A^* and R, respectively, exceed the rate at which the excitation is transferred to other species. It must also exceed the rate

$$k(\mathrm{A,R})[\mathrm{A}][\mathrm{R}]$$

and the rate at which thermal excitation leads to the A* + R reaction. A further complication is that any isotopic enrichment reached in the A* + R reaction may be reduced by subsequent undesired reactions.

Isotopically selective photochemistry with excited atoms has been extensively studied using conventional excitation techniques. An example can be found in the experiments of Desnoyer *et al.* (1963) and Morand and Nief (1968). The photochemical reaction of Hg(6 ^3P) with O_2 was used to enrich ^{202}Hg to 96.8% in HgO. This technique has not been extensively used in laser isotope separation.

Laser-induced reaction (LIR) was used in one of the earliest experiments on laser-induced isotope separation [Tiffany (1968)]. In this experiment a temperature tuned ruby laser was used to induce the reaction

$$Br_2 + C_4F_8 \rightarrow C_4F_8\,Br_2$$

However, subsequent reactions under the conditions of the experiment led to isotopic scrambling so that no net enhancement occurred.

The first successful separation of isotopes using LIR also involved Br_2 but in this case as a scavenger. Mayer *et al.* (1970) irradiated a mixture of CH_3OH and CD_3OD with lines from an HF laser in the wavelength range 2.64–2.87 μm. This produced excitation of the OH stretching vibration in CH_3OH and led to reaction of these molecules with Br_2. The resulting mixture was enriched to 95% CD_3OD from an original 1:1 mixture.

Other isotope separation experiments performed with CW laser sources include the LIR separation of ^{37}Cl from ICl using CW Ar^+ laser [Liu *et al.* (1975)] or a CW dye laser [Brenner *et al.* (1977)], the separation of ^{79}Br and ^{81}Br from CH_3Br with CW CO_2 radiation [Manuccia *et al.* (1976)], the separation of D from CH_2D_2 with CW CO_2 radiation [Hsu and Manuccia (1978)], and the isotope-selective photobromination of $^{12,13}CH_3F$ using CW CO_2 laser radiation.

The experiment of Brenner *et al.* (1977) illustrates many of the salient features of isotope separation using LIR. A CW dye laser was used to irradiate $^{35,37}ICl$ in a cell containing bromobenzene as a scavenger. It was shown that the following reaction led to enrichment of ^{37}Cl in $C_6H_5\,^{37}Cl$

$$I^{37}Cl^* + C_6H_5Br \rightarrow I + C_6H_5\,^{37}Cl + Br$$

Selective excitation of $I^{37}Cl$ was accomplished by having an intracavity absorption cell with $I^{35}Cl$ in the dye laser to suppress oscillation at the wavelengths of lines of $I^{35}Cl$.

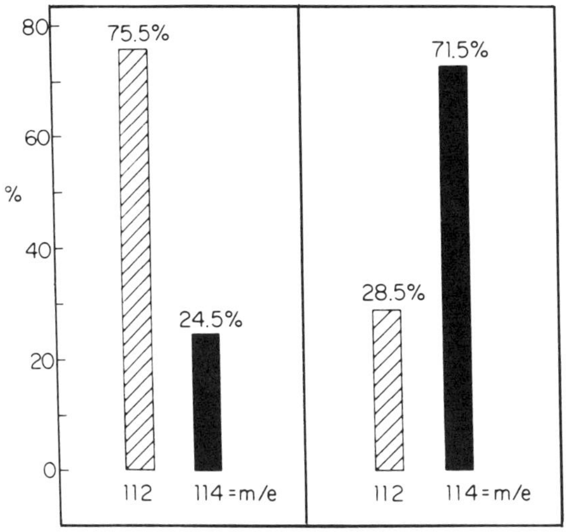

Figure 4.1. Percent abundance of parent ions from $C_6H_{65}{}^{35}Cl$ (112) and $C_6H_5{}^{37}Cl$(114): (a) natural abundance; (b) from reaction of ICl (11.6 μmol) + C_6H_5Br (306 μmol) irradiated at λ6053 Å for 1 h [Brenner *et al.* (1977)].

Separation of ^{37}Cl is effective because the laser excites $I^{37}Cl$ from the X $^1\Sigma^+$ ground state to the A $^3\Pi_1$ state which then exhibits enhanced reactivity toward bromobenzene. Brenner *et al.* show in this reaction that $I^{37}Cl^*$ behaves kinetically like a Cl atom. The reaction leads to an excited radical intermediate $(C_6H_5{}^{37}Cl\ Br)^*$ that rapidly eliminates a Br atom yielding $C_6H_5{}^{37}Cl$.

The quantum yield of $C_6H_5{}^{37}Cl$ is ~ 0.1, while enrichment factors of up to 9 were observed. An example of the enrichment produced is shown in Figure 4.1. Limitations to enrichment are provided by competing reactions that tend to scramble ^{35}Cl and ^{37}Cl. One of these is initiated by the energy transfer

$$I^{37}Cl^* + I^{35}Cl \rightarrow I^{37}Cl + I^{35}Cl^*$$

which then participates in the reaction

$$I^{35}Cl^* + C_6H_5Br \rightarrow C_6H_5{}^{35}Cl + I + Br$$

Another competitive reaction involves ^{35}Cl

$$^{35}Cl + C_6H_5Br \rightarrow C_6H_5{}^{35}Cl + Br$$

where ^{35}Cl can be formed from the exchange reaction

$$^{37}Cl + I^{35}Cl \rightarrow I^{37}Cl + {}^{35}Cl$$

Chlorine atoms can be formed in a variety of ways

$$I^{37}Cl^* + M \rightarrow I + {}^{37}Cl + M$$
$$I^{37}Cl^* + C_6H_5Br \rightarrow C_6H_5Br + I + {}^{37}Cl$$

or

$$\begin{aligned} Br + ICl &\rightarrow IBr + Cl \\ &\rightarrow BrCl + I \end{aligned}$$

with

$$BrCl + h\nu \rightarrow Br + Cl$$

These reactions would all inhibit isotopic enhancement were it not for the fact that the deactivation of $(C_6H_5Br^{37}Cl)^*$ is rapid compared to energy transfer or chlorine reactions.

Stuke and Marinero (1979) examined the selective photoaddition of ICl to acetylene

$$ICl^* + C_2H_2 \rightarrow C_2H_2\ ICl$$

ICl* was obtained by irradiation with a CW dye laser having an intracavity absorption cell containing one isotomer. Enrichment factors of ~ 50 were observed. It was found that an increase in buffer gas pressure reduced β. This was attributed to collision-induced dissociation of ICl* with attendant scrambling of isotopes due to reaction with Cl atoms.

The reaction between vibrationally excited HCl and Br has been shown to be isotopically selective [Leone *et al.* (1975) and Arnoldi *et al.* (1975)]. This reaction is ordinarily endothermic when HCl is in the $v'' = 0$ state. When HCl is vibrationally excited, however, the reaction is catalyzed at a rate that may exceed the rate for $v'' = 0$ molecules by a factor $\sim 10^{12}$ [Moore (1973)]. The reaction

$$H^{35}Cl\ (v'' = 2) + Br \rightarrow HBr + {}^{35}Cl$$

followed by

$$^{35}Cl + Br_2 \rightarrow {}^{35}ClBr + Br$$

has resulted in ^{35}Cl enhancements of ~ 3.

The enrichment of deuterium by CW CO_2 laser excitation of CH_2D_2 has been reported by Hsu and Manuccia (1978). A mixture of CH_2D_2 + Cl was irradiated with a line at 9.2 μm from a CW CO_2 laser. This acted to vibrationally excite CH_2D_2

$$CH_2D_2 + h\nu \rightarrow CH_2D_2^{v}$$

Then

$$\begin{aligned} CH_2D_2^{v} + Cl &\rightarrow HCl + CHD_2 \\ &\rightarrow DCl + CH_2D \end{aligned}$$

The reaction

$$CH_4 + Cl \rightarrow HCl + CH_3$$

occurs at the thermal rate which is substantially less than that for the vibrationally enhanced reaction. Using this technique, a deuterium enhancement $\beta = 1.72$ was obtained.

Molin *et al.* (1978) have studied the effect of vibrational excitation on the reaction of $^{12}CH_3F$ and $^{13}CH_3F$ molecules with Br. Molecules were excited vibrationally by irradiation with lines from a tunable CW CO_2 laser and Br atoms were produced photolytically from Br_2 using a conventional source of ultraviolet radiation. The reactions were

$$\begin{aligned} CH_3F + Br &\rightarrow HBr + CH_2F \\ CH_2F + Br_2 &\rightarrow CH_2FBr + Br \end{aligned}$$

Isotopic enhancements of 2.5 and 1.4 were observed on irradiating $^{13}CH_3F$ and $^{12}CH_3F$, respectively. The CW CO_2 laser power was in the 10–100 W range. Irradiations were performed in a large excess of inert gas. This is reported to reduce the scrambling effect of V–V energy transfer between $^{12}CH_3F$ and $^{13}CH_3F$.

A TEA-type CO_2 laser has been used to induce reactions between Cl_2CF_2 and Br_2CF with isobutylene, propylene, and ethylene. The radiation chemistry of these systems is complex although some enrichment of ^{13}C was observed in the reaction products.

Similar complexities were observed in the selective excitation of Cl_2CS in mixtures with diethoxyethylene [Lamotte *et al.* (1975)]. This excitation resulted in the separation of ^{35}Cl or ^{37}Cl with β = 1.2 and 1.7, respectively.

Irradiation of mixtures of $U^{16}O_2F_2$ and $U^{18}O_2F_2$ in a methanol solution with added HF with pulses from a flashlamp pumped dye laser has been shown

[Rofer-Depoorter and Depoorter (1979)] to result in the separation of ^{16}O and ^{18}O. The reaction

$$UO_2F_2 + 2HF + CH_3OH + h\nu \rightarrow UF_4 + \text{Organic products}$$

was shown to occur more readily for the ^{18}O isotomer than with the ^{16}O isotomer when irradiation occurred at 455 or 588 nm.

4.5. SINGLE-PHOTON PREDISSOCIATION

The mechanism of the single-photon predissociation is shown schematically in Figure 4.2. Absorption leads to a transition to a bound state that predissociates via overlap with a repulsive state. Alternatively, the dissociation may be collision induced as shown in part (b) of the figure. Here a transition is made to a bound rovibronic level near the dissociation limit. Collision with another species then dissociates the molecule.

The advantage of both methods derives from the fact that only a single photon is required. However, this is often offset by one or more of the following disadvantages:

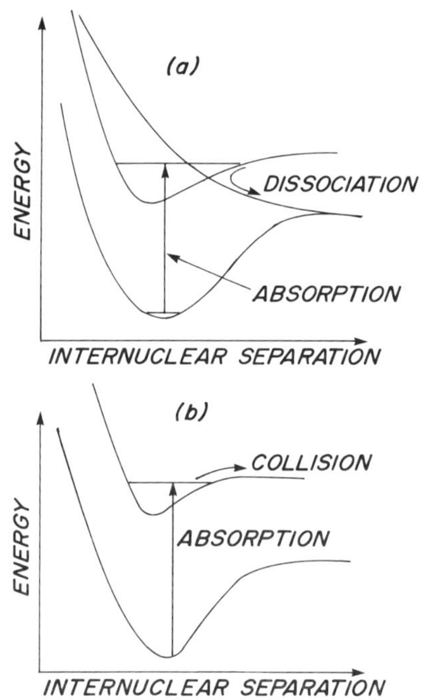

Figure 4.2. (a) Single-photon predissociation; (b) collisional dissociation following absorption.

1. The rate of predissociation must be small enough that it does not lead to substantial broadening of the final state. If broadening is substantial isotopic selectivity may be reduced.
2. The rate of predissociation must be large enough to allow efficient conversion to products in competition with other reaction routes.
3. The products of dissociation must be readily removed from the system or must be stable molecules that do not participate in potential isotopic scrambling reactions. This is particularly true when dissociation yields atoms or radicals which may react with undissociated molecules.

Isotope separation by single-photon predissociation (SPP) was observed in Br_2 by Leone and Moore (1974). The ground state in Br_2 is $X^1\ \Sigma_{0+g}$ deriving from two $Br(^3P_{3/2})$ atoms. A predissociating excited state, $^3\Pi_{0+u}$ derives from $Br(^2P_{1/2}) + Br(^2P_{3/2})$ and lies $\sim$ 16,000 cm^{-1} above the minimum of the X state. The $^3\Pi_{0+u}$ state dissociates via crossing with the dissociative $^1\Pi_{1u}$ $(^2P_{3/2} + {}^2P_{3/2})$ state. Therefore, energy levels are almost identical to those shown in Figure 4.2(a). Photolytically produced Br atoms are scavenged by reaction with HI. The relevant reactions are

$$^{n}Br_2 + h\nu \rightarrow {}^{i}Br_2^* \tag{4.1}$$
$$^{i}Br_2^* \rightarrow {}^{i}Br + {}^{i}Br \tag{4.2}$$
$$^{i}Br + HI \rightarrow H^{i}Br + I \tag{4.3}$$

where $^{n}Br_2$ is the normal mixture of ^{79}Br and ^{81}Br isotomers, while $^{i}Br_2$ is the selected isotomer. Competitive reactions are

$$^{i}Br_2^* \rightarrow {}^{i}Br_2 + h\nu \tag{4.4}$$
$$^{i}Br_2^* + M \rightarrow {}^{i}Br + {}^{i}Br + M \tag{4.5}$$
$$\rightarrow {}^{i}Br_2 + M \tag{4.6}$$
$$^{i}Br_2^* + {}^{n}Br_2 \rightarrow {}^{i}Br_2 + {}^{n}Br_2^* \tag{4.7}$$

Furthermore, one can have

$$^{i}Br + {}^{n}Br_2 \rightarrow {}^{i}Br\,{}^{n}Br + {}^{n}Br \tag{4.8}$$

Rates are

$$k_2 \sim 10^6\text{–}10^7\ s^{-1}$$
$$k_3 \sim 10^{11}\ cm^3\ s^{-1}$$
$$k_4 \sim 5 \times 10^4\ s^{-1}$$
$$\sigma_5 \sim 10^{-14}\ cm^2\ s^{-1}$$

$$\sigma_6 < 10^{-14}\ \text{cm}^2$$
$$k_8 = 4 \times 10^{-11}\ \text{cm}^3\ \text{s}^{-1}$$

Reactions (4.4)–(4.6) lower the efficiency with which separation is produced while reactions (4.7) and (4.8) lead to scrambling. Overall enhancements of ^{81}Br of ~ 5 were produced using this method.

Radiation from an ArF laser at 193.5 nm has been used to excite O_2 molecules to predissociating levels [Sander *et al.* (1977)]. The output of the ArF laser that extends over a range of ~ 1 nm was filtered by operation in air so that lines from the $^{16}O_2$ isotomer would not be excited in the sample cell. The remaining radiation induced transitions in the (4–0) band of the $X^3\Sigma_g^- \rightarrow B\ ^3\Sigma_u^-$ Schumann–Runge system only in ^{17}O and ^{18}O bearing molecules. The final states predissociate to yield O atoms. These then add to the O_2 to produce isotopically enriched O_3.

$$^i\text{O} + {}^{16}\text{O}_2 + \text{M} \rightarrow {}^i\text{O}_3 + \text{M}$$

This must compete with the two-body reaction

$$^i\text{O} + {}^{16}\text{O}_2 \rightarrow {}^{16}\text{O}^i\text{O} + {}^{16}\text{O}$$

Another mechanism which leads to scrambling involves the production of O atoms from the nonselective process

$$\text{O}_2(X\ ^3\Sigma_g^-) + h\nu \rightarrow \text{O}_2(A\ ^3\Sigma_u^+) \rightarrow \text{O} + \text{O}$$

which occurs over the same wavelength range. Under the conditions of the experiment, the rate for the three-body reaction was ~ 0.25 that of the competitive two-body scrambling reaction. Some enhancement of ^{17}O and ^{18}O in the O_3 product was nevertheless observed.

Formaldehyde, H_2CO, is known to predissociate in its first excited singlet state at energies in excess of 28,188 cm^{-1}. This has led to much interest in isotope separation from H_2CO, HDCO, D_2CO, and $^{13}CH_2O$. The photodissociation of H_2CO results in the formation of either H_2 + CO or H + HCO depending on the energy of the predissociating state. Some dissociative states are shown schematically in Figure 4.3.

The first exploitation of this effect to yield isotope separation was reported by Yeung and Moore (1972). In this study a frequency-doubled ruby laser was used to excite a 1:1 mixture of H_2CO and D_2CO at 3 torr in a gas cell. After ~ 25, 15-ns pulses a deuterium enrichment of ~ 6 was obtained in the products. A similar result was obtained with frequency-tripled Nd–YAG radiation by Ambartzumyan *et al.* (1975a).

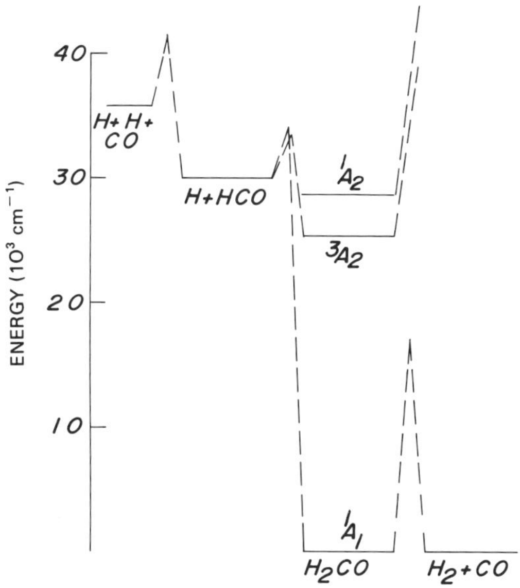

Figure 4.3. Energy levels in H_2CO and separated states. Excitation of H_2CO to the $^1A_2$ state can result in the generation of H + HCO or H_2 + CO due to predissociation.

Marling (1975) found that a coincidence between the 325.0-nm line of the CW He–Cd laser and an absorption feature of HDCO could be used to produce separation of D from naturally occurring H_2CO. Irradiation of a sample of H_2CO with a normal D:H ratio for 4 h resulted in a single-step deuterium enrichment factor $\beta \gtrsim 15$. Subsequent experiments with a Xe-ion laser yielded $\beta = 58$ [Marling (1977)].

These early studies were limited by the lack of a tunable laser source in the 300–350 nm region. Mannik *et al.* (1979) used a frequency-doubled N_2 pumped dye laser to selectively excite HDCO at 345.608 nm in a multipass cell. The resulting enrichment factor for D was 254 $\pm$ 30. This was obtained from samples of natural formaldehyde at pressures of $\sim$ 4 torr.

A portion of the spectrum of $H_2{}^{12}C^{18}O$ and $H_2{}^{17}C^{16}O$ is shown in Figure 4.4 [Marling (1976)]. A coincidence with the 332.375-nm line of the Ne^+ laser is also shown. Marling found that this coincidence could be used to selectively dissociate the $H_2{}^{12}C^{18}O$ isotomer. An ^{18}O enhancement of 8 was obtained.

Doubling of the output of a flashlamp pumped dye laser was used by Clark *et al.* (1975) to excite a mixure of $^{12}CH_2O$ and $^{13}CH_2O$ with NO. The role of NO was uncertain but it may have acted as a scavenger for radical products. Irradiation at an absorption line of $^{12}CH_2O$ was found to produce a ^{12}C enrichment of $\sim$80 in the photolytically produced CO.

SPP has also been used to separate C and N isotopic species from gas-phase *s*-tetrazine ($C_2H_4N_4$) [Karl and Innes (1975) and Boesl *et al.* (1979)] and *s*-tetrazine in various matrices at low temperatures [Hochstrasser and

King (1975, 1976) and Dellinger *et al.* (1977)]. Karl and Innes found that enrichments of ^{13}C and ^{15}N were obtained in the irradiation of *s*-tetrazine with a 10-mW CW tunable laser. The maximum enrichment factor was 72 and a mass separation of 17 moles per kWh of laser energy was obtained. Boesl *et al.* (1979), using an excitation wavelength optimized to produce maximum dissociation, found that enrichment factors for ^{15}N and ^{13}C could be increased to $\gtrsim 10^3$.

In the matrix isolation technique a sample of the desired species is condensed together with an excess of inert gas on a substrate cooled to a cryogenic temperature. Dellinger *et al.* (1977) deposited a mixture of *s*-tetrazine with Ar at 4.2 K. At a dilution of 1000, the origin of the *s*-tetrazine $^1A_g \rightarrow {}^1B_{3u}$ transition shows clearly resolved features due to ^{15}N and ^{13}C isotomers. By irradiation with pulses from an N_2 pumped dye laser, absorption features due to individual isotomers could be removed leaving the remaining *s*-tetrazine enriched

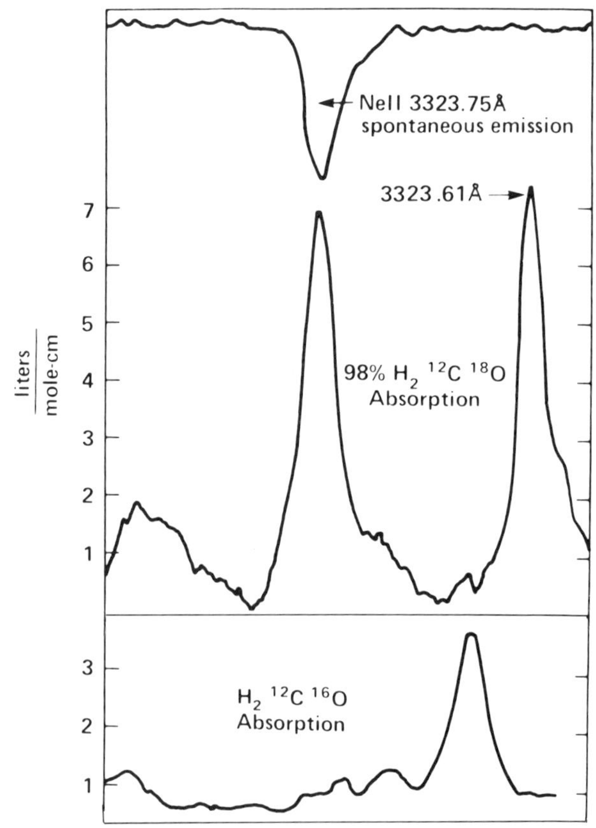

Figure 4.4. Comparison of Ne II 3323.75 Å spontaneous emission with absorption spectra of normal and 0–18 enriched formaldehyde [Marling (1976)].

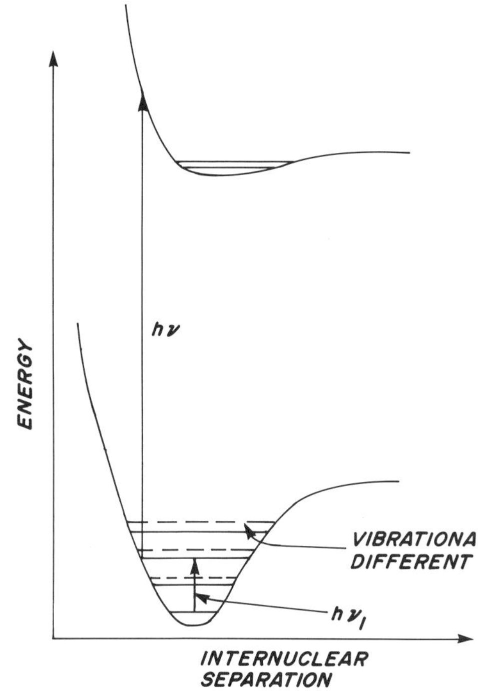

Figure 4.5. Ground and excited states involved in IR + UV TPD. The IR photon causes a $v = 0 \rightarrow v = 2$ transition in the desired isotomer. The UV photon leads to dissociation.

in the other isotopes. The dissociation products HCN and N_2 are stable against recombination at the matrix temperature.

Separation of ^{235}U and ^{238}U from the gaseous uranium complex $[UO_2(hfacac)_2]_2$† under irradiation with lines from a CW CO_2 laser has been demonstrated by Cox and Maas (1980). Enrichment factors were typically 1.2–1.5.

4.6. TWO-PHOTON DISSOCIATION

In two-photon dissociation (TPD) the first laser photon serves to label the isotomer of interest while the second photon produces a transition that leads to dissociation. Both photons may have comparable energy or one photon may be of high energy while the other has low energy (IR–UV photodissociation). The energy levels involved in TPD are shown schematically in Figure 4.5.

†hfacac = hexafluoroacetylacetonate.

Letokhov and Moore (1977) have identified the following three limitations to the selectivity of the TPD for isotope separation.

1. *Thermal excitation of vibrational levels.* This results in the nonselective excitation of all isotomers. Since molecules in these excited vibrational levels can often be photodissociated with the same wavelength that leads to selective dissociation, the population of these levels reduces selectivity. This effect can be particularly noticeable when vibrational–translational (V–T) energy transfer dominates so that vibrational excitation rapidly degrades to gas heating. This effect can be minimized if the pulse at ν_2 is of shorter duration than the V–T transfer time.

2. *Diffuseness of photoabsorption band edge.* Since the second photon results in a transition to a continuum state there is no well defined energy threshold for photodissociation. This implies that similar dissociating transitions may be possible from unexcited molecular states (Figure 4.6). The effect of overlapping dissociation continua can be minimized by tuning the second laser to the extreme red wing of the $\sigma(A^*)$ vs. ν curve. However, this is accompanied by reduced excitation efficiency. Overlap can be reduced by initially exciting the molecule to a higher vibrational state.

3. *Rotational relaxation bottleneck.* While the transition at ν_1 is shown in Figure 4.5 as occurring between two discrete vibrational states, this actually is a rovibrational transition. Since ν_1 corresponds to the frequency difference between two rotation–vibration levels, the intensity of this transition depends on the population difference between two rotational energy levels: one in the excited vibrational level and one in the ground vibrational level. Because of inefficient energy transfer between individual rotational energy levels, saturation of this transition can occur at relatively low incident laser intensity. This means that only some small fraction, q, of available molecules participates in the excitation. This fraction may be as small as 0.01 but usually lies in the range $q \sim 0.1$–0.01. q increases at high pressures due to the overlapping of

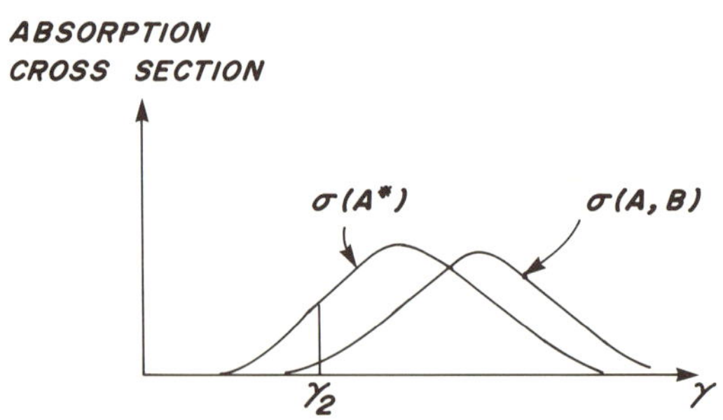

Figure 4.6. Photodissociation cross sections for ground-state molecules $\sigma(A,B)$ and vibrationally excited isotomer $\sigma(A^*)$. ν_2 is the frequency of the second laser photon. For good selectivity $\sigma(A^*)\nu_2 \gg \sigma(A,B)\nu_2$.

rotational lines as the result of pressure broadening. Attempts to increase q via collision often also result in gas heating since the time scale for attainment of rotational equilibrium is often comparable to that for V–T transfer.

The first experiments on isotope separation via TPD were reported by Ambartzumyan *et al.* (1972). Pulses from a line tunable TEA-type CO_2 laser were used to vibrationally excite $^{15}NH_3$ molecules. These were then photodissociated with UV light from a spark source synchronized in time with the TEA laser pulse. A $^{14}NH_3$ filter was used to remove radiation at the $^{14}NH_3$ wavelengths. The result of photodissociation was to produce isotopically enriched N_2. The overall reaction is

$$4\ ^{15}NH_3 \rightarrow 2\ ^{15}NH_3 + 3\ H_2 + {}^{15}N_2$$

An enrichment factor for ^{15}N of $\beta \sim 2.5$–6 was obtained.

Ambartzumyan and Letokhov (1972) studied the TPD of HCl without an attempt to separate isotopes. An elegant excitation system involving a tunable Q-switched Nd–glass laser, frequency shifted via stimulated Raman scattering and frequency quadrupled to 265.0 nm, was used to excite a mixture of HCl and NO. The TPD-induced reaction was

$$\begin{aligned} &HCl + h\nu_1 \rightarrow HCl(v = 3) \\ &HCl(v = 3) + h\nu_2 \rightarrow H + Cl \\ &Cl + NO \rightarrow NOCl \end{aligned}$$

Rockwood and Rabideau (1974) irradiated $BCl_3 + O_2$ in a quartz tube with radiation from a pulsed CO_2 laser synchronized with UV light from a Xe flashlamp. A ^{10}B enrichment of 1.1 was obtained.

Two infrared lasers have been used to dissociate gaseous OsO_4 yielding β $(^{192}Os/^{187}Os) = 1.15$. Both TEA-type CO_2 lasers had pulse energies in the 1–3 J range with pulse durations of 90–250 ns. Optimum results were obtained when ν_1 was tuned to the CO_2 $P(6)$ line while ν_2 was tuned to the CO_2 $P(20)$ line. Scavengers used were OCS, C_2H_4, and NO.

4.7. PHOTOISOMERIZATION

The use of both single-photon and multiphoton-induced photoisomerization has been reported in the literature. When a given quantity of vibrational energy is given to a molecule, the molecule may oscillate between various isomeric structures in the absence of quenching collisions. Collisional quenching of part of this internal energy can result in deactivation to a product of the original molecule that has the same chemical composition but different struc-

Table 4.1. Isotope Selective Photoisomerization of CH_3NC^a

Vibration excited	Wavelength (nm)	Pressure (torr)	Percent conversion to CH_3CN	β^b
6ν(CH)	621.4	35	25	0.81
6ν(CH)	626.0	15	25	1.35
5ν(CH)	730.0	1.3	30	1.31
5ν(CH)	739.2	8.0	40	1.48

^{a}After Reddy and Berry (1980); $^b\beta$ is the enrichment factor for ^{13}C.

ture. Since the chemical properties of this isomer can be quite different from those of the original molecule it may be separated from the final mixture. When the original excitation has been isotope selective, then the isomer will contain a higher proportion of the desired isotopic species.

Glatt and Yogev (1976) have reported the photoisomerization of D-labeled, 1,5 hexadiene using a line tunable TEA-type CO_2 laser. An enhancement factor of $\sim$ 3.3 was obtained.

Reddy and Berry (1980) observed an isotope-sensitive photoisomerization in gaseous methyl isocyanide, CH_3NC, using intracavity absorption to excite a vibrational overtone in the visible region of the spectrum. The vibrations excited corresponded to 5ν(CH) and 5ν(CH) combined with various other fundamentals, where ν(CH) is the frequency of CH fundamental stretching vibration. The product of isomerization was CH_3CN (methyl cyanide). A summary of enrichment efficiences is given in Table 4.1.

Methyl and ethyl isocyanide have been shown to undergo isomerization to the corresponding nitriles when irradiated with pulses from a focused TEA CO_2 laser [Hartford and Tuccio (1979)]. While thermal isomerization was present at pressures in excess of 2 torr, no evidence of this effect was observed at low pressures. It was proposed that under these conditions isomerization occurred when the molecules absorbed sufficient infrared energy to exceed the activation energy for this process. The enhancement factors for ^{15}N separation were 1.2 $\pm$ 0.03 for the isomerization of CH_3NC and 1.16 $\pm$ 0.03 for C_2H_5NC. These enhancements were obtained without any evidence for laser-induced dissociation or isotope scrambling collisional effects. As such, the process appears to have many attractive features with few disadvantages.

Another study of multiphoton-induced isomerization has been reported by Yogev and Benmair (1979). ^{13}C separation was obtained by the laser-induced isomerization of hexafluorocyclobutene to hexafluorobutadiene using pulses from a CO_2 laser.

In general, laser-induced isomerization offers a potentially attractive method of isotope separation involving, in principle, only one excitation step and fredom from competing reactions. However, the applicability of this technique is limited in scope by the availability of suitable candidate molecules.

4.8. TWO-STEP PHOTOIONIZATION

The atomic analogue of selective two-photon dissociation is provided by two-step photoionization. This process has been widely used in laser isotope separation in atomic vapors. Several possible selective two-step photoionization schemes are shown in Figure 4.7. All processes are isotope selective in the first transition where the radiation is tuned to excite a single isotope.

Figure 4.7(a) shows the transition involved in resonant two-step photoionization. ν_1 selects the isotope of interest and ν_2 produces a transition from the selected state to the ionization continuum. Note that one can often have $\nu_1 = \nu_2$ so that the net process is a resonant two-photon ionization. Application of this straightforward technique to particular atomic systems is limited by the need to have $E(\mathrm{IP}) \lesssim (h\nu_1 + h\nu_2)$.

Resonant three-photon ionization [Figure 4.7(b)] can be used for atoms with larger ionization potentials. The cross section for three-photon ionization scales as $\sigma_I \propto I(\nu_1)I(\nu_2)I(\nu_3)$.

When the first photon leads to population of an atomic level within $\simeq kT$ of the ionization energy [Figure 4.7(c)], then ionization can be produced by collision. Alternatively, the transition to a continuum state can be induced by infrared or microwave photons.

Efficient ionization can be produced via autoionizing states [Figure 4.7(d)]. The autoionizing level is usually connected to lower energy levels via strongly allowed transitions, permitting efficient excitation and producing a large increase in photoionization cross section.

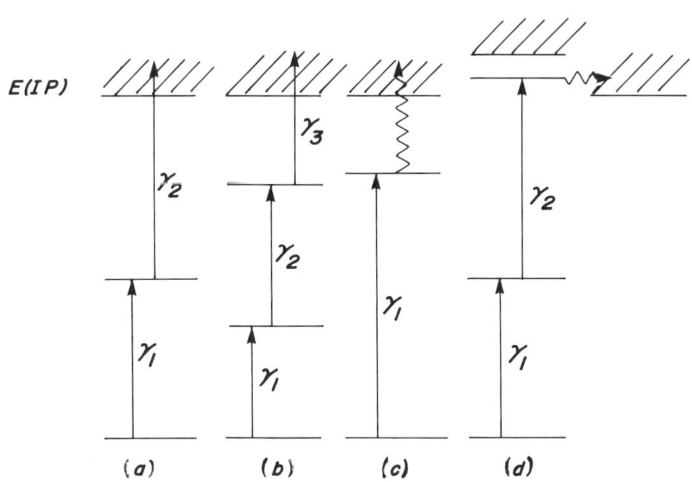

Figure 4.7. Selective photoionization of an atomic species: (a) two-step photoionization; (b) three-step photoionization; (c) collision-induced photoionization; (d) ionization by autoionization.

If we consider a simple two-step photoionization [Figure 4.7(a)] then an estimate of laser powers required for efficient isotope separation can be easily obtained. Saturation at ν_1 requires a laser intensity

$$I_s(\nu_1) = \frac{h\nu_1}{2\sigma_1\tau}$$

where σ_1 is the cross section for excitation and τ is the lifetime of the excited state. Typical values for these parameters would be $h\nu_1 = 5 \times 10^{-19}$ J, $\sigma_1 = 10^{-12}$ cm^2, and $\tau = \tau(\text{radiative}) = 10^{-8}$ s. This yields $I_s(\nu_1) \sim 25$ W cm^{-2}. Saturation at ν_2 requires

$$\begin{aligned} I_s(\nu_2) &= \frac{h\nu_2}{\sigma_2\tau} \\ &= \frac{2\nu_2}{\nu_1}\frac{\sigma_1}{\sigma_2} I_s(\nu_1) \end{aligned}$$

Taking $\nu_2 \sim \nu_1$ it is evident that $I_1(\nu_2) \sim (\sigma 1/\sigma_2)I_s(\nu_1)$. Since σ_2 is typically $\sim 10^{-18}$ cm^2, one has $I_s(\nu_2) \sim 10^6\, I_s(\nu_1)$. Thus requirements on the powers of the two laser sources at ν_1 and ν_2 are quite different; in the present example $I_s(\nu_1) = 25$ W cm^{-2} while $I_s(\nu_2) \sim 5 \times 10^7$ W cm^{-2}.

Isotope separation via two-step photoionization is usually carried out in a beam configuration. Such a system is shown in Figure 4.8. A beam of neutral atoms produced by thermal evaporation is intersected by the two laser beams at the point shown. This produces selective excitation and photoionization. The ions generated are then extracted with a strong electric field (several kV cm^{-1}). Since the region of interaction between the laser beams and the atomic beam is of limited spatial extent, much of the incident laser radiation may not be

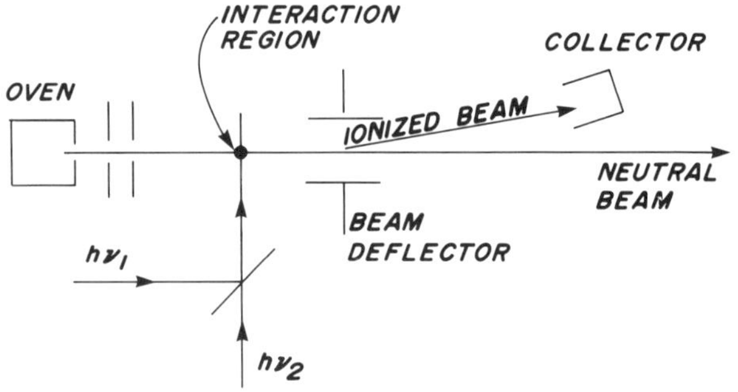

Figure 4.8. Schematic diagram of system for isotope separation via two-step photoionization.

Table 4.2. Data on Isotope Separation via Two-Step Photoionization

Isotopic species	Laser	Wavelength (nm)	β	Reference
^6Li	CW Ar^+	Various	180	Rothe *et al.* (1978)
^7Li	CW dye	Intracavity		Yamashita *et al.* (1978)
^{40}Ca	CW Ar^+/dye	616.2		Brinkmann *et al.* (1974)
	+ CW Ar^+	488.8		
^{85}Rb	Ruby/dye	794.8		Ambartzumyan and Letokhov (1972), and Ambartzumyan *et al.* (1976)
	Double dye	347.1		
Nd	CW Ar^+/dye N_2	588.79	50–96	Karlov *et al.* (1977, 1978)
Sm	CW Ar^+/dye N_2	531.636	51–86	Karlov *et al.* (1977, 1978)
Gd	CW Ar^+/dye N_2	585.622	11.5–143	Karlov *et al.* (1977, 1978)
Dy	CW Ar^+/ dye N_2	538.856	17–726	Karlov *et al.* (1977, 1978)
Er	CW Ar^+/dye	582.679	46–100	Karlov *et al.* (1977, 1978)
^{235}U	N_2/dye	426.6	1.5	James *et al.* (1976)
		360.9		
^{235}U	CW Xe^+	378.1	4	Böhm *et al.* (1978)
	CW Kr^+	356.4		
		350.7		
^{235}U	Basic spectroscopic data			Böhm *et al.* (1978) and Radziemski *et al.* (1975)

utilized. In fact, under conditions in which the ν_1 transition is saturated, the fraction of the incident beam absorbed is $\sigma_1 NL/2$ for the beam at ν_1 and $\sigma_2 NL/2$ for the beam at ν_2. In this expression N is the number of atoms per cm^3 and L is the transverse atomic beam dimension. Thus, even when the atomic beam is optically thick at ν_1 ($\sigma_1 NL \simeq 1$) only a small fraction of the power in the beam at ν_2 will be absorbed. This provides a severe limit to the efficiency of the overall separation process. This limitation can be eliminated in part by the use of collisional ionization [Figure 4.7(c)] or by excitation to autoionizing states [Figure 4.7(d)]. Alternatively, a microwave or IR radiation field can be used to produce ionization.

A summary of data on laser isotope separation using two-step photoionization is given in Table 4.2. These experiments show that enhancements of 10^2–10^3 can be readily obtained. All experiments, with the exception of that on ^6Li, involve photoionization of atomic states. Rothe *et al.* (1978) found that significant isotropic enrichment of ^6Li could be obtained from the two-photon ionization of Li_2.

4.9. PHOTODEFLECTION

Isotope separation by radiation pressure was first suggested by Ashkin (1970). When a light beam is tuned to a resonance line of an atom in an atomic beam, each absorbed photon transfers momentum h/λ to the absorber. For $\lambda = 600$ nm, this momentum is approximately 10^{-27} kg m s^{-1}. With an atomic mass of $\sim 10^{-25}$ kg, this corresponds to a perpendicular velocity of 1 cm s^{-1} per absorption.

A thermal velocity for a heavy atom effusing from an oven at 1000 K is typically 500 m s^{-1}. The angular divergence of an atomic beam with $v_{\|}$ in the direction of beam travel and $v_{\perp}$ perpendicular to this direction is $\theta = v_{\perp}/v_{\|}$. For $\theta = 10^{-3}$ rad; $v_{\perp} = 10^{-3} \times 500 = 0.5$ m s^{-1}. Thus, if one assumes perfect collimation in the initial beam a divergence of 1 mrad after irradiation would be obtained if each atom participated in ~ 50 transitions while in the laser beam.

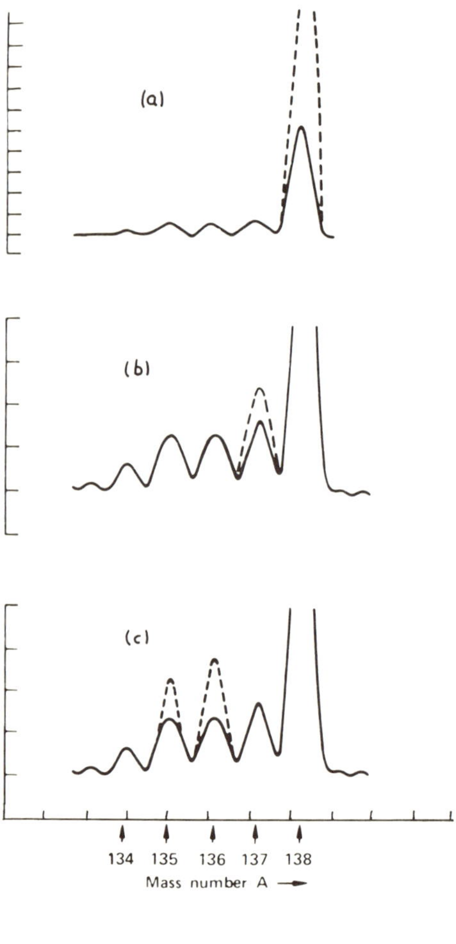

Figure 4.9. Superimposed mass spectrometer traces: dashed line indicates change in spectrometer output where laser was tuned to the center of a given isotope's absorption line. (a) Laser tuned to ^{138}Ba line; (b) laser tuned to ^{137}Ba line; and (c) laser tuned to overlapping ^{135}Ba and ^{136}Ba resonance lines [Bernhardt *et al.* (1974)].

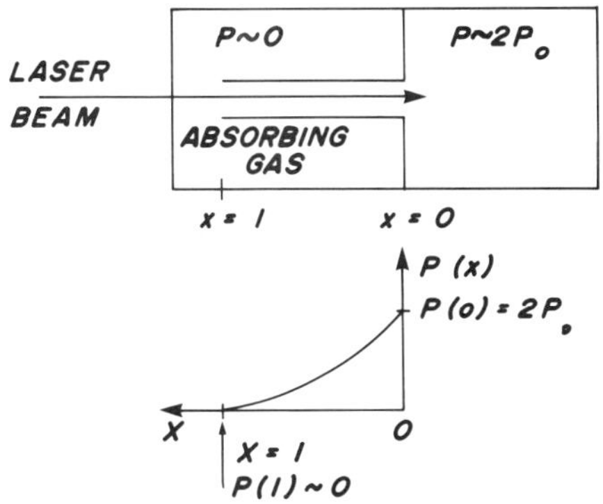

Figure 4.10. Pressure variation in absorbing gas due to radiation pressure. If the laser is tuned to excite only one isotope of a mixture, only the isotopic component is separated.

If we assume that the laser beam irradiates a 1-cm length of the atomic beam, then each atom spends 0.2×10^{-4} s in the laser beam. During this period it must make 50 transitions each involving $\sim$ 3 eV or 5×10^{-19} J of energy. Thus the energy required to separate each atom is 250×10^{-19} J or 1.5×10^7 J mol^{-1}. This rather large amount of energy is exclusive of losses involved in generating the requisite laser power. Hence photodeflection is inherently a rather inefficient means of generating appreciable quantities of specific isotopes.

Bernhardt *et al.* (1974) have studied the separation of various Ba isotopes in an atomic beam using photodeflection with the beam from a narrow line-width CW dye laser. The laser was tuned to the $6s^2\ ^1S_0 \rightarrow 6s6p\ ^1P_1^0$ resonance line of individual Ba isotopes and emitted $\sim$ 50 mW. Figure 4.9 shows results obtained when the Ba beam was sampled through a slit by a quadrupole mass spectrometer downstream of the point where irradiation occurred. Enhancement of mass peaks due to individual isotopes of Ba are evident. It was found, however, that the overall efficiency was limited by the trapping of population in the metastable $6s5d\ ^1D_2$ state. Atoms trapped in this state were unable to participate entirely in the full number of radiation driven absorption–emission cycles. The significance of this effect was subsequently reduced by the use of a second CW dye laser to excite atoms out of the $6s5d\ ^1D_2$ state [Bernhardt *et al.* (1976a)]. With this modification the overall efficiency of photodeflection of individual Ba isotopes could be considerably enhanced. The use of laser photodeflection in the measurement of basic spectroscopic data has been described by Bernhardt *et al.* (1976b, c, d).

Separation of isotopes by light pressure in a static as opposed to a beam system was also suggested by Ashkin (1970). In a system where the optical depth is small, the force exerted on an atom with absorption cross-section σ is $F = I_0\,\sigma/c$, where I_0 is the light intensity. The pressure variation with distance, x, in a closed cell (Figure 4.10) will then be

$$p(x) = p_0 \exp(-Fx/kT)$$

Ashkin calculates that $p(20 \text{ cm}) = 3.4 \times 10^{-9}$ torr, where Na vapor at a pressure $p_0 = 10^{-3}$ torr is exposed to a laser beam with intensity $I_0 = 7.6 \times 10^4$ W cm^{-2}.

Gelbwachs and Hartwick (1975) calculated the efficiency of the separation of molecular species with light pressure at various wavelengths. It was found that the theoretical separation achieved with IR or microwave radiation was several orders of magnitude smaller than that obtained with visible or UV radiation. Experimental studies have subsequently demonstrated the existence of pressure variation in SF_6 gas excited at 10.6 μm [Rinehart *et al.* (1976)] and in Na vapor excited in the visible [Bjorkholm *et al.* (1975)].

4.10. MULTIPHOTON DISSOCIATION

When a polyatomic molecule is exposed to an intense IR radiation field under collisionless conditions, the molecule stores absorbed energy in a variety of internal modes. As excitation continues, this energy may accumulate to the point where bond breaking can occur. This is often accompanied by luminescence at visible and UV wavelengths. Isenor and Richardson (1971) were the first to report the existence of this fluorescence.

Multiphoton dissociation occurs at field strengths that are significantly lower than those that produce gas breakdown. The luminescence observed under these conditions is several orders of magnitude weaker than light produced during gas breakdown. The time dependence of this luminescence has two components as shown in Figure 4.11. The prompt component coincides with the exciting laser pulse at both high and low gas pressure while the broad component is resolved from the prompt conponent only at low gas pressure. The origin of the prompt component is attributed to multiphoton dissociation (MPD) under collisionless conditions, while the delayed component arises from vibrational energy (V–V) transfer. The time delay τ of the second component is found to satisfy the following relation

$$p\tau = \text{constant}$$

where p is the gas pressure [Letokhov and Moore (1977)].

The intensity threshold for the production of visible luminescence is typically 10^8 W cm^{-2} or $\sim 10^{-2}$ of that required to induce optical breakdown. In this intensity regime excitation of a specific isotomer results in the highly selective dissociation of this isotomer. In fact, such selectivity occurs even if the exciting radiation is not closely resonant in wavelength with a line of the particular isotomer dissociated.

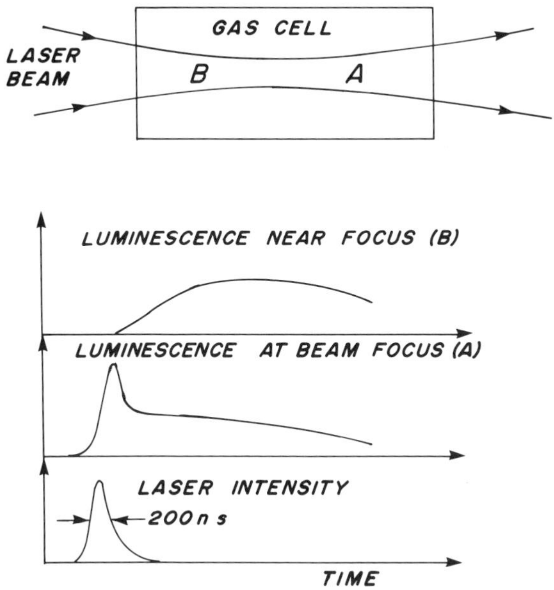

Figure 4.11. Time dependence of luminescence from near-focus (B) and at focus (A) from polyatomic gas excited with laser pulse as shown.

Under these conditions, the degree of dissociation increases with the number of laser pulses absorbed by the sample. β depends exponentially on the total number of pulses and increases with a decrease in gas pressure. The overall effect of this absorption of laser light by a particular isotomer is to store energy selectively in molecules of this type leading to unimolecular decomposition when a certain stored energy threshold is reached.

The mechanism of MPD has been elucidated by the elegant experiments of Ambartzumyan *et al.* (1976 a,b,c). Initial absorption of photons from the IR radiation field occurs via rovibrational transitions until $\sim$ 0.5 eV of energy is stored in the molecule. During this stage, anharmonicity is compensated by shifts from the $P \rightarrow Q \rightarrow R$ rotational branches with increasing vibrational excitation (Figure 4.12). When a high vibrational state is reached (v = 3 or 4), overlapping of vibrational states due to different modes provides a quasicontinuum of states. This leads to rapid redistribution of energy stored in the selected vibrational mode with a subsequent reduction in constraints on further photon absorption.

Figure 4.13 shows how the density of molecular vibrational states increases with increasing excitation energy in several polyatomic molecules. When ρ_{vib} exceeds 1–10 per cm^{-1} one has a quasicontinuum of vibrational

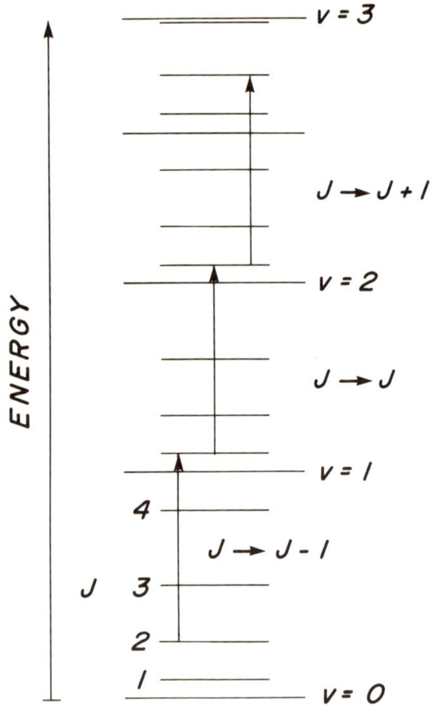

Figure 4.12. Example of sequential IR photon absorption in a series of rovibrational transitions maintaining energy coincidence in the presence of anharmonicity of the vibrational levels.

states. For SF_6 this point is reached when the excitation energy exceeds ~ 3000 cm^{-1}.

Figure 4.14 shows the frequency dependence of the multiphoton dissociation rate in $^{32}SF_6$ [Letokhov and Moore (1977)]. It shows that the peak dissociation rate obtained with single wavelength excitation occurs at a frequency that is shifted by ~ 10 cm^{-1} from the peak of the IR absorption cross section (curve 4). This shift corresponds to the anharmonicity of the selected vibrational mode. Curves 2 and 3 show the dissociation produced by irradiation at two frequencies, ν_1 = 940 cm^{-1} and ν_2 = 1084 cm^{-1}. These irradiations were performed with $I(\nu_1)$ = 4 MW cm^{-2} and $I(\nu_2)$ = 58 cm^{-2} and show the effect of initial excitation within a manifold of discreet states followed by a transition to nonselective excitation once the vibrational quasi continuum has been reached. The lower power beam at ν_1 is used to increase population in the first few excited vibrational levels, i.e., to raise average molecular excitation energies to ~ 0.5 eV. This step yields the initial selectivity that can be utilized for isotope separation.

A large number of studies have been carried out on isotope separation via MPD. We summarize the data obtained in the following section.

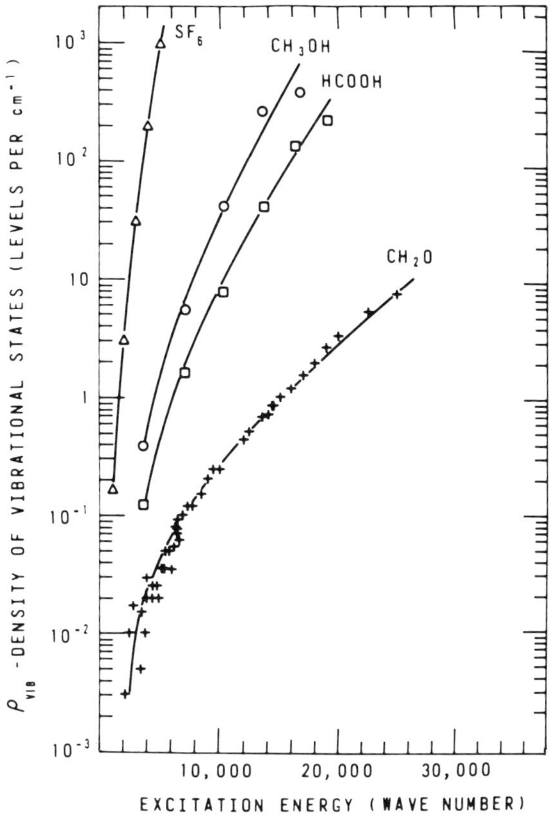

Figure 4.13. The density of states (ρ_{vib}) vs. excitation energy for SF_6, CH_3OH, HCOOH, and H_2CO [Evans *et al.* (1979)].

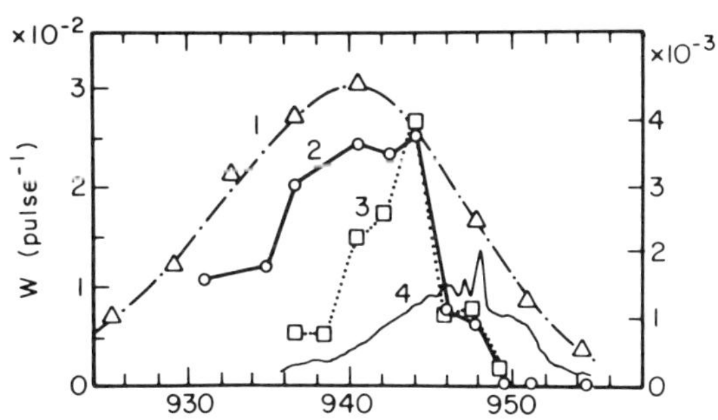

Figure 4.14. Resonance characteristics of the rate of SF_6 dissociation in an IR laser field: curve 1, single-frequency irradiation $P_{\text{avg}} = 31$ MW cm^{-2}, $T = 300$ K; curve 2, two-frequency irradiation $I(\nu_1) = 4$ MW cm^{-2}, $P_{\text{avg}}(\nu_2) = 58$ MW cm^{-2}, $\nu_2 = 1084$ cm^{-1}, $T = 300$ K; curve 3, two-frequency irradiation with the same laser parameters but at $T = 193$ K; curve 4, linear IR absorption of SF_6 at 300 K. The ordinate, curve 1 left and curves 2 and 3 right, gives fractional decomposition per pulse averaged over the entire sample cell [Letokhov and Moore (1977)].

4.10.1. Deuterium

One of the first separations of D and H using MPD was reported by Yogev and Benmair (1975). A mixture of CD_2Cl_2 and CH_2Cl_2 was selectively decomposed by excitation with 100-ns pulses from a TEA CO_2 laser at an energy of 943 cm^{-1}. An appreciable effect was only observed when the radiation was strongly focused in the cell. β was found to be 1.2.

Separation of D from formaldehyde has been reported by Letokhov *et al.* (1972) and by Koren *et al.* (1976). A list of some wavelength coincidences between lines of $^{12}C^{16}O_2$ and H_2CO, HDCO and D_2CO is given in Table 4.3. It is evident that lines of the 9.4-μm band are well suited for excitation of HDCO and D_2CO. Koren *et al.* used the *P*(20) line of this band to irradiate formaldehyde gas with 4-J, 100-ns pulses. The enrichment factor was ~ 40 after irradiation with 300 laser pulses. A similar result was obtained by Letokhov *et al.* (1972). Evans *et al.* (1979) have studied the MPD of H_2CO/D_2CO mixtures with pulses from a DF laser. Fundamental studies of MPD in formaldehyde have been reported by Koren *et al.* (1977) and Koren and Oppenheim (1978).

An enrichment factor for H in HCOOH/HCOOD mixtures of ~ 25 was obtained by Evans *et al.* (1978) using a HF laser. β was found to increase dramatically as the gas pressure was lowered (Figure 4.15). Dissociation of HCOOH required 260 kJ mol^{-1} of laser energy. The same laser has also been used in a study of the isotopic-selective decomposition of methanol [McAlpine *et al.* (1979)]. H was enriched by a factor ~ 60 in a low-pressure (2 torr) mixture of CH_3OH and CH_3OD. A similar increase in β was observed at low pressure.

Table 4.3. Coincident Frequencies for $^{12}C^{16}O_2$ Laser Lines and Formaldehyde Isotomers[a]

Band (μm)	Laser line	Frequency (cm^{-1})	H_2CO (cm^{-1})	HDCO (cm^{-1})	D_2CO (cm^{-1})
10.4	*P*(26)	938.8	—	—	938.5
	P(14)	949.5	—	—	949.1
	P(12)	951.2	—	951.6	—
	R(18)	974.6	—	—	974.5
9.4	*P*(28)	1039.4	—	1039.7	—
	P(24)	1043.2	—	—	1043.4
	P(16)	1050.4	1051	—	—
	P(12)	1053.9	—	—	1053.8
	P(10)	1055.6	—	1055.6	—
	R(16)	1076	1075.9	—	—
	R(26)	1082.3	—	1082	—

[a]Following Koren *et al.* (1976).

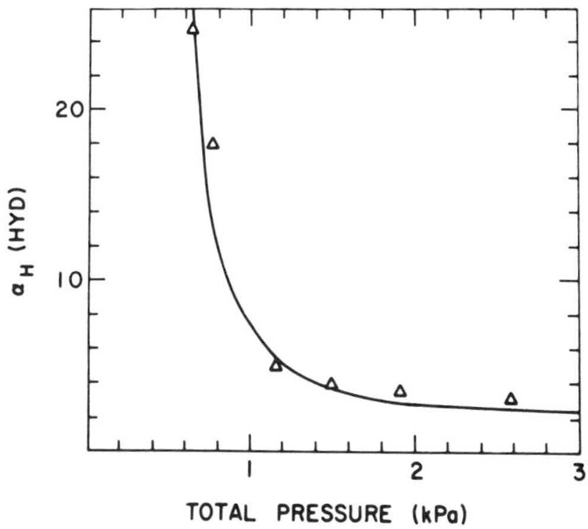

Figure 4.15. Separation factors for selective irradiation of HCOOH in an equimolar mixture of HCOOH/HCOOD at various total pressures [Evans *et al.* (1978)].

MPD of fluoroform-d has been demonstrated to yield large D separation efficiencies [Tuccio and Hartford (1979) and Herman and Marling (1979)]. CF_3D absorbs strongly at the frequency of several lines in the *R* branch of the CO_2 10.4-μm band ($\sigma \sim 3 \times 10^{-19}$ cm^2), while CF_3H is almost transparent in this region ($\sigma \sim 4 \times 10^{-23}$ cm^2). Irradiation at these frequencies ($\nu \sim 970$ cm^{-1}) therefore is effective in promoting the MPD

$$CF_3D + nh\nu \rightarrow :CF_2 + DF$$

followed by

$$2{:}\, CF_2 \rightarrow C_2F_4$$

Figure 4.16 shows a comparison of the infrared spectrum of a CF_3H–CF_3D mixture before and after irradiation at 970 cm^{-1}. The decrease in the strength of features due to CF_3D is evident as is the appearance of features due to C_2F_4. The deuterium enhancement factor derived from these observations is $\beta \sim$ 5000.

Herman and Marling (1979) find that decomposition of CF_3D is greatly enhanced by the addition of the Ar buffer gas over a limited pressure range. This was attributed to the effect of rotational relaxation in increasing the fraction of molecules that can absorb IR photons in the initial stage of MPD excitation. They report that with the optimum Ar pressure, 100% of CF_3D molecules in the laser focus decompose when the laser fluence is 30 J cm^{-2}. Under collision-free conditions 180 J cm^{-2} is required to achieve the same effect while 30 J cm^{-2} produces only $\sim$ 2% decomposition.

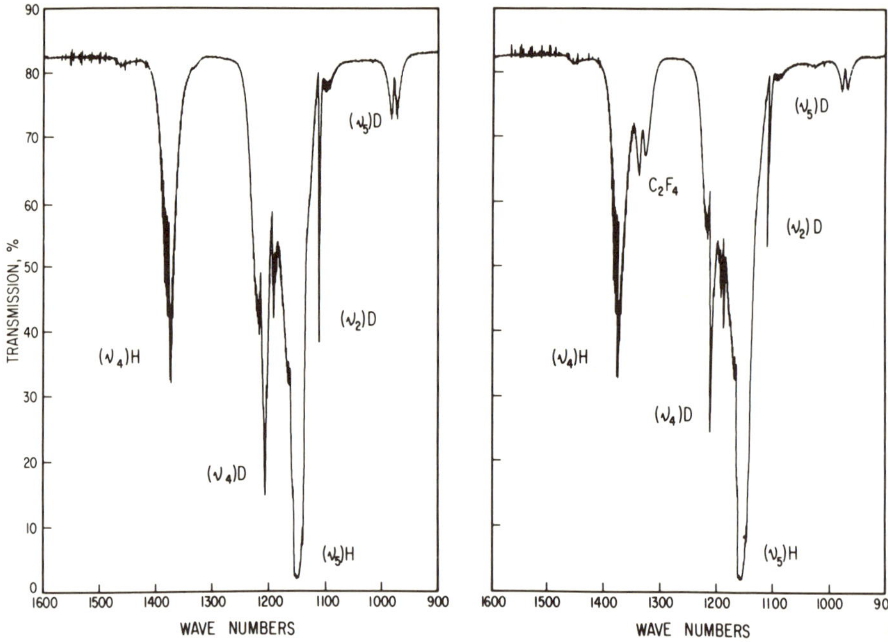

Figure 4.16. Infrared spectra of a 4:1 mixture of CF_3H:CF_3D before (left spectrum) and after (right spectrum) CO_2 laser irradiation. No change in the CF_3H band intensities is detectable following irradiation, while a substantial decrease in the CF_3D band intensities occurs. Also, following irradiation C_2F_4 formation is indicated by the appearance of the band at 1337 cm^{-1} [Tuccio and Hartford (1979)].

The large enhancements possible with MPD of CF_3D have been discussed in relation to possible applications of this technique to large scale H–D separation [Herman and Marling (1979)].

4.10.2. Boron

Several groups have studied the separation of ^{10}B and ^{11}B isotopes from BCl_3 under high-power irradiation with CO_2 laser pulses [Freund and Ritter (1975), Lyman and Rockwood (1976), and Ambartzumyan *et al.* (1976d)]. Pulse energies were in the range 0.1–2 J and the pulse duration was typically 100–200 ns. Under these conditions BCl_3 undergoes MPD. Reactions with scavengers such as H_2S, O_2, HBr, and H_2 were used to remove decomposition products. The largest value of β obtained was $\sim$ 20. Figure 4.17 shows infrared spectra of a $^{10}BCl_3$ and $^{11}BCl_3$ mixture before and after irradiation at the $^{10}BCl_3$ frequency [part (a)] and at the $^{11}BCl_3$ frequency [part (b)]. These data were produced with an intensity of $\sim 10^9$ W cm^{-2} in the beam focus and dry air was used as a scavenger.

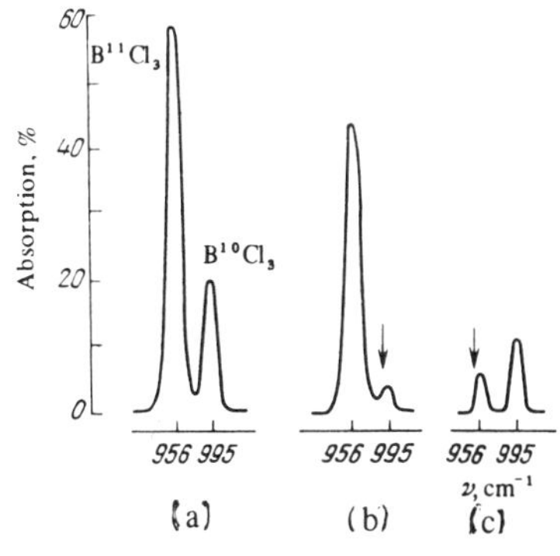

Figure 4.17. Absorption spectra of (a) BCl_3 before and after 30-min irradiation causing excitation of (b) $B^{10}Cl_3$ and (c) $B^{11}Cl_3$ (c). The BCl_3 pressure was 0.44 mm Hg and the pressure of air was 11 mm Hg. The arrows indicate the excitation frequency [Ambartzumyan *et al.* (1976d)].

4.10.3. Carbon

The first report of ^{13}C separation via MPD appeared in 1976 [Lyman and Rockwood (1976)]. Pulses of 1 J energy and 200-ns duration from a TEA CO_2 laser were focused in mixtures of $^{13}CF_2{}^{35}Cl$ and $^{12}CF_2{}^{35}Cl$. The resulting value of β was 1.13 after 2000 pulses. It was estimated that 25 laser photons were required to break the bond that leads to product formation. Figure 4.18 shows the dependence of β on the total number of pulses.

Subsequent work has centered on the separation of ^{13}C from CF_3X compounds [Drouin *et al.* (1978) and Gauthier *et al.* (1980a, b)]. Borsella *et al.* (1980) have also reported on ^{13}C separation from the MPD of C_2F_5Cl.

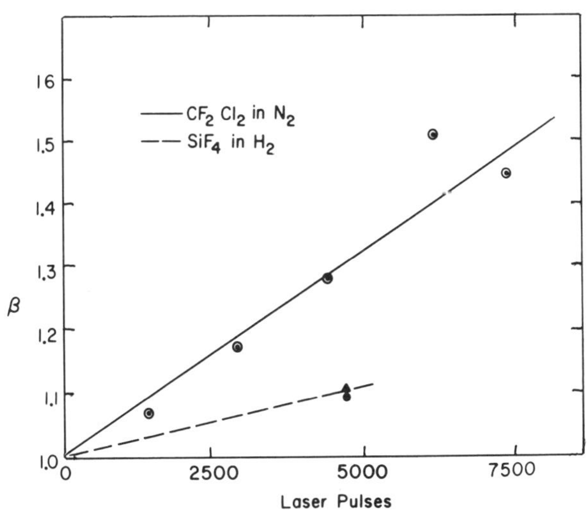

Figure 4.18. Enrichment parameter β for CF_3Cl_2 (solid line) and SiF_4 (dashed line) as a function of the number of laser pulses. For SiF_4, (▲) indicates $^{29}Si/^{28}Si$ ratio and (●) indicates $^{30}Si/^{28}Si$ ratio [Lyman and Rockwood (1976)].

Drouin *et al.* (1978) studied the effect of laser wavelength on the selectivity of CF_3X dissociation. The highest selectivities were found for CF_3Cl and CF_3Br at $-50\,^\circ C$ using the $R(8)$ and $P(24)$ lines, respectively, of the 9.4-μm band. The selectivity of ^{13}C production from CF_3I was significantly smaller even at a temperature of $-50\,^\circ C$. This was attributed to an overlapping of ^{12}C and ^{13}C IR spectral features in CF_3I, together with an enhancement in the scrambling reaction

$$^{13}CF_3 + {}^{12}CF_3I \rightarrow {}^{12}CF_3 + {}^{13}CF_3I$$

The gas pressure was 0.2 torr.

In subsequent work, Gauthier *et al.* (1980a) found that the selectivity of ^{13}C production from CF_3Br was sensitive to CF_3Br pressure. The maximum value $\sim$ 150 was obtained at a pressure of 20–30 torr but was accompanied by a decrease in the overall quantum efficiency for MPD from that observed at lower pressures. These results were obtained when CF_3Br was irradiated with a laser line that was 28 cm^{-1} to the red of the $^{13}CF_3Br$ absorption frequency and 53 cm^{-1} to the red of that for $^{12}CF_3Br$. Irradiation at a frequency that is closer to the $^{13}CF_3Br$ band yields the normal result that β decreases with increasing pressure [Gauthier *et al.* (1980b)]. An explanation for the anomalous pressure dependence of the isotopic selectivity for off-resonance irradiation has been proposed by Gauthier *et al.* (1980).

4.10.4. Silicon

Lyman and Rockwood (1976) have reported the separation of ^{29}Si and ^{30}Si from SiF_4 via MPD with pulses from a TEA CO_2 laser. The laser was operated on the $P(36)$ line of the 9.4-μm band to excite a resonance in $^{28}SiF_4$. Elimination of the dominant ^{28}Si isotope resulted in the enhancement of ^{29}Si and ^{30}Si in the residual gas. β was found to be $\sim$ 1.1. Dissociation products were removed by reaction with H_2.

4.10.5. Sulfur

The separation of the various isotopes of S from SF_6 via MPD has been extensively studied. Figure 4.19 [Letokhov (1977)] shows low-resolution infrared spectra of $^{32}SF_6$ and $^{34}SF_6$ in the region of the ν_3 fundamental together with experimental curves of the laser-induced MPD rate as a function of frequency. This figure also shows the frequency dependence of enrichment factors for ^{34}S and ^{32}S over this energy range. As expected, these show that $\beta(32)$ is largest when excitation occurs in the $^{32}SF_6$ band.

At the spectral resolution shown in Figure 4.19 rotational structure in SF_6 is unresolved. McDowell *et al.* (1976) have obtained high-resolution spectra of

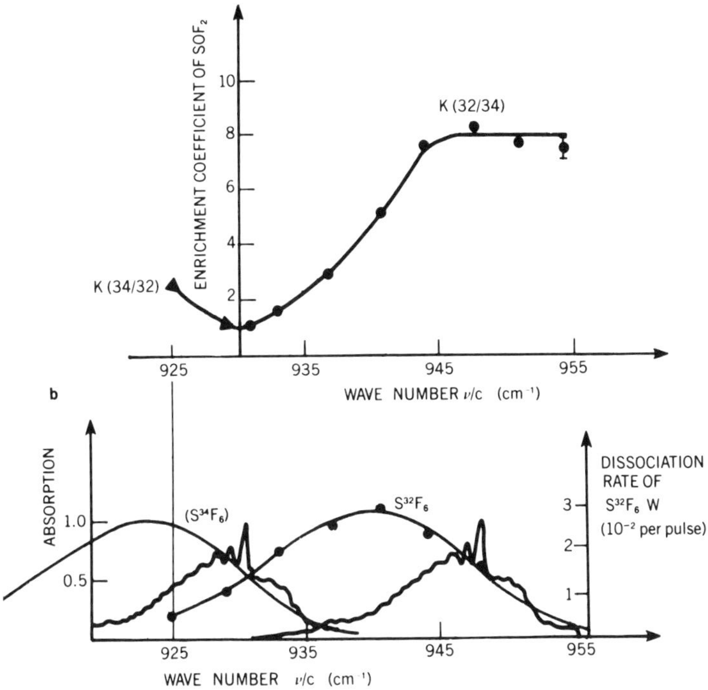

Figure 4.19. Frequency dependence of multiphoton-selective dissociation of SF_6 molecules by infrared radiation (Measurements made in focused beam; average radiation power density 31 MW cm^{-2}, SF_6 pressure 0.2 torr.) Curve (a) shows variation of enrichment coefficients $K(32/34)$ and $K(34/32)$ as points (●) and (▲), respectively. Frequency dependence of $S^{34}F_6$ dissociation rate is an assumption [Letokhov (1977)].

SF_6 in the vicinity of several CO_2 laser lines which show a complex rotational structure. The spectrum of SF_6 in the vicinity of the $P(18)$ CO_2 laser line at 945.9802 cm^{-1} is shown in Figure 4.20. It shows that the $P(18)$ laser line pumps the ${}^1A_2$ component of the $P(33)$ line in SF_6.

The first investigations of $^{34,36}S$ separation from SF_6 using CO_2 laser-induced MPD showed that enormous enhancements could be readily obtained [Lyman *et al.* (1975) and Ambartzumyan *et al.* (1975)]. The largest value of β that has been reported in these studies is $\sim$ 4000 [Letokhov (1977)]. These results were obtained by irradiating SF_6 in H_2.

A comprehensive parametric study of enrichment in SF_6 has been reported by Gower and Billman (1977). Their data on the pressure dependence of enrichment factor ($K \equiv \beta$) for ^{34}S are shown in Figure 4.21 for SF_6 and $SF_6 + H_2$ mixtures. Both curves exhibit maxima between 0.1 and 0.01 torr and decreasing β for $p \gtrsim 0.1$ torr. Gower and Billman (1977) derive a disso-

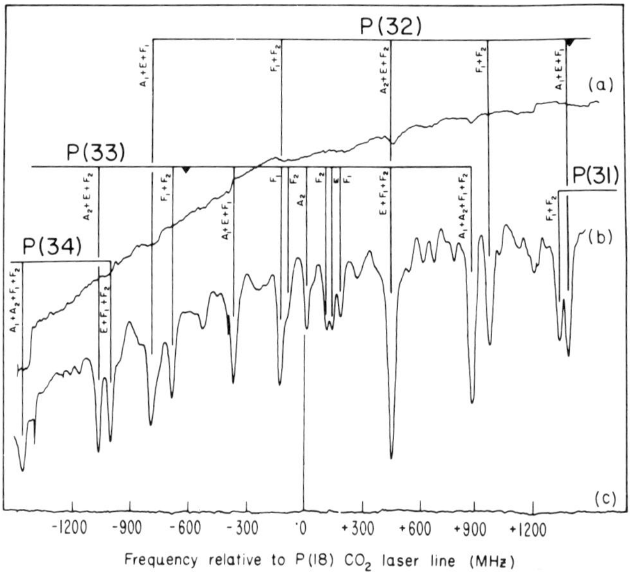

Figure 4.20. Absorption spectrum of SF_6 near the $P(18)$ line of the CO_2 laser at 945.9802 cm^{-1}. Cell length, 10 cm; temperature 300 K; SF_6 pressure, (a) $\sim$ 0 torr, (b) 0.1 torr; trace (c) is the zero [McDowell *et al.* (1976)].

ciation cross section σ_{32} which eliminates the effect of geometrical factors on enrichment. σ_{32} is defined as follows:

$$\sigma_{32} = \frac{h\nu V}{El}(1 - K^{1/q})$$

where V is the cell volume, E the pulse energy, l the length of the cell, K the enrichment factor, and q the number of laser pulses.

The effect of fluence (J cm^{-2}) on the ^{34}S enrichment factor is shown in Figure 4.22. This shows a distinct threshold for enrichment at 4.6 J cm^{-2} in SF_6 and at 6 J cm^{-2} in SF_6:H_2 mixtures. Little variation in K is observed for different pulse lengths over the 40–200 ns range providing the same energy flux is delivered to the sample. Work with much shorter pulses (0.5–1 ns) has shown that MPD occurs with increasing probability [Kolodner *et al.* (1977)] and that the energy threshold for dissociation is maintained regardless of pulse duration.

Experiments with long excitation pulse lengths ($\sim$1 μs) [Lin *et al.* (1978) and Koren and Oppenheim (1979)] show that dissociation can be induced by collisions following the initial collisionless dissociation. This effect arises when collisions act to redistribute the population of rotational states bleached by

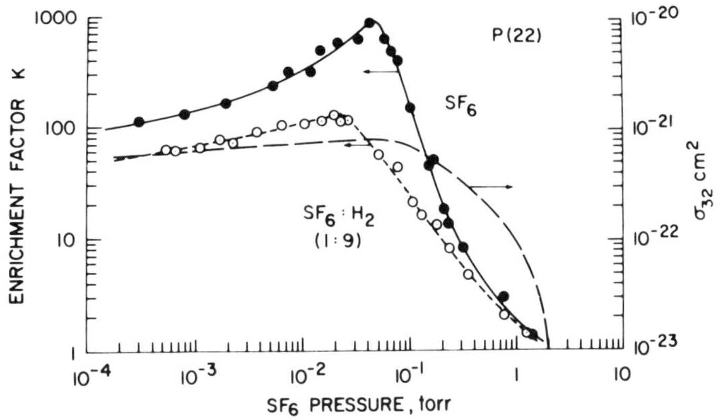

Figure 4.21. Enrichment factor K vs. SF_6 partial pressure for SF_6 only and an SF_6:H_2(1:9) mix.---, curve is the dissociation cross section σ_{32}; $P(22)$ laser line. Energy = 1.2 J per pulse for q = 100 pulses [Gower and Billman (1977)].

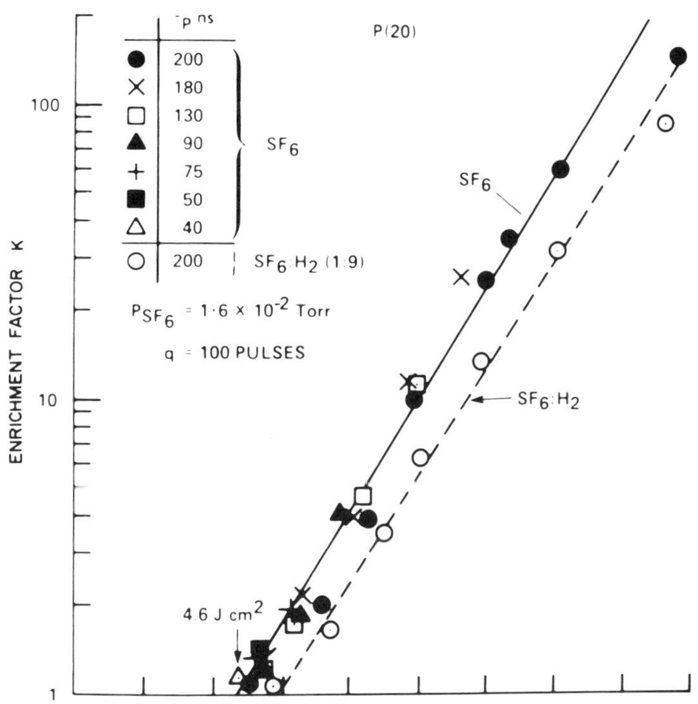

Figure 4.22. Enrichment factor K vs. laser pulse energy density for the $P(20)$ laser line using different pulse lengths; number of pulses, q = 100; SF_6 pressure = 1.6×10^{-2} torr [Gower and Billman (1977)].

absorption under collisionless conditions. Some data obtained by Lin *et al.* illustrating this effect are given in Table 4.4. This shows that increasing gas pressure to 0.2 torr has little effect on $^{32}SF_6$ dissociation while $^{32}SF_6$ dissociation increases with the inert gases and decreases with H_2. At higher pressures the $^{32}SF_6$ dissociation rate decreases substantially. The dissociation rate of $^{34}SF_6$ is a strong function of perturber mass at high pressure, with the greatest reduction in dissociation efficiency occurring with H_2.

The effect of H_2 at high pressures is attributed to enhanced collisional deactivation of excited vibrational SF_6 states. With low H_2 pressure ($\sim$0.2 torr) the dissociation rate of $^{32}SF_6$ is little affected, while that of $^{34}SF_6$ is reduced by 50%. Since $^{34}SF_6$ is always "off resonance" with respect to excitation with the $P(20)$ line while $^{32}SF_6$ is "on resonance," $^{34}SF_6$ is less highly excited than is $^{32}SF_6$ with given laser fluence. As a result, collisional deactivation of $^{34}SF_6$ is more important because it has a large effect on the probability that many photons can be absorbed from the radiation field. A larger collisional loss term can be sustained by $^{32}SF_6$ since the rate of absorption of photons by this isotomer is high.

The dependence of the dissociation rate in SF_6 on laser intensity is shown in Figure 4.23 [Letokhov (1977)]. Data are given for excitation in the ν_3 fundamental and the $\nu_2 + \nu_6$ intercombination band. A clear threshold is seen at $I \sim 20$ MW cm^{-2}. Above threshold the dissociation rate is linearly related to I, although the slopes of the ν_3 and $\nu_2 + \nu_6$ curves differ.

It should be noted that most S enrichment experiments have been performed by removing the abundant ^{32}S isotope with MPD. Thus ^{34}S (and other isotopes) are enriched by depleting the gas in ^{32}S not by directly driving ^{34}S into a separable compound. This represents an inefficient use of laser photons

Table 4.4. Dissociation of $^{32}SF_6$ and $^{34}SF_6$ in the Presence of Various Gases[a]

Mixture (torr)	Percent $^{32}SF_6$	Dissociation[b] $^{34}SF_6$	β
0.2 SF_6	89.9	40.2	5.92
1.0 SF_6	78	34	3.0
+ 0.2 Ar	90.8	48.7	5.61
+ 19.3 Ar	81.6	44.8	3.00
+ 0.2 He	91.9	43.2	7.01
+ 20.3 He	60.0	36.2	1.60
+ 0.2 H_2	78.2	23.5	3.51
+ 20.6 H_2	42.0	15.4	1.46

[a] Data were taken following 500 laser pulses. The pulse duration was 1-μs [Lin *et al.* (1978); [b] Using $P(20)$ line of 10.4-μm band to excite $^{32}SF_6$.

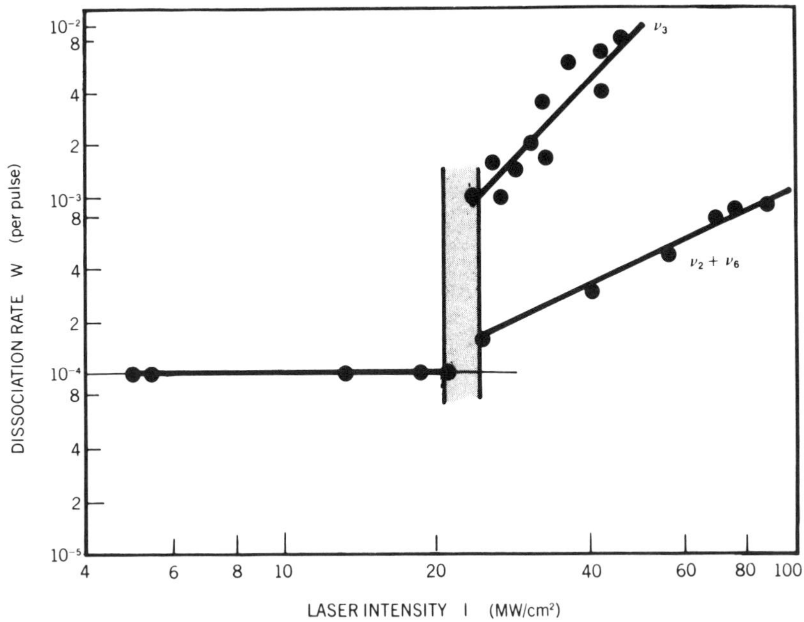

Figure 4.23. Dissociation threshold in SF_6. The graph shows the dependence of dissociation rate W on laser intensity when the laser pulse excites the ν_3 fundamental band or the $\nu_2 + \nu_6$ compound band of SF_6. Laser pulse length is 90 n, SF_6 is 0.2 torr [Letokhov (1977)].

(overall quantum efficiency $\sim 10^{-8}$) and represents a limitation to the possible application of MPD in sulphur enrichment on a large-scale basis.

4.10.6. Chlorine

MPD of CF_2Cl_2 leading to Cl isotopic enhancements have been reported by Huie *et al.* (1978). The largest value of β observed was ~ 1.8. MPD of CF_2Cl_2 leads to the formation of Cl_2. A plot of the $^{35,35}Cl_2/^{37,37}Cl_2$ ratio observed in this product is plotted in Figure 4.24 as a function of CF_2Cl_2 pressure. An interesting extension of isotopic selectivity to very high pressures is evident in this plot. Huie *et al.* attribute this effect to the existence of isotopically specific interactions occurring with vibrationally excited molecules following the termination of the laser pulse.

Figure 4.24 shows that the addition of SiF_4 followed by irradiation with the $P(30)$ (9.64-μm) line, which is absorbed by SiF_4 but not by CF_2Cl, leads to no isotopic enhancement in Cl_2. This shows that collisional effects with excited SiF_4 can produce dissociation but that no isotopic selectivity results.

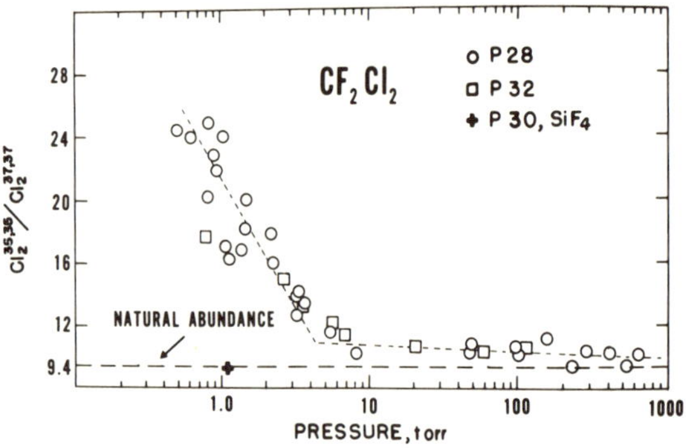

Figure 4.24. The ratio of the monoisotopic forms of Cl_2 produced in the CO_2 TEA laser photolysis of pure CF_2Cl_2 as a function of pressure [Huie *et al.* (1978)].

4.10.7. Molybdenum

A weak intercombination band of MoF_6 overlaps with several lines in the 9.4-μm band of the CO_2 laser. This band exhibits no resolvable isotopic structure when observed at moderate resolution [Freund and Lyman (1978)]. Furthermore, many low-lying vibrational levels are thermally populated at 300 K adding to the complexity of infrared spectra and to the overlap of lines due to various isotopic species. Despite these complications, Freund and Lyman have demonstrated that MPD leads to significant enhancements in the abundance of certain Mo isotopes when MoF_6 is irradiated with pulses from a TEA CO_2 laser. Some of their data are shown in Figure 4.25.

4.10.8. Osmium

The first study of isotope separation in OsO_4 via MPD was reported by Nakarov *et al.* (1975). ^{187}Os and ^{192}Os were separated from OsO_4 by irradiating with 3-J, 90-ns duration pulses. Enhancements were 1.1–1.2. Later work has centered on the elucidation of dissociation processes [Ambartzumyan *et al.* (1978a,b)].

4.10.9. Uranium

With the knowledge gained from MPD studies on isotope separation from SF_6, the analogous uranium compound UF_6 would appear to be an ideal source of photochemically produced U isotopes. Unfortunately, the UF_6 molecule does not absorb in the wavelength range available from the CO_2 laser and thus the most useful infrared laser source is not available.

Since UF_6 absorbs near 16 μm, much work has been done on the development of laser sources for use in this spectral region. A summary of candidate laser types has been published [*Laser Focus,* March (1977), p. 66]. The most promising of these involved Raman shifting the output of a CO_2 laser using H_2 gas. This produced 1.6-J, 20-MW pulses.

Studies of UF_6 MPD with pulses from an optically pumped CF_4 laser [Tiee and Wittig (1978) and Rabinowitz *et al.* (1978)] show a threshold for dissociation of ~1 J cm^{-2} or 15 MW cm^{-2}. Similarities to SF_6 MPD are noticeable although isotopic enhancement has not yet been reported. In an interesting development, Tiee and Wittig (1978) show that the photodissociation yield is enhanced when both CF_4 and CO_2 laser radiation is incident on the sample. This enhancement (which amounts to a factor of 10–100) is attributed to the fact that UF_6 molecules that have been preexcited by absorption of CF_4 laser radiation to the vibrational quasicontinuum can absorb CO_2 laser radiation.

Another approach to U isotope separation has been suggested by Miller *et al.* (1979). It was shown that it is possible to "tailor" U compounds such that they absorb at ~10 μm. As an example, Miller *et al.* report on the synthesis of uranium hexamethoxide, $U(OCH_3)_6$, and demonstrate that measurable enrichments of ^{235}U can be obtained from the MPD of this compound with pulses from a TEA laser operating on lines of the 10.4-μm band. A subsequent report by Cox *et al.* (1979) showed that somewhat greater isotopic enhancements could be obtained on irradiation of uranyl hexafluoroacetylacetonate tetrahydrofuran with CO_2 laser radiation. CW radiation was also effective in yielding isotope separation.

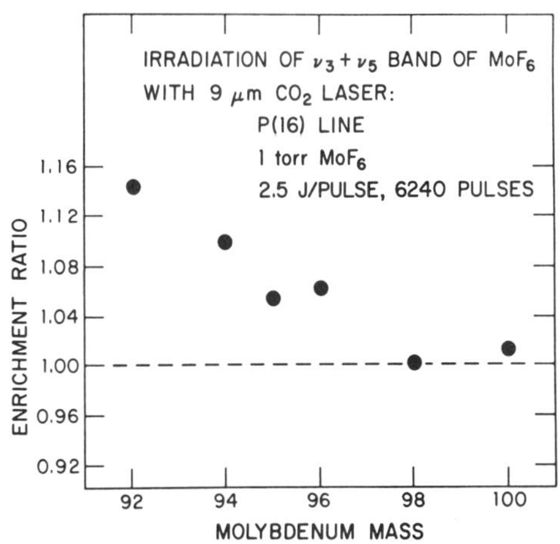

Figure 4.25. β_m vs. m after irradiation with $P(16)$ line at 1050.5 cm^{-1}, 30% of MoF_6 remains in all after irradiation [Freund and Lyman (1978)].

4.11. SELECTIVE RAMAN EXCITATION

This method has not been widely applied to isotope separation. It involves the utilization of the population of excited vibrational levels induced by selective Raman scattering to enhance the reaction between a particular isotomer and a scavenger molecule.

This technique has been used by Basov *et al.* (1973, 1974a) and Arkhipov *et al.* (1972) to induce the isotopically selective reaction

$$N_2 + O_2 \rightarrow 2NO$$

Typical enhancements for ^{15}N of ~100 were obtained.

4.12. ECONOMICS OF LASER ISOTOPE SEPARATION

Much attention has been given to potential commercial application of laser isotope separation [an excellent review has been given by Letokhov and Moore (1977)]. The greatest economic benefits would derive from the development of methods to separate isotopes needed for nuclear reactors. In other areas, laser isotope separation may be efficient but the costs of commercial development are unlikely to be justified given the small usage of these isotopes.

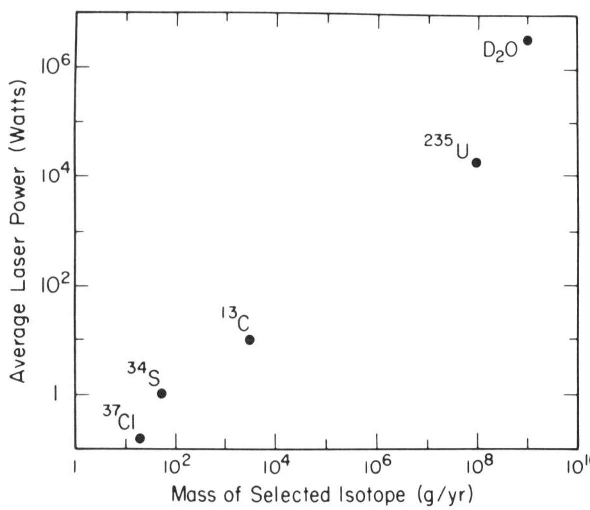

Figure 4.26. Average laser power required to reproduce output of largest present day production facilities for several isotopes [Letokhov and Moore (1977)].

The two most important isotopes from an economic point of view are ^{235}U (for use in light-water reactors) and D (for use in the heavy-water CANDU reactor). An increase in the production efficiency of either isotope could offer a potential saving of many $\$10^9$.

The requirements that such processes place on laser sources are severe: Average laser powers of $\sim 10^4$ W would be required to compete with current technology for the production of ^{235}U. To produce 10^9 g yr^{-1} of D_2O would require a laser with almost 10^7 W average power. Letokhov and Moore (1977) have provided a useful summary of the relation between average laser power and isotope generation rate for rates that would make laser isotope separation competitive with conventional technology. This graph is shown in Figure 4.26.

4.13. LASER-INDUCED REACTIONS

In this section we review work done on the laser enhancement of chemical reactions exclusive of those that lead to isotope separation. Laser modification of reaction rates can occur when a component involved in a chemical reaction is modified by the absorption of radiation from the laser field. The absorber may be a reactant, a reaction intermediate, or a product. The absorbed energy may be in the form of infrared, visible, or UV photons.

Since the effect of electronic excitation on reaction rates has been the subject of much work using conventional light sources [cf. Calvert and Pitts (1966)], while almost no data on the effect of infrared excitation on reaction rates were available prior to the development of laser sources, much emphasis has been placed on the new field of *vibrational photochemistry*. This emphasis will be continued in the present review.

Excellent surveys of work in this area have been published by Basov *et al.* (1974b), Weitz and Flynn (1974), Berry (1974), Ambartzumyan and Letokhov (1977), Kimel and Speiser (1977), Panfilov and Molin (1978), and Yardley (1979).

4.13.1. Infrared Photochemistry—Basic Mechanisms

If we consider an isolated polyatomic molecule in a gas under collisionless conditions, the absorption of infrared radiation will produce nonequilibrium populations within the manifold of vibrational and rotational levels of the lowest electronic level. In particular, at high field intensity the population of these rotational and vibrational levels will be as shown schematically in Figure 4.27. The effect of irradiation is to transfer population to higher vibrational states. In the absence of collisions absorption will "burn a hole" in the rotational energy distribution of the lower vibrational level and will enhance the population of a single rotational state in the excited vibrational state. While several

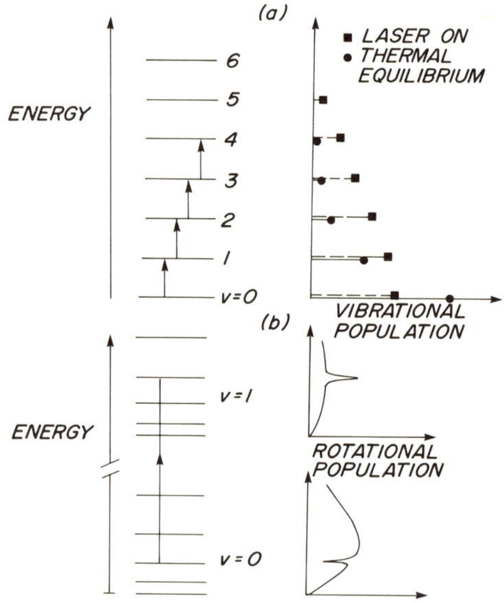

Figure 4.27. Population in (a) vibrational and (b) rotational levels with and without laser field.

rotational states may participate in transitions at high laser intensity due to power broadening, generally only a few rotational levels participate in absorption at most intensities of interest.

If collisional processes are now "switched-on," their effect will be (1) to exchange vibrational energy between excited and unexcited molecules (V–V transfer) and (2) to exchange vibrational energy with translational modes (V–T transfer). The latter process leads to gas heating and is the source of a major complication: laser-driven thermal reactions.

Collisions also reestablish equilibrium within the manifold of rotational energy levels removing any "bottleneck" that may exist under collisionless conditions. An inert gas is often used to ensure that the rotational energy distribution is basically thermal. The pressure should be such that rotational relaxation occurs on a faster time scale than V–T transfer if thermal heating effects are to be avoided.

Unlike thermal excitation, the absorption of a limited amount of laser energy excites only a single vibrational mode of the absorber. If this vibrational mode is involved in the route to a particular reaction channel, then excitation in this mode can increase the probability of occurrence of the reaction. The effect of stored energy in such a mode is to reduce the activation energy.

Energy may be accumulated in higher vibrational states by multiphoton absorption or by V–V energy transfer. The rapid accumulation of vibrational energy can lead to unimolecular dissociation which can occur on shorter time scales than reaction when the incident laser flux is high (10^8–10^9 W cm^{-2}). V–V transfer increases the population of higher vibrational states via

$$M(v = 1) + M(v = 1) \rightarrow M(v = 0) + M(v = 2) + \text{Energy}$$

followed by

$$M(v = 2) + M(v = n) \rightarrow M(v = 0) + M(v = n + 2) + \text{Energy}$$

Excitation of an absorbing gas with a CW laser source provides a shift in the equilibrium population of vibration/rotation states. However, since V–T transfer is important in such a system, much of this shift in equilibrium will be due to thermal excitation. In order to separate the effect of nonequilibrium populations in higher vibrational states, such molecules must react before thermalization occurs for these levels. Generally, this will only occur when appreciable energy is stored in high vibrational states. This in turn implies high pumping rates and hence high laser powers. It is not surprising, therefore, that reactions induced with low-power CW laser sources often mimic thermal reactions.

As an example of the ambiguity that often surrounds the role played by thermal reactions, we consider the laser-induced conversion of CF_2Cl_2 into $C_2F_4Cl_2$ [Zitter *et al.* (1975)] [also see Slater and Parks (1979)]. A mixture of CF_2Cl_2 and SF_6 was irradiated with ~5 W of CW radiation from a line tunable CO_2 laser. Since SF_6 absorbs at 948 cm^{-1} while CF_2Cl_2 absorbs at 931 cm^{-1}, it is possible to irradiate the mixture to excite either SF_6 or CF_3Cl_2 but not the other. If both components absorb equal amounts of radiation, then equal thermal heating should be produced in both circumstances. Zitter *et al.* (1975) found that no reaction was observed when the mixture was irradiated in the SF_6 band while a strong reaction was observed when irradiation occurred in the CF_2Cl band. Under the latter conditions the reaction rate was found to vary as the eleventh power of the laser intensity. This led to the justifiable conclusion that selective vibrational excitation under irradiation was responsible for the observed enhancement in reaction rate.

However, it has been shown, [Panfilov and Molin (1978)] that the case for selective nonthermal infrared photochemistry has not been conclusively demonstrated even in this system. Detailed analysis of the temperature dependence of the relevant infrared absorption bands shows that a higher temperature could be present in the beam focus when the laser is tuned to the CF_2Cl band. This could produce the necessary enhancement via a thermal effect.

Thermal effects can be reduced by irradiation at low gas pressure. An experiment conducted under these conditions was reported in 1972 by Yogev *et al.* (1972). Limonene and isoprene were irradiated with a CW CO_2 laser. Two pressure regions (<3 torr and >3 torr) were delineated. At low pressure a wide variety of reaction products were observed with the concentration of these products maximizing at ~0.05 torr. No reaction was observed between 3 and 30 torr, but above 30 torr dimeric products of uncertain structure were formed. Detailed analysis of activation and deactivation showed that at low pressures sequential absorption of several laser photons could be obtained and the reaction was vibrationally induced. At high pressures normal thermal chemistry was obtained.

4.13.2. Vibrationally Enhanced Chemical Reactions

In this section we review some representative data on the role of infrared laser radiation in the enhancement of chemical reactions. We consider only reactions that are influenced by the vibrational state of one reactant. Furthermore, the role of vibrational excitation on unimolecular processes such as isomerization and unimolecular decomposition is not dealt with here but will be discussed in separate sections.

A quantitative analysis of the enhancement of bimolecular reaction rates by vibrational excitation has been published by Birely and Lyman (1975). An attempt was made to modify the standard Arrhenius rate equation

$$k = A \exp(-E_a/RT)$$

to describe enhancements produced with vibrational excitation of one of the reactants. In the above equation k is a rate constant, A the preexponential factor, E_a the activation energy, R the gas constant, and T is in K. The modified equation was taken to be

$$k^+ = A^+ \exp(-E_a^+/RT)$$

i.e., the Arrhenius form was assumed. Birely and Lyman further assumed that $A^+ = A$ and that E_a^+ could be represented by

$$E_{\mathrm{a}}^+ = E_a - \alpha E_{\mathrm{vib}}$$

where E_{vib} is the vibrational energy stored in reactants and α is the empirically determined factor. They concluded from an analysis of available experimental data on the effect of vibrational excitation on reaction rates that α is typically $\lesssim 0.6$. Furthermore, α was found to be uncorrelated with E_a, E_{vib}, or with exothermicity. Thus the use of a modified Arrhenius equation to describe the effect of vibrational energy on reaction rates was found to be convenient. However, the parameter α was concluded to be of little physical significance.

Orel and Miller (1978) have shown that the presence of an intense laser field can lower the activation energy for a chemical reaction even when the field is nonresonant for vibration modes of individual reactants (or even if these modes are IR inactive). The enhancement occurs because coupling can occur to IR-active collisional modes. An example was given of the $H + H_2 \rightarrow H_2 +$

H reaction in which the $\vec{H} \ldots \overleftarrow{H} \ldots \overleftarrow{H}$ mode is IR active even though the H_2 fundamental is not. Extra energy therefore appears directly in the reaction coordinate most effective in promoting the reaction.

A good example of the effect of vibrational energy on bimolecular reaction rates can be seen in experiments on the reaction of various atomic species with vibrationally excited HCl [Arnoldi and Wolfrum (1974), Brown *et al.* (1975, 1976), Odiorne *et al.* (1971), Karny *et al.* (1975), Arnoldi *et al.* (1975), Leone *et al.* (1975), and Karny and Katz (1976)]. Arnoldi and Wolfrum (1974) measured rate constants for the reactions

$$\mathrm{O} + \mathrm{HCl}(v = 1) \xrightarrow{k_1'} \mathrm{OH} + \mathrm{Cl}$$
$$\mathrm{O} + \mathrm{HCl}(v = 1) \xrightarrow{k_1''} \mathrm{HCl}(v = 0) + 0$$
$$\mathrm{H} + \mathrm{HCl}(v = 1) \xrightarrow{k_2'} \mathrm{H_2} + \mathrm{Cl}$$
$$\mathrm{H} + \mathrm{HCl}(v = 1) \xrightarrow{k_2''} \mathrm{HCl}(v = 0) + \mathrm{H}$$

where $\mathrm{HCl}(v = 0) \rightarrow \mathrm{HCl}(v = 1)$ was excited by pulses from an HCl laser. They conclude that

$$k_1' + k_1'' = (2.2 \pm 0.7) \times 10^{12}\ \mathrm{cm^3\ mol^{-1}\ s^{-1}}$$
$$k_2' + k_2'' = (3.9 \pm 1.3) \times 10^{12}\ \mathrm{cm^3\ mol^{-1}\ s^{-1}}$$

These should be compared to the room temperature rate constants $k_1 = (5.6 \pm 1.9) \times 10^8\ \mathrm{cm^3\ mol^{-1}\ s^{-1}}$ and $k_2 = (6.7 \pm 0.4) \times 10^{10}\ \mathrm{cm^3\ mol^{-1}\ s^{-1}}$ for the reactions

$$\mathrm{O} + \mathrm{HCl} \xrightarrow{k_1} \mathrm{OH} + \mathrm{Cl}$$
$$\mathrm{H} + \mathrm{HCl} \xrightarrow{k_2} \mathrm{H_2} + \mathrm{Cl}$$

Vibrational deactivation is important in both reactions and hence it is difficult to estimate directly the enhancement of k_1 and k_2 due to vibrational excitation alone. However, enhancements of $\sim 10^2$ for the O + HCl reactions are likely.

Brown *et al.* (1976) and Leone *et al.* (1975) have studied the removal of vibrationally excited HCl by Br atoms. Reactions between $\mathrm{HCl}(v = 2)$ and Br gives

$$\mathrm{HCl}(v = 2) + \mathrm{Br} \rightarrow \mathrm{HBr}(v = 0) + \mathrm{C}$$
$$\mathrm{HCl}(v = 2) + \mathrm{Br} \rightarrow \mathrm{HCl}(v = 1) + \mathrm{Br}$$
$$\mathrm{HCl}(v = 2) + \mathrm{Br} \rightarrow \mathrm{HCl}(v = 0) + \mathrm{Br}$$

The rate constant for the total removal of $HCl(v = 2)$ was found to be $k = 1.1 \times 10^{12}$ cm^3 mol^{-1} s^{-1}. This includes the effect of all three reaction channels, although the first reaction (yielding HBr) probably proceeds at a large fraction of the above rate. This reaction has been used in the separation of ^{35}Cl (cf. Section 4.4). A theoretical calculation of dynamical reaction mechanisms in this and Br + DCl (v = 2,3) reactions has been reported by Smith (1977).

The effect of CW CO_2 laser radiation on the reaction of ground-state ($^2P_{3/2}$) Br atoms with CH_3F has been investigated by Krasnopyrov *et al.* (1979). The reaction is

$$Br + CH_3F \rightarrow HBr + CH_2F$$

followed by

$$CH_2F + BR_2 \rightarrow CH_2FBr + Br$$

Figure 4.28 shows a plot of the log of the bromination rate, W, with and without CW CO_2 laser irradiation. As one would expect, the effect of laser radiation is greatest at low temperature.

The enormous enhancement in the rate of this reaction at low temperature is only partly due to direct irradiation: effects due to V–V transfer are likely to be important. Surprisingly, vibrational excitation in CH_3F was found not to influence the reaction rate with excited ($^2P_{1/2}$) Br atoms.

The overlap of the asymmetric stretching vibrational mode of O_3 with lines from the CO_2 laser means that O_3 molecules are easily excited vibration-

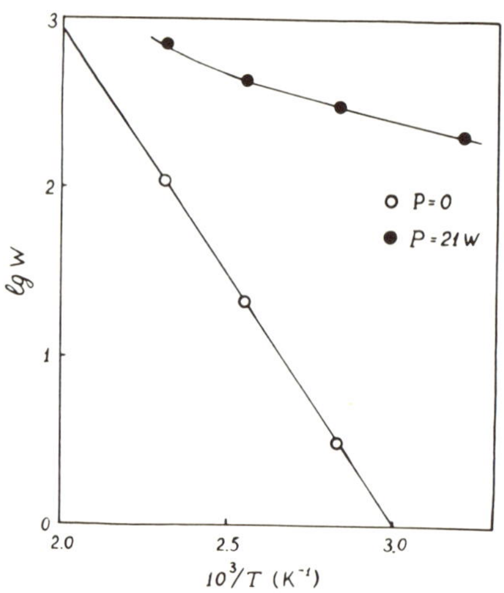

Figure 4.28. Plots of log W vs $1/T$ under CO_2 laser irradiation with 21-W power and without it: $[CH_3F] = 4 \times 10^{15}$ cm^{-3}; $[Br_2] = 8 \times 10^{15}$ cm^{-3}; $[Ar] = 2.3 \times 10^{17}$ cm^{-3} [Krasnopyrov *et al.* (1979)].

ally. Thus, much attention was focused on the reactions of vibrationally excited O_3 ($O_3^\dagger$) with various atomic and molecular species [Gordon and Lin (1973, 1976), Braun *et al.* (1974), Kurylo *et al.* (1975, 1976), Manning *et al.* (1976), Herman *et al.* (1976), and Stephenson and Freund (1976)].

Several papers have appeared on the enhancement of O_3 + NO reactions produced by vibrational excitation in O_3. The overall reactions are

$$O_3 + NO \xrightarrow{k_1} NO_2\ (^2A_1) + O_2 \tag{4.9}$$

$$O_3 + NO \xrightarrow{k_2} NO_2^*\ (^2B_1) + O_2 \tag{4.10}$$

$$O_3^\dagger + NO \xrightarrow{k_3} NO_2\ (^2A_1) + O_2 \tag{4.11}$$

$$O_3^\dagger + NO \xrightarrow{k_4} NO_2^*\ (^2B_1) + O_2 \tag{4.12}$$

where $O_3^\dagger$ is the vibrationally excited O_3 molecule and $^2A_1$ is the ground state of NO_2 (vibrational excitation can occur) While $^2B_1$ is an electronically excited state. Rate constants for the first two reactions [(4.9) and (4.10)] are

$$k_1 = 4.3 \times 10^{11} \exp(-2300/RT) \text{ cm}^3 \text{ mol}^{-1} \text{ s}^{-1}$$
$$k_2 = 7.6 \times 10^{11} \exp(-4200/RT) \text{ cm}^3 \text{ mol}^{-1} \text{ s}^{-1}$$

Chemiluminescence from reactions (4.10) and (4.12) lies in the visible and can be used to monitor the progress of the reaction. Braun *et al.* (1974) conclude from a measure of this chemiluminescence that the rate enhancement $k_4/k_2 = 5.6 \pm 1.0$. A modulated CW CO_2 laser was used to enhance the reaction.

The above ratio of k_4/k_2 was derived on the assumption that only O_3 molecules vibrationally excited in the ν_2 mode contribute to the reaction. As O_3 is excited in the asymmetric ν_3 mode by absorption of laser radiation, a rapid energy interchange between $\nu_3(001)$, $\nu_2(010)$, and $\nu_1(100)$ modes is required. Kurylo *et al.* (1975) discuss this process and conclude that $k_4/k_2 = 4.1 \pm 2$ and $k_3/k_1 = 17.1 \pm 4.3$.

The formation of NO_2 from O_3 and NO involves the breaking of the O_2–O bond in O_3 and hence should be sensitive to the vibrational state of O_3. Since the NO bond is not broken, no similar effect should be observed if this bond is excited. Stephenson and Freund (1976) have studied the analogous reaction

$$NO^\dagger + O_3 \xrightarrow{k_5} NO_2(^2A_1) + O_2 \tag{4.13}$$

$$NO^\dagger + O_3 \xrightarrow{k_6} NO_2(^2B_1) + O_2 \tag{4.14}$$

where $NO^{\dagger} \equiv NO(v = 1)$ and NO was excited by light from a CO laser. They found that

$$k_6/k_2 = 5.7\ ^{+2.6}_{-1.4} \quad \text{and} \quad (k_5 + k_{NO\text{-}O_3})/k_1 \leq 22$$

where $k_{NO\text{-}O_3}$ is the rate constant for collisional deactivation of $NO^{\dagger}$ by O_3. The excitation of NO is therefore about as effective in promoting the reaction with O_3 as is the excitation of O_3 even though the energy involved for NO^+ is approximately twice that available from $O_3^{\dagger}$. Stephenson and Freund suggest that vibrational excitation in NO is transferred to the O_2–O bond in the $(NO.O_3)^+$ collision complex.

Other reactions with vibrationally excited $O_3^{\dagger}$ that have been studied include $O_3^{\dagger}$–SO [Kaldor *et al.* (1974)], $O_3^{\dagger} + O_2$ [Kurylo *et al.* (1976)], and $O_3^{\dagger}$ + Pb [Kurylo *et al.* (1976)].

Manning *et al.* (1976) have reported on the vibrational enhancement of the following reactions using CW laser excitation

$$O + OCS \rightarrow CO + SO \tag{4.15}$$
$$O + OCS^{\dagger} \rightarrow CO + SO \tag{4.16}$$
$$O + C_2H_4 \rightarrow CH_3 + CHO \tag{4.17}$$
$$O + C_2H_4^{\dagger} \rightarrow CH_3 + CHO \tag{4.18}$$

where O atoms are in the ^3P state. They find that while some enhancement does occur in both reactions, the observed effect could be readily attributed to thermal heating. They conclude that little coupling of vibrational energy occurs to the reaction coordinate in these systems.

The reaction

$$H_2^{\dagger} + D_2 \rightarrow 2HD$$

has been investigated by Bauer *et al.* (1973). Vibrationally excited H_2 molecules were generated by stimulated Raman scattering of ruby laser radiation in H_2 gas

$$H_2(v = 0) + (h\nu = 14{,}404\ \text{cm}^{-1}) \rightarrow H_2(v = 1) + 10{,}251\ \text{cm}^{-1}$$

This is followed by rapid V–V transfer so that the reaction likely involves molecules in a variety of excited vibrational levels.

A summary of data on the enhancement of simple biomolecular reactions via vibrational excitation of one reactant is given in Table 4.5.

Farneth *et al.* (1976) have reported the photosensitized chemical decomposition of tetramethyl-1-2 dioxetane (M). This reaction was initiated by the

able 4.5. Enhancement of Simple Reactions via Vibrational Excitation of Reactants

Reaction	ΔH (kcal mol^{-1})	E (kcal mol^{-1})	$h\nu$ (kcal mol^{-1})	$(k^{\dagger}/k)^a$	Reference
$Cl^{\dagger}$ ($v = 1$) + O → OH + Cl	0.8	4.5	8.2	300	Karny *et al.* (1974)
$_3^{\dagger}$ + SO → SO_2 ($^1B_1$) + O_2	−25	—	3.0	2.5	Kaldor *et al.* (1974)
$_3^{\dagger}$ + O_2 ($^1\Delta g$) → $2O_2$ + O	1.4	—	3.0	38	Kurylo *et al.* (1974b)
$_3^{\dagger}$ + NO → NO_2 ($^2B_1$) + O	−20.0	4.2	3.0	4.1	Kurylo *et al.* (1974)
$_3^{\dagger}$ + NO → NO_2 ($^2A_1$) + O	−48.0	2.3	3.0	17.1	Kurylo *et al.* (1974)
$O^{\dagger}$ + O_3 → NO_2 ($^2B_1$) + O_2	−20.0	4.2	5.4	5.7	Stephenson and Freund (1976)
$H_4^{\dagger}$ + Cl → HCl + CH_3	−1.0	3.8	3.7	<30	Panfilov and Molin (1978)
$H_3F^{\dagger}$ + (Br($^2P_{3/2}$) → HBr + CH_2F	12.0	14.6	3.0	30	Krashopyrov *et al.* (1979)
$H_3F^{\dagger}$ + Br($^2P_{1/2}$) → HBr + CH_2F	1.5	4.5	3.0	1	Krashopyrov *et al.* (1979)

atio of reaction rate with vibrational excitation to that in normal thermal reaction at 300 K.

absorption of 9.6-μm CO_2 laser radiation by CH_3F, followed by VV transfer to M.

$$CH_3F + (h\nu = 9.6\ \mu m) \rightarrow CH_3F^{\dagger}$$
$$CH_3F^{\dagger} + M \rightarrow CH_3F + M^{\dagger}$$
$$M^{\dagger} \rightarrow A^{\dagger} + A$$

where A is acetone, C_3H_6O, and possible V–T routes have not been included. Electronically excited acetone molecules radiate in a band near 410 nm. The time dependence of this fluorescence together with IR fluorescence from CH_3F is shown in Figure 4.29. The long tail on the visible fluorescent signal is attributed to enhanced reaction due to thermal heating. The estimated temperature rise was 100–150 K.

A number of studies have reported the initiation of explosive reactions with pulsed CO_2 laser radiation [Lyman and Jensen (1973), Bauer *et al.* (1978), and Bauer and Haberman (1978)]. An example is the laser-initiated reaction between SF_6 and SiH_4

$$SF_6 + SiH_4 \rightarrow S_2^{*} + SiF_4 + HF + H_2$$

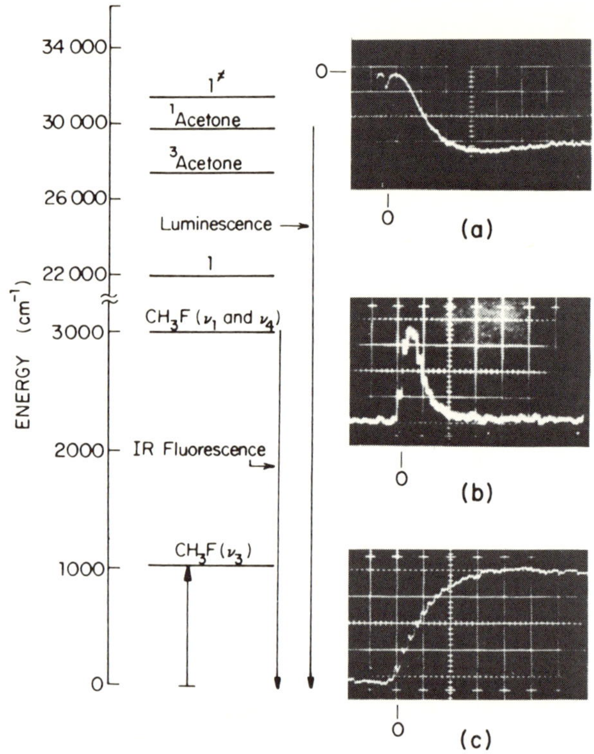

Figure 4.29. A partial energy level diagram for CH_3F, 1, and acetone up to about 32000 cm^{-1}. Energy scale referenced to CH_3F and acetone both in ground vibrational state 1* denotes the transition state for the decomposition of 1. Also displayed are: (a) Time resolved visible luminescence ($\lambda_{max} \simeq 410$ nm). Signal is inverted with respect to (b) and (c). The small spike at short time is an instantaneous spark reflected off the front of the reaction cell. Zero on the absicca represents the initiation of the laser pulse. Zero on the ordinate represents the ambient signal level. (b) Time resolved 3-μm infrared fluorescence. Zero on the abscissa represents the initiation of the laser pulse. (c) Time resolved translation heating. Zero on the abscissa represents the initiation of the laser pulse. Time base is 5 μs per major division along the abscissa in (a), (b), and (c) measured with P_{CH3F} = 5 torr and $P_{C6H12O2} \simeq 1$ torr. The translational heating was measured via the thermal lens effect. 1 ≡ tetramethyl-1,2-dioxetane [Courtesy of Farneth *et al.* (1976) Reprinted with permission; copyright 1976 American Chemical Society].

[Bauer and Haberman (1978)]. This reaction occurs via collisions with vibrationally excited SF_6 and SiH_4. The overall reaction is isotope selective although the detailed mechanism is uncertain.

4.13.3. Vibrationally Induced Decomposition

In this section we consider reactions that are initiated by the absorption of several IR laser photons leading to dissociation. Multiphoton infrared laser photochemistry has been extensively reviewed by Ambartzumyan and Letokhov (1977). A discussion of the multiphoton dissociation (MPD) process has been given in Section 4.10.

One of the simplest systems in which MPD has been observed involves the dissociation of NH_3 with high-power CO_2 laser radiation [Campbell *et al.* (1976a, b) and Avouris *et al.* (1979)]. The absorption of radiation leads to the decomposition

$$NH_3 + nh\nu \rightarrow NH_2(^2B_1) + H$$

or

$$NH_3 + nh\nu \rightarrow NH_2(^2A_1) + H$$

where $^2A_1$ is an electronically excited state. Figure 4.30 shows the variation of NH_2 fluorescence (excited by a CW dye laser) with frequency in the region of the 9.4- and 10.4-μm bands. The largest signals are observed for lines in the 10.4-μm band to the red of the NH_3 band heads. Such a red shift for optimum yield was predicted theoretically by Mukamel and Jortner (1976a, b).

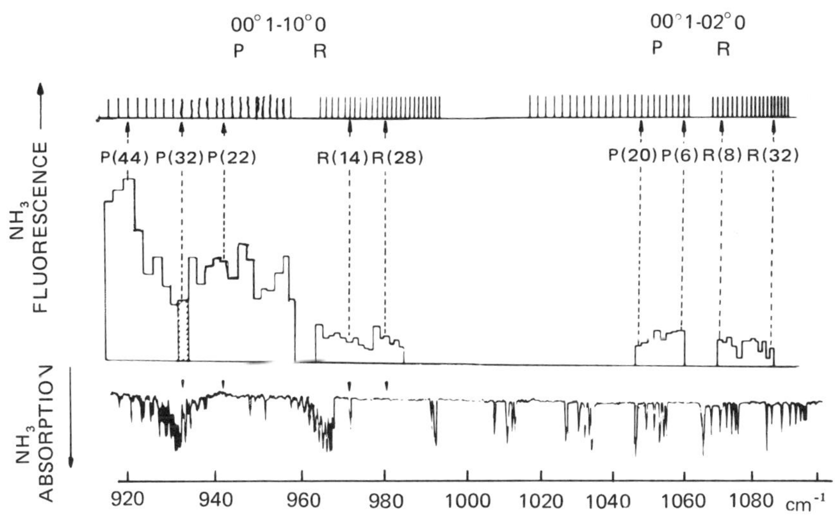

Figure 4.30. NH_2 laser excited fluorescence intensities as a function of CO_2 laser line. The upper part of the figure indicates the CO_2 laser line positions, with a normal infrared spectrum of NH_3 shown below on the same wavenumber scale. The central part of the figure shows the NH_2 laser-excited fluorescence intensities measured at a total pressure of 1 torr NH_3, and normalized with respect to CW laser power and CO_2 laser energy. Measurements were made using the following CO_2 lines: $P(4)$ to $P(48)$, $R(4)$ to $R(34)$ of the (00°1–10°0) band, $P(6)$ to $P(20)$, $R(8)$ to $R(32)$ of the (00°1–02°0) band [Campbell *et al.* (1976b)].

Table 4.6. Some Representative Studies of MPD-Induced Chemical Reactions

Absorber	Reaction[a]	Comments	Reference
$CClF_3$	$M \rightarrow C_2F_6, CF_4, CCl_2F_2$		Dever and Grunwald (1976)
	$M + H_2 \rightarrow CHF_3, C_2F_4, C_2H_2$		
CF_3Br	$M \rightarrow C_2F_6, CF_4, CF_2Br_2, Br_2$		Jalenak and Nogar (1979)
CF_3I	$M \rightarrow CF_3 + I$	Titration of products with NO_2 or Br_2	Rossi *et al.* (1979)
CH_3X ($X = F, Cl, Br$)	$M + Cl_2 \rightarrow CH_2XCl + HCl$		Earl and Ronn (1976)
CH_3NH_2	$M \rightarrow CH_3 + NH_2^*$		Filseth *et al.* (1979)
CH_3NO_2	$M \rightarrow H_2CO^\dagger, CH_3OH^\dagger, NO, N_2O$	IR luminescence seen from $H_2CO^\dagger$ and $CH_3OH^\dagger$	Hartford (1978)
CH_3OH	$M \rightarrow CO, CH_4, C_2H_6, C_2H_4, C_2H_2$	Products independent of laser frequency	Bhatnagar *et al* (1979)
CF_3CH_3	$M \rightarrow CF_2CH_2^\dagger + HF^\dagger$	Similar reaction with CF_2ClCH_3 and C_2H_5Cl	West *et al.* (1978)
CH_2CO	$M \rightarrow CO, H_2, C_2$	Collision-assisted dissociation	Jalenak *et al.* (1979)
$C_2H_2Cl_2$	*trans*-M $\rightarrow$ *cis*-$C_2H_2Cl_2$		Ambartzumyan *et al.* (1976)
C_2F_3Cl	$M \rightarrow C_2F_4, ClFC = CFCl$ (*trans*)		Nagai *et al.* (1979)
C_2H_4	$M + Cl_2 \rightarrow HCl^\dagger + Cl + C_2H_3$		Quick *et al.* (1980)

C_3H_6	M →	Strong wall effect	Karny and Zare (1977)
trans-butene, C_4H_8	M → CH_4, C_2H_4, *cis*-2 butene, organics		Yogev and Loewenstein-Benmair (1973)
i-C_4H_9X (X = Cl, Br, I)	M → *i*-C_4H_8		Braun and Tsang (1976)
tetramethyldioxetane	M → acetone		Haas and Yahav (1977)
cyclo-C_4F_8	M → tetrafluoroethylene		Preses *et al.* (1977)
N_2F_4	M → 2 NF_2		Lyman and Jensen (1972)
O_3, OCS		First triatomic dissociated via IR MPD	Proch and Schröder (1979)
N_2H_4	M → 2 NH_2		Filseth *et al.* (1979)
UO_2(hfacac)$_2$THF	M → UO_2(hfacac)$_2$ + THF		Kaldor *et al.* (1979)
cyclo-propane	M → C_2H_2, CH_4, C_2H_4, C_2HCH_3	Analogous flame reaction	Lesiecki and Guillory (1977)
vinylcyclopropane	M + NO → M + O_2 M → organics		Farneth *et al.* (1979)
ethyl vinyl ether	M → aldehydes, C_2H_4	Depends on pulse duration	Brenner (1978)
cis-3-4-dichlorocyclobutene	M → *cis*-; *trans*-1, 4; dichloro-1,3; butadiene	Only product	Danen *et al.* (1979)

[a]These are not necessarily in stoichiometric form.

An experimental study of the dependence of dissociation on pulse energy [Avouris *et al.* (1979)] shows that two distinct thresholds are observed—one at ~ 50 mJ per pulse and the other at ~ 0.4 J per pulse. The second threshold is characterized by an increase in the slope of a plot of [NH_2] vs. E. It is suggested that this effect may arise from the onset of power broadening.

Other studies of the luminescence accompanying MPD have been reported by Isenor and Richardson (1971), Ambartzumyan *et al.* (1974, 1975). The work of Ambartzumyan *et al.* (1975) showed that many polyatomic molecules dissociate to simple radical fragments following MPD by CO_2 laser radiation. For example, methanol was found to yield electronically excited CH, OH, and C_2 while the MPD of CH_3NO_2 generated emission from CN, CH, OH, and NH or O_2. The variety of species produced suggests that absorbed IR laser energy is distributed over many vibrational modes within an excited molecule and that more than one of these modes may lead to decomposition.

A list of representative chemical reactions induced by MPD with high-power pulsed O_2 laser sources is given in Table 4.6. Some aspects of MPD-induced chemistry can be obtained from studies that have been carried out on the trifluorohalomethanes [Earl and Ronn (1976), Jalenak and Nogar (1979), Rossi *et al.* (1979)]. These studies show that the product yield is a sensitive function of laser wavelength. Furthermore, this dependence also is a function of laser pulse length. An example of these effects taken from the work of Rossi *et al.* is shown in Figure 4.31.

Jalenak and Nogar (1979) have obtained the following parametric dependence for CO_2 laser-induced MPD of CF_3Br. The primary products of this photolysis were CF_4, C_3F_6, and CF_2Br_2. It was found that the yield of CF_4 and C_3F_6 increased linearly with number of laser pulses up to $\simeq$ 75 pulses, at which point a weaker increase was observed. The yield of C_2F_6 per laser pulse increased linearly with CF_3Br pressure up to ~ 50 torr. Addition of NO or N_2 resulted in a decrease in yield with increasing pressure. The absorbed energy per pulse was found to increase linearly with CF_3Br pressure up to ~ 40 torr in the absence of foreign gases. With the addition of NO and N_2, no such increase was observed. Decrease observed with increasing foreign gas pressure can be attributed to collisional deactivation of $CF_3Br^{\dagger}$.

Prior to the experiments of Proch and Schröder (1979), MPD had been observed only in molecules with more than three atoms. This limitation arises because the density of available vibrational states at low total system energy is small for molecules with only a few atoms. Proch and Schröder showed that MPD could be produced in the triatomics O_3 and OCS if the laser fluence was increased sufficiently. Their curve of dissociation probability for OCS vs. fluence is shown in Figure 4.32. This result was obtained with the $P(24)$ line at 1043.2 cm^{-1}. A clear energy threshold of 50 J cm^{-2} was obtained. Under the conditions used, the power broadening was only 0.3 cm^{-1}. Detuning the laser frequency eliminated MPD.

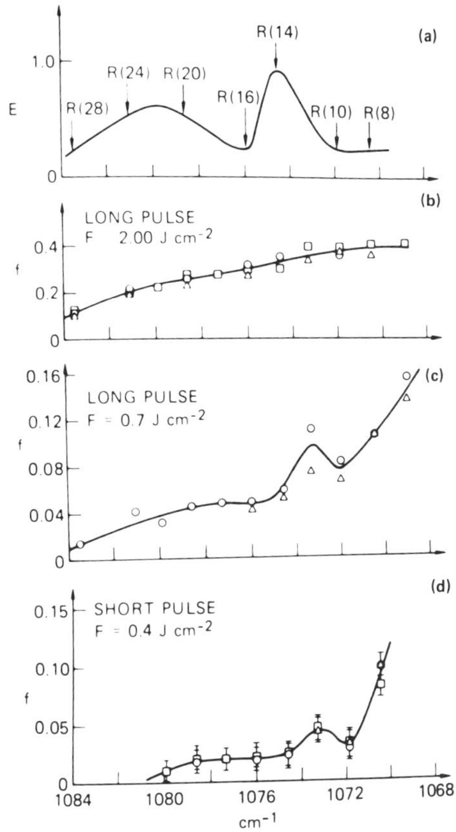

Figure 4.31. Wavelength dependence of dissociation yield of CF_3I upon irradiation in the 9.6-μm transition. (a) small signal absorption spectrum [8]; (b) long pulse (2 μs), $F = 2.0$ J cm^{-2}; (c) long pulse (2 μs), $F = 0.7$ J cm^{-2}; (d) short pulse (150 ns), $F = 4$ J cm^{-2} [Rossi *et al.* (1979)].

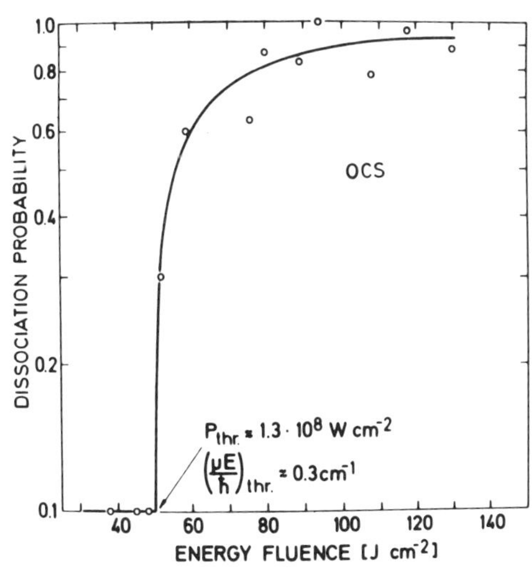

Figure 4.32. Dissociation probability of OCS in a supersonic molecular beam. Threshold intensity and the corresponding power broadening are indicated on the diagram [Proch and Schröder (1979)].

Laser decomposition induced by CW CO_2 laser radiation are summarized in Table 4.7. The lower excitation rates available with CW laser sources ensure that MPD is not an important process. Instead, one has rapid energy transfer via V–V processes that lead to the population of highly excited vibrational levels and can produce dissociation.

The effect of energy redistribution by collision can be seen by examining the pressure dependence of a factor Q which is equal to the number of photons absorbed per number of molecules dissociated. Under conditions at which collisions are unimportant, Q should be independent of pressure. Figure 4.33 shows a plot of Q vs. laser power for several pressures [Zitter and Koster (1978)]. These data were obtained from the CO_2 laser-induced decomposition of CH_3CF_2Cl. It indicates that fewer absorbed photons are required to produce dissociation as the pressure increases. Zitter and Koster have provided evidence that $Qh\nu$ should approach ΔH for the reaction at high laser power and high pressure. This is 5.2 photons per molecule for CH_3CF_2Cl. The corresponding value for E_a, the activation energy, is 22 photons per molecule. The Zitter and Koster data indicate that the excess energy $E_a - \Delta H$ is almost totally converted to excitation in the reactants. This is likely via V–V transfer from HCl†.

The trimerization of tetrachloroethylene sensitized by the excitation of BCl_3 with CW CO_2 laser radiation is a good example of a reaction that pro-

Table 4.7. Reactions Induced by CW CO_2 Laser Radiation

Absorber	Laser power (W)	Reaction	Reference
BCl_3	600	$M + H_2 \rightarrow B + Cl_2$	McAfee *et al.* (1976)
B_2H_6	1.5	$M \rightarrow B_{20}H_{16}$	Bachmann *et al.* (1974)
	8	$M \rightarrow B_{10}H_{14}$	Riley *et al.* (1978)
	8	$M \rightarrow$ various	Shatas *et al.* (1978)
$B_2H_6 + SF_6$	6	$M \rightarrow B_{10}H_{14}$ + others	Riley *et al.* (1979) Riley and Shatas (1979)
H_3BPF_3	100	$2M \rightarrow B_2H_6 + 2PF_3$	Lory *et al.* (1975)
D_3BPF_3		$2M \rightarrow B_2D_6 + 2PF_3$	Chien and Bauer (1976)
CF_2Cl_2	400	$M \rightarrow CF_3Cl + C + Cl_2$	Freeman *et al.* (1974)
	40		Slezak *et al.* (1978)
$C_2H_2 + SF_6$	300		Farneth *et al.* (1976)
$C_2H_6 + SF_6$	340	$M \rightarrow C$	Tardieu de Maleissye *et al.* (1976)
$C_2Cl_4 + BCl_3$	6	$M + BCl_3 \rightarrow C_6Cl_6$	Bachmann *et al.* (1977)
$C_2F_4Cl_2$	4	$M \rightarrow C_2F_4$	Zitter and Koster (1976)
CH_3CF_2Cl	10	$M \rightarrow HCl + CH_2CF_2$	Zitter and Koster (1978)
$C_{10}H_{16}$	5	$M \rightarrow$ isoprene	Yogev *et al.* (1972)

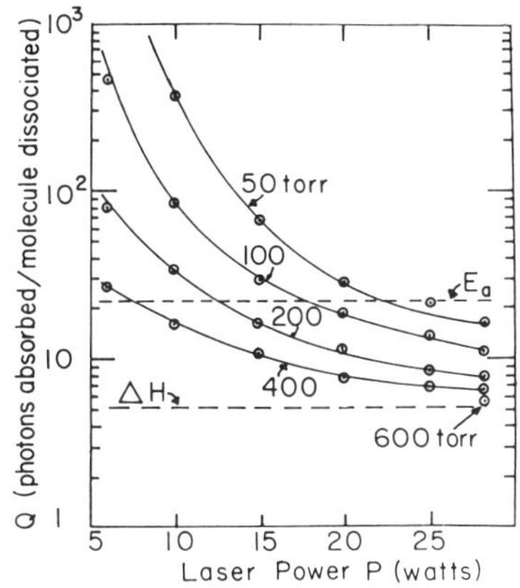

Figure 4.33. Quantum requirement Q as a function of laser beam power P for several gas pressures. The horizontal dashed lines indicate the activation energy and enthalpy in units of photons per molecule. The laser frequency is 953 cm^{-1} [Zitter and Koster (1978). Reprinted with permission; copyright 1978 American Chemical Society].

ceeds by V–V transfer [Bachmann *et al.* (1977)]. C_2Cl_4 is transparent to 940-cm^{-1} laser radiation while this is strongly absorbed by BCl_3. BCl_3 is therefore excited directly via absorption of laser photons while C_2Cl_4 is converted to C_6Cl_6 only as the result of collisional transfer of this excitation. It is noteworthy that the same reaction does not occur when SF_6 is added to C_2Cl_4, indicating that thermal heating is unlikely to be the source of the trimerization reaction.

Many experiments have been carried out on reactions induced in diborane, B_2H_6, with CW CO_2 laser radiation [Bachmann *et al.* (1974), Riley *et al.* (1978, 1979), Shatas *et al.* (1978), and Riley and Shatas (1979)]. Both neat B_2H_6 and B_2H_6–SF_6 mixtures have been irradiated. Initial experiment found that icosoborane, $B_{20}H_{16}$, was a major reaction product, but this has not been confirmed in subsequent work [Shatas *et al.* (1978)]. Controversy exists over the mechanism of these reactions: some experimenters conclude that thermolysis is dominant [Riley *et al.* (1978)] while others favor a V–V mechanism [Bachmann *et al.* (1974)].

The large absorption coefficient of SF_6 at CO_2 laser wavelengths means that SF_6 can be used to couple thermal energy into gaseous mixtures through V–T transfer. Shaub and Bauer (1975) have used this effect to induce laser-driven pyrolysis reaction in a variety of molecular gases. The addition of 10–100 torr of SF_6 was shown to be sufficient to produce a temperature rise of $\sim 10^3$ K along the path of a CW CO_2 laser beam with $I \sim 10$–10^2 W cm^{-2}. It was found that a 1-min irradiation of a system M + SF_6 was sufficient to yield almost complete decomposition of compounds with $E_a \sim 50$ kcal mol^{-1} and 10% decomposition of compounds with $E_a \sim 100$ kcal mol^{-1}. An advantage of this technique is that rapid heating is obtained without the influence of wall effects.

4.14. ISOMERIZATION

Figure 4.34 shows potential curves for *cis* and *trans* forms of a polyatomic molecule together with some processes that can lead to *cis–trans* conversion. The simplest is single-photon isomerization. Here, absorption of a photon lowers the activation energy for isomerization when it results in a transition to a higher vibrational state.

Miltiphoton absorption can lead to excitation of a high vibrational state which can mix with a mode that leads to isomerization. The deactivated species is then the isomer of the original molecule.

Laser heating via V–T transfer can also produce thermal isomerization as shown in the figure. Such a process would not differ from normal thermal isomerization.

Ambartzumyan *et al.* (1976) were one of the first groups to report multiphoton isomerization using a pulsed IR laser. Their data for $C_2H_2Cl_2$, shown in Figure 4.35 relates the rate of isomerization to that of dissociation as laser pulse energy increases. It is evident that MP isomerization scales almost identically with MPD over an energy range $E_{\text{pulse}} \lesssim 0.75$ J. Fluorescence is also observed due to the formation of CH* and C_2^* following MPD.

The activation energy for trans–cis isomerization in dichloroethylene is 1.78 eV and the *cis* form is more stable than the *trans* form by 0.02 eV. It was found that a mixture of 80% *trans*- and 20% *cis*-$C_2H_2Cl_2$ was converted to 30% trans and 50% cis after $\sim 2 \times 10^4$ pulses at a fluence of 100 J cm^{-2}.

Some other studies of laser-induced isomerization have been reported by Buechele *et al.* (1979) (isomerization of hexadienes), Yogev and Benmair (1977) (electrocyclic isomerization of hexafluorocyclobutene), Cheng and Keehn (1977) (isomerization of allene and methylacetylene), and Nagai and

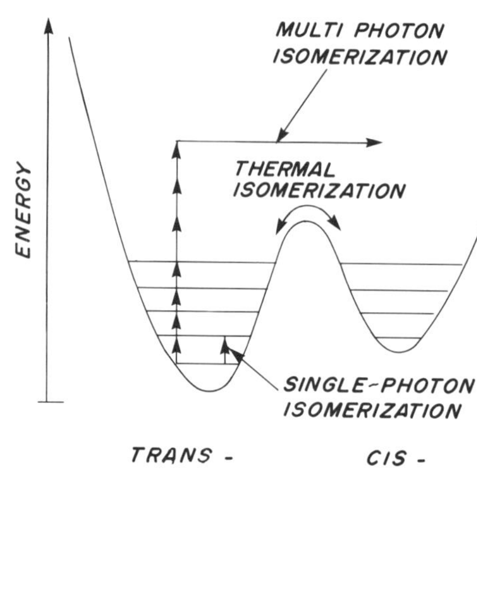

Figure 4.34. Conversion from *trans-cis* forms via absorption of photons.

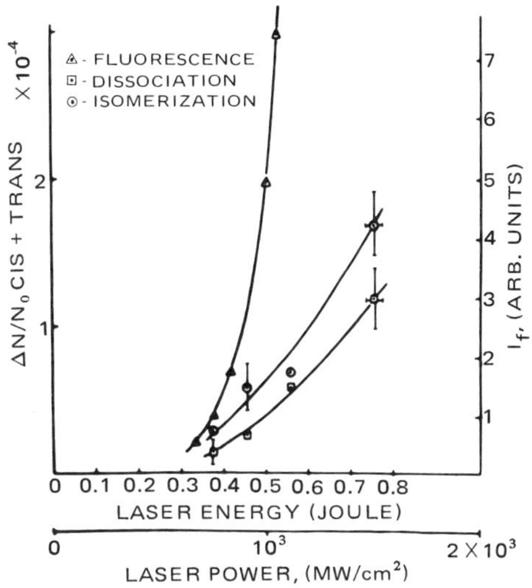

Figure 4.35. The dependence of the relative rates of dissociation, isomerization, and intensity of visible luminescence vs. intensity and energy density of the incident laser radiation; ν_{laser} = 980.9 cm^{-1}. $C_2H_2Cl_2$ pressure is 1 torr. The luminescence signal was picked up from the whole luminous region [Ambartzumyan *et al.* (1976e)].

Katayama (1977) (1,2 dichloroethylene). Some studies have also been carried out in matrices [Davies *et al.* (1977) and Tasumi *et al.* (1979)]. The isomerization of 1,2-diiodoethylene in the Ar^+ laser-induced reaction of I_2 and C_2H_2 has also been observed [Kushawaha (1980)].

Enhancement of isomerization rates by the addition of a selective absorber has been reported by Cheng and Keehn (1977) and Nagai and Katayama (1977). The latter show that the rate of isomerization of *trans*-1,2-dichloroethylene is increased sixfold by the addition of 1 torr of SF_6 to 10 torr of the isomer before irradiation with pulses from a TEA laser. A similar result was obtained under irradiation with a 10-W CW CO_2 laser. This effect is presumably due to thermal isomerization.

4.15. LASERS IN CATALYSIS

It has been shown that the excitation of gaseous molecules with infrared laser radiation can enhance reaction rates and alter reaction routes. Such effects should also occur for adsorbed molecules. This suggests that new chemical effects may be observable if laser radiation is used to selectively excite specific vibrational modes of adsorbed species. In particular, it should be possible to influence the course of catalytic reactions occurring on the surface of a solid catalyst by exposing the surface to laser light of the correct frequency.

Some anticipated effects of laser light on adsorbed species are the following:

1. *Photodesorption.* Absorption of laser radiation increases the probability that the adsorbate will leave the surface. This may be the result of direct excitation of the adsorbate–surface bond or by excitation of a vibrational state within the adsorbate that can exchange energy with the surface–adsorbate bond. Alternatively, excitation of an electronic transition in the absorbate may lead to desorption of excited molecules or to desorption following reemission to an excited vibrational level in the ground state. Photodesorption can also occur when an electronic transition within the substrate creates an exciton that migrates to the surface and is destroyed at a surface site. The resulting energy then acts to desorb an adsorbed molecule.

2. *Photocatalysis.* We can examine some potential photocatalytic effects with the aid of Figure 4.36. Part (a) shows that the absorption of IR radiation corresponding to an internal vibrational mode of AB can enhance the reaction AB + C, leading to the desorption of the product. Part (b) shows how an intense radiation field can alter the products of a heterogeneous catalytic reaction by exciting particular modes of an intermediate complex formed during the catalytic reaction of AB + C. By exciting specific vibrational modes of this complex, the reaction may be driven into novel reaction channels.

Figure 4.36(c) shows an example of laser-enhanced diffusion. Absorption of infrared radiation by AB can reduce the activation energy for surface diffusion and lead to reaction with C some distance away.

In part (d) the excitation of AB occurs in the gas phase prior to adsorption. A vibrationally excited molecule therefore arrives at the surface. This may react in a different way with other adsorbed species or may undergo an enhanced reaction along with a normal reaction route.

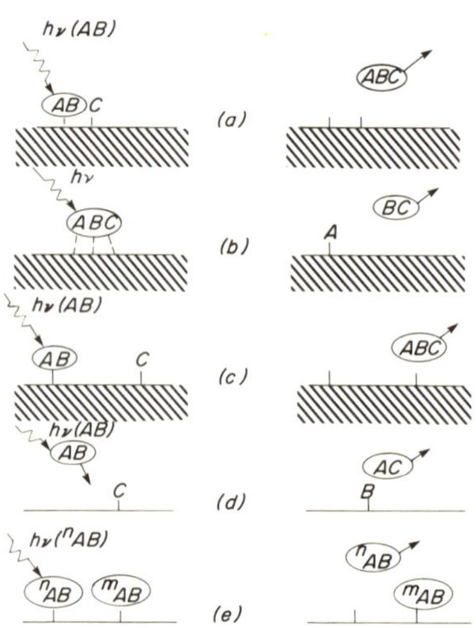

Figure 4.36. Some photocatalytic processes (see text for description).

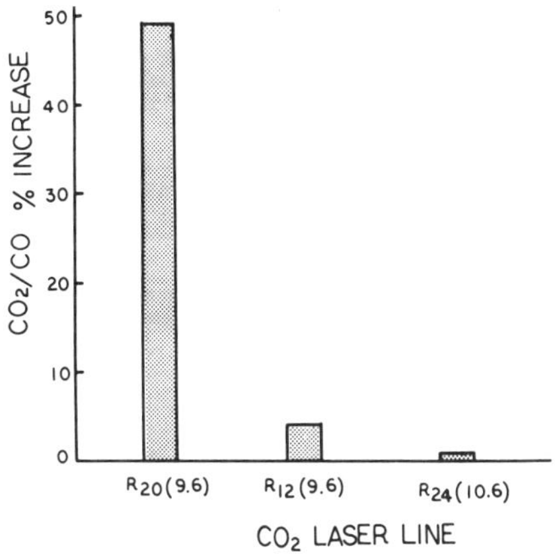

Figure 4.37. Change in CO_2/CO ratio effected by laser radiation: 9.6 μm, $R(20)$, strongly absorbed; 9.6 μm, $R(12)$, weakly absorbed; 10.6 μm, $R(24)$, nonabsorbed [Umstead and Lin (1978)].

The final process, shown in Figure 4.36, leads to isotope separation. The IR radiation field is tuned to excite a particular isotomeric surface species which then either desorbs directly or participates in an enhanced surface reaction before desorbing. In either case, the gas phase is enhanced in ^{n}AB while the surface preferentially retains ^{m}AB.

While there has been considerable theoretical interest in these processes [Askaryan and Namiot (1975a,b), Goldanskii *et al.* (1976), Gochelashvili *et al.* (1975), and Gangwer (1978)], few experimental studies have been carried out [Bass and Fanchi (1976) and Umstead and Lin (1978)]. Experimental difficulties involve the separation of gas-phase and surface-phase contributions. Furthermore, surface heating may mask selective laser-induced effects due to thermal reactions.

These problems have been addressed by Bass and Fanchi in their investigation of the enhancement of the N_2O–Cu surface interaction by vibrational excitation of N_2O and by Umstead and Lin in their study of the catalytic decomposition of formic acid on Pt.

Bass and Fanchi conclude that the reaction probability for an N_2O molecule excited to the (001) state is ~ 5000 times greater than that of ground state, (000), N_2O molecules colliding with Cu surfaces at 363 K. The overall reaction is

$$N_2O + Cu \rightleftharpoons N_2 + CuO$$

Since the reaction involves severing the N–O bond in N_2O, excitation of this band (the ν_3 mode) should enhance the reaction. Such an effect was indeed found as mentioned above.

The catalytic conversion of HCOOH to $CO_2 + H_2$ and $CO + H_2O$ was shown by Umstead and Lin (1978) to be sensitive to the presence of certain wavelengths of CW CO_2 laser radiation. The resulting change in the $[CO_2]/[CO]$ ratio under laser irradiation is shown in Figure 4.37. A large enhancement in this ratio was observed when the laser was tuned to the 9.6-μm $R(20)$ line. This resulted from a decrease in the amount of CO produced. No change in CO_2 yield was observed while the increase in the $[CO_2]/[CO]$ ratio was independent of pressure over the range 6–20 torr. The mechanism of the observed changes are uncertain in the absence of further empirical data. However, some radiation-specific interaction with the catalytic system does appear to be likely.

Isotope-sensitive evaporation from mixed $^{10}BCl_3$ and $^{11}BCl_3$ films has been reported by Karlov *et al.* (1977). A thin film of mixed composition was frozen on a transparent substrate at 77 K. This was then exposed to radiation within the $^{11}BCl_3$ band of the frozen solid. As a result, $^{11}BCl_3$ was selectively evaporated from the film. An enhancement of $\beta \sim 3$ was observed in the gas phase.

The potential use of photodesorption for isotope separation can be illustrated by means of an example [Gangwer (1978)]. We consider the adsorption of H_2O, HDO, and D_2O on a transparent substrate. The OH fundamental of H_2O adsorbed on silica lies in the 2.6–2.8 μm range while the OD fundamental in HDO absorbs between 3.5 and 3.7 μm. Thus irradiation of an H_2O/HDO–silica system with laser radiation near 3.6 μm would selectively excite HDO and could lead to desorption of this molecule. The problem is to estimate the excitation rate that would lead to significant enhancement of HDO in the gas over a selectively irradiated surface.

The residence time, t, of a molecule adsorbed with binding energy E_b on a surface at temperature T (K) is

$$t = \nu^{-1} \exp(E_b/k_B T)$$

where ν is a surface vibrational frequency ($\sim 10^{13}$ Hz) while k_B is the Boltzmann constant. If we take $E_b = 50$ kJ mol^{-1} $= 8 \times 10^{-20}$ J per molecule then $t = 2.5 \times 10^{-5}$ s when $T = 300$ K. The saturation intensity is

$$I_{\text{sat}} = \frac{h\nu}{2\sigma t}$$

where $h\nu = 2780$ cm^{-1} $= 5.5 \times 10^{-20}$ J, and σ is the excitation cross section ($\sim 10^{-20}$ cm^2). Then

$$I_{\text{sat}} = 1.1 \times 10^5 \text{ W cm}^{-2}$$

We now assume that the effect of vibrational excitation is to change the residence time for HDO to

$$t^* = \nu^{-1} \exp(E_b/h\nu)$$

Then, under these conditions, $t^* = 10^{-13} \exp(8 \times 10^{-20}/5.5 \times 10^{-20}) = 4.3 \times 10^{-13}$ s. This result is extremely sensitive to the value of $h\nu$ used in the calculation. For example, if we assume that only 10% of $h\nu$, on average, is retained by each adsorbed HDO molecule, then t^* becomes 2×10^{-7} s. Obviously a measurable effect will be observed only if the average energy retained per molecule is $\gtrsim k_B T = 4 \times 10^{-21}$ J in this example. However, when this constraint is satisfied, the desorption rate of HDO subject to laser excitation is significantly greater than the normal thermal rate.

If $t^* \ll$ (laser pulse duration) then essentially all selectively excited molecules will be desorbed during each laser pulse. Then, if R is the laser repetition frequency, p the partial pressure of desired isotomer in initial gas, and A the beam area the rate of isotomer separation, F is

$$F = 0.4AR\left(\frac{pn}{m}\right)^{1/2}$$

where n is the number of isotomer molecules in the gas and m is the mass of one molecule. This equation assumes that p is large enough that all requisite surface adsorption sites can be filled with the desired isotomer between laser pulses. This may be a problem as the surface becomes enhanced in the other isotomer since the requisite adsorption sites will not be available. Taking $p = 10^{-3}$ torr $= 0.13$ N m^{-2}, $m = 19 \times 1.6 \times 10^{-27}$ kg, $A = 10^{-4}$ m^2, and $R = 10$ Hz, one obtains $F \sim 10^{-5}$ mol min^{-1}. It is apparent that the throughput of an isotope separation system based on photodesorption is potentially large.

4.16. LASER-INDUCED REACTIONS: UV–VIS EXCITATION

In this section we briefly review some experiments that have been carried out on the stimulation of chemical reactions with the radiation from lasers operating in the UV or visible. Since much work has been done with incoherent radiation sources, the availability of laser sources has not opened a new field as in IR photochemistry. However, laser sources have been readily adopted as convenient high-intensity sources for photochemical studies in this region.

A selection of representative experiments that outline the capability of laser sources in photochemical studies is given in Table 4.8.

Table 4.8. Some Photochemical Studies with UV–VIS Laser Sources

Reaction	Laser	Features	Reference
$Na(^2P) + H_2O \rightarrow NaOH + H$	Flashlamp pumped dye	Flame study	Muller et al. (1978)
$Na(^2P) + H_2 \rightarrow NaH + H$			
$NO_2^* + CO \rightarrow NO + CO_2$	Ar^+-dye	Reaction induced at room temperature	Herman *et al.* (1976)
SiH_4	ArF	Removal of PH_3, AsH_3, and B_2H_6 via photolysis	Clark and Anderson (1978)
ortho $I_2 \rightarrow 2I$	Ar^+	Enrichment of para-I_2 in ortho–para mixture	Balykin *et al.* (1976)
$CF_2Cl_2 \rightarrow$	ArF	Multiphoton dissociation to yield CF^*, CF_2^*	Tiee *et al.* (1979)
$UF_6 \rightarrow UF_5 + F$	KrF	Measure IR spectrum of UF_5	Kim *et al.* (1979)
$CH_3NH_2 \rightarrow$	ArF	Two-photon dissociation to yield NH, CH, CH, and NH_2 via superexcited state	Nishi *et al.* (1980)
Cyclobutanone $\rightarrow$ C_2, C_3 products	Flashlamp pumped dye	Measures product distribution vs. excitation wave length	Yang and Lee (1976)
$Br_2 + C_4F_8 \rightarrow C_4F_8Br_2$	Tunable ruby	Early study	Tiffany (1968)
$S(solid) + C_2H_4 \rightarrow C_2H_4S$	KrF	Nonthermal reaction	Betteridge and Yardley (1979)

4.17. PROCESSING VIA THERMAL HEATING

Laser heating can be used directly in chemical processing. The experiments of Shaub and Bauer (1975) on the pyrolysis of SF_6–hydrocarbon mixtures is an example. Many other studies have been reported in the literature. Some of the first of these involved the initiation of reactions between C_n species and organic gas-phase molecules by the evaporation of C_n from graphite using laser heating [Namba *et al.* (1969), Schaeffer and Pearson (1969), and Blaxell *et al.* (1972)]. The dominant product was acetylene, C_2H_2.

The economic feasibility of generating C_2H_2 from coal with high-power CW CO_2 laser radiation has been discussed by Asmus (1971). Earlier studies showed that coals could be pyrolyzed with high-power pulsed or CW laser radiation [Sharkey *et al.* (1964), Karn *et al.* (1967, 1969, 1970), Shultz and Sharkey (1967), Karn and Singer (1968), Karn and Sharkey (1968), and Joy *et al.* (1968, 1970)]. While no economic feasibility has been demonstrated, laser pyrolysis may offer some advantage over conventional thermal volatilization of coals in certain situations. One of these is that the high temperatures reached in laser heating could enhance the methane–acetylene conversion $2CH_4 \rightarrow C_2H_2 + 3H_2$.

Laser heating has been used to characterize oil shales [Biscar (1971) and Hanson *et al.* (1975, 1976)]. This has been suggested [Biscar (1971)] as a method of measuring the hydrocarbon content of oil shales.

Reactions between sulfur and hydrocarbons have been studied by Miknis and Biscar (1971). The ejecta from the S surface under irradiation with pulses from a *Q*-switched ruby laser was shown to react with the products of gas breakdown at the surface to form CS_2.

The main advantage of lasers in pyrolysis lies in the ability to rapidly heat small sample areas and volumes to high temperature. The rapid heating and cooling can "freeze" in high-temperature chemical equilibrium leading to the formation of different chemical species from those observed in conventional pyrolysis.

4.18. POLYMERIZATION

Since polymerization can be initiated by a variety of radiation sources, it is not surprising that similar effects are observed with laser radiation. The primary process in the initiation of photopolymerization is the absorption of light by a sensitizer molecule [Frigerio and Stefanini (1971), Bukin and Kosyahov (1967), and Kuhlmann *et al.* (1977)] or directly in the monomer [Aleksandrov *et al.* (1977)]. Figure 4.38 shows the quantum yield (defined as the ratio of the number of monomers linked in the chain to the number of quanta absorbed by the solution) for laser-induced polymerization of acrylamide in solutions containing methylene blue and other dyes [Frigerio and Stefanini (1971)]. Evidently the yield is a sensitive function of the laser source used and the nature of the dye sensitizer. Aleksandrov *et al.* (1977) concluded from an experimental and theoretical study of the polymerization of methyl methacrylate that it should be possible to control the molecular weight distribution function of the resulting polymer by changing laser irradiation conditions.

In a related application, Kal'vina *et al.* (1975) found that the use of resonant frequency laser radiation significantly increased the degree of polymerization of alkyd resin-based enamel and lacquer coatings and also decreased drying times. A laser whose frequency corresponded to specific infrared bond resonances was used.

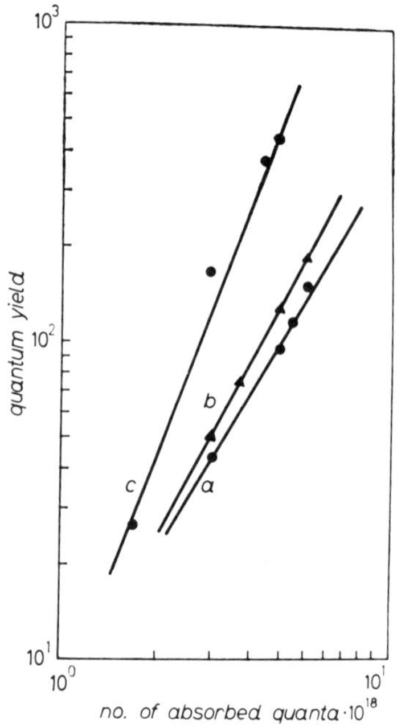

Figure 4.38. Quantum yields with various dyes: (a) solution containing 1.7×10^{-5} mol methylene blue radiated by ruby laser; (b) solution containing 5×10^{-5} mol methylene blue and 1×10^{-6} mol cosine Σ radiated by ruby laser; (c) solution containing 1.7×10^{-5} mol methylene blue radiated by He–Ne gas laser [Frigerio and Stefanini (1971)].

Polymerization can also be initiated by laser-induced breakdown. The breakdown plasma within a solution of the monomer presumably acts as a local source of heat and ultraviolet radiation as well as a source of free radicals which act to polymerize material adjacent to this breakdown region [Chin and Bedard (1971)]. A limitation to this and other laser polymerization processes is provided by the requirement that the monomeric solution be sufficiently transparent at the laser wavelength to permit penetration of laser radiation. A large absorption coefficient would limit polymerization to a thin surface layer.

Chapter 5

Lasers in Chemical Analysis

5.1. INTRODUCTION

By way of introduction to the use of lasers in chemical analysis Tables 5.1 and 5.2 summarize many of the laser-based analytical techniques discussed in this chapter together with primary areas of utility. While several of these represent significant advances over conventional technology others are novel but have not yet found wide application. Other laser applications relate to the improvement of standard analytical devices, e.g., the use of laser sampling in gas chromatography.

5.2. ABSORPTION SPECTROSCOPY

In a conventional absorption experiment the relation between incident intensity I_0 and transmitted intensity I is (Figure 5.1)

$$I = I_0 e^{-\alpha L}$$

where L is the length of the absorber and α is the absorption coefficient. For absorbers with cross section σ (cm^2) present at density N (cm^{-3}) this expression becomes

$$I = I_0 e^{-\sigma N L}$$

The sensitivity of direct absorption measurements is limited by the requirement that the absorption must be recorded against a large background. This can be seen as follows: For small absorption $e^{-\sigma NL} \simeq 1 - \sigma NL$. Then

$$I = I_0(1 - \sigma NL)$$

Table 5.1. Use of Lasers in Gaseous Sampling

Technique	Utility
Absorption	IR sampling of atmospheric molecules
Intracavity absorption	Flame diagnostics, isotope studies
Laser-induced fluorescence	Trace element analysis in flames
Multiphoton ionization	Detection of polyatomic molecules
Coherent anti-Stokes Raman spectroscopy	Combustion and plasma diagnostics
Laser Stark spectroscopy	Trace analysis of organic molecules
Laser photoacoustic spectroscopy	Gaseous pollutant, aerosol absorption

and defining percentage absorption as $100(I_0 - I)/I_0$

$$\text{Absorption} = 100\ \sigma NL$$

An absorption of 0.1% therefore corresponds to $\sigma NL = 10^{-3}$. With $\sigma = 10^{-12}$ cm^2, as typical of an atomic transition at 500 nm, $NL = 10^9$ cm^{-2}. For a 10-cm path length this yields N(detectable) $\sim 10^8$ cm^{-3}. While these numbers are reasonably conservative, they are valid to within an order of magnitude. This implies that the minimum detectable partial pressure of an absorber present in a gas cell with a 10-cm path length is $\sim 10^{-9}$ torr. The primary limitation to the determination of trace quantities of compounds is therefore the difficulty of measuring small values of $(I_0 - I)/I_0 = \Delta I/I_0$.

This limitation can be overcome in part by increasing L or performing measurements over long path lengths. However, this in often not practical, particularly in laboratory analysis.

The development of tunable laser sources in the visible region of the spectrum has done little to improve detection limits for gaseous species *using conventional absorption spectrometry*. Reasons for this are twofold: First, the use

Table 5.2. Use of Lasers in Sampling Liquids and Solids

Technique	Utility
Laser-induced fluorescence	Fluorimetry, phosphorimetry
Selective excitation of probe ion luminescence	Trace analysis of lanthanides
Laser microprobe	Microsampling of solid samples for trace analysis, atomic absorption spectrometry
Laser microprobe mass spectrometer	Microsampling
Laser photoacoustic spectroscopy	Analysis of trace molecules in liquids, biological and medical samples

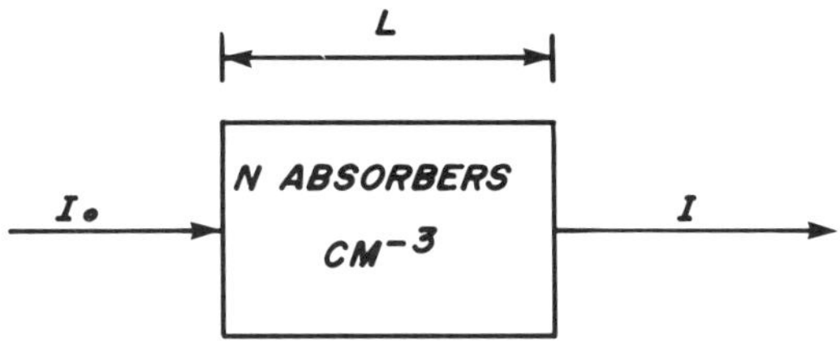

Figure 5.1. Incident and transmitted intensities in absorption.

of a narrow-band tunable source does not eliminate the need to measure ΔI in the presence of a large background, I_0. Second, under most conditions of interest, pressure broadening increases the linewidth of individual spectral lines to the point where σ at the center of the line is reduced significantly. Under these conditions, a narrow laser linewidth is of no advantage.

While these considerations are also valid in the infrared, the complexity of infrared spectra is such that individual rotational lines of vibrational bands are not normally resolved. A major advance in the study of gases by infrared spectroscopy has involved the development of tunable diode lasers with extremely narrow linewidth [Hinkley and Freed (1969)]. With linewidths of less than 10^{-5} cm^{-1}, complex spectra can be readily resolved. Furthermore, lines due to particular species can be sampled in the presence of other absorbers.

Figure 5.2 shows the diode laser spectrometer developed by Reid *et al.* (1978) for the measurement of trace absorption in gases. The White cell used

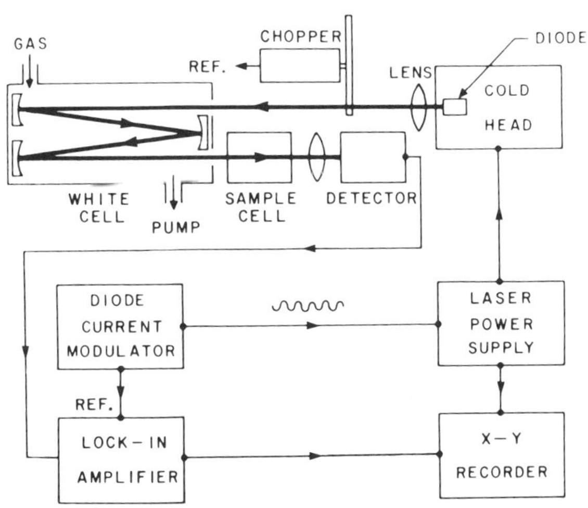

Figure 5.2. Schematic diagram of laser diode measurement system [Reid *et al.* (1978)].

in these measurements could be used to achieve pathlengths of up to 500 m. A typical spectral scan in the 1142-cm^{-1} region of SO_2 absorption is shown in Figure 5.3. It was found that frequency modulation of the laser output was superior to amplitude modulation in the detection of weak absorptions. The performance of this system was sufficient to permit detection of SO_2 at the $\lesssim$

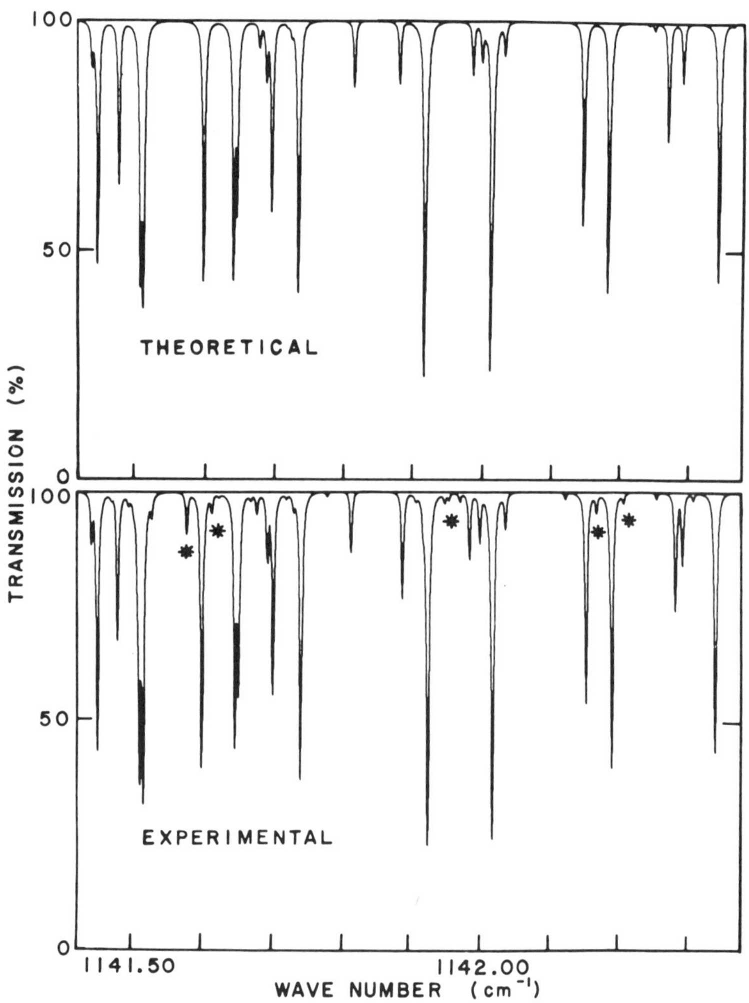

Figure 5.3. Experimental and theoretical transmission spectra for SO_2 in the 1142-cm^{-1} region. A path length of 60 m was used, with a 1000-ppm mixture of SO_2 in N_2 at a total pressure of 10 torr. Many weak SO_2 absorption lines which were obtained experimentally are not predicted by the theory; a few of these are indicated by an asterisk [Reid *et al.* (1978)].

Table 5.3. Minimum Detectable Concentration for Many Molecules of Atmospheric Importance[a]

Molecule	Absorption coefficient of strong lines (m^{-1} ppb^{-1})	Approximate wavenumber of lines (cm^{-1})	Sensitivity (ppb)
SO_2	3.5×10^{-8}	1140	3
O_3	2×10^{-7}	1050	0.5
N_2O	5×10^{-8}	1150	2
CO_2	3.5×10^{-10}	1075	300
H_2O	2×10^{-9}	1135	50
NH_3	2×10^{-6}	1050	0.05
PAN	$\sim 3 \times 10^{-7}$	1150	0.03
CH_4	3×10^{-6}	1300	0.03
SO_2	3.5×10^{-7}	1370	0.3
NO_2	5×10^{-6}	1600	0.02
NO	3×10^{-6}	1800	0.03
CO	10^{-5}	2120	0.01
CO_2	10^{-4}	2350	0.001

[a]The results are quoted for a sensitivity of 10^{-7} m^{-1} [Reid *et al.* (1978)].

1-ppb level. This was equivalent to the measurement of absorption coefficients $\alpha \sim 10^{-9}$ cm^{-1}. Table 5.3 gives a list of representative detection limits for a variety of gaseous molecules. It should be noted that sensitivities of this order can only be attained at low pressure ($\lesssim$ 10 torr total pressure): at higher pressure, broadening eliminates structure due to individual spectral lines. The sensitivity of discrete structure to pressure is shown in Figure 5.4 [Hinkley *et al.* (1976)].

Despite some problems associated with the operation of diode lasers [Dubs and Günthard (1978)] the sensitivity and selectivity of this technique has contributed greatly to infrared spectroscopy as a diagnostic technique. As one example of the utility of this method, Forrest (1980) reports on the use of diode laser absorption spectrometry in the detection and characterization of molecules in cigarette smoke (see Figure 1.9).

5.2.1. Absorption vs. Other Techniques

Detection limits can be lowered significantly by measuring the *effect* of absorption rather than absorption itself. In this case, the effect of delivering $I - I_0$ W cm^{-2} of optical power to the sample is measured by some other means than the reduction of transmitted intensity. This approach is the basis for some of the most sensitive analytical techniques in trace analysis. Some examples are:

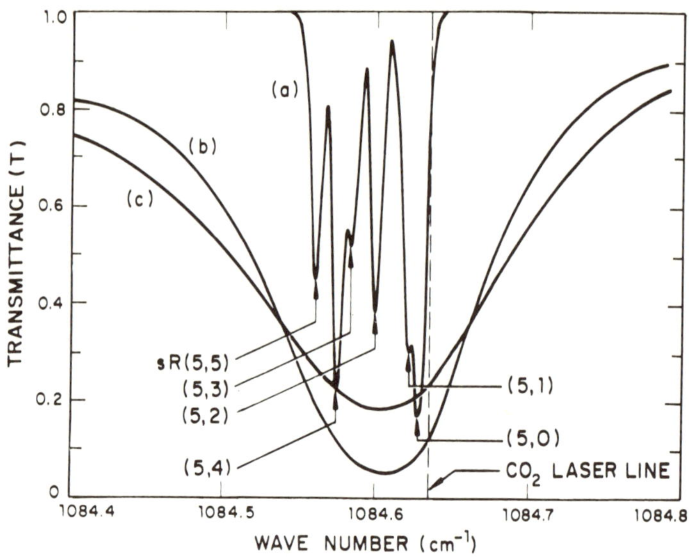

Figure 5.4. Laser spectroscopy of NH_3 using diode laser in closed-cycle refrigerator, operating at approximately 90 K. Trace (a) represents 0.1-torr NH_2; (b) is 0.46-torr NH_3 with 360-torr air added; and (c) is 0.46-torr NH_3 with 750-torr (1 atm) air added. The CO_2 laser line is $R(30)$ at 1084.635 cm^{-1} [Hinkley *et al.* (1976)].

Fluorescent analysis
Photoacoustic measurements
Laser-enhanced ionization
Photoionization

All of the above techniques enable the presence of absorption to be detected by its effect on some other parameter of the system. Since the measurement in this case can be made against a small background, great increases in sensitivity are possible.

The intracavity absorption technique represents an interesting hybrid process involving the measurement of absorption by its effect on the gain of a tunable laser cavity. While this involves a measurement of $\Delta I/I_0$, the quenching of oscillation within the absorption line profile enhances ΔI and often improves sensitivity over conventional absorption spectrometry.

5.2.2. *Intracavity Absorption*

The use of intracavity absorption techniques in the determination of weak absorption was first reported by Peterson *et al.* (1971). This method relies on the sensitivity of laser oscillation to cavity losses. When an absorber is introduced into the cavity of a dye laser, absorption resonances in the absorber introduce loss at wavelengths characteristic of the absorber. This quenches oscillation at these wavelengths. If the dye laser is operated in a broadband emission mode, the spectrum of the output will show decreased emission at these resonant wavelengths. Figure 5.5 shows how the presence of an absorption line in a material introduced into the laser cavity modifies the output emission.

In the first experiments on this effect by Peterson *et al.* (1971) the dye laser output was recorded photographically on a spectrograph. Na or I_2 vapor was placed in a 10-cm cell within the cavity. It was found that the minimum detectable column density of Na was $NL \sim 1 \times 10^{11}$ cm^{-2} when the cell was within the cavity and $NL \sim 8 \times 10^{12}$ cm^{-2} when absorption was measured with the cell outside the cavity. The resulting enhancement of absorption, while impressive, was limited by the detection method used—notably photographic spectrometry. The use of a spectrometer introduces broadening of the absorption resonance and can reduce sensitivity by many orders of magnitude.

Hänsch *et al.* (1972) showed that the full potential for an increase in sensitivity using the intracavity absorption technique can be realized by the use of *CW* laser excitation and fluorescent detection. This is shown schematically in Figure 5.6.

In the experiments by Hänsch *et al.* (1972) the absorption and fluorescence cells contained I_2 vapor. The detection of 3.3×10^{12} I_2 molecules cm^{-2} was reported with intracavity absorption. This represented an enhancement by a factor $\sim 10^5$ of detection limits from conventional absorption spectroscopy

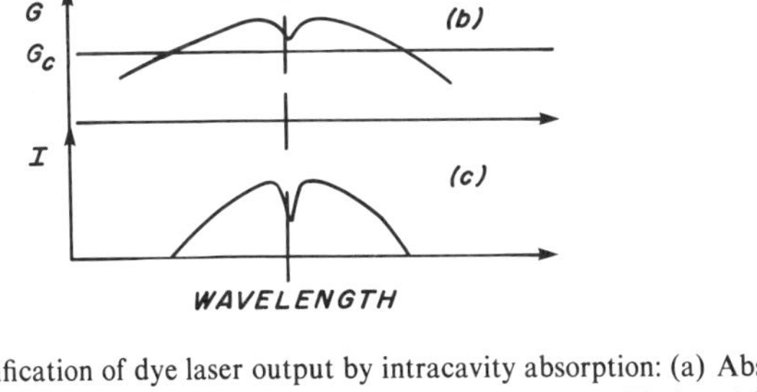

Figure 5.5. Modification of dye laser output by intracavity absorption: (a) Absorption cross section vs. wavelength; (b) gain curve—G_c is the critical gain for oscillation; (c) spectrum of dye laser output showing decreased emission at position of absorption line.

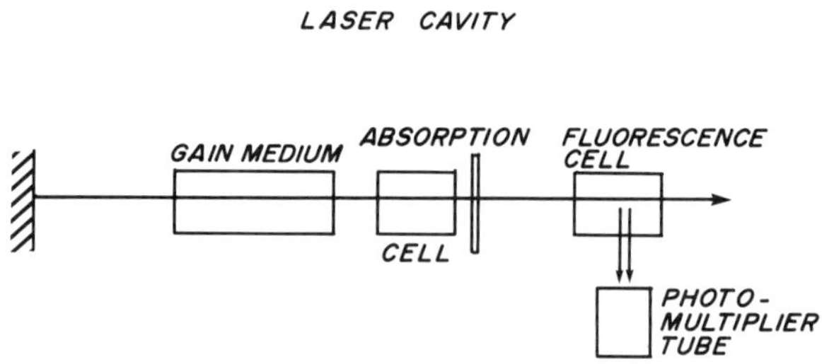

Figure 5.6. Intracavity absorption with fluorescent detection.

using the same laser and fluorescent cell detection. The gain in sensitivity from fluorescent detection derives from the high spectral resolution possible when the laser emission exactly overlaps that of absorption lines in the fluorescence cell. This effect can be seen clearly in Figure 5.7 which shows the fluorescent emission from cells containing $^{127}I_2$ and $^{129}I_2$ with and without intracavity absorption by $^{127}I_2$. When $^{127}I_2$ is placed in the laser cavity emission is quenched at the wavelengths of this isotope. This suppresses fluorescent emission from the external cell containing $^{127}I_2$. The use of intracavity techniques in the detection of $^{127}I_2$ and $^{129}I_2$ has been discussed by Hohimer and Hargis (1979).

There have been several theoretical studies of the mechanisms involved in intracavity absorption [Keller *et al.* (1972), Hänsch *et al.* (1972), and Brunner and Paul (1974)]. Enhancement has been shown to arise from a combination of increased effective path length due to multiple passes through the sample within the cavity and the threshold nature of laser emissions together with mode competition.

A simple calculation by Hänsch *et al.* (1972) valid near threshold shows that

$$\frac{dN_i}{N_i} = -d\eta_i\, 2j\, \frac{\alpha}{\eta(\alpha - \eta)}$$

where dN_i is the decrease in photon number in ith cavity mode, N_i the total number of photons in ith mode, α the gain parameter, η the loss parameter, j the number of oscillating modes within the spectral emission profile of the laser, and $d\eta_i$ the selective increase in loss of the ith cavity mode (due to the absorber). The enhancement factor therefore derives from the large value of j together with the near equality of α and η. For example, Hänsch *et al.* take $j = 10^4$ and $\alpha - \eta = 10^{-3}$ and find that enhancements of 10^7 are possible.

Brunner and Paul (1974) have generalized this analysis to the region far from threshold. Further discussion on the mechanism of intracavity absorption can be found in the articles by von Weyssenhoff and Rehling (1974), Holt

(1975), Tohma (1975, 1976), Stepanov and Rubinov (1975), Brunner and Paul (1976, 1978). Brink (1980), Lewellen and Brink (1981), Morgan *et al.* (1978), and Schröder *et al.* (1975) have studied the time development of interactivity absorption in pulsed dye lasers. The effect of saturation on the sensitivity of the intracavity absorption method has been discussed by Mironenko and Pack (1979).

A summary of analytical applications of the intracavity absorption technique is given in Table 5.4. The list of experiments is not exhaustive but has been chosen to be representative of typical studies.

A block diagram of a system used to detect intracavity absorption with CW laser is shown in Figure 5.8. Detection of absorption features superimposed on the dye laser output is usually via a photomultiplier tube on a grating spectrometer. With a repetitively pulsed or CW dye laser the spectrum can be obtained by scanning the monochromator. For a pulsed laser operating at a low repetition rate, point by point or photographic recording is often necessary. When the full spectrum is required on real-time basis, the photomultiplier tube can be replaced by an array of solid-state photodiodes [Horlick and Codding (1974)] or an optical multichannel analyzer [Atkinson *et al.* (1977)]. In the system developed by Atkinson *et al.*, a spectral resolution of 0.025 nm could be obtained. The intracavity absorption spectrum of NO_2 in a low-pressure cell was recorded by operating the optical multichannel analyzer in the difference mode. Spectra obtained with 15 pulses from the dye laser with and without the intracavity absorber were stored and differenced to yield the absorption due to

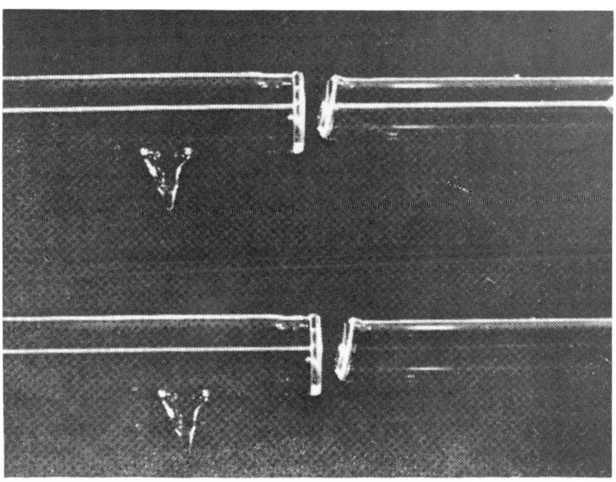

Figure 5.7. Fluorescence light from I_2^{127} (right) and I_2^{129} (left): (a) without intracavity absorption; (b) with a trace of I_2^{127} vapor inside the laser cavity [Hänsch *et al.* (1972) Reproduced with permission; copyright 1972 *IEEE*].

the sample. Integrated absorption under individual spectral features was found to be linearly related to NO_2 pressure for pressures $\lesssim$80–100 mtorr.

Intracavity absorption has been used to measure trace elements in flames [Thrash *et al.* (1971), Green and Latz (1974), Maeda *et al.* (1975), and van Dijk *et al.* (1978)] and free radicals in the flash photolysis of organic compounds [Atkinson *et al.* (1973)] as well as in conventional absorption cells [Antonov *et al.* (1975a), and Atkinson *et al.* (1977)].

A useful variation on the basic intracavity absorption technique has been reported by Crescenzi and Shirk (1977), who show that modification of the laser resonator permits the recording of spatial variations in the magnitude of intracavity absorption. In principle, this method could be used to construct a double beam (reference/absorber) intracavity absorption spectrometer analogous to conventional dual beam spectrometers.

The use of intracavity absorption to measure weak spectral transitions has been reported by Antonov *et al.* (1975b) and Bray *et al.* (1977). The relation between absorption coefficient α and the absorption cross section σ of an atom or molecule present at density N is

$$\alpha = \sigma N$$

For a strong electronic transition in the visible $\sigma = 10^{-12}$ cm^2. The minimum detectable absorption is such a gas using intracavity techniques corresponds to

Table 5.4. Summary of Absorption Measurements Using Intracavity Spectroscopy

Element/compound	Source[a]	Detection limit	Reference
Na	FLDL	0.2 ppb	Maeda *et al.* (1975)
	CWDL	5×10^5 cm^{-3}	Maeda *et al.* (1977)
Li	FLDL	5 ppb	Maeda *et al.* (1975)
Cs	FLDL	7×10^8 cm^{-3}	Childs *et al.* (1974)
Ba	FLDL	5000 ppm	Green and Latz (1974)
Sr	FLDL	—	Thrash *et al.* (1971)
Pr^{3+}	FLDL	1×10^{-4} *M*	Spiker and Shirk (1974)
Eu^{3+}	RLPDL	150 ppm	Horlick and Codding (1974)
Ho^{3+}	FLDL	3×10^{-4} *M*	Spiker and Shirk (1974)
I_2	CWDL	6×10^{10} cm^{-3}	Hänsch *et al.* (1972)
NH_2	FLDL	—	Atkinson *et al.* (1973)
HCO	FLDL	—	Atkinson *et al.* (1973)
NO_2	FLDL	$\sim 4 \times 10^{13}$	Atkinson *et al.* (1977)
O_2, H_2O	CWDL	—	Antonov *et al.* (1975a,b)
Anthracene, stilbene	—	—	Kleinschmidt *et al.* (1974)

[a]FLDL, flashlamp pumped dye laser; CWDL, CW dye laser; RLPDL, ruby laser pumped dye laser (frequency doubled).

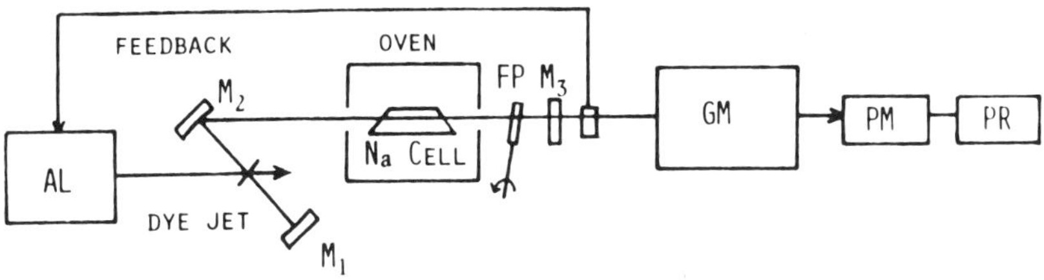

Figure 5.8. System for intracavity absorption measurement using a CW dye laser [Maeda *et al.* (1977).

$N \sim 10^6\ cm^{-3}$ [Maeda *et al.* (1977)]. Hence $\alpha_{min} \sim 10^{-6}$ cm. With $N = 2.7 \times 10^{19}\ cm^{-3}$ as appropriate for a gas at a pressure of 1 atm and 273 K, this value of α corresponds to a minimum detectable value of $\sigma = 10^{-6}(27 \times 10^{19}) = 0.37 \times 10^{-25}\ cm^2$. Such a small value of σ is characteristic of highly forbidden electronic transitions.

Bray *et al.* (1977) and Antonov *et al.* (1975b) have succeeded in recording absorption spectra of such highly forbidden transitions as the (2,0) band of the $b \leftarrow X$ transition of O_2 using intracavity absorption and a long path length cell. This transition has an oscillator strength of $\sim 10^{-12}$ and $\sigma \sim 10^{-24}\ cm^2$.

Reilly and Pimentel (1976) have investigated the use of intracavity techniques in the measurement of small values of gain in various laser media and fluorescent flames.

Applications of the intracavity absorption method in the field of laser isotope separation have been discussed by Datta *et al.* (1975), Stuke and Schäfer (1977), and Yamashita *et al.* (1978). The advantages derive from the fact that when the intracavity absorber is composed of a gas enriched in one isotope, the laser output can be used to selectively excite the other isotope component. This can be accomplished without detailed information for the compounds involved. Datta *et al.* (1975) have used a cell with ^{35}ICl inside the cavity of a CW dye laser to quench emission lines of this isotopic molecule. The output was used to irradiate a mixture of ICl + BrHC = CHBr. The resulting *trans*-Cl HC–CH Cl photoproduct was significantly enhanced in ^{37}Cl.

Further discussion of the role of intracavity absorption in laser isotope separation can be found in Chapter 4.

5.3. LASER-INDUCED FLUORESCENCE

In this section we consider the use of laser-induced fluorescence (LIF) as an analytical and diagnostic technique. The advantages of LIF over atomic absorption for the detection of trace species are many. Some notable contributions to the usefulness of this method are summarized below:

1. Only small sample volumes are required.
2. Increased signal-to-noise ratio because measurement of the fluorescence signal can usually be made against a low background.
3. At low light levels the intensity of fluorescence is directly proportional to emitter concentration.

The processes involved in the generation of fluorescence can be seen in Figure 5.9. When an atom or molecule absorbs a photon from a radiation field it makes a transition from its initial state to one of higher energy (transition A in the figure). This can lead to emission at the same wavelength (resonance fluorescence) or to emission at some other wavelength. (E_s and E_p in the figure). Prior to this emission, the molecule may make a transition to a more long-lived state. The subsequent phosphorescent emission is shown as E_p in part (a) of the figure.

Collisional excitation may lead to a redistribution of excited-state population following absorption. This can result in emission at a wavelength shorter than that used to excite the atom or molecule. This anti-Stokes emission is shown as E_{AS} in the figure. Further discussion of possible absorption/emission processes in atomic fluorescence can be found in Section 5.3.2. In this section

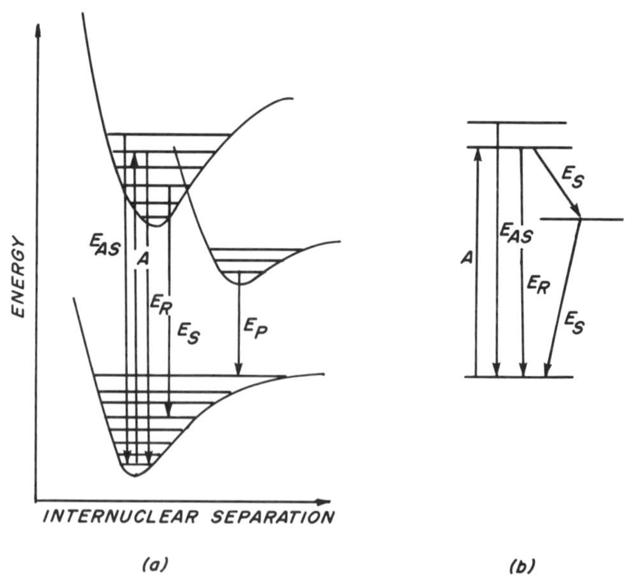

Figure 5.9. Transitions involved in (a) molecular and (b) atomic fluorescence. A, absorption; E_e, resonance fluorescence; E_S, shifted or Stokes fluorescence; E_p, phosphorescence; E_{AS}, anti-Stokes fluorescence.

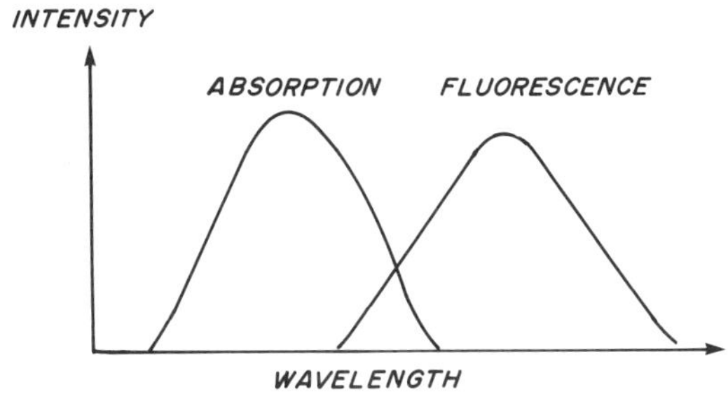

Figure 5.10. Relation between absorption and emission band profiles for a molecular transition.

we examine applications of LIF to the detection of molecular species in a variety of environments.

Typical absorption and fluorescence spectra for a large molecule are shown in Figure 5.10. In many cases the emission and absorption bands are essentially mirror images of one another. Excitation within the absorption band will lead to emission within the fluorescence band. A fluorescence experiment therefore involves monitoring the emission within the fluorescence band. This is accomplished by tuning a monochromator to a wavelength within the fluorescence band. If this system can respond to rapid changes in intensity, the fluorescent signal can be resolved temporally. This may be useful, particularly when two fluorescence bands overlap spectrally but have different time dependence.

Since the fluorescence band profile is often not dependent on the wavelength of excitation, LIF can be used to obtain information on the absorption bands of the fluorescent molecules by scanning the laser wavelength while monitoring the intensity of fluorescence at a particular wavelength within the fluorescence band. This yields the photoexcitation spectrum of the absorber. Since LIF is a sensitive measurement technique, weak absorption features may be seen in the photoexcitation spectrum even when the column density of absorbers is small.

A schematic diagram of a LIF system incorporating tunable wavelength excitation and time-resolved detection is shown in Figure 5.11. The sample is contained in a light-tight box together with optics to collimate the fluorescence and focus an image on the slit of the monochromator. Filters act to reduce Raman and Rayleigh scattered components in the light focused on the slit of the monochromator. The output of the photomultiplier tube is fed to a boxcar integrator. The boxcar window can be adjusted to sample the fluorescence signal some time after the trigger pulse provided by the laser. Since the position of this window is variable in time, it may be set to optimize the signal-to-noise ratio. The boxcar integrator can also be used to obtain the time dependence of the fluorescence signal.

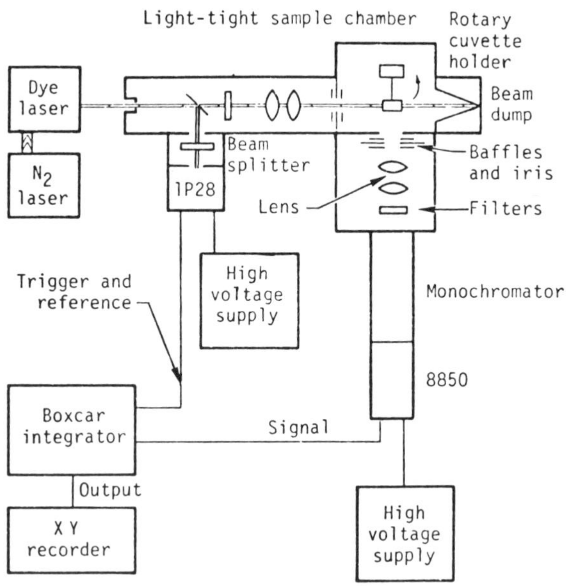

Figure 5.11. Schematic of a LIF system incorporating tunable wavelength excitation and time-resolved detection [Richardson and Ando (1977) Reprinted with permission; copyright 1977, American Chemical Society].

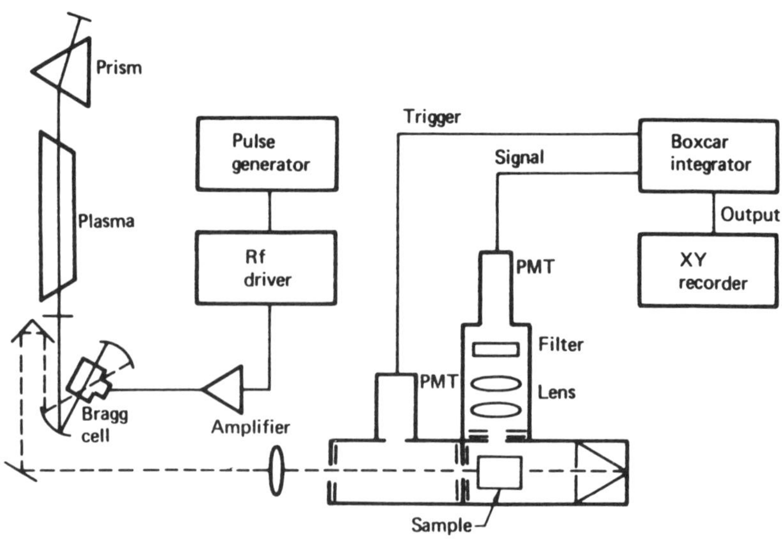

Figure 5.12. LIF system with excitation by pulses from a cavity-dumped Ar^+ laser together with boxcar detection [Richardson and George (1978) Reprinted with permission; copyright 1978, American Chemical Society].

A similar system utilizing cavity dumped pulses from an Ar^+ laser to excite fluorescence is shown in Figure 5.12. This resulted in the generation of pulses with a width ~ 1 MHz. Peak power was 5 W. This should be compared to the 1–10 kW peak power available from an N_2 pumped dye laser at a pulse repetition rate of 10–100 Hz.

Richardson and George (1978) have compared the performance of an N_2 pumped dye laser and a cavity dumped Ar^+ laser in the detection of rhodamine B via LIF. While both systems perform acceptably and provide a detection limit of ~ 1 ppt (2×10^{-12} *M*) for the rhodamine B in water, the flexibility of the dye laser system is preferred because it permits the excitation wavelength to be varied over a wider range. This allows detection to be optimized for a given solute–solvent combination.

Evaluation of a photon counting system under similar conditions and using pulse selection from a mode-locked Ar^+ laser showed that sensitivity was considerably reduced over gated analog detection. The detection limit for rhodamine B in water using this system was 15 ppt (3×10^{-11} *M*). However, a pulse-gated photon counting system has been used with a dye laser to yield sub part per trillion detection of fluorescein and riboflavin [Ishibashi *et al.* (1979)].

LIF has been used as a diagnostic technique in many areas of physics and chemistry. Major areas of application are discussed in the following sections.

5.3.1. Laser-Induced Fluorescence: Theory

In this section we estimate the fluorescence signal of a sample using a simple calculation based on the properties of the fluorescent system. Performing such a calculation requires that some simplifying assumptions be made about the electronic structure of the fluorescent atom or molecule. We adopt the model of Piepmeier (1972) and perform calculations for a three-level system as shown in Figure 5.13. This analysis is appropriate for a homogeneously broadened line in which the spectral linewidth exceeds the laser linewidth.

The notation used is as follows:

A_{ij} = radiation decay rate between levels i and j

Q_{ij} = collisional relaxation rate between levels i and j.

b_{ij} = rate of absorption or stimulated emission between levels i and j

$= \rho(\nu) B_{ij} f(\nu_0 - \nu)$

where $\rho(\nu)$ = laser energy density

B_{ij} = Einstein coefficient for stimulated emission

$f(\nu_0 - \nu)$ = lineshape function

ν_0 = absorber frequency

ν = laser frequency

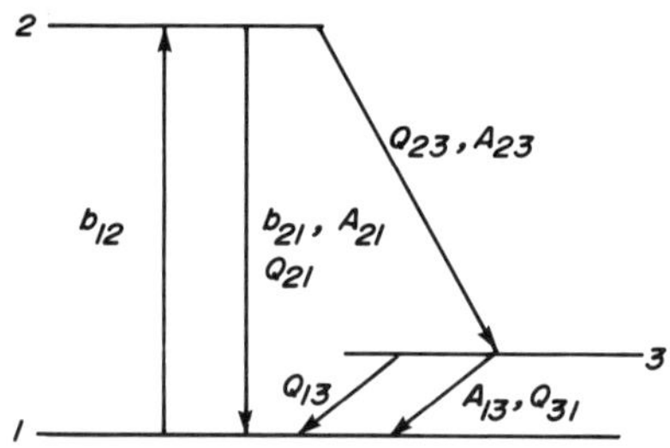

Figure 5.13. Energy levels used in calculation of fluorescence signal. Level 3 is coupled strongly to level 1 by collisional relaxation. Laser excitation occurs from level 1 to level 2.

If one assumes the linewidth is Lorentzian, then

$$f(\nu_0 - \nu) = \frac{A_{21} + Q}{4\pi^2(\nu_0 - \nu)^2 + [(A_{21} + Q)/2]^2}$$

where Q is the total quenching rate of level 2 due to collisional or other effects.

In equilibrium, excitation and deactivation balance for each level. Then, if N_i is the population of the ith level,

$$\frac{N_2}{N} = \frac{D\rho B_{12}}{(1 + L)[4\pi^2(\nu_0 - \nu) + [(A_{21} + Q)/2]^2 + D\rho\{B_{21} + [(1 + \mathrm{K})/(1 + \mathrm{L})]B_{12}\}}$$

where $N = N_1 + N_2 + N_3$ = total number of fluorescent atoms or molecules. Also

$$K = \frac{A_{23} + Q_{23}}{Q_{31} + A_{31}}$$

$$L = \frac{Q_{13}}{Q_{31} + A_{31}}$$

$$D = \frac{A_{21} + Q}{A_{21} + Q_{21} + A_{23} + Q_{23}}$$

If Doppler broadening is assumed so that the $1 \rightarrow 2$ transition has width $\Delta\nu_d$, then it can be shown that

$$\frac{N_2}{N} = F(a,R)\,\frac{R}{(1 + R)(1 + L)}\,\frac{1}{[g_1/g_2 + (1 + K)/(1 + L)]}$$

where

$$F(a,R) = \pi^{1/2}a(1 + R)^{1/2} \exp [a^2(1 + R)] \operatorname{erfc} [a(1 + R)^{1/2}]$$

$$a = 0.13 \frac{A_{21} + Q}{\Delta\nu_d}$$

$$R = \frac{4D\rho\{B_{21} + [(1 + \mathrm{K})/(1 + L)]B_{12}\}}{(A_{21} + Q)^2}$$

and erfc is the error function, while g_1 and g_2 are the degeneracies of levels 1 and 2, respectively.

The parameters a and R are, respectively, measures of the ratio of the homogeneous linewidth to the Doppler width and the saturation of the $1 \rightarrow 2$ transition. The parameter $F(a,R)$ approaches unity as either a or R become large. When this occurs and R is large (i.e., saturation of $1 \rightarrow 2$ transition) the fractional population in level 2 becomes

$$\lim_{R\rightarrow\infty} F(a,R) \rightarrow 1; \left(\frac{N_2}{N}\right) = \frac{1}{(1 + L)[g_1/g_2 + (1 + K)/(1 + L)]}$$

Note that in this limit the population in level 2 is independent of ρ or laser intensity.

Fluorescent photons are emitted from the excited volume, V, at the rate

$$\eta = A_{21}VN_2$$

The energy of each photon is $h\nu$, where ν is the laser frequency (we are assuming resonant fluorescence) and h is Planck's constant. Then the fluorescent power collected by an optical system that accepts light over a solid angle Ω is

$$P_F = h\nu \frac{\Omega}{4\pi} N_2A_{21}V$$

This is a general result and applies to fluorescence from level 2 regardless of the number of other levels in the system.

For a two-level system, the population in level 2 is

$$\frac{N_2}{N} = \frac{1}{(g_1/g_2 + 1)} \frac{\mathrm{R}}{(1 + R)} F(a,R)$$

with $R = 4D\rho(B_{21} + B_{12})/(A_{21} + Q)^2$. In this case, saturation yields $N_2/N = [(g_1/g_2) + 1]^{-1}$ and so knowledge of the degeneracies of levels 1 and 2 is sufficient to obtain a value for N_2/N. If P is then measured under controlled conditions

$$N = \frac{4\pi P}{h\nu\Omega A_{21} V}\left(\frac{g_1}{g_2} + 1\right)$$

and N the emitter density can be obtained directly.

The laser intensity required to achieve saturation in a two-level system can be estimated from

$$B_{12}\rho(\nu_0) \gg Q + A_{21}$$

i.e., the constraint that the pumping rate due to laser excitation exceeds the collisional and spontaneous radiative relaxation rates from level 2. In this expression $\rho(\nu_0)$ is the laser energy density when tuned to the center of the absorption line. Since

$$B = \frac{A\lambda^3}{8\pi h}$$

where λ is wavelength, taking $A = 10^8 \text{ s}^{-1}$, $\lambda = 500$ mm, and $h = 6.6 \times 10^{-34} \text{ J s}^{-1}$ gives $B = 7.5 \times 10^{20} \text{ m}^3 \text{ J}^{-1} \text{ s}^{-2}$.

The collisional quenching rate at atmospheric pressure for an atom such as Na in air would be $Q \sim 10^{11} \text{ s}^{-1}$. Thus the condition for saturation becomes

$$\rho(\nu_0) \gg \frac{10^{11} + 10^8}{7.5 \times 10^{20}}$$
$$= 1.3 \times 10^{-10} \text{ J m}^{-3} \text{ Hz}^{-1}$$

This can be related to laser intensity through

$$I_{\nu 0} = \rho(\nu_0) c \, \Delta\nu$$

where $c = 3 \times 10^8 \text{ m s}^{-1}$, the speed of light, and $\Delta\nu$ is the linewidth in Hz. A linewidth of 0.1 cm^{-1} corresponds to $\Delta\nu = 3 \times 10^9$ Hz, hence

$$I_{\nu 0} \text{ (saturation)} = 1.2 \times 10^8 \text{ W m}^{-2}$$
$$= 1.2 \times 10^4 \text{ W cm}^{-2}$$

This is a relatively modest intensity. For example, focusing the output of a dye laser with 1-kW peak power pulses to a 1-mm^2 spot would give $I = 10^5 \text{ W cm}^{-2}$.

An interesting property of performing a laser fluorescence experiment in the saturated mode is that, in principle, the effect of quenching can be eliminated or at least reduced. This can be seen most clearly for a two-level system. We adopt the analysis of Daily (1977) and Baronavski and McDonald (1977), which assumes that the laser linewidth exceeds the spectral linewidth. In this case

$$\frac{N_2}{N} = \frac{1}{[(1 + g_1)/g_2]} \frac{r}{(r + 1)}$$

where

$$r = \frac{(B_{12} + B_{21})\rho}{Q + A_{21}}$$

If the laser pulse is tailored to saturate the $1 \rightarrow 2$ transition and then to switch off abruptly, the time dependence of P_F is

$$P_F(t) = P_{F(0)}\exp\left[-(Q + A_{21})t\right].$$

Thus a measurement of $P_F(t)$ yields a value for Q, the quenching rate of level 2, if A_{21} is known. Then the total population in level 2 before the laser pulse was switched off is

$$N_2(\text{total}) = \frac{4\pi}{\Omega h\nu A_{21} V} \int_0^\infty P_F(t)\, dt$$

A measurement of the decay curve for fluorescence then provides direct information on N_2(total) and hence N(total).

In practice, this method can be applied only when quenching is not large. One requires that the laser pulse terminate in a time $\Delta t \ll (Q + A_{21})^{-1}$ for accurate measurement. With $Q = 10^{11}\ \text{s}^{-1}$ this requirement implies $\Delta t \sim 1$ ps. Thus the use of this method with dye lasers for which $\Delta t \sim 1$ ns is possible only at low pressures ($\sim 10^{-5}$ atm).

Development of the theory to generate a simple analytic relation between fluorescence signal and laser intensity in the region where saturation is not fully achieved has been reported by Baronavski and McDonald (1977). In the region near saturation

$$\frac{N_2}{N} = \frac{g_2}{(g_1 + g_2)} \frac{(B_{12} + B_{21})\rho}{[A_{21} + Q + \rho(B_{12} + B_{21})]}.$$

Then

$$P_F = h\nu \frac{\Omega}{4\pi} A_{21} V \frac{g_2}{(g_1 + g_2)} \frac{(B_{12} + B_{21})\rho N}{[A_{21} + Q + \rho(B_{12} + B_{21})]}$$

With $g_1 = g_2$ and $B_{12} = B_{21}$ this becomes

$$P_F = h\nu \frac{\Omega}{4\pi} A_{21} V \frac{\rho N B}{A_{21} + Q + 2\rho B}$$

This can be expanded in a Taylor's series to give

$$P_F \simeq h\nu \frac{\Omega}{8\pi} A_{21} V N \left(1 - \frac{Q + A_{21}}{2\rho B} \right)$$

Baronavski and McDonald (1977) show that in the region where this approximate expression is valid (i.e., near saturation) a plot of P_F vs. ρ^{-1} should yield a straight line with negative slope. This slope is $[(Q + A_{21})/2B]\, N$ from which Q can be obtained. The intercept of this plot gives N directly. Application of this technique has been investigated quantitatively by Baronavski and McDonald in a measurement of C_2 density in a oxyacetylene flame.

A full discussion of saturation in multilevel systems is given in the review by Eckbreth *et al.* (1977), who have performed a detailed analysis of saturation for some simple molecules in flames. Some of their data are summarized in Table 5.5. It is evident that OH is nearly saturated under these conditions while NO is far from saturation.

Table 5.5. Saturation for Diatomic Molecules in a Flame at Atmospheric Pressure[a]

Molecule	Fluorescent wavelength (nm)	Saturation parameter
OH	306.4	0.89
CN	388.3	0.64
CH	431.5	0.45
NH	336.0	0.32
NO	227.0	0.09

[a]Spectra are homogeneously broadened by 120 torr of H_2O at 2600 K. The laser linewidth is assumed to be 2.5 cm^{-1}. [Eckbreth *et al.* (1977)]. The laser intensity is 10^8 W cm^{-2}.

Table 5.6. Laser Intensity to Achieve Saturation in Diatomic Molecules in a Flame at Atmospheric Pressure[a]

Molecule	Fluorescent wavelength (nm)	Laser intensity [W(cm^2 − cm^1)$^{-1}$)]	
		12 torr H_2O	120 torr H_2O
NO	227.0	4×10^7	4×10^8
NH	336.0	1×10^7	1×10^8
CH	431.5	4×10^6	4×10^7
CN	388.3	3×10^6	3×10^7
OH	306.4	6×10^5	6×10^6

[a]From Eckbreth *et al.* (1977).

Values of laser intensity required to saturate the same transition in some diatomic flame species are listed in Table 5.6. [Eckbreth *et al.* (1977)]. Here, as in Table 5.5, the laser linewidth is 2.5 cm^{-1} and $T = 2600$ K.

A useful comparison of cross sections for laser-excited fluorescence of atoms and molecules in air at atmospheric pressure in the absence of saturation has been published by Wang (1976). These data are given in Table 5.7. The scattering cross section for resonance fluorescence is defined as

$$\frac{d\sigma}{d\Omega} = \frac{\sigma_{\text{abs}}}{4\pi} \Omega q$$

where Ω is the solid angle of the detector system, σ_{abs} the cross section for absorption of laser light, and q the Stern–Volmer quenching ratio defined as

$$q = \frac{1}{1 + (\tau_r/\tau_c)(1/z)}$$

ble 5.7. Resonance Scattering Cross Sections for a Selected Number of Atoms and Molecules[a]

cies	Wavelength λ(nm)	σ_{abs} (cm^2)	τ_r (s)	τ_c (s)	$1/z$	q	$d\sigma/d\Omega^b$ (cm^2 sr^{-1})
	589.6	3.1×10^{-12}	1.6×10^{-8}	2.4×10^{-11}	0.14	1.1×10^{-2}	2.6×10^{-16}
	253.7	2.5×10^{-13}	1.1×10^{-7}	3.6×10^{-11}	0.31	1.1×10^{-3}	2.1×10^{-18}
2	400.0	1.1×10^{-17}	4.4×10^{-5}	1.4×10^{-10}	0.13	2.5×10^{-5}	2.2×10^{-24}
[illegible]	290.0	1.4×10^{-17}	4.2×10^{-5}	1.4×10^{-10}	0.28	1.2×10^{-5}	1.3×10^{-24}
I_6	250.0	1.3×10^{-18}	6.2×10^{-7}	$\sim 10^{-10}$	~ 1	1.6×10^{-4}	1.7×10^{-24}
	564.0	1.0×10^{-17}	5×10^{-7}	$\sim 10^{-10}$	~ 0.1	$\sim 10^{-3}$	8.0×10^{-23}
	5300.0	5.2×10^{-19}	8×10^{-2}	3×10^{-5}	~ 1	3.8×10^{-4}	1.6×10^{-24}

m Wang (1976); [b]Assuming $\Omega = 0.1$ and atmospheric pressure air background.

where τ_r is the radiative lifetime, τ_c the collisional lifetime, and $1/z$ the probability for quenching at each collision.

While the cross sections given in the last column of Table 5.7 are useful in estimating the relative strength of fluorescent signals under standard conditions, the detectability of these species in a given source will depend on many other parameters. For example, in flames background emission arises from Mie and Rayleigh scattering from particles as well as from laser-induced incandescence. Raman scattering may also be important at high laser intensities. Factors such as turbulence and vibration of optical components may also be important but are difficult to incorporate in any estimate of net defectivities for particular species. Some of these factors have been discussed in depth by Wang (1976) and Eckbreth *et al.* (1977).

5.3.2. Laser-Excited Atomic Flame Fluorescence

While atomic flame fluorescence spectrometry is a sensitive and useful analytical technique [Winefordner and Vickers (1970)], the use of laser sources to excite fluorescence has led to significant improvements and impressive detection limits for certain elements [Weeks *et al.* (1978)]. The advantages of laser excitation can be understood immediately from a simplified expression for the fluorescent intensity I_F from an atomic flame species

$$I_F = K\Omega I\phi NL$$

where K is a factor depending on wavelength, the Doppler width of the absorbing line, and its transition probability, Ω is a function of the solid angle of incident radiation, ϕ is an efficiency factor, N is the vapor concentration, L is the length of the excited region, and I is the exciting source intensity. This expression shows that

$$N \propto \frac{I_F}{I}$$

Then if the minimum detectable fluorescent intensity I_F (min) is fixed

$$N(\min) \propto \frac{I_F(\min)}{I}$$

i.e., the minimum detectable concentration is inversely proportional to the intensity of the exciting source I. Since laser radiation is characterized by high intensity and monochromaticity compared to conventional light sources, it is evident that reductions in N(min) should be possible by using laser excitation.

The potential of this modification of the standard atomic fluorescence analytical method was first realized in 1971 [Denton and Malmstadt (1971) and Kuhl and Marowsky (1971)]. Kuhl and Marowsky reported a detection limit for Na of 3×10^{-12} g cm^{-3} (8.5×10^{10} atom cm^{-3}) in a sodium absorption cell excited with a flashlamp pumped dye laser. It was also found that the ratio I_F/I remained linear over a variation of I of five orders of magnitude. A comparison of I_F vs. N for different detection techniques (Figure 5.14) showed that I_F vs. N was also linear and illustrated the advantages of laser excitation in fluorescent detection. The detection limits of 33 ng cm^{-3} with hollow cathode lamp (hcl) excitation and 0.5 ng cm^{-3} for excitation with a sodium vapor lamp were reduced to 0.003 ng cm^{-3} with laser excitation. This limit has now been lowered by many orders of magnitude [Gelbwachs *et al.* (1977)].

Both pulsed and CW lasers have been used in atomic fluorescence studies. Figures 5.15 and 5.16 show schematic diagrams of pulsed and CW systems, respectively. In both cases, the laser is tuned to an electronic transition of an atomic species in the flame. When resonance is achieved, the volume excited by the laser is seen to brighten as resonance fluorescence is emitted. This emis-

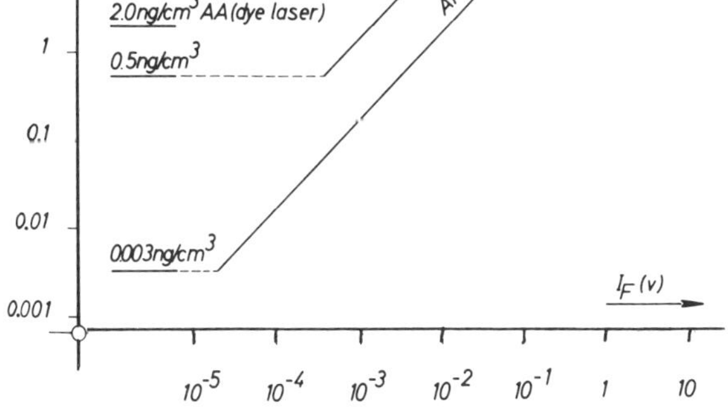

Figure 5.14. Working curves for different light sources for atomic absorption (AA) and atomic fluorescence (AF). I_F, fluorescence signal in volts; c, sodium density in ng cm^{-3}; hcl, hollow cathode lamp; svl, sodium vapor lamp [Kuhl and Marowsky (1971)].

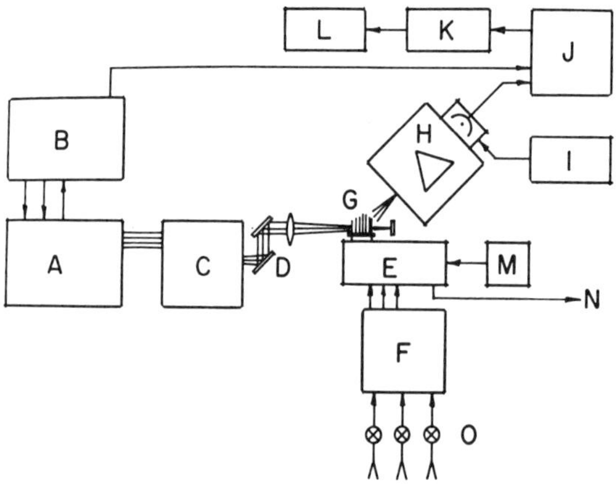

Figure 5.15. Block diagram of experimental system for pulsed laser-excited atomic fluorescence flame spectrometry (experimental conditions) [Fraser and Winefordner (1972)]: *A*, N_2 laser; *B*, power supply; *C*, dye laser; *D*, output optics; *E*, mixing chamber; *F*, flow system; *G*, flame; *H*, monochromator; *I*, photomultiplier tube power supply; *J*, boxcar integrator; *K*, analog integrator; *L*, recorder; *M*, sample supply; *N*, excess sample drain; *O*, gas tanks and regulator. (Reproduced with permission; copyright 1972, American Chemical Society).

sion is detected by a photomultiplier tube mounted on a monochromator. The monochromator is tuned to select one of the fluorescence lines emitted from the sample. This signal is then processed with a boxcar integrator (pulsed excitation) or a lock-in amplifier (CW excitation). One advantage of laser excitation is that different elements can be sampled simply by changing the dye laser wavelength. Signal-to-background ratios are enhanced by use of a narrow exciting line. For this reason best results are obtained when the laser output is narrowed by an intracavity etalon. Typical source conditions for pulsed and CW laser sources are given in Table 5.8.

The parameters listed for the pulsed laser refer to an N_2 pumped dye laser system with frequency doubling to achieve UV emission. Flashlamp pumped

Table 5.8. Source Conditions for Pulsed and CW Lasers

	Pulsed	CW
Power	1–100 kW	1–300 mW
Linewidth	0.1–0.01 nm	0.003 nm
Repetition rate/chopping rate	10–30 Hz	2 kHz
Wavelength range	250–800 nm	420–800 nm

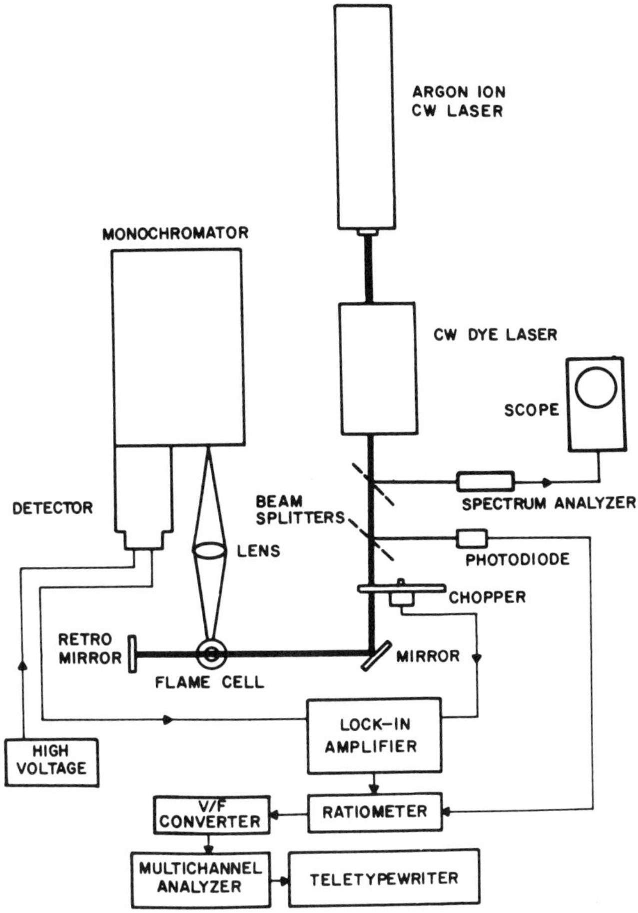

Figure 5.16. Block diagram of CW atomic flame fluorescence configuration. Laser beam indicated by heavy solid line [Green *et al.* (1976) Reproduced with permission; copyright 1976, American Chemical Society].

dye lasers can also be used [Kuhl and Spitschan (1973)]: In this case the laser pulse length is increased from $\sim 1\ \mu s$ and the repetition rate may be reduced to $\lesssim 10\ s^{-1}$.

A wavelength-modulated CW dye laser system for use in atomic fluorescence has been developed by Goff and Yeung (1978). The advantage of wavelength modulation in synchronous detection of atomic fluorescence lies in the discrimination against background scattering. Since Mie or Rayleigh scatter-

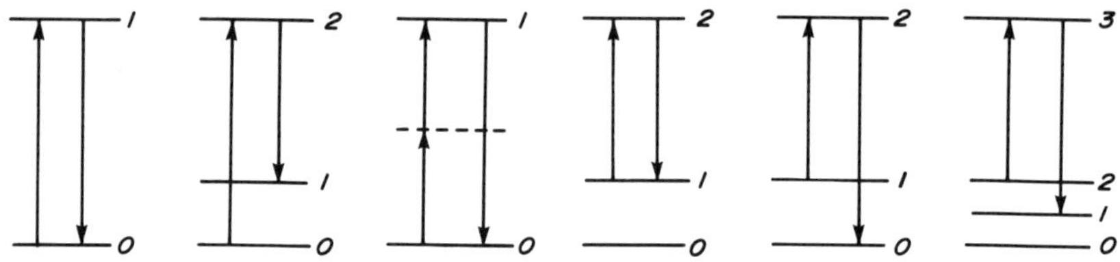

Figure 5.17. Various excitation/fluorescence transitions seen in laser-excited atomic fluorescence.

ing can be considered to be independent of wavelength over a narrow wavelength range, the scattered flux does not contain a time-varying component. Hence the time-varying fluorescent signal can be extracted from a large dc background with lock-in detection.

A variety of excitation–fluorescence combinations can be used in laser-excited atomic fluorescence. Several of these are shown in Figure 5.17.

Part (a) shows the energy levels involved in resonance fluorescence from the ground state (level 0). If the fluorescence is to an excited state then one has the situation shown in part (b); this is often called *Stokes direct line fluorescence.* A direct two-photon transition (c) sums the energies of two long-wavelength photons to make a transition from $0 \rightarrow 1$. The emission may occur as a UV photon although selection rules for two-photon transitions normally require that the parities of states 0 and 1 are the same. This would preclude a single-photon transition between 1 and 0.

Resonance fluorescence can also be observed from excited states [part (d)] as long as $(E_1 - E_0)/k_BT$ is not large ($k_BT = 1200\ \text{cm}^{-1}$ at 1800 K). The fluorescent photon can be emitted to the ground state [part (e)] or to another low-lying state [part (f)]. These last transitions are referred to as *excited-state anti-Stokes direct line fluorescence.*

Table 5.9 summarizes data obtained on various elements by Fraser and Winefordner (1972) together with the types of transitions involved. In most cases, the detection limits quoted have been reduced in more recent experiments. It is evident from an inspection of these data that substantial improvements in sensitivity are achieved when the fluorescent photons have different energy from those used to excite fluorescence. This effect arises in part from improved discrimination against scattered light when detection can be made at a wavelength separated from that of the exciting source.

A variety of possible sources of spectral interference have been identified in laser atomic flame fluorescence. Omenetto *et al.* (1973) illustrate these by considering the detection of an element A in the presence of a dominant matrix constituent B. The processes identified are:

1. Scattering from flame gases and species at λ.
2. Scattering from solvent at λ.
3. Scattering of A at λ.
4. Scattering from B at λ.
5. Fluorescence from A at λ excited by λ.
6. Fluorescence from A at λ excited by $\lambda_x \neq \lambda$.
7. Fluorescence from B at λ excited by λ.
8. Fluorescence from B at λ excited by $\lambda_x \neq \lambda$.

In this list λ refers to the wavelength of detection, while λ_x is the excitation wavelength. If $\lambda_x \neq \lambda$ then process 6 gives the desired signal, while other processes can be considered to generate interference of one type or another. Some

Table 5.9. Laser-Excited Atomic Fluorescence[a]

Element	Type of excitation/ fluorescence	Energy levels (cm^{-1})	Detection limit ($\mu g\ ml^{-1}$)
Pulsed excitation			
Al	a	0–25348	0.1
	d	112–25348	0.03
	b	0–25348, 112–25348	0.005
Cr	a	0–27935	0.02
Co	d	816–28777	50
	e	0–28777, 816–28777	0.2
Ga	a	0–24788	0.1
	b	0–24788, 826–24788	0.02
In	a	0–24373	0.01
	b	0–24373, 2213–24373	0.002
Ni	d	880–28569	2
	f	880–28569, 205–28569	0.05
Tl	a	0–26478	1
	b	0–26478, 7793–26478	0.02
CW excitation			
Ba	a	0–18060	0.04
	f*	9597–25957, 25704–9216	0.5
Cu	e	13245–30535, 30535–0	100
Li	d	14904–31282	0.5
Mo	e	11859–28924, 28924–0	14
Na	a	0–16956	0.0001
Nd	a	0–17787	2
Rh	e	12723–29431, 29431–0	0.3
Sc	e	11677–29190, 29190–11610	50
U	a	0–16900	500
V	d	8716–23131	0.3

[a]Data from Fraser and Winefordner (1972) and Smith *et al.* (1977).

of these interferences can be removed by utilizing offset wavelengths for excitation and detection [Smith *et al.* (1977)] or by frequency doubling the dye laser output to permit excitation in the ultraviolet [Kuhl and Spitschan (1973), Weeks *et al.* (1978), and Bolshov *et al.* (1976)]. With frequency doubling, Neumann and Kriese (1974) were able to detect Pb at the 10^{-4} μg ml^{-1} level in a nonflame atomic fluorescence spectrometer. Bolshov *et al.* (1976) used a similar technique to detect 2.5×10^{-6} μg ml^{-1} in Pb and 25×10^{-6} μg ml^{-1} in Fe.

In these and other laser fluorescence experiments, the dependence of the fluorescence signal I_F on analyte concentration is seen to follow a curve similar to that shown in Figure 5.18 for Na [Green *et al.* (1976)]. Although the I_F vs. N curve is linear at small N, some saturation is apparent as N increases. At large concentrations, the curve turns over so that I_F actually decreases with a further increase in N. When this occurs, the fluorescent line shows evidence of self-absorption at the line core.

A similar saturation is seen at constant analyte concentration as I, the laser intensity, is increased, although the curve does not bend over completely

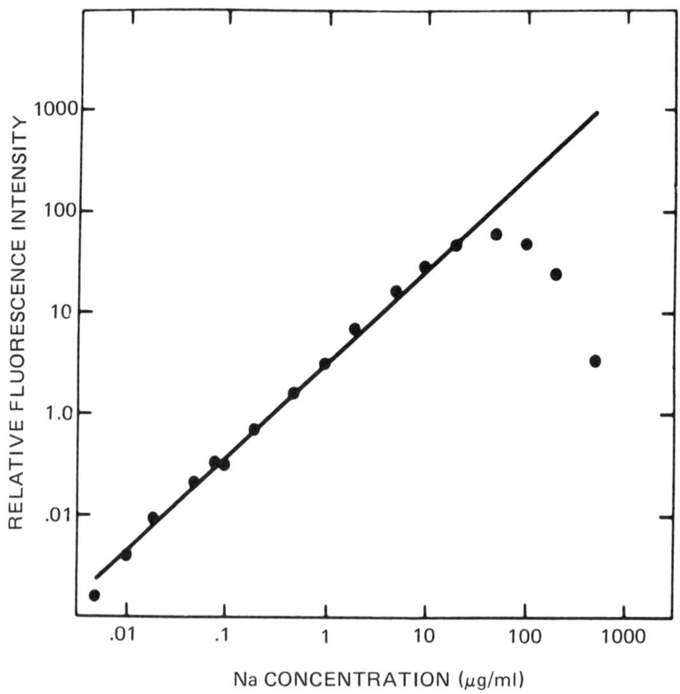

Figure 5.18. Analytical curve for sodium fluorescence showing the dynamic range of the laser fluorescence technique [Green *et al.* (1976) Reproduced with permission; copyright 1976, American Chemical Society].

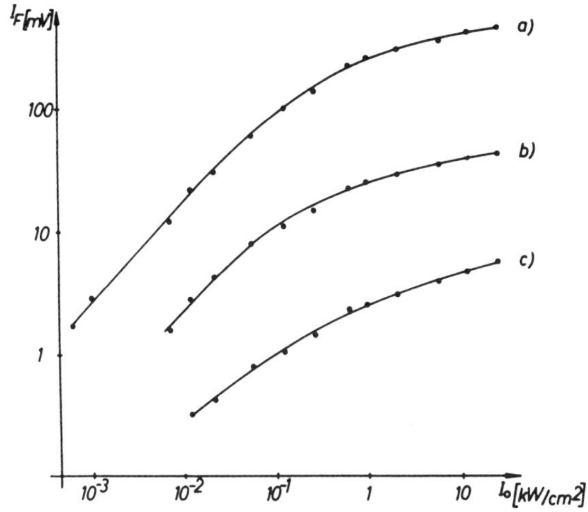

Figure 5.19. Fluorescence signals observed in the Na D_2 line as a function of the exciting laser radiation intensity for various Na concentrations: (a) 10 μg ml^{-1}, (b) 1 μg ml^{-1}, and (c) 0.1 μg ml^{-1} [Kuhl *et al.* (1973)].

[Figure 5.19; Kuhl *et al.* (1973)]. Saturation at high laser intensity provides an effective limit to detectability because an increase in laser intensity is not accompanied by an increase in fluorescent photons or an attendant decrease in concentration detected. The laser intensity required for saturation is a function of the intensity distribution within the focused beam and the size of the beam relative to that of the flame [Sharp and Goldwasser (1976), Rodrigo and Measures (1973), and Daily (1978)]. When the beam diameter is smaller than the flame no saturation is observed.

An analysis of saturation in a Gaussian beam by Daily (1978) shows that N_k, the population of the upper level of the atom being observed, can be expressed.

$$N_k = \beta N_T \frac{I_\nu}{I_\nu + I_\nu^s}$$

where

$$\beta = \left[1 + \frac{g_i}{g_k} + \sum_{i \neq k} \left(\frac{N_i}{N_k}\right)^*\right]^{-1}$$

and $N_T = \Sigma N_i$ is the total population in all levels, I_ν the specific intensity in the laser beam I_ν^s the saturation intensity, g_i the degeneracy factor for the initial level (from which absorption occurs), g_k the degeneracy of the final level, and $(N_i/N_k)^*$ the quasiequilibrium populations for all levels except the initial level. As an example, for a three-level system (Figure 5.20) one obtains the following expression for the equilibrium population in level 3 assuming steady pumping from 1 → 3

$$N_3 = \frac{N_1 + N_2 + N_3}{[(1 + g_1)/g_3] + (N_2/N_3^*)} \frac{I_\nu}{(I_\nu^s + I_\nu)}$$

These expressions follow the observed relation between I_ν and N over the range from small N (no saturation) to large N (saturation).

The effect of increasing optical depth on the transmission of incident laser light through a flame and the subsequent emission of fluorescent photons has been discussed by Kuhl *et al.* (1973) and Brod and Yeung (1976). These effects are not large in the concentration range where the laser atomic fluorescence technique is of the most value (concentration $\lesssim 1\ \mu g\ ml^{-1}$).

Resonance fluorescence has been used in a detection of trace quantities of Na inside a quartz furnace at 1000°C [Mayo *et al.* (1976)]. A CW dye laser was used with wavelength modulation to repetitively scan over the Na spectral line. Data from 100 scans were stored in a multichannel analyzer and used to synthesize the Na resonance. An estimated detection limit of 5×10^5 atoms cm^{-3} was quoted. This represented an improvement in detection limit of $\sim 10^3$ over the obtained using atomic absorption.

The detection limit for Na atoms in a flame was further reduced by Gelbwachs *et al.* (1977) using a technique known as saturated optical nonresonant emission spectroscopy (SONRES). This method involves the saturation of an atomic transition under temperature and pressure conditions that produce an exchange of population between the upper excited state and another nearby state. In the Na atom, the two states involved correspond to the D_1 and D_2 transitions at 589.6 and 589.0 nm, respectively. A flashlamp pumped dye laser was used to saturate the D_1 transition of Na atoms in a flame. Collisional redistribution led to an increase in the population of the other member of the Na doublet with subsequent emission of the D_2 line.

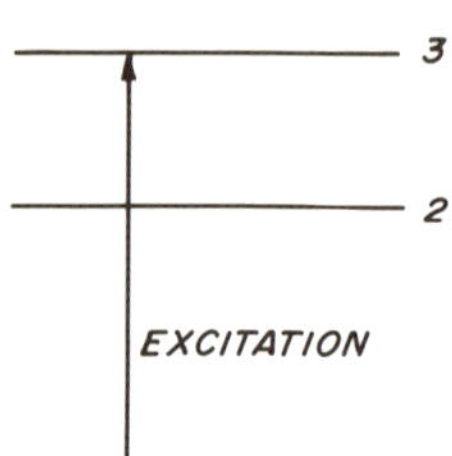

Figure 5.20. Levels in three-level system with laser excitation.

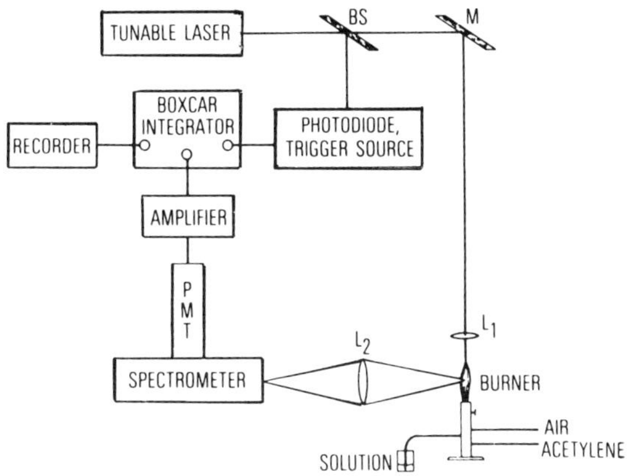

Figure 5.21. Schematic representation of experimental apparatus in laser fluorescent detection of Na in a flame [Gelbwachs *et al.* (1977)].

Figure 5.21 shows a block diagram of the experimental apparatus used. A detection limit of 10^4 Na atoms cm^{-3} was reported. When the flame was replaced by a seated cell containing metallic Na, 200 Na atoms cm^{-3} were detected. When geometrical factors were taken into account, the system detected the emission from 0.2 Na atoms with a signal-to-noise ratio of 25.

Complete flexibility of a laser atomic flame fluorescence trace analysis system requires that the wavelength of excitation can be easily varied so that various elements can be sampled. Perry *et al.* (1977) have developed a microprocessor controlled dye laser that allows rapid and reproducible scanning of laser wavelength over the 360–650 nm range. Figure 5.22 shows this system. When used in an atomic fluorescence application, information about the wavelength to be scanned were loaded manually into the microprocessor.

5.3.3. Laser-Excited Molecular Flame Fluorescence

Several of the simple diatomic molecules present in flames at atmospheric pressure have been detected by LIF. However, detection of molecular species through LIF is inherently more difficult than detection of atomic species. This arises in part because of the relative weakness of individual vibronic lines compared to the electronic transitions of atoms. As an example, $\sigma_{abs}(Na) = 3.1 \times 10^{-12}$ cm^2 at the peak of the Na resonance line at 589.6 nm, while $\sigma_{abs}(NO_2) = 1.1 \times 10^{-17}$ cm^2 for lines of the $X \rightarrow A$ band of NO_2 at 400 nm (see Table 5.7). When quenching in air at 1 atm is taken into account, the corresponding differential cross section for resonance fluorescence are

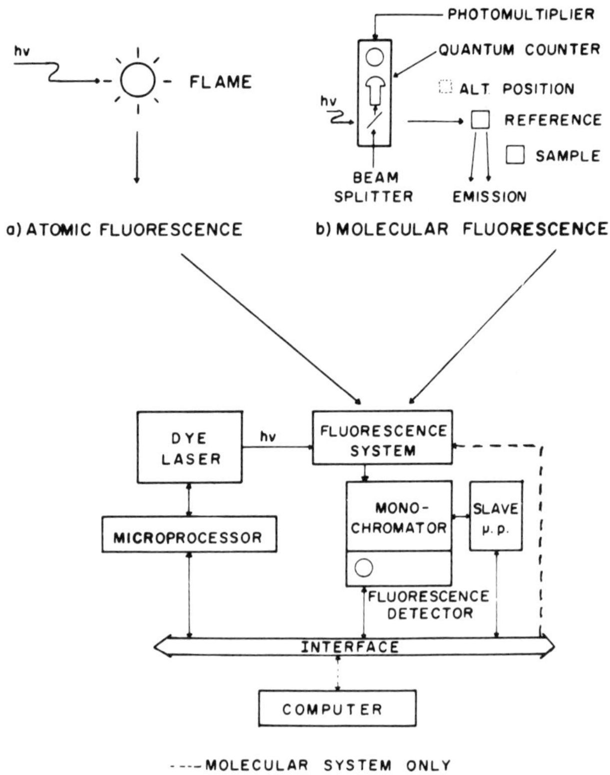

Figure 5.22. Experimental configurations for collecting (a) atomic flame fluorescence with microprocessor and (b) molecular fluorescence and background under minicomputer control [Perry *et al.* (1977) Reproduced with permission; copyright 1977, American Chemical Society].

$$\left.\frac{d\sigma}{d\Omega}\right|_{\mathrm{Na}} = 2.6 \times 10^{-16}\ \mathrm{cm}^2\ \mathrm{sr}^{-1}$$

$$\left.\frac{d\sigma}{d\Omega}\right|_{\mathrm{NO_2}} = 2.2 \times 10^{-24}\ \mathrm{cm}^2\ \mathrm{sr}^{-1}$$

The first measurement of LIF of a molecule in a flame was published by Barnes *et al.* (1973). A nitrogen pumped dye laser was used to excite emission from CH in a oxyacetylene welding torch. Spectra are shown in Figure 5.23. Emission was excited by tuning the laser wavelength to coincide with the Q-branch at 431.2 nm. Excitation spectra could be obtained by detecting emission from the flame over a broad band while scanning the wavelength of the dye laser.

OH fluorescence in a flame was measured by Wang and Davis (1974b). It was found possible to obtain information concerning the spatial distribution of the OH radical with much higher resolution than previously possible using conventional absorption techniques. Wang and Davis (1974a) were able to detect OH in air via LIF with a detection limit of 5×10^6 cm^{-3}. These measurements utilized a frequency-doubled dye laser to excite the OH absorption near 280 nm.

Low concentrations of C_2 in flames have been observed by LIF using the 514.53-nm line from an Ar^+ laser [Jones and Mackie (1976)] and the output of a flashlamp pumped tunable dye laser [Baronavski and McDonald (1977)]. Excitation with the Ar^+ laser is possible because of a chance coincidence between the 514.53-nm line and two rotational lines of the (0,0) band of the Swan system. Phase sensitive detection was used to detect the C_2 fluorescence in the presence of a large flame background.

Jones and Mackie performed a useful calculation of the relative importance of various emission terms in the detection of LIF with this system. These

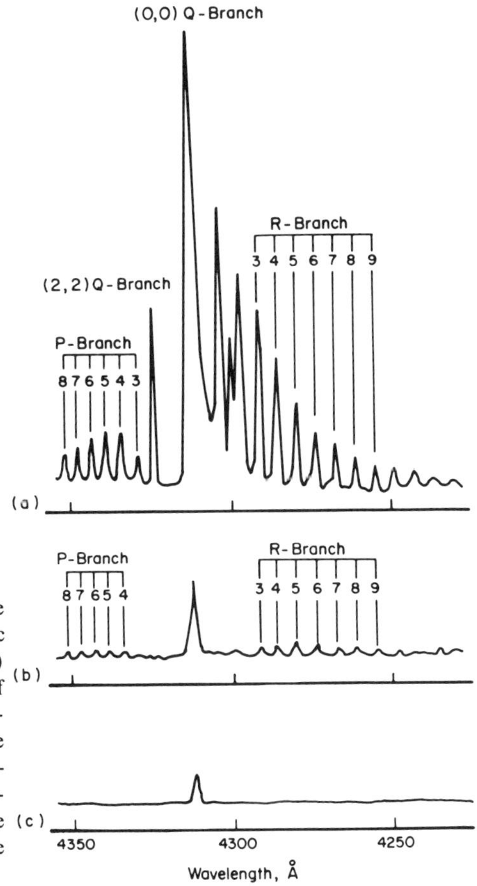

Figure 5.23. (a) Emission spectrum of the (0,0) band of the $A^2\Delta \rightarrow X^2\Pi$ electronic transition of CH in oxyacetylene flame. (b) Fluorescence spectrum of the (0,0) band of CH excited inside the reaction zone surrounding inner cone of oxyacetylene flame using 4312-Å dye laser radiation. (c) Spectrum of 4312-Å dye laser radiation scattered from region of oxyacetylene flame outside the inner cone reaction zone [Barnes *et al.* (1973)].

Table 5.10. Emission Terms for Various Processes

Process	I/I (fluorescence)
Flame emission	70
Molecular Rayleigh scattering	21
Rotational Raman scattering	0.01
Resonance fluorescence	1.0

terms are expressed relative to the fluorescent signal and were calculated for a laser power of 0.125 W and a focal volume of 0.8×10^{-2} mm^3. They are given in Table 5.10. The experimentally determined background was actually somewhat smaller than that that would be expected from the table.

Baronavski and McDonald (1977) investigated the role of saturation in quantitative determination of C_2 in flames via LIF. The theory describing the system response in the saturated or nearly saturated mode has been developed in Section 5.3.1. For a two-level system the relation between the fluorescent power P_F and laser energy density ρ is given by

$$P_F = XA_{21}\left(1 - \frac{Q + A_{21}}{2\rho B}\right)$$

where X is a constant for a particular transition and collection geometry. ρ is proportional to the laser intensity (W cm^{-2}) or fluence (J m^{-2}). Hence a plot of P_F vs. ρ^{-1} should be linear. Such a plot for C_2 is shown in Figure 5.24 [Baronavski and McDonald (1977)]. The linear relation predicted from the theory was observed, although experimental uncertainties were found to limit the final accuracy of the resulting values of Q and N. Nevertheless, the validity of the saturation technique in LIF of flame species has been demonstrated.

The detection of NO_2 at the ppm level in flames via LIF has been reported by Barnes and Kircher (1978).

5.3.4. Beam Diagnostics

The sensitivity of LIF and the ease with which fluorescence can be excited from small subvolumes of space have resulted in the use of LIF as a sensitive detector in beam diagnostics. As such, LIF has been found to be useful in the study of the products of chemical reactions and their state of excitation. The properties of atoms, molecules, and complexes in single molecular beams and nozzle beams are also easily studied with LIF.

Possible applications of LIF to beam diagnostics can be seen by reference to possible scattering geometries (Figure 5.25). The first of these involves the

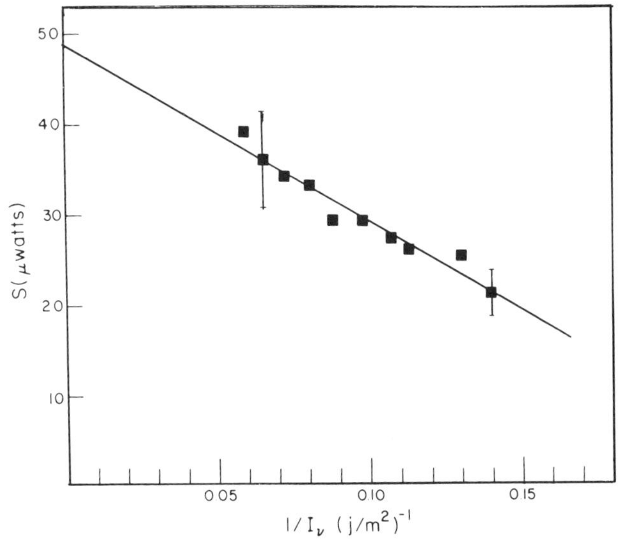

Figure 5.24. A plot of the C_2 fluorescence emission signal vs. $1/I_p$. The fluorescence signal has been fully corrected for efficiency factors. The slope of this plot gives the total quenching rate for the $^3\Pi_g$ state while the intercept is used to evaluate the $^3\Pi_u$ state number density [Baronavski and McDonald (1977)].

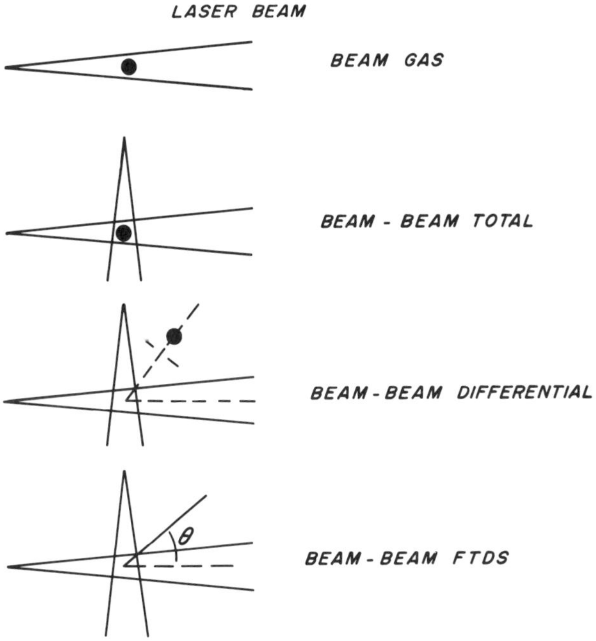

Figure 5.25. Various geometries in LIF beam diagnostics [after Kinsey (1977)].

use of a laser beam to probe the conditions at some point within a molecular beam. An early example of an experiment of this type was the observation of the vibrational and rotational state of Na_2 molecules after expansion through a nozzle [Sinha *et al.* (1973)].

Beam–beam total scattering involves the use of LIF to probe the total internal state distribution in the products of reaction of two separate beam species. In this case, fluorescence is excited from all products of the beam reaction, regardless of the initial angle at which scattering was produced. One of the first experiments in this area was reported by Schultz *et al.* (1972). This involved a study of the LIF from BaO produced in the beam–beam reaction.

$$Ba + O_2 \rightarrow BaO + O$$

The beam–beam differential technique (Figure 5.25) uses LIF to probe the products of beam–beam scattering at a particular angle. Variation of the position of the slit–laser beam combination over a range of angles permits measurement of the angular distribution function for reactive scattering. Smith and Zare (1976) have studied the angular distribution of BaCl in the reaction

$$Ba + KCl \rightarrow BaCl + K$$

The final technique shown as *beam–beam FTDS* in Figure 5.25 involves the measurement of the Doppler profile of fluorescence as a function of angle of incidence of the exciting laser beam. This can be converted to information about the scattering distribution function. As such, it represents a Fourier transform Doppler spectroscopic (FTDS) measurement of scattering amplitudes. Further discussion of this process can be found in the review by Kinsey (1977).

Multiphoton ionization (MPI) techniques can also be used to probe atomic and molecular beam species. Potential advantages of this approach have been discussed by Feldman *et al.* (1977). It appears that the high signal-to-noise ratio possible with MPI may permit measurements to be made at lower densities than those feasible with LIF. On the other hand, Feldman *et al.* find that the spectral resolution attainable in their preliminary MPI experiments was significantly less than that commonly available in LIF.

A diagram of the components of a MPI beam diagnostic system is shown in Figure 5.26. Zandee *et al.* (1978) used a system similar to this to investigate the MPI spectra of I_2 in a molecular beam. By limiting detection to either I^+ or I_2^+ ions, excitation spectra of I_2 could be obtained by scanning the wavelength of a tunable dye laser. It was found that these spectra showed significant differences.

Feldman *et al.* (1977) studied the beam reaction $Ba + HCl \rightarrow BaCl +$

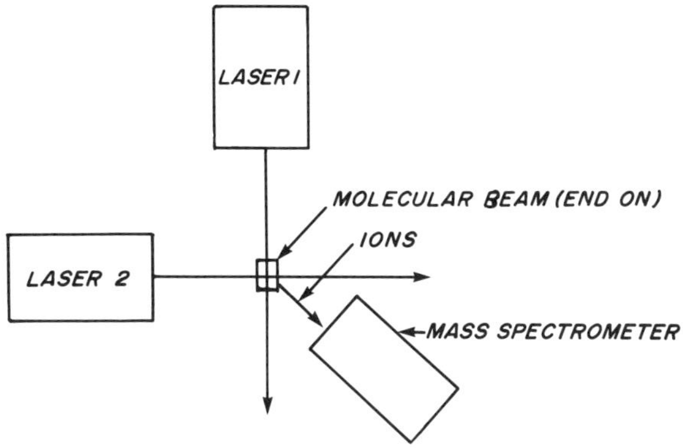

Figure 5.26. Schematic diagram of MPI beam diagnostic system. The mass spectrometer samples ions generated from the interaction region. Either laser 1 or laser 2 is tunable. The system may also use only one laser, in which case two or more photons of the same energy produce MPI.

H using MPI without mass selection. Spectra of BaCl were overlaid by strong atomic lines due to MPI of Ba atoms. In addition, the fine scale structure of BaCl bands was not observable in MPI.

In view of the rapid development of LIF beam diagnostic techniques and the burgeoning literature on this subject, no attempt will be made to give a complete review of work in this field at this time. Rather, the power of this method in physical and chemical analysis will be illustrated by reference to several experiments chosen to be representative of work currently being carried on in this area.

5.3.4.1. Fundamental Molecular Data from LIF. The spectra of heavy molecules produced in conventional sources (furnaces, discharge lamps) are often complicated by emission from many levels or absorption from thermally populated states. The effect of this excitation is to produce complex overlapping spectra whose analysis is difficult or impossible. Beam techniques offer the possibility of producing molecules in specific, well characterized states under collisionless conditions. In addition, by the use of supersonic expansion, extremely low translational and rotational temperatures are possible. Thus, molecular beams probed by LIF have been demostrated to be an almost ideal source of data on molecular constants and the lifetime of molecular states.

An example of an experiment that yields data on both molecular constants and the lifetime of electronic states is shown in Figure 5.27 [Pruett and Zare (1975)]. Pruett and Zare used LIF to study the spectroscopic states of BaO produced by the reaction

$$Ba + CO_2 \rightarrow BaO + CO$$

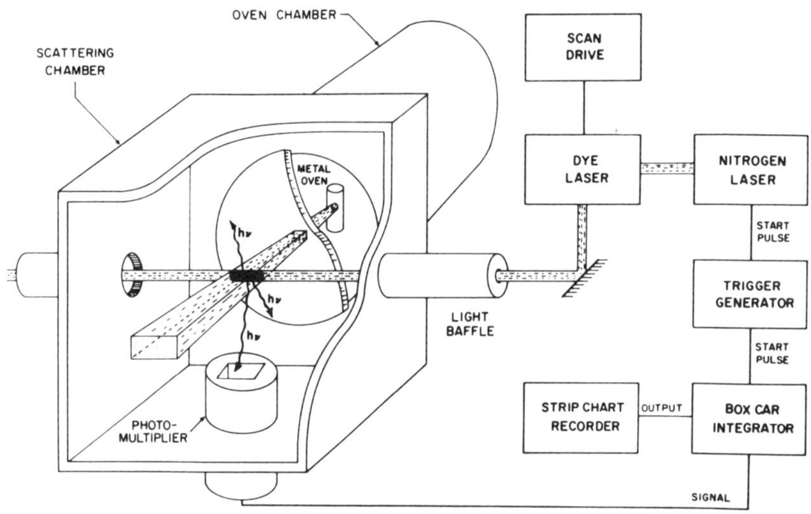

Figure 5.27. Experimental setup: Perspective of the "beam + gas" scattering apparatus and schematic of the laser fluorescence detection system [Pruett and Zare (1975)].

BaO generated by this reaction is predominantly in the ground state ($X\,'\Sigma^+$, $v'' = 0$). A dye laser is used to excite the following transitions by absorption from the X state

$$\begin{aligned} X\,{}^1\Sigma^+ &\xrightarrow{h\nu} A\,{}^1\Sigma^+ \\ &\rightarrow A'\,{}^1\Pi \end{aligned}$$

The A' and A states have different radiative lifetimes and hence can be separated by gating the detection system. With BaO in an environment in which single collisions dominate, radiative deactivation of both states can be followed for long times. This allows the lifetimes of both states to be obtained from fluorescent decay curves (Figure 5.28). Since the lifetime of the A state ($\sim$300 ns) is much shorter than that of the A' state ($\sim$10 μs), emission at times exceeding several microseconds is predominantly due to the $A' \rightarrow X$ system. Vibrational constants were obtained for the $A'\,{}^1\Pi$ state and several bands were rotationally analyzed to yield rotational constants.

Dagdigian (1976) has used LIF to produce fluorescence spectra of LiH and NaH molecules in a supersonic molecular beam. Measurement of the fluorescent decay rate following excitation with $\sim$10-ns pulses from a dye laser was used to obtain an accurate estimate of radiative lifetimes of several ($v'J'$) levels of the $A\,{}^1\Sigma^+$ state. The rotational intensity distribution indicated that rotational temperatures as low as 230 ± 50 K were produced in the supersonic expansion of NaH.

The cooling of internal degrees of freedom in supersonic beams greatly simplifies molecular spectra, especially when this is combined with the high spectral selectivity possible with LIF. This method has been used to simplify the complex visible spectrum of NO_2 [Smalley *et al.* (1974, 1975)]. More recently, Breford and Engelke (1979) have obtained high-resolution spectra of NaK molecules in a supersonic jet and have used these spectra to characterize several molecular states of this molecule. Large molecules have also been studied using LIF to probe a supersonic beam. An example is free base phthalocyanine [Fitch *et al.* (1979)].

Internal conversion is a radiationless process that results in the deactivation of excited states of large organic molecules. The effect can be seen in the $S_0 \rightarrow S_1$ transition of aromatic molecules where it results in a radiationless deactivation $S_1 \rightarrow S_0$. While internal conversion has been frequently invoked in studies of organic molecules in condensed phases, direct observation of this process has not been possible until gas-phase measurements could be made on isolated molecules. This was accomplished by Sander *et al.* (1976) by LIF of pentacene molecules in a molecular beam. Radiationless deactivation $S_1 \rightarrow S_0$ results in this temporary population of highly excited levels, S_0^*, of the S_0 state. These can be detected by resonance fluorescence $S_0^* \xrightarrow{h\nu} S_1 \xrightarrow{h\nu} S_0^*$. Sander *et al.* have succeeded in measuring this fluorescence and have obtained an excitation spectrum of the states associated with internal conversion.

5.3.4.2. Weakly Bound Complexes. Supersonic beams are well suited to the study of fragile molecular complexes because of the low internal temperatures achieved during supersonic expansion as a free jet. In addition, molecules can be examined in a collision-free environment. The result is that weakly bound molecules can be formed and studied. This method is particularly well suited to the observation of the electronic properties of van der Waals molecules in which bond energies may be $\sim 10^{-2}$ eV (80 cm^{-1}).

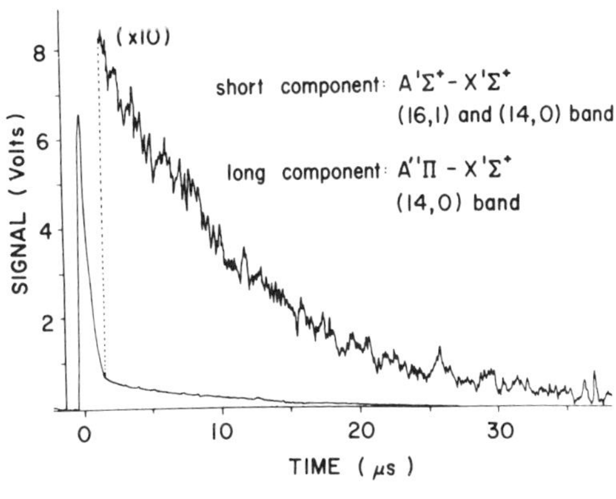

Figure 5.28. Fluorescence signal as a function of time using excitation at 4364 Å [Pruett and Zare (1975)].

I_2He is an example of such a complex. It can be studied by expanding a mixture of I_2 in He at 100 atm through a supersonic nozzle [Smalley *et al.* (1976) and Sharfin *et al.* (1979)]. The resultant cooling was shown to quench the rotational temperature of I_2 to 0.5 K and its vibrational temperature to 50 K. Similar cooling of the internal states of I_2He was also observed.

A schematic diagram of the system developed by Smalley *et al.* (1976) for a spectroscopic study of these and other I_2–He complexes is shown in Figure 5.29. The complexes were probed with a beam from a CW tunable dye laser having a linewidth of 1 GHz. An example of a fluorescence excitation spectrum of I_2He is shown in Figure 5.30. Major spectral features are due to the (10,0) and (12,1) bands of the $BO_u^+(^3\Pi) \leftarrow XO_g^+(^1\Sigma)$ transition of I_2. Other features (labeled A, B, and C) are due to I_2He and I_2He_2 molecules. The narrow width of the features due to I_2–He complexes implies that the complex is relatively long-lived even when electronically and vibrationally excited. This in turn suggests that the coupling between the I–I vibration and the I_2–He vibration is weak so that high vibrational excitation in the I_2 molecule is not easily transferred to the I_2–He band.

5.3.4.3. Reactive Scattering. One of the first comprehensive studies of the product state distribution in reactive scattering was published by Dagdigian *et al.* (1974). LIF was used to excite fluorescence from BaO produced in the reactions

$$Ba + O_2 \rightarrow BaO + O$$

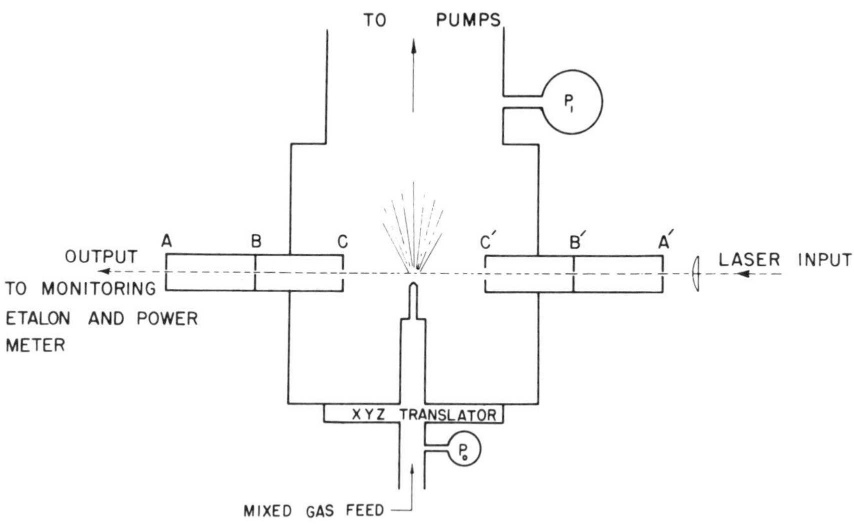

Figure 5.29. Schematic of supersonic jet apparatus [Smalley *et al.* (1976)].

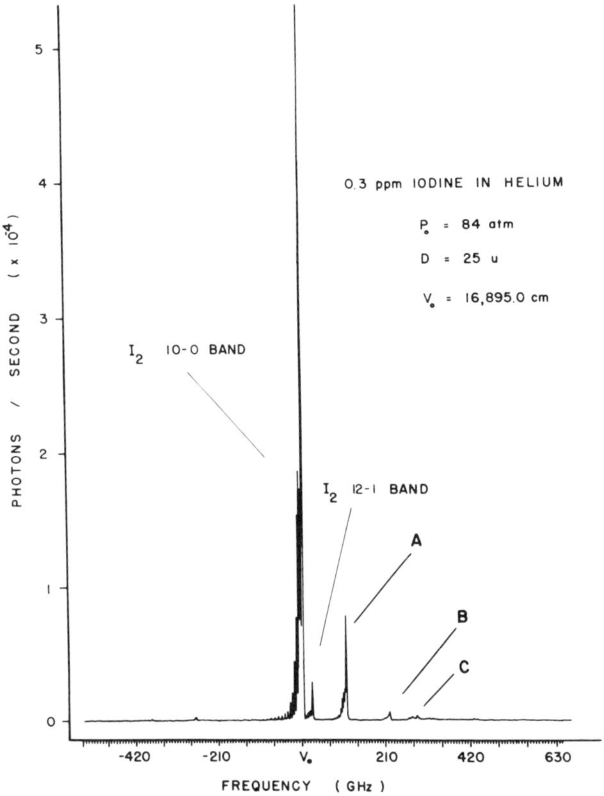

Figure 5.30. Fluorescence excitation spectrum of a supersonic jet of I_2 in He in the spectral region near the 10–0 vibronic band of the $\tilde{B} \rightarrow \tilde{X}$ transition of I_2. Laser bandwidth was 1-GHz FWHM [Smalley *et al.* (1976)].

and

$$Ba + CO_2 \rightarrow BaO + CO$$

The reactions occur when Ba atoms from a heated cell enter a chamber filled with O_2 or CO_2. As such, the reaction occurs in a "beam + gas" rather than a "Beam + beam" configuration.

The Ba + O_2 reaction was found to produce BaO in a variety of excited vibrational and rotational levels. The internal excitation of BaO produced from the Ba + CO_2 reaction was more complex. In both cases, rotational distributions were nonthermal. However, in the Ba + CO_2 reaction the rotational distribution in the $v'' = 0$ level was reasonably well fit by a Boltzmann distribution with a temperature $T_R \sim 2000$ K. The corresponding vibrational distribution was given by $T_V \sim 1300$ K.

The reaction

$$Ba + CCl_4 \rightarrow BaCl + CCl_3$$

has been studied by Schmidt *et al.* (1976) using LIF to probe the internal state distribution of the BaCl product. This reaction occurred in the region of overlap between Ba and CCl_4 beams. LIF was excited from the $X\,^2\Sigma \rightarrow C\,^2\Pi_{1/2}$ and $^2\Pi_{3/2}$ transitions. The similarity of the potential curves in the C and X states resulted in the restriction of strong fluorescence to sequences with small values of $|v' - v''|$ (v' is the vibrational quantum number of lower energy level and v″ the quantum number for upper level). The vibrational distribution of BaCl was found to be strongly peaked at $v'' \sim 44$. This was taken as evidence that much of the reaction energy of $Ba^+ + Cl^-$ finally ends up as vibrational excitation of BaCl. Significant energy was also predicted for the CCl_3 fragment. A summary of energy partition is as follows

Reaction energy	1.70 eV(100%)
E_{vib}^{BaCl}	75 ± 3%
E_{rot}^{BaCl}	4 ± 2%
$E_{internal}^{CCl_3}$	15%
E_{trans}	6%

The model developed to describe this reaction was based on an electron jump from Ba to form a complex $CCl_3 \cdots Cl^-$ that subsequently dissociates. The subsequent reaction $Ba^+ + Cl^-$ forms BaCl*.

Reaction of electronically excited atoms have been studied by Pasternack and Dagdigian (1978) and Solarz and Johnson (1979). Pasternack and Dagdigian used LIF to probe CaO produced from

$$Ca(4s4p\,^3P) + O_2 \rightarrow CaO + O$$
$$Ca(4s4p\,^3P) + CO_2 \rightarrow CaO + CO.$$

Ca* was generated by an electrical discharge. This technique can be rather nonselective. However, in this case, the other likely metastable species [$Ca(4s3d\,^1D)$] represented about 1% of the ^3P concentration.

Analysis was complicated by the complexity of the fluorescence spectrum of CaO which shows features due to at least three transitions. These data showed that only some 5–6% of available reaction energy ends up stored in CaO internal excitation. It was suggested that reaction occurs from $Ca(^3P)$ via an electron jump mechanism forming (for the Ca + O_2 reaction) a relatively long lived Ca^+–O_2^- complex.

Reactions between electronically excited Ba and Sr and halomethanes have been investigated by Solarz and Johnson (1979). Excited-state atoms were produced by excitation with pulses from a flashlamp pumped dye laser. These atoms were then reacted as follows:

$$\mathrm{Ba}\ ({}^3D) + \mathrm{CH_3X} \rightarrow \mathrm{BaX} + \mathrm{CH_3}$$
$$\mathrm{Sr}\ ({}^3P_1) + \mathrm{CH_3X} \rightarrow \mathrm{BaX} + \mathrm{CH_3}$$

where X = F, Cl, Br. Reactions with CCl_3 and CCl_4 were also studied. It was found that the state of vibrational excitation of the BaX reaction product was mostly a function of the nature of the halomethane reactant and was not strongly dependent on the excitation state of the atomic reaction. On the other hand, electronic excitation of Ba and Sr was found to significantly affect reaction rates.

The effect of *collisional* energy on the spin-forbidden reaction

$$\mathrm{Ba}({}^1S) + \mathrm{SO_2}\ ({}^1A_1) \rightarrow \mathrm{BaO}(X\ {}^1\Sigma^+) + \mathrm{SO}(X\ {}^3\Sigma^-)$$

has been studied by Dirscherl and Lee (1979). This energy could be varied over the range 0.1–1 eV. It was found that the extra collisional energy appears mainly as rotational energy in BaO although only some 20% of available reaction energy appears as internal energy in BaO; an example of the data obtained is given in Table 5.11.

A classic experiment on the role of the state of vibrational excitation in a diatomic reactant on the internal state of a reaction product was reported by Pruett and Zare (1976). The reaction studied was

$$\mathrm{Ba} + \mathrm{HF} \rightarrow \mathrm{BaF} + \mathrm{H}$$

Table 5.11. Data for the Effect of Collisional Energy on Energy Terms in Reaction of Ba with SO_2

Collisional energy (eV)	Vibration energy (eV)	Rotational temperature (K)	$E_{int}(BaO)/E_{total}$ (%)
0.05	0.014	400 ± 100	20 ± 5
0.11	0.007	600 ± 200	20 ± 5
0.41	0.006	1000 ± 300	16 ± 5
0.60	0.003	1500 ± 400	17 ± 5
0.89	0.001	2000 ± 500	17 ± 5

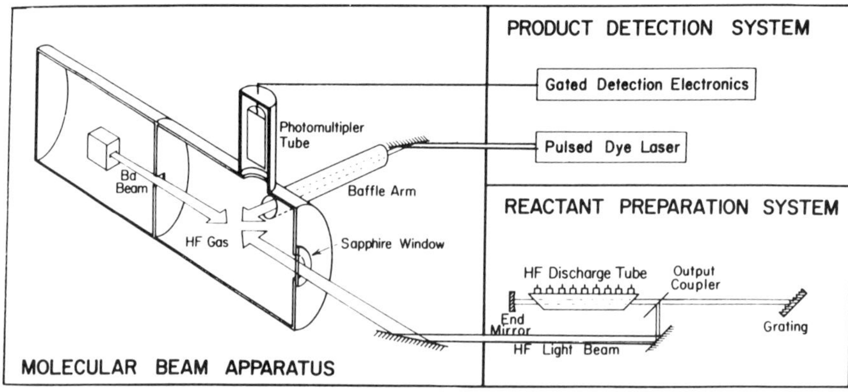

Figure 5.31. Schematic of LIF experimental setup showing the molecular beam apparatus (cutaway drawing), the product detection system, and the reactant preparation system [Pruett and Zare (1976)].

The vibrational state of HF could be modified by irradiation of HF with an HF laser. A diagram showing the irradiation/LIF system is shown in Figure 5.31. BaF excitation spectra recorded with the HF laser on or off were differenced to yield a spectrum characteristic of the enhancement due to HF excitation to the $v'' = 1$ state. This spectrum could then be used to obtain an estimate of the enhancement in BaF ground-state vibrational excitation that could be attributed to HF($v'' = 1$). This calculation was complicated by the fact that the results obtained with HF laser excitation were due to the reaction of both HF($v'' = 0$) and HF($v'' = 1$). Thus the excitation of HF to $v'' = 1$ was accompanied by a decrease in HF($v'' = 0$) population. The resulting relative state-to-state reaction rates are shown in Figure 5.32 [Pruett and Zare (1976)]. It is evident that vibrational excitation of HF dramatically enhances BaF populations in levels with $v'' \geq 4$. The BaF product was found to retain some 64% of the initial vibrational excitation.

The angular distribution of individual BaCl vibrational states in the

$$\mathrm{Ba} + \mathrm{KCl} \rightarrow \mathrm{BaCl} + \mathrm{K}$$

reaction has been reported by Smith and Zare (1976). It was found that molecules with low v'' were scattered with strong forward and backward amplitudes in the laboratory reference frame. Some 58% of BaCl was found to be forward scattered. This was suggested to be the result of the formation of a long-lived collision complex. Figure 5.33 shows a plot of BaCl v'' fluxes for several v'' at various laboratory scattering angles.

When a reactant is pumped with polarized light from a laser before encountering another reactant, then the effect of orientation on reaction rate

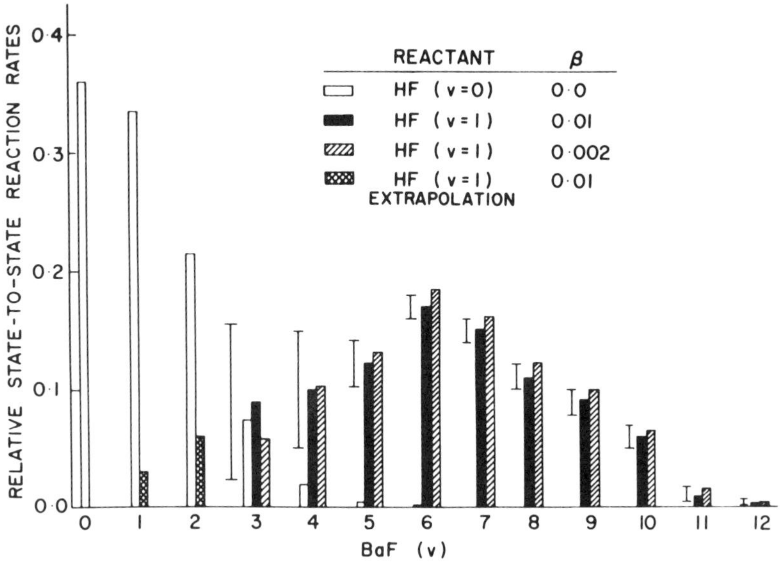

Figure 5.32. Bar graph showing the state-to-state reaction rates for Ba + HF(v = 0,1) → BaF(v) + H. Two calculations for Ba + HF(v = 1) are displayed, corresponding to two different possible fractions of HF vibrational excitation (as measured by β). The error bars indicate the uncertainty in the $k(1,v)$ ratio [Pruett and Zare (1976)].

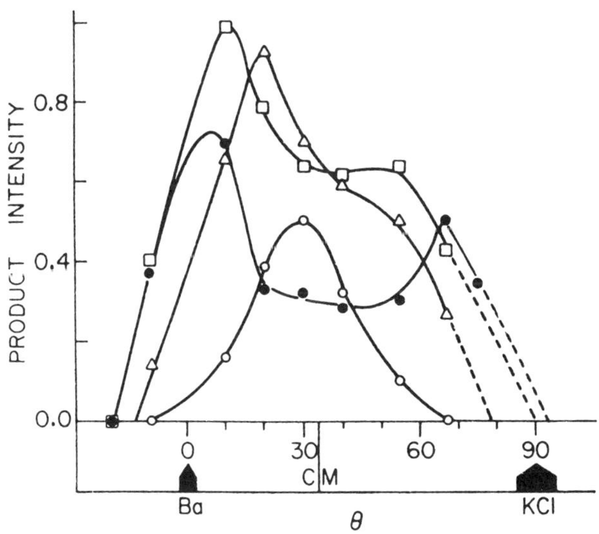

Figure 5.33. Plot of the BaCl vibrational state fluxes as a function of the laboratory scattering angle θ [Smith and Zare (1976)].

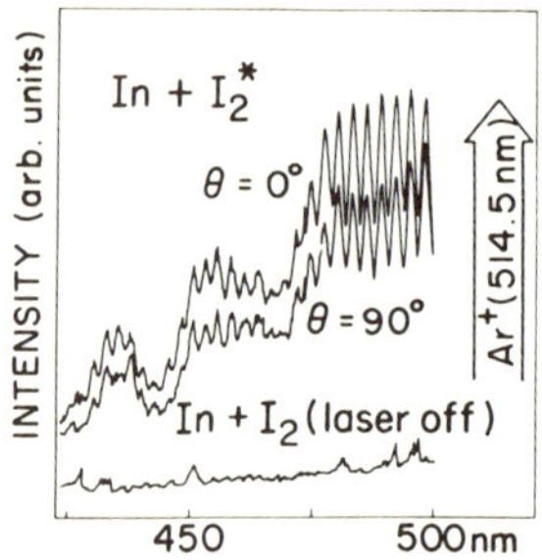

Figure 5.34. Laser-induced chemiluminescence spectra of In + I_2^* (left) and Tl + I_2^* (right) for different orientations of I_2^* ($\theta = 0°$, parallel; $\theta = 90°$, perpendicular) with respect to the metal atom approach direction [Zare (1979)].

and product state distributions can be determined [Zare (1979)]. This occurs since the polarization direction of incident light determines the probability that a particular molecule will absorb a photon. As a result, when absorption does occur, the orientation of the absorber is determined. This orientation can be carried to another beam where a reaction can proceed. By studying the degree to which this reaction occurs as a function of the incident polarization, the effect of steric factors on reaction rates can be readily studied. Figure 5.34 shows laser-induced emission spectra from InI and TlI prepared from electronically excited I_2 as a function of the angle of I_2^* relative to the beam of metal atoms.

5.3.5. Fluorimetry and Phosphorimetry

Fluorimetry is a well established and useful technique in analytical chemistry. While this method is characterized by its selectivity and sensitivity [Guilbault (1973)] the use of laser excitation has brought about many new instrumental developments that have, in turn, led to increased selectivity and sensitivity. The tunability of laser sources permits optimization of absorption by the fluorescent species and discrimination against competitive species. Pulsed laser sources also lend themselves naturally to the use of time-resolved detection.

An indication of the improvement in detection limits possible with laser excitation can be seen in Table 5.12 [Richardson (1977)]. Richardson used a N_2 pumped dye laser system with frequency-doubling capability to excite fluorescence from a variety of amino acids and vitamins in water. Nonfluorescent compounds, such as aniline and arginine, were labeled by fluorescamine. Detection limits for all compounds were reduced by at least a factor 10^2 below those attainable with conventional fluorimetry. Plots of fluorescent intensity vs. solute concentration were linear over many orders of magnitude.

Frequency doubling has been used by Richardson and Ando (1977) to excite fluorescence from several polycyclic aromatic hydrocarbons in water or water–methanol solutions. Detection limits for the compounds studied were

Table 5.12. Limits of Detection of Biochemicals in Water[a]

Compound	Excitation wavelength, λ_{ex} (nm)	Absorptivity at λ_{ex} (M^{-1} cm^{-1})	Fluorescence wavelength, λ_{em} (nm)	Quantum yield	Limits of detection: LIMF, M	LIMF, μg/1(ppb)	Conventional, M	Conventional, μg/1(ppb)
Vitamin A acetate	337.1	1.8×10^4	500	—	3×10^{-9}	1.0	—	100
Vitamin B_2 (riboflavin)	375.0	1.1×10^4	540	0.26	1.25×10^{-12}	0.47×10^{-3}	1×10^{-10}	0.038
FAD[b]	267.0	3.8×10^4	525	0.04	5×10^{-11}	39×10^{-3}	—	—
Vitamin B_6 (pyridoxine)	337.1	1.6×10^3	410	0.10	2.5×10^{-10}	50×10^{-3}	—	1.5
Vitamin B_{12} (cyanocobalamin)	260.0	1.2×10^4	310	—	ca 1×10^{-7}	1.4×10^2	—	—
Tryptophan	270.0	5.4×10^3	358	0.13	2.5×10^{-10}	50×10^{-3}	—	3
Aniline/fluram	390.0	1.7×10^4	500	0.28	3.8×10^{-11}	3.5×10^{-3}	—	> 40
Arginine/fluram	390.0	1.7×10^4	500	~ 0.25.	5×10^{-11}	10×10^{-3}	—	300

[a]From Richardson (1977); [b]Flavin adenine dinucleotide.

benzene—19 ppb, napthalene—1.3 ppt, anthracene—$<$ 4.4 ppt, fluoranthene—1.0 ppt, and pyrene—0.5 ppt. The relatively high detection limit for benzene was largely due to limitations provided by frequency doubling the dye laser to wavelengths $<$ 260 nm. At the wavelength used for excitation of benzene fluorescence (259.95 nm), the absorption of benzene was small, leading to inefficient coupling.

An interesting comparison of various source-detection systems in fluorimetry has been published by Ishabashi *et al.* (1979). The optimization of the signal-to-noise ratio (S/N) by variation of excitation and detection wavelength was also discussed. This is shown in Figure 5.35 for fluorescein in water. The

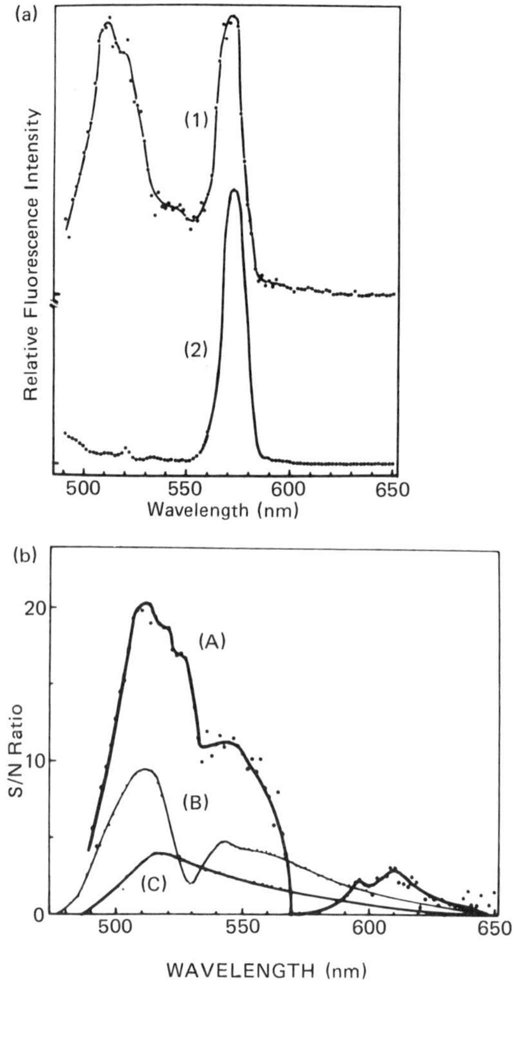

Figure 5.35. (a) Comparison of fluorescence spectrum of fluorescein in water (1) with that of water (2): excitation wavelength, 480 nm; linewidth 0.3 nm. (b) S/N ratio vs. wavelength of fluorescein excited at (A) 480 nm, (B) 450 nm, and (C) 420 nm [Ishibashi *et al.* (1979)].

Table 5.13. Detection Limits for Fluorescein

Source	Detector	Excitation detection limit				
		λ (nm)	Δλ (nm)	λ (nm)	Δλ (nm)	ppt
DL[a]	Pulse gated photon counting	480	0.3	510	0.5	0.02
DL[b]	Averager	470	10	514	1.6	2
DL[c]	Boxcar	480	0.4	514	—	100
DL[d]	Boxcar	470	0.1	514	—	40
N_2[d]	Boxcar	337	<0.1	514	—	30
XE lamp[b]	dc amplifier	470	10	514	1.6	70

[a]Ishibashi *et al.* (1979); [b]Imasaka *et al.* (1977); [c]Smith *et al.* (1974); [d]van Geel and Winefordner (1976).

advantages to be gained by optimization of the detector wavelength–laser wavelength combination is obvious. Best results for fluorescein are obtained with laser excitation at 480 nm and detection at 510 nm. A summary of detection limits for fluorescein with this and other methods is given in Table 5.13.

Fluorescence can also be excited via direct or sequential two-photon absorption. Studies of the fluorescence of aromatic hydrocarbons excited by sequential two-photon absorption have been reported by Orner and Topp (1975), Lin and Topp (1977), and Topp and Lin (1977). A review of this technique has been published by Topp (1978).

It may be possible to fully exploit the high resolution available from dye laser excitation by using the matrix isolation technique [Meyer (1971)] for the production of samples. Maple *et al.* (1980) have shown that condensation of polynuclear aromatic hydrocarbons with *n*-heptane at cryogenic temperatures results in samples in which the aromatic molecules are frozen in well-defined sites. This reduces solute–solute and solute–solvent interactions to the point where narrow spectral lines are observed from the solute. The dye laser can then be tuned to excite individual lines to produce fluorescence from particular aromatic compounds in a complex mixture. Despite the complications involved in the preparation of samples and the need for cryogenic equipment, the LIF matrix isolation technique shows promise in the analysis of multicomponent samples.

Applications of time-resolved laser-excited fluorimetry and phosphorimetry in the analysis of aromatic hydrocarbons have been discussed by Lytle and Kelsey (1974), Brown *et al.* (1974), Wilson and Miller (1975), Boutilier and Winefordner (1979), and Ghiggino *et al.* (1979). The approach can be illustrated by the experiments of Brown *et al.* (1974) on the separation of prompt fluorescence from total emission as observed from solutions of the aromatic carbonyls. In these compounds, the weak singlet–singlet prompt fluorescence

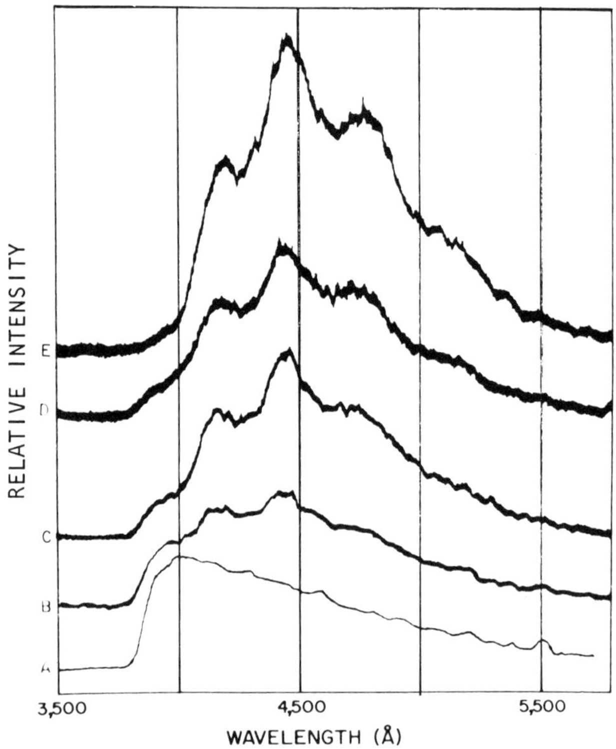

Figure 5.36. Time-resolved emission spectra from 0.1 *M* 4,4′-dimethoxybenzophenone in benzene, 25 °C, boxcar aperture t_g = 100 ns: (a) t_g = 0 ns; (b) t_d = 200 ns, r_s = 5; (c) t_d = 500 ns, r_s = 10; (d) t_d = 1 μs, r_s = 10; and (e) t_d = 5 μs, r_s = 50 [Brown *et al.* (1974) Reproduced with permission; copyright 1974, American Chemical Society].

is overlaid, in part, by strong phosphorescence due to triplet–singlet emission. Spectral analysis shows that the prompt fluorescence occurs on the high-energy side of the phosphorescence band. Brown *et al.* found that the prompt fluorescence was characterized by a short lifetime (<10 ns). This was superimposed on a phosphorescent emission that had lifetimes of ~1 ms at 77 K to ~1 μs at 300 K.

By using a boxcar integrator, the intensity of fluorescence delayed by specific times following the 10-ns laser pulse could be recorded as a function of wavelength. An example of spectral data recorded in this way is shown in Figure 5.36. The sample was a 0.1 *M* solution of 4,4′-dimethoxybenzophenone in benzene at 25 °C. Individual spectra were recorded with the boxcar aperture set at 100 ns. In the bottom spectrum, this aperture was placed to include the laser pulse. The top spectrum has the aperture open 5 μs after the laser pulse. It is evident that the use of time resolution enables the short-wavelength prompt fluorescence to be extracted from the dominant phosphorescent emission.

Picosecond pulses extend the time scale for the study of fluorescent phenomena in solutions to $\sim 10^{-12}$ s. On a time scale of 10^{-12} s, the role of the solvent and solvent–solute interactions become important and lead to new time-dependent phenomena. A review of fluorescence with picosecond pulses is given by Topp (1978).

5.3.6. Selective Excitation of Probe Ion Luminescence

Analysis of trace inorganic ion concentrations in solids by the selective excitation of probe ion luminescence (SEPIL) technique has been developed by Wright and coworkers [Miller *et al.* (1977), Wright (1977), Gustafson and Wright (1977, 1979), and Johnston and Wright (1979)]. This technique relies on the fact that the fluorescence lines of many lanthanide ions remain narrow in solids, even when these solids are in microcrystalline form. While these lines are narrow, their energy is sensitive to the nature and number of coordinating ions in the crystal lattice. Thus the spectrum of an ion such as Eu^{3+} in a solid may show structure due to emission from Eu^{3+} sites with neighboring defects or other ions. By monitoring the emission from a particular site while scanning the excitation wavelength with a tunable dye laser, an excitation spectrum of that center is obtained. This can be used in a determination of the concentration of that particular lanthanide ion–(defect, analyte, ion) combination.

A simplified representation of the energy levels involved in SEPIL is shown in Figure 5.37. X represents an ion such as Eu^{3+} in substitutional sites in a lattice such as CaF_2. The spectrum of ions in this site consist of sharp lines as shown. When X ions are in sites with adjacent impurity ions or defects produced by these ions, the energies of X levels are modified. By monitoring the $3 \rightarrow 0$ transition while scanning E, spectral features due to ions in these sites (and only these sites) can be recorded.

A block diagram of the spectroscopic system developed by Wright *et al.* to study SEPIL is shown in Figure 5.38. Samples in the form of precipitates that have been fired at elevated temperatures are placed in a variable temperature cryostat for cooling to ~ 10 K. Cooling to cryogenic temperatures results in sharp lines and suppresses absorption from excited states. The cooled solid is excited with pulses from an N_2 pumped dye laser. The resulting fluorescent intensity is monitored at a single wavelength as the laser wavelength is scanned. Figure 5.39 shows excitation spectra obtained from a $BaSO_4$:Eu^{3+}, PO_4^{3-} solid while monitoring various fluorescent lines. It can be seen that using a broadband detector for fluorescence results in a complex excitation spectrum that has contributions from ions at many sites. By narrowing down the detection bandwidth to center on a particular fluorescent line (at longer wavelengths than those shown in the figure), spectra due to ions at different sites can be

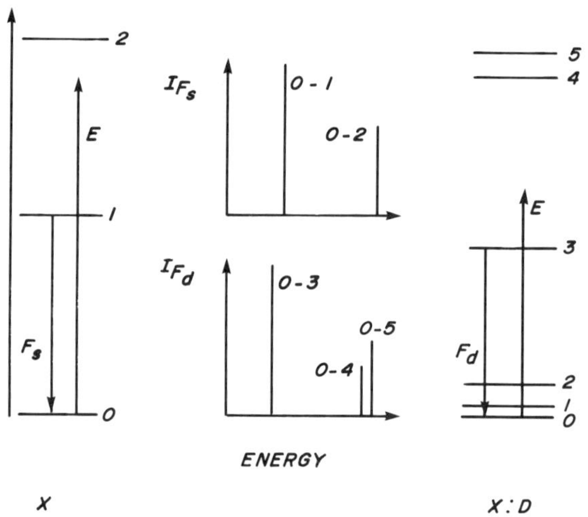

Figure 5.37. Excitation spectra of substitutional and defected sites for X-type ions in a solid. X, substitutional site of high symmetry; X:D, defect site of low symmetry. *F*, fluorescence; *E*, excitation; I_{FS}, fluorescent intensity at $1 \rightarrow 0$ transition; I_{FD}, fluorescent intensity at $3 \rightarrow 0$ transition.

obtained. Part (d) of the figure corresponds to sites having neighboring PO_4^{3-} ions. This spectrum can then be used for quantitative measurements of PO_4^{3-} concentration [Wright (1977)].

The detection of fluorescent and nonfluorescent lanthanide ions using this method has been discussed by Gustafson and Wright (1979) and Johnston and Wright (1979). Direct detection of lanthanide emission or lanthanide-pair emission is used to provide analysis at the 10^{-12} g ml^{-1} level.

The sites of importance in the analysis of coprecipitated samples fired in air involve association with interstitial O atoms introduced to achieve charge compensation or lanthanide $-Er^{3+}$ pair sites. These latter sites can be used to detect the presence of nonfluorescent ions such as La^{3+}, Ce^{3+}, Gd^{3+}, and Lu^{3+}. A list of detection limits for these and fluorescent lanthanides is given in Table 5.14. Detection is possible over a wide range of lanthanide concentrations for both fluorescent and nonfluorescent ions.

5.4. LASER-ENHANCED IONIZATION SPECTROSCOPY

The newly developed technique of laser-enhanced ionization (LEI) spectrometry promises to be an important analytical tool. In LEI spectrometry, the laser wavelength is tuned to an electronic transition of an atomic or molecular species in a flame or other source of excitation. In a flame, the conductivity,

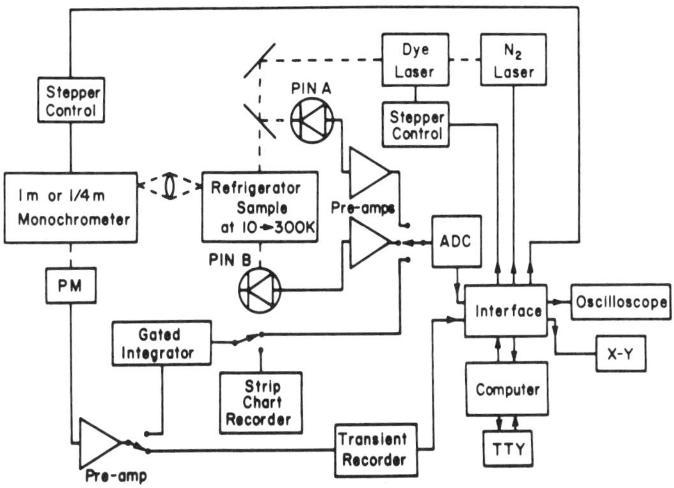

Figure 5.38. Block diagram of experimental apparatus for *SEPIL*. Dashed lines indicate light path [Wright (1977) Reproduced with permission; copyright 1977, American Chemical Society].

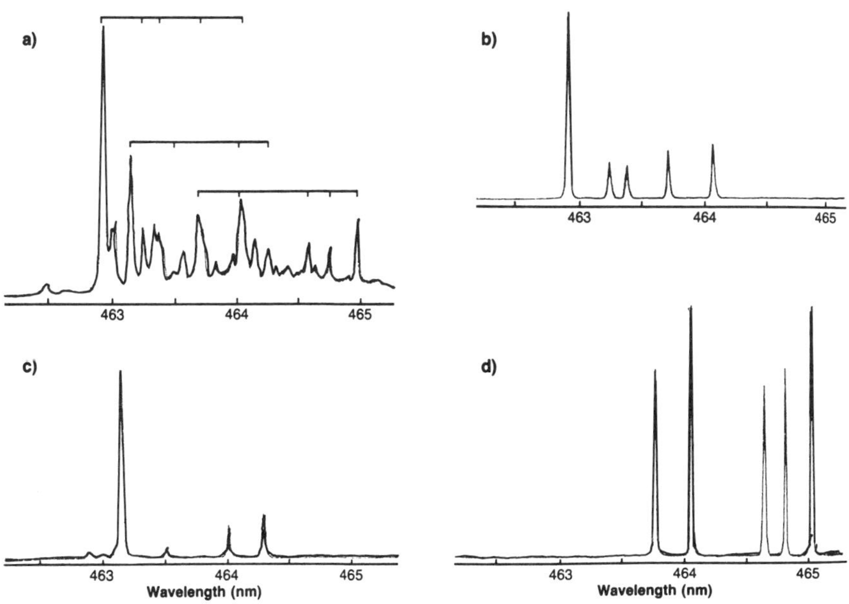

Figure 5.39. Excitation spectrum of $BaSO_4{:}Eu^{3+}$, PO_4^{3-} obtained by monitoring: (a) all fluorescence; (b) fluorescence line at 576.8 nm; (c) fluorescence at 577.0 nm; (d) fluorescence line at 578.2 nm corresponding to Eu ions with nearby PO_4 [Wright (1977) Reproduced with permission; copyright 1977, American Chemical Society].

Table 5.14. Detection Limits for Lanthanides Using the Selective Excitation of Probe Ion Luminescence Method

Lanthanide	Limit for S/N = 1		Practical limit[c] (mol %)
	mol %[a]	pg ml^{-1}[b]	
La	4×10^{-8}	8	10^{-6}
Gd	2×10^{-7}	45	10^{-6}
Lu	6×10^{-8}	16	10^{-6}
Y	3×10^{-8}	4	10^{-6}
Cl	2×10^{-8}	4	10^{-6}
Th	2×10^{-8}	8	$<5 \times 10^{-7}$
Pr	9.7×10^{-9}	2.4	—
Nd	1.8×10^{-6}	450	—
Sm	2.5×10^{-7}	63	—
Eu	1.6×10^{-9}	0.4	—
Tb	1.5×10^{-7}	38	—
Dy	5.8×10^{-8}	15	—
Ho	4.4×10^{-8}	11	—
Er	2.4×10^{-9}	0.6	—
Tm	8.0×10^{-9}	2	—

[a]In solid; [b]In solution used to prepare solid; [c]Provided by sample contamination.

which can be measured with external electrodes, is a sensitive fuction of ion and electron concentration. If these concentrations are disturbed, the conductivity is seen to vary. When a laser is tuned to an electronic transition of an absorber in the flame, the resulting excited species may be more easily ionized by collisional processes. Thus laser excitation of the flame influences the concentration of ionic species and can be detected via flame conductivity. Since this effect is only produced when the laser is tuned to a transition of an atomic or molecular flame constituent, it provides a measure of the concentration of such constituents in the flame. Since this technique involves the direct measurement of an electrical signal produced by the fraction of laser light absorbed, it is not affected by the many noise and interference sources present in conventional flame atomic spectrometry. The result is an analytical method with sensitivity for atomic metals in the 10^{-3} ng ml^{-1} range.

Figure 5.40 shows a schematic diagram of an LEI system [Schenck *et al.* (1978)]. The output from a tunable dye laser is passed through the flame near two electrodes based at -1000 V relative to the burner. Current pulses due to LEI are superimposed on a large dc background due to the static conductivity. These pulses are amplified and detected with a boxcar integrator (for pulsed laser excitation) or by a lock-in detector (for chopper CW laser excitation). When the laser wavelength is tuned through a spectral resonance in a constituent of the flame, an increase (or decrease) in current similar to that shown in Figure 5.41 for Na is observed [Green *et al.* (1976)].

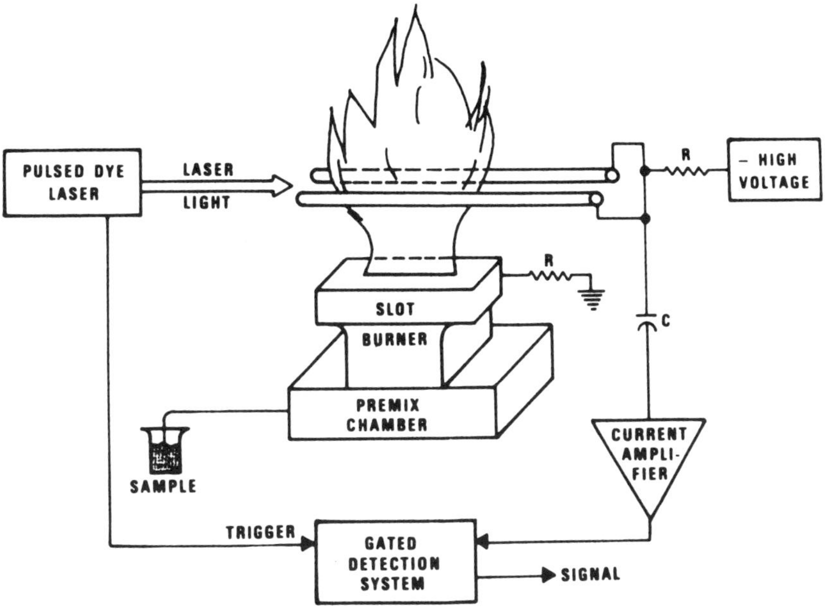

Figure 5.40. Schematic diagram of LEI apparatus [Schenck *et al.* (1978)].

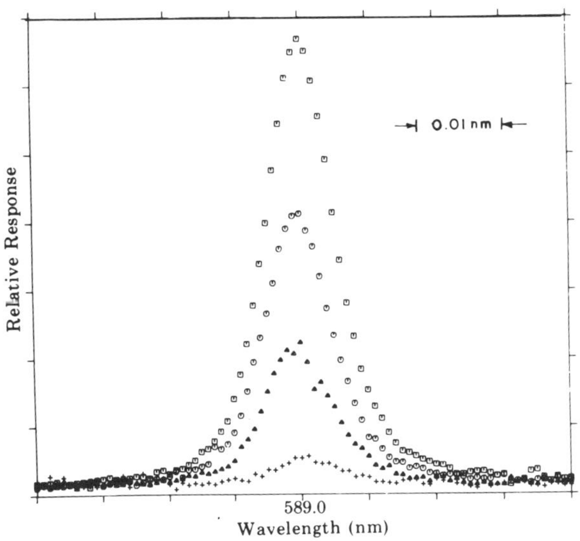

Figure 5.41. Optogalvanic (LEI) signal as a function of wavelength (~0.05 nm total scan) for several concentrations of sodium: □, 10 ppb; ⊙, 5 ppb; △ 2 ppb; +, blank [Green *et al.* (1976a) Reproduced with permission; copyright, 1976, American Chemical Society].

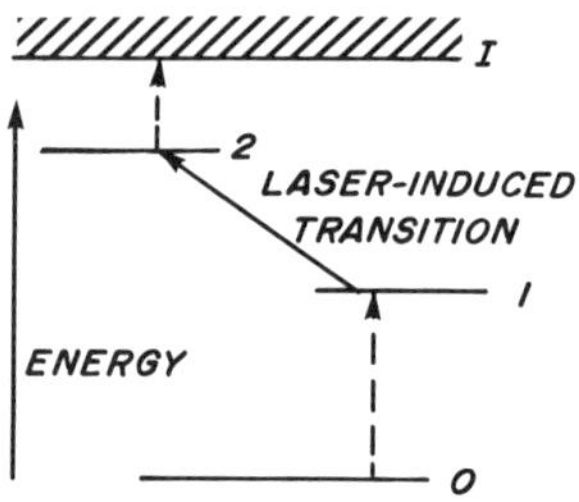

Figure 5.42. Energy levels for atomic LEI spectroscopy: 0—ground state; 1, 2—excited states with energy difference equal to energy of laser photon; *I*, ionization limit.

Figure 5.42 shows the atomic energy involved in an LEI system. Dashed lines represent collisional excitations that occur in thermal equilibrium in the flame. One of these excites the atom from its ground state (level 0) to an excited state (level 1). The laser excites an atom in state 1 to state 2, where it can ionize collisionally making a transition to the set of levels *I*. Since the efficiency for ionization due to collisions is greater in level 2 than level 1, the laser-induced transition results in an enhancement of the ion concentration in the flame. Travis *et al.* (1979) have shown that the sensitivity of LEI for a particular electronic transition in a given atom is

$$S_T(A) = C_1 X_B(A)\beta(\mathrm{A})\{\exp[-E_i^*(A)/k_BT]B_{12}(A)E_\nu^p(A)\}^{C_2}$$

where $S_T(A)$ = sensitivity nA ng^{-1} ml^{-1}
C_1 = empirical constant
$X_B(A)$ = Boltzmann population factor for level 1
= $\exp[-(E_1 - E_0)/k_BT]$
$\beta(A)$ = atomization efficiency for element and flame combination
$E_i^*(A)$ = $E_I - E_2$
k_B = Boltzmann constant
T = flame temperature (K)
$B_{12}(A)$ = Einstein probability factor for transition $1 \rightarrow 2$
$E_\nu^p(A)$ = peak laser spectral irradiance
C_2 = empirical constant

Turk *et al.* (1979a) have derived the following values for C_1 and C_2 from an analysis of data for 23 different optical transitions in 14 elements

$$\log C_1 = -3.3 \pm 0.3$$
$$C_2 = 0.65 \pm 0.05$$

These apply for elements introduced via standard procedures into an air–acetylene flame. Representative detection limits for several elements are listed in

Tables 5.15. These data were obtained using a flashlamp pumped dye laser with frequency doubling where applicable and show clearly that LEI sensitivity depends significantly on E_i^*, the energy difference between the terminal photoexcited level and the ionization limit. Although other factors can influence these data (notably a reduction in sensitivity due to low laser power when doubling is required), it can be seen that low detection limits tend to occur with elements that have small E_i^*. This result arises from the Boltzmann factor $\exp(-E_i^*/k_BT)$ in the expression for $S_T(A)$.

A comparison of detection limits between LEI in flames and such standard techniques as flame atomic absorption/emission and flame atomic fluorescence shows that for many elements LEI flame spectrometry leads to a reduction in effective detection limits by 1–2 orders of magnitude.

The sensitivity of LEI spectrometry for detection of a particular element is affected by the presence of interferences from neighboring lines and from background ionization due to matrix material. Since the wings of resonance lines can extend over many nanometers, particularly at atmospheric pressure and at elevated temperature, the presence of small quantities of elements such as Na or K may interfere with the detection of other elements with nearby

Table 5.15. Detection Limits for Elements Measured with LEI in Air–Acetylene Flame[a]

Element	Wavelength (nm)	Energy levels (cm^{-1})	E_i^* (cm^{-1})	Detection limit ($ng\ ml^{-1}$)
Ag	328.1	0–30473	30633	1
Ba	307.2	0–32547	9488	0.2
Bi	306.8	0–32588	26202	2
Ca	300.7	15316–48564	741	0.1
Cr	298.6	8308–41782	12788	2
Ca	324.8	0–30784	31533	100
Fe	302.1	0–33096	30604	2
Ga	287.4	0–34782	13606	0.07
In	303.9	0–32892	13788	0.006
K	294.3	0–33973	1037	1
Li	610.4	14904–31283	12204	0.012
	670.8	0–14904	28583	0.001
Mg	285.2	0–35051	26620	0.1
Mn	279.5	0–35770	24200	0.3
Na	285.3	0–35042	6407	0.05
	589.0	0–16973	24476	0.1
Ni	300.2	205–33501	28078	7
Pb	280.2	10650–46329	13491	0.6
Sn	284.0	3428–38629	20603	3–6
Tl	391.8	7793–43049	7218	0.09

[a]From Turk *et al.* (1979a,b).

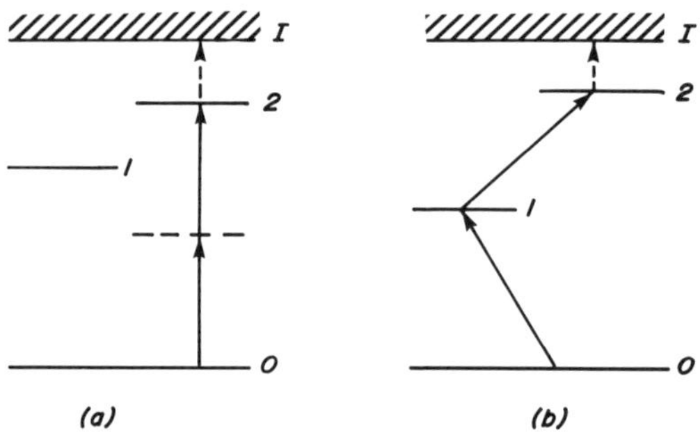

Figure 5.43. Energy levels involved in (a) direct two-photon transition 0 → 1; (b) stepwise two-photon transition 0 → 1 → 2.

transitions. As an example of this effect, Turk *et al.* (1979a) find that the introduction of 20 μg ml^{-1} of Na into a flame with Pb interferes strongly with the detection of Pb through the line at 280.2 nm. This interference is due to the wing of the Na line at 285.3 nm.

The high intensity obtainable from laser sources makes it possible to observe LEI from direct two-photon [van Dijk *et al.* (1978)] and stepwise two-photon transitions [Turk *et al.* (1979b)]. The real and virtual states involved in such transitions are shown in Figure 5.43. In part (a) of the figure the laser energy is $h\nu = (E_2 - E_0)/2$, where h is Planck's constant and ν is the frequency of the laser output. The two-photon transition occurs via a virtual intermediate state. The transition probability for such a two-photon transition scales as F^2, where F is the laser intensity.

Part (b) of the figure shows a stepwise two-photon transition via a real intermediate state. This transition requires two lasers of different wavelengths synchronized in time to yield overlapping excitation pulses. The transition probability for this process scales as $F_{0\to1}F_{1\to2}$.

Since two-photon transitions permit access to highly excited states, an enhancement of LEI efficiency may be possible using this technique. Turk *et al.* (1979b) illustrate this with an example of LEI detection of Cu using one- and two-photon excitation. With one-photon excitation no Cu signal could be detected from a 1000-ppm solution when $E_i^* = 31{,}533$ cm^{-1}. The use of a second laser to excite atoms in the first level to within 9468 cm^{-1} of the ionization limit allowed detection of Cu from a 1000-ppm solution with a signal to blank noise ratio of 5700.

Two-photon excitation may reduce interferences from matrix species since the state observed can be chosen to lie in a different spectral region from the interfering line. However, at high laser intensity new interferences due to multiphoton resonances may occur.

While atomic species are relatively easily detected in flames using LEI, common molecular species such as C_2, OH, or CN are not observed. The reason for this appears to be related to the high ionization potential of such molecules which inhibits thermal ionization from laser-excited states. Despite the failure to observe intrinsic molecular entities in flames via LEI, Schenck *et al.* (1978) have seeded flames with salts of several heavy metals and subsequently detected spectral features due to diatomic oxides of these metals. Transitions were observed in ScO, YO, and LaO, suggesting that LEI detection may be a useful spectroscopic source for refractory molecules.

LEI or optogalvanic spectroscopy can also be studied with discharge sources [Green *et al.* (1976b)]. For spectroscopic investigations hollow cathode sources are preferred because of their stable emission characteristics and narrow linewidth. Several galvanic optospectroscopic studies have been reported on elements in hollow cathode sources [Smyth and Schenck (1978), Zalewski *et al.* (1979), and Bridges (1978)].

5.5. MULTIPHOTON IONIZATION

A sensitive method of detecting trace quantities of gas-phase molecules using laser-excited, two-photon ionization has recently been described [Brophy and Retner (1979a,b) and Frueholz *et al.* (1980)]. Potential detection limits with the two-photon ionization technique may be 1 molecule cm^{-3}.

Energy levels involved in two-photon resonant molecular ionization are shown in Figure 5.44. The first photon is absorbed resonantly by the molecule on a line of the $S_0 \rightarrow S_1$ band. At high laser intensity, a second photon can be

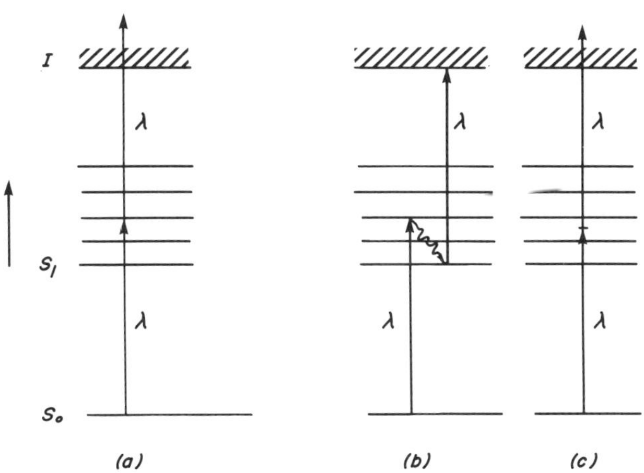

Figure 5.44. $S_0 \rightarrow S_1$ transition as intermediate resonance in two-photon ionization: (a) no relaxation; (b) relaxation to lower vibrational level of S_1 state; (c) nonresonant two-photon ionization.

absorbed while the molecule is in the S_1 state. This is also a resonant process when the photon energy is sufficient to promote a transition to the ionization continuum. Since the cross section for ionization is large only when the transition occurs via real intermediate states, the ionization current shows resonances due to bound molecular states when the photon energy is scanned through the $S_0 \rightarrow S_1$ band. Conversely, by tuning the laser wavelength to a sharp molecular resonance, the detectability of that molecule can be enhanced.

The electrical detection system for ions produced by two-photon absorption can be a simple pair of electrodes [Brophy and Rettner (1979b)] biased to ~50 V. Enhanced sensitivity is obtained, however, by creating ions inside a proportional counter [Frueholz *et al.* (1980)]. Alternatively, the ions produced can be sampled by a mass spectrometer [Antonov *et al.* (1978)].

Two-photon ionization need not involve two photons of the same energy. Antonov *et al.* (1978) use a pulsed H_2 laser operating at 160 nm to ionize molecules excited by an N_2 pumped dye laser. Since the photoionization continuum extends over many electron volts, resonant ionization is still possible even when the state reached by absorption of the second photon lies much higher than the ionization limit.

Figure 5.45 shows a schematic diagram of a two-photon ionization system in which the two photons have different energies [Brophy and Rettner (1979b)]. In this case, the output from each laser was frequency doubled to the 290–300 nm range. This results in transitions to continuum states with energies ~8.5 eV. Since the ionization potential of the molecule studied (aniline) is 7.7 eV, efficient ion formation is possible. By scanning the wavelength of one of the lasers, spectral features associated with the $^1B_2$–$^1A_1$, O_0^0 band appeared as variations in the ion signal. An analysis of the intensity of the two-photon ionization signal showed that $\sigma_1 = (1.0 \pm 0.2) \times 10^{-17}$ cm^2 and $\sigma_2 = (3.5 \pm 0.8) \times 10^{-17}$ cm^2 for the first and second transitions, respectively.

Antonov *et al.* (1978) studied the two-photon ionization of a variety of inorganic and organic molecules. Figure 5.46 shows a comparison between the two-photon ionization spectrum of NO_2 with its absorption spectrum. The ionization spectrum was recorded by monitoring the appearance of NO_2^+ mass spectrometrically. It is evident that major features of the one-photon absorption spectrum are well reproduced in the photoionization spectrum.

Spectra were also obtained for benzaldehyde, C_6H_5CHO. Here it was found that benzaldehyde fragmented to form $C_6H_5CO^+$ ions. The yield of this ion was twice as large with two-photon excitation as with one-step photoionization.

The sensitivity of this technique in the measurement of trace quantities of gas-phase organic molecules has yet to be thoroughly evaluated. However, Antonov *et al.* report that absorption coefficients as small as 10^{-10} cm^{-1} can be measured for ground-state molecules. The minimum detectable absorption for molecules that have already absorbed a photon was found to be 10^{-14} cm^{-1}.

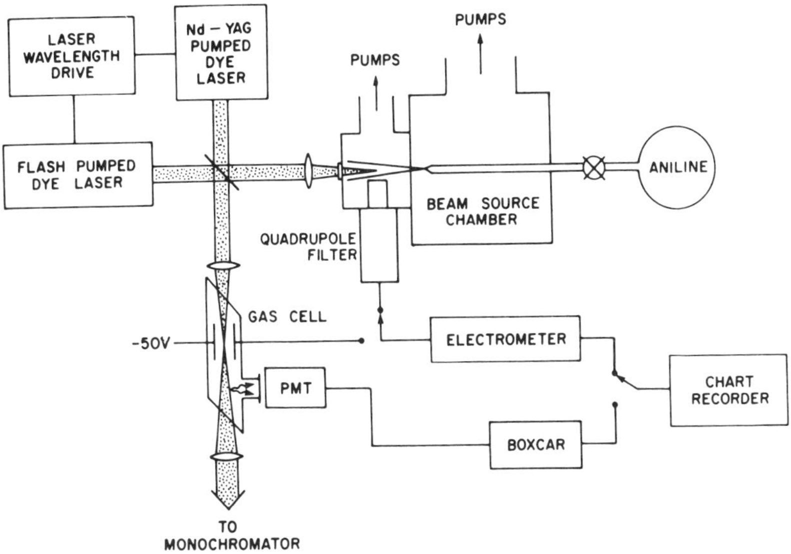

Figure 5.45. Schematic of a laser multiphoton ionization apparatus [Brophy and Rettner (1976b)].

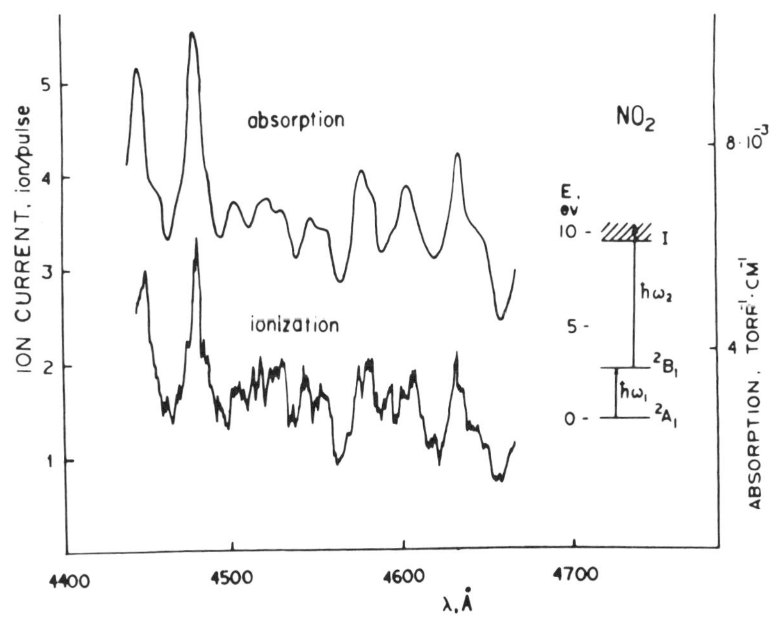

Figure 5.46. Dependence of the two-step photoionization yield and the absorption of NO_2 on the dye laser wavelength [Antonov *et al.* (1978)].

5.6. RAMAN SPECTROSCOPY

5.6.1. Theory and Physical Principles

When a sample of molecular gas is placed in an optical radiation field two fundamental interactions with this field are possible—absorption and scattering. At low intensity, scattering falls into two basic categories

1. *Rayleigh scattering*—no change in wavelength in scattered light.
2. *Raman scattering*—changes in wavelength of scattered light with the molecule either gaining or losing energy.

If a monochromatic source of frequency ω irradiates the sample, Rayleigh scattering will be responsible for a strong scattered component at ω. Raman scattering generates new frequencies in the scattered beam. These are

$$\omega_S = \omega - \omega_R, \quad \text{Stokes components}$$
$$\omega_{AS} = \omega + \omega_R, \quad \text{Anti-Stokes component}$$

where ω_R is a Raman frequency in the molecule. Raman frequencies correspond to those of vibrational and rotational modes of the molecule. These frequencies are not necessarily those seen in infrared absorption spectra of the molecule. In fact, for many molecules, infrared and Raman spectra provide complimentary information about fundamental vibrational modes. This is particularly true for symmetric molecules where selection rules for the excitation of infrared and Raman modes may be quite different. Figure 5.47 shows the energy levels participating in Rayleigh and Raman scattering together with resulting spectra.

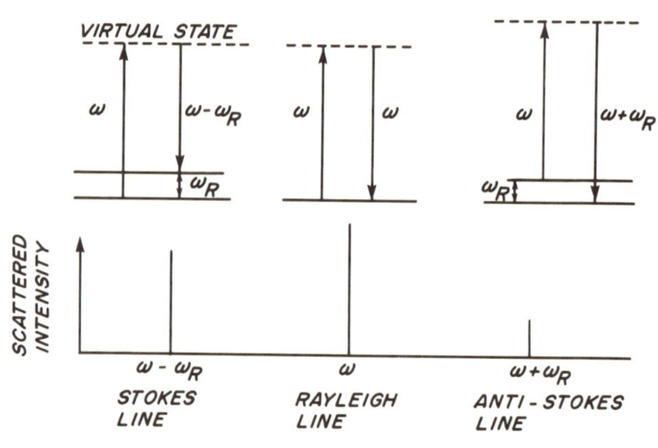

Figure 5.47. Energy levels and spectra for Rayleigh and Raman scattering.

The response of an isolated molecule to an applied electric field, E, can be treated in terms of an induced dipole moment μ, where

$$\mu = \alpha E$$

and α is the polarizability. This expression actually involves the tensor polarizability. However, by using orthogonal coordinates x, y, and z within the molecule and assuming that the polarizability tensor is symmetric, one can write

$$\mu_x = \alpha_{xx} E_x$$
$$\mu_y = \alpha_{yy} E_y$$
$$\mu_z = \alpha_{zz} E_z$$

The x, y, and z axes are therefore identified as the principal axes of polarizability. The polarizability can be calculated for any direction referred to these axes and defines a polarizability ellipsoid. Since the polarizability of a molecule

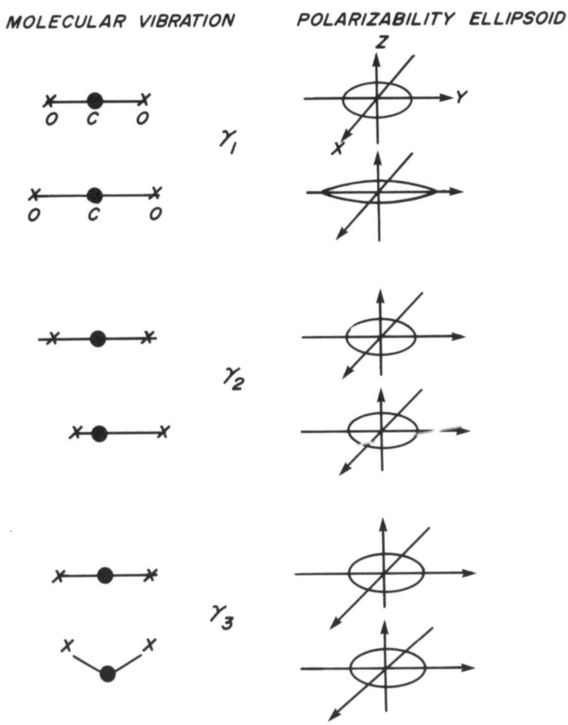

Figure 5.48. Fundamental vibrational modes of CO_2 and corresponding polarizability ellipsoids.

Table 5.16. Fundamental Vibrational Modes for CO_2 Molecule

Mode	Frequency	IR	Raman
Symmetric stretch	ν_1	Inactive	Active
Asymmetric stretch	ν_2	Active	Inactive
Bending	ν_3	Active	Inactive

depends on the relative positions of atoms within the molecule it may change during a vibration. When a molecular vibration leads to a change in the shape, size, or orientation of the polarizability ellipsoid, μ changes and a Raman transition is possible. Figure 5.48 shows the shape of the polarizability tensor for the CO_2 molecule. The three fundamental vibrational modes for this molecule and their IR/Raman activities are given in Table 5.16.

Figure 5.48 shows how the polarizability tensor varies with each of these vibrational modes. It can be seen that only the ν_1 mode leads to a change in this ellipsoid. Hence only the ν_1 mode is Raman active.

The magnitude of the Raman effect and its relation to that of other scattering effects can be obtained from a comparison of differential scattering cross sections, $d\sigma/d\Omega$. The power scattered, P_s, from a monochromatic beam of power P_i is

$$P_s = NLP_i\left(\frac{d\sigma}{d\Omega}\right)\Omega$$

where N is the density of the absorber (cm^{-3}), L the length of the scattering region, and Ω the solid angle over which scattered radiation is detected. Typical values for $d\sigma/d\Omega$ for various resonant and nonresonant scattering processes are given in Table 5.17.

As an example, for a gas at 1 atm and with $L = 1$ cm and $P_i = 1$ W, P_s(Raman) $= 8.5 \times 10^{-12}$ W, where $\Omega = \pi$ sr. This corresponds to $\sim 10^7$ photons s^{-1} at 500 nm. The Rayleigh power under these conditions is P_s(Rayleigh) $= 8.5 \times 10^{-9}$ W. Since P_s(Rayleigh) $\gg P_s$(Raman) the Raman signal would be difficult to detect were it not for the fact that it is shifted in

Table 5.17. Typical Values for $d\sigma/d\Omega$

Process	$d\sigma/d\Omega$ (cm^2 sr^{-1})
Resonance fluorescence (atomic)	10^{-16}
Resonance fluorescence (molecular)	10^{-23}
Rayleigh scattering	10^{-28}
Raman scattering	10^{-31}

frequency from the Rayleigh line. Even so, good sensitivity in Raman measurements requires that the detector have high discrimination against the Rayleigh line.

Enhancemennt in $d\sigma/d\Omega$ occurs when the incident frequency approaches or overlaps an electronic transition in the scattering molecule. If ω_E is the frequency of an electronic transition, then three ranges can be defined in terms of the relation of ω to ω_E:

$\omega \ll \omega_E$	Raman effect	$\epsilon \lesssim 1$
$\omega < \omega_E$	Preresonance	$\epsilon = 1\text{–}10^2$
$\omega > \omega_E$	Raman effect	
$\omega \sim \omega_E$	Resonance Raman effect	$\epsilon > 10^2$

where ϵ is the molar extinction coefficient (liter mol^{-1} cm^{-1}). When in the resonant range, enhancements of the Raman cross section by several orders of magnitude are possible. Resonance Raman effects have been extensively exploited since the advent of frequency-doubled dye laser sources. A fine example of resonant enhancement can be seen in the frequency dependence of the 1338- and 1484-cm^{-1} bands of adenosine 5′-monophosphate (AMP) shown in Figure 5.49 [Blazej and Peticolas (1977)]. A review of resonant Raman scattering has been published by Rousseau *et al.* (1979).

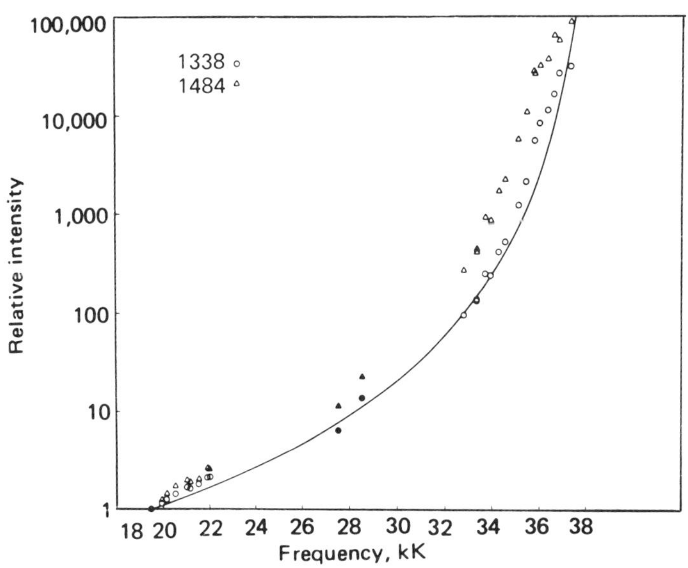

Figure 5.49. Resonant Raman effect in two bands of AMP [Blazej and Peticolas (1977)].

5.6.2. Experimental Techniques

Laser Raman spectroscopy (LRS) is a well established technique of great versatility. As a result, there have been many reviews on this subject (a list appears at the end of this section). For this reason, the present discussion will be limited in scope.

The three major components of a Raman spectrometer are (i) the laser source, (ii) the irradiation cell, and (iii) light gathering optics and spectrometer/spectrograph. We consider each of these in turn.

5.6.2.1. Laser Sources. The basic requirement for a Raman source is that it provides high-intensity monochromatic radiation free from spectral interferences due to other lines within $\sim +3000\ \text{cm}^{-1}$. The laser linewidth should be small enough so that it does not appreciably broaden Raman lines at the spectrometer resolution used. This may be $\sim$0.01 cm^{-1} although larger values ($\sim$0.1 cm^{-1}) are also acceptable for most applications.

These requirements are well satisfied by the Ar ion laser which can oscillate on a variety of lines in the 450–520 nm region and several in the 350-nm region. A list of these wavelengths is given in Table 5.18. The Kr^+ laser is also useful for Raman spectroscopy, as is the He–Ne laser with its output at 632.8 nm. Most commercial units use either Ar^+, Kr^+, or He–Ne lasers as the excitation source.

5.6.2.2. Irradiation Cells. The optimum irradiation configuration for a particular sample must be devised with the collection optics and spectrometer response in mind. With relatively transparent samples it may be possible to place the sample in the laser cavity thereby benefiting from increased laser intensity. With most solid and liquid samples, however, the laser beam is customarily focused in (or on) the sample and the scattered light is taken from

Table 5.18. Lines from Ar Ion Laser in the 450–520 nm Wavelength Range

λ_{air} (nm)	λ_{vac} (nm)	Energy (cm^{-1})
454.50579	454.63261	21995.7826
457.93495	458.06327	21831.0451
465.79012	465.92053	21462.8879
472.68683	472.81906	21149.7396
476.48646	476.61969	20981.0886
487.98635	488.12264	20486.6548
496.50795	496.64650	20135.0456
501.71628	501.85621	19926.0261
514.53083	514.67417	19429.7686

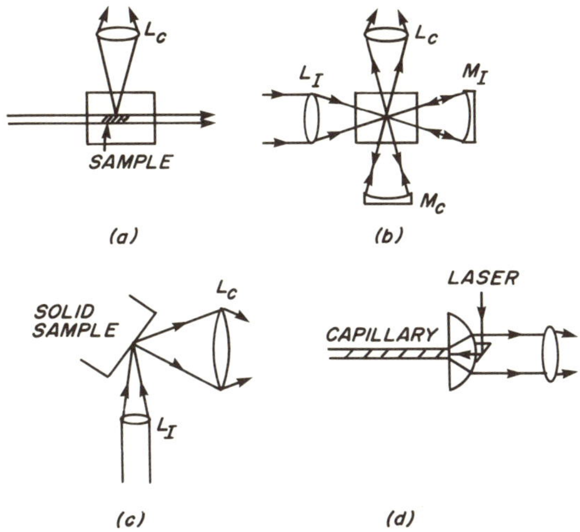

Figure 5.50. Some sample irradiation geometries. L_I, L_C—irradiation and collection lenses; M_I, M_C—irradiation and collection mirrors.

the focal volume with a suitable collecting lens. Figure 5.50 shows several irradiation geometries. The configurations shown in (a) and (b) are typical of those used for liquid or gaseous samples. In (b) the irradiation/collection system is optimized with M_I and M_C placed to enhance irradiation/collection efficiency. A useful system for the irradiation of liquid samples is shown in part (d).

Multipass cells can also be used both intra- and extracavity. Since only a small fraction of the laser beam is scattered on each pass, in principle, many passes are possible (with suitable optics) before the laser intensity is significantly reduced.

5.6.2.3. Collection Optics and Spectrometers. Usually a single lens is used to image scattered light from the sample on the spectrometer slit. In choosing this lens attention must be paid to the speed (f/number) of the spectrometer used relative to that of the lens. Optimum results are obtained when the lens matches the speed of the spectrometer. A filter may be placed between the collecting lens and the spectrometer to suppress the Rayleigh scattered component. Alternatively, the scattered radiation may be prefiltered with a low dispersion monochromator.

The basic requirements of the spectrometer used to analyze the Raman scattered light are that it have (i) sufficient spectral resolution to study the system of interest and (ii) good rejection of the Rayleigh component. The requisite resolution may vary from several cm^{-1} for solid or liquid samples to ~ 0.01 cm^{-1} in studies of rotational Raman spectra.

Rejection of Rayleigh scattered light often requires that two monochro-

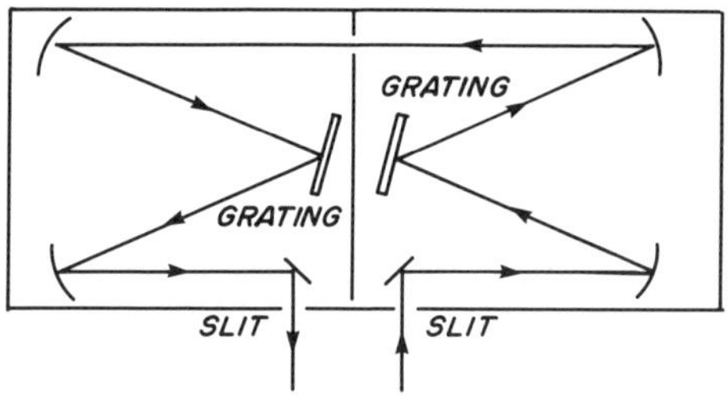

Figure 5.51. Double monochromator based on back-to-back Czerny–Turner configuration.

mators be operated in tandem. The double monochromator configuration can result in rejection ratios of 10^{-14} only a few tens of cm^{-1} from the Rayleigh line. A popular configuration is the back to back Czerny–Turner monochromator shown in Figure 5.51.

5.6.3. Experimental Results

In view of the wealth of data available on Raman spectra of molecules in various states of aggregation and the range of reviews published on all aspects of this subject, the reader is referred to the following bibliography.

Overviews	Gilson and Hendra (1970) Hendra (1975) Anderson (1971, 1973) Freeman (1974)
Rotational Raman	Weber (1979)
Solids	Hendra (1975)
Biological materials	Spiro (1974)
Adsorbed molecules	Van Duyne (1979)
Atmospheric molecules	Inaba (1976)
Resonant Raman	Rousseau *et al.* (1979)
Theory	Koningstein (1972) Herzberg (1945)

5.6.4. Coherent Anti-Stokes Raman Spectroscopy

In the technique of coherent anti-Stokes Raman spectroscopy (CARS) two laser beams with frequencies ω_1 and ω_2 are focused together in a sample. Mixing in the sample generates a coherent beam at a frequency $2\omega_1 - \omega_2 = \omega_3$. The intensity of this beam is highly dependent on the existence of molecular vibrations at $\omega_R \sim \omega_3$. When $\omega_3 = \omega_R$ an enhancement of the coherently scat-

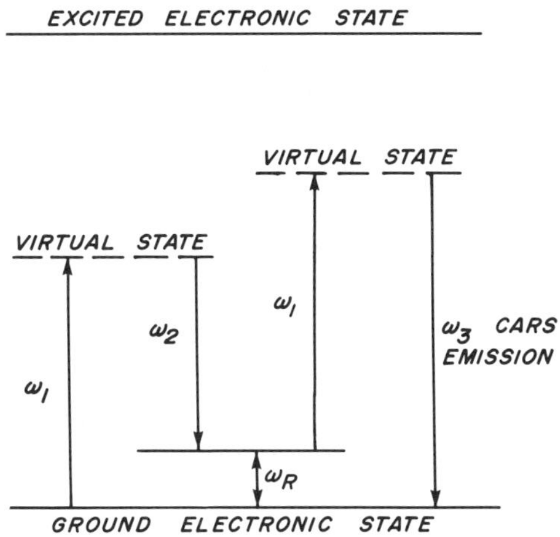

Figure 5.52. Molecular levels involved in CARS emission process.

tered beam is seen. Thus as ω_2 is varied, the beam at ω_3 scans the molecular Raman modes ω_R. The energy levels involved in CARS are shown in Figure 5.52.

Since linear momentum must be conserved, the wave vectors of incident and scattered beams must seem as shown in Figure 5.53. Thus the CARS beam will be emitted into a specific angle ϕ. This represents one of the advantages of the CARS technique because photons are emitted only into a narrow beam instead of into 4π steradians as in conventional Raman spectroscopy.

CARS is a third-order effect in the optical electric field as can be seen from the following simple analysis. The relation between electric field **E** and poliarization **p** can be expressed in the series

$$\mathbf{p} = \chi^1\mathbf{E} + \chi^2\mathbf{E}^2 + \chi^3\mathbf{E}^3 + \cdots$$

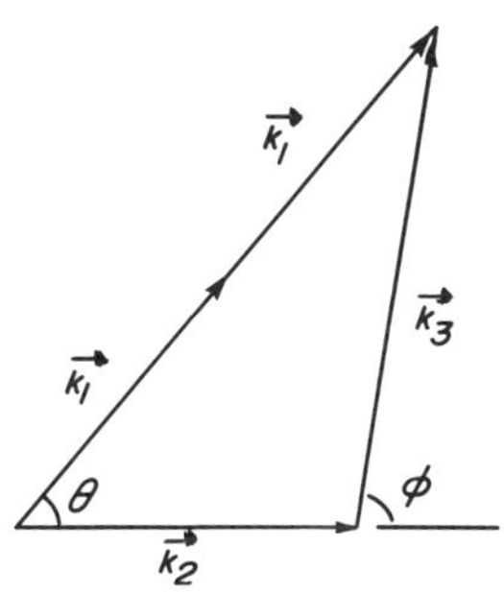

Figure 5.53. Phase-matching diagram in CARS. $\mathbf{k}_1$, wave vector at ω_1; $\mathbf{k}_2$, wave vector at stimulated Stokes emission frequency; $\mathbf{k}_3$, wave vector at CARS frequency.

where χ^i is the dielectric susceptibility tensor of rank $i + 1$. Each term χ^i is responsible for a variety of interactions between the system and the applied electric field. For example, the first term, χ^1, is responsible for classical Raman scattering. The second term, χ^2, gives rise to second-order effects, notably second harmonic generation and sum and difference frequency generation. The term in χ^3 yields third harmonic terms and the CARS effect.

For isotropic media such as gases, liquids, and some solids, the term χ^2 can be shown to vanish, leaving χ^3 as the first nonlinear term. CARS and some other coherent Raman techniques are shown schematically in Figure 5.54. Table 5.19 summarizes these processes and their properties.

With focused Gaussian beams the power in the CARS beam at ω_3 is given by

$$P_3 = \left[\frac{2}{\lambda n_3}\tan^{-1}\left(\frac{2L\lambda}{\pi d_0^2}\right)\right]^2 \frac{4\pi^2\omega_3}{c^2}\,|3\chi^3|^2\,P_1^2 P_2$$

where P_1 is the power in the beam at ω_1, P_2 the power in the beam at ω_2, $\lambda \sim \lambda_1$, n_3 is the refractive index of the medium at ω_3, L the coherence length, and d_0 the Gaussian beam diameter. Maximum power conversion occurs over a length $L \simeq 2\pi d_0^2/\lambda$. For $d_0 = 0.01$ cm, and $\lambda = 5 \times 10^{-5}$ cm and $L \sim 10$ cm.

It is evident that the third-order electric susceptibility χ^3, which is a function of ω, contains all the spectroscopic information about the system under examination. χ^3 can be written

$$\chi^3 = \chi^R + \chi^{NR}$$

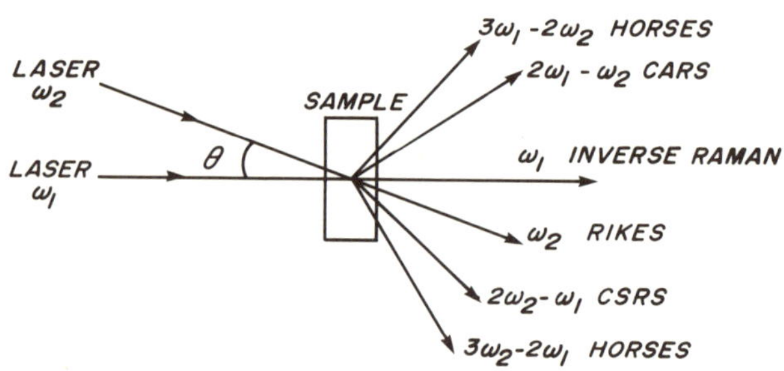

Figure 5.54. Some coherent Raman scattering geometries. See Table 5.19 for an explanation of the acronyms.

Table 5.19. Coherent Raman Scattering Processes

Acronym	Name	Process
CARS	Coherent anti-Stokes Raman spectroscopy	$2\omega_1 - \omega_2 = \omega_3$
CSRS	Coherent Stokes Raman spectroscopy	$2\omega_2 - \omega_1 = \omega_3$
HORSES	Higher-order Raman spectral excitation studies	$3\omega_1 - 2\omega_2, 3\omega_2 - 2\omega_1 = \omega_3$
RIKES	Raman-induced Kerr effect	Polarized ω_2

where χ^R is a term that arises because of resonances in the sample while χ^{NR} is a background term (basically frequency independent) arising from the sample environment. Then it can be shown that

$$\chi^3 = \frac{\chi^{NR}}{3} + \frac{1}{3}\sum_j \frac{2Nc^4}{h\omega_2^4}\left(\frac{d\sigma}{d\Omega}\right)_j \left\{\frac{\omega_R(j)\,\Delta_j}{[\omega_R(j)]^2 - (\omega_1 - \omega_2)^2 - i\Gamma_j(\omega_1 - \omega_2)}\right\}$$

where N is the density of scatterers, $(d\sigma/d\Omega)_j$ the Raman cross section for the jth Raman resonance at frequency $\omega_R(j)$, Γ_j the Raman linewidth, and Δj the factor measuring population difference between the two levels involved in the jth Raman transition. It can be seen from this expression that $\chi^3 \sim \chi^{NR}$ when $\omega_1 - \omega_2$ is far from any $\omega_R(j)$. When $\omega_1 - \omega_2 \rightarrow \omega_R$, χ^3 increases dramatically and P_3 likewise increases. Some salient features of the CARS method are apparent from the fact that $P_3 \propto |\chi^3|^2$ and the functional dependence of χ^3 on frequency. These are as follows:

1. The output at frequency ω_3 is coherent.
2. $P_3 \propto P^3$, i.e., small increases in laser power can give appreciable increases in P_3.
3. Signals are greatly enhanced near Raman lines due to $\omega_1 - \omega_2$ term.
4. $P_3 \propto N^2\,(d\sigma/d\Omega)^2(\Gamma_j)^{-1}$. Thus lines with narrow widths are enhanced. This makes CARS particularly useful for gas-phase studies.
5. Electronic resonances will produce a background χ^3. This limits the ultimate sensitivity of the CARS technique.
6. Since χ^3 appears as $|\chi^3|^2$ in P_3, interacting terms may appear and can interfere with particular resonances.
7. P_3 is sensitive to small changes in the population of Raman levels.

An example given by Tolles *et al.* (1977) serves to illustrate the advantages of CARS over normal Raman spectroscopy. We consider the detection of a Raman line in gaseous H_2 using conventional Raman spectroscopy and CARS. The Raman power is

$$P(\text{Raman}) = NL\Omega \frac{d\sigma}{d\Omega} P_1$$

where N is the H_2 density and L is the length of the focal region. The corresponding expression for power in the CARS beam is

$$P_3 = \left(\frac{16\pi\lambda_2^4}{hc\lambda_1\lambda_3}\right)^2 \left(\frac{N\Delta}{\Gamma}\frac{d\sigma}{d\Omega}\right)^2 P_1^2 P_2$$

The following parameters are assumed: Units in the expression for P_3 are g, cm, s, and ergs and Γ is in radians. $N = 3.6 \times 10^{15}$ cm^{-3}, $L = 1$ cm, $\Omega = 1$ sr, $d\sigma/d\Omega = 2 \times 10^{-30}$ cm^2 sr^{-1}, $P_1 = 2$ MW (peak power), $P_2 = 0.2$ MW (peak power), $\lambda_1 = 532$ nm $\simeq \lambda_2$, $\Gamma = 0.03$ cm^{-1}, and $\Delta = 0.1$.

Then

$$P(\text{Raman}) = 1.4 \times 10^{-8}\ \text{W}$$

and

$$P_3(\text{CARS}) = 6.5 \times 10^3\ \text{W}$$

Thus, on a pulse by pulse basis, CARS represents an increase in signal strength of $\sim 10^{11}$. Similar enhancements are found when the analysis is performed for CW laser excitation. When one also notes that the CARS signal involves a spectral resolution (0.03 cm^{-1} in our example) that is not easily obtained in conventional Raman spectra, the gain in signal strength given above is seen to be somewhat conservative.

Tolles *et al.* (1977) have provided a useful overview of the advantages and disadvantages of the CARS technique. Following Tolles *et al.* these are:

Advantages

1. High Raman conversion efficiencies are obtained.
2. Excellent collection efficiency is possible since CARS is generated as a beam.
3. Narrow spectra are obtained without the need for a monochromator.
4. Linewidths of 10^{-3} cm^{-1} can be obtained.

5. High brightness of CARS beam permits detection in bright sources—plasmas, flames, and fluorescent sources.
6. Technique gives good sensitivity at low pressure.
7. Since CARS is performed with pulsed laser sources it can be easily used to study time-dependent phenomena in photochemistry, chemical kinetics, and fluid dynamics.

Disadvantages

1. CARS generates its own nonresonant background that provides a practical limit to detectivity of ~1% in liquids and 10 ppm in gas.
2. It is not useful for strongly absorbing or scattering media.
3. It has a high capital cost.
4. There is a limited tuning range with practical systems.
5. Small Raman shifts not easily obtained.
6. Occasional sample damage occurs and there is the requirement of high laser power.
7. The technique is sensitive to changes in laser power, sample density fluctuations, and variations in linewidth.
8. Interferences due to neighboring resonances are relatively common.

Systems for the production of CARS spectra and the use of CARS in a variety of analytic and diagnostic roles have been extensively reviewed [Levenson (1977), Tolles *et al.* (1977), Eckbreth *et al.* (1977), Harvey and Nibler 1978, and Druet and Taran (1979)]. Only a brief overview will therefore be presented here and the reader is referred to the above articles and references in them for a full account of CARS methodology.

Figure 5.55 shows a schematic diagram of the system developed by Nibler and Knighten (1979) for recording CARS spectra of gases. The output from a pulsed Nd–YAG laser is doubled to provide 10 MW of power at ω_1. Part of the beam at 1060 nm is split off and doubled to pump a dye laser oscillator–amplifier combination. The output beam at ω_2 is mixed with that at ω_1 in the sample to produce the CARS output at ω_3. This is monitored with a photomultiplier–boxcar system. The dye laser output at ω_2 can also be monitored with an etalon to provide fringes for wavelength calibration. Figure 5.56 shows CARS spectra for CH_4 recorded under similar conditions [Tolles *et al.* (1977)]. The spectral resolution and good signal-to-noise ratio possible with the CARS technique is well demonstrated in this spectrum.

While the system shown in Figure 5.55 uses two lasers with narrow linewidth to generate CARS spectra, useful measurements can also be made if one laser operates as a broadband oscillator. If a broadband dye laser is used for

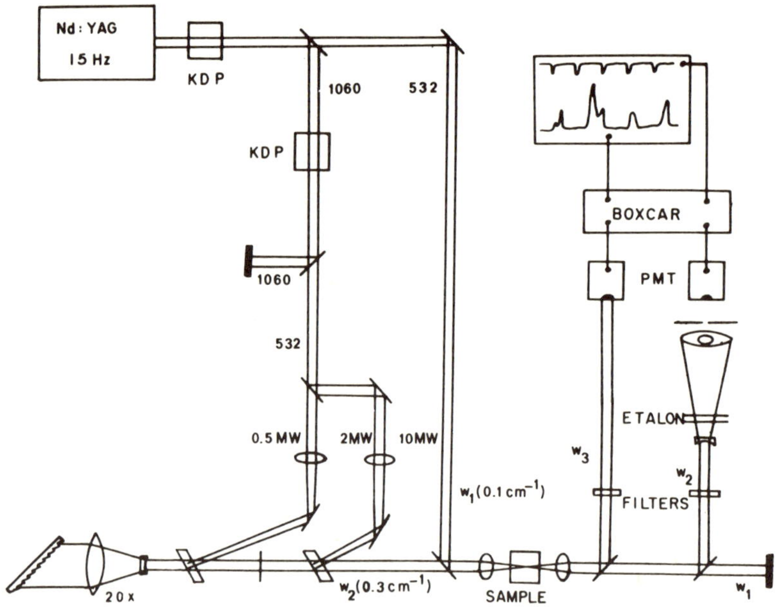

Figure 5.55. Apparatus to record CARS spectra of gases [Nibler and Knighten (1979)]. The diffraction coupled YAG laser has a linewidth of 0.1 cm^{-1} in a donut mode. The dye laser provides a linewidth of ~0.3 cm^{-1} in a similar mode. Part of the ω_2 beam is passed through an etalon to provide an accurate relative frequency mark.

ω_2 then the CARS output will appear as a complete Raman spectrum superimposed on the broadband emission. In this system a monochromator or spectrograph is required to obtain this spectrum. This limits spectral resolution to that of the spectrometer but does allow real time measurements to be made of the entire spectrum. The use of an optical multichannel analyzer is particularly advantageous.

CW lasers can also be used for CARS measurements although with much reduced sensitivity. However, this is compensated for in part by the extra resolution and stability available from CW laser sources. Two configurations utilizing CW excitation to record CARS spectra are shown in Figure 5.57(a) and (b).

The high brightness of the CARS beam makes this technique extremely useful in diagnostic studies of molecules in sources that have high intrinsic brightness. As a result, the CARS technique has been extensively exploited in the study of combustion and plasma diagnostics. Figure 5.58 shows how CARS can be effectively used to obtain spatially resolved spectra in a high-temperature source.

Since CARS data can be obtained with pulsed sources, the technique has been extensively used to study photochemical processes in flash photolysis systems. In a novel experiment Nibler *et al.* (1976) used a doubled 532-nm beam to photolyze benzene vapor in a flow cell. The CARS spectrum in the CH stretching region of the resulting products of this photolysis was complex (Figure 5.59) indicating that a variety of fragments had been produced.

The nonresonant background (χ^{NR}) present in all samples, but most noticeable in solids and liquids, provides the practical detection limit in most CARS measurements. Begley *et al.* (1974) showed, for example, that detection

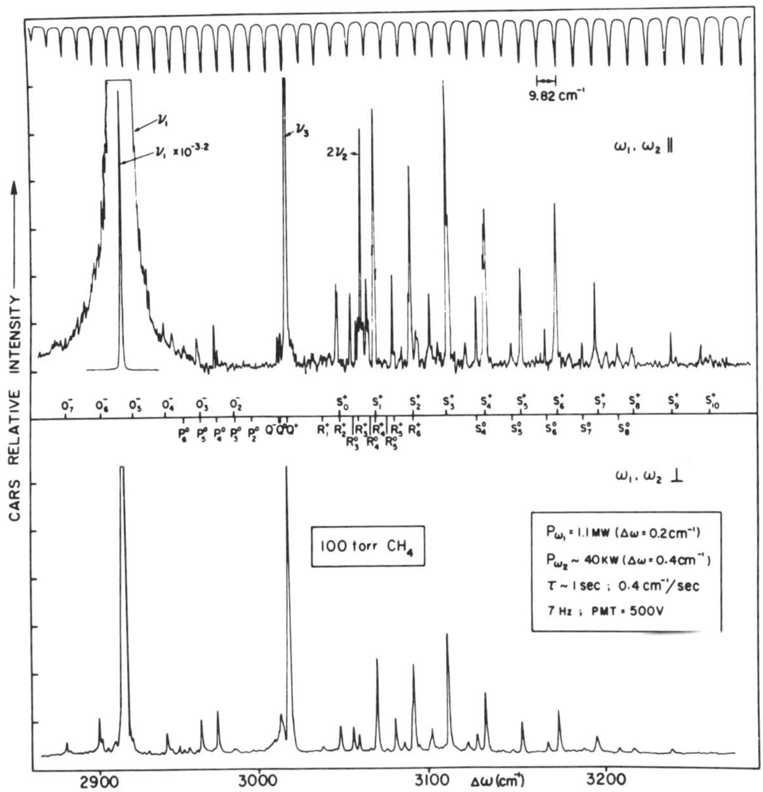

Figure 5.56. A CARS spectrum of methane near 3000 cm^{-1}. The upper trace is recorded with parallel polarized laser beams, the lower with perpendicularly polarized beams. Note the sharp reduction in signal intensity for the strongly polarized bands, p_1 and $2p_2$. As a means of identification an incomplete assignment based on literature information is provided at the center of the figure. At the top of the figure is a trace of the Fabry–Perot fringes of the dye laser output. This trace serves as both an accurate relative frequency marker and an indicator of dye laser amplitude [Tolles *et al.* (1977)].

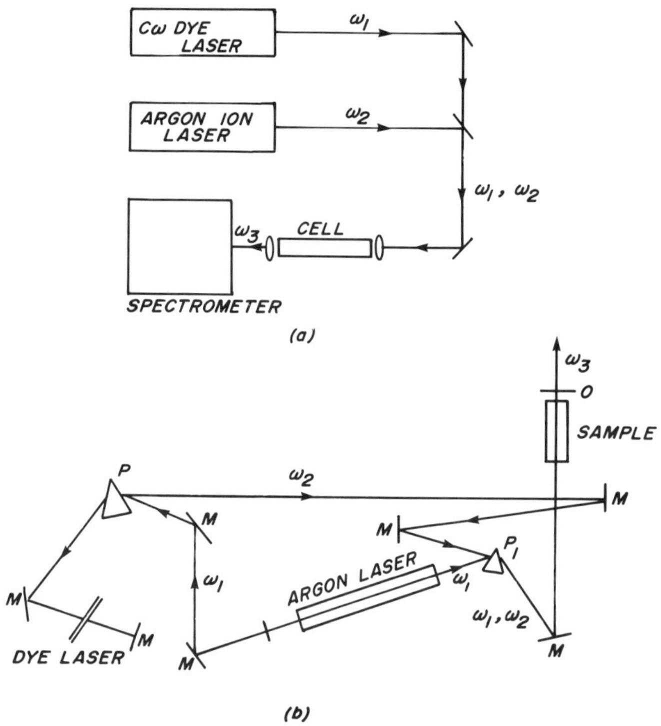

Figure 5.57. (a) CW CARS system without sample in cavity; (b) system with sample in cavity. Light from an Ar^+ laser pumps CW dye laser. This is combined with the dye laser radiation in P_1 and passes through the sample before exciting the cavity at O.

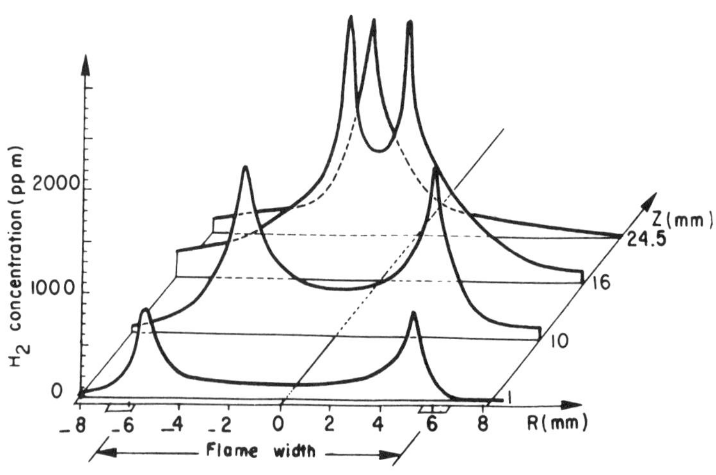

Figure 5.58. H_2 distribution in a horizontal natural gas flame. R is the distance from the burner axis; Z is the distance along the axis; coordinate R is vertical, pointing downward [Regnier and Taran (1973)].

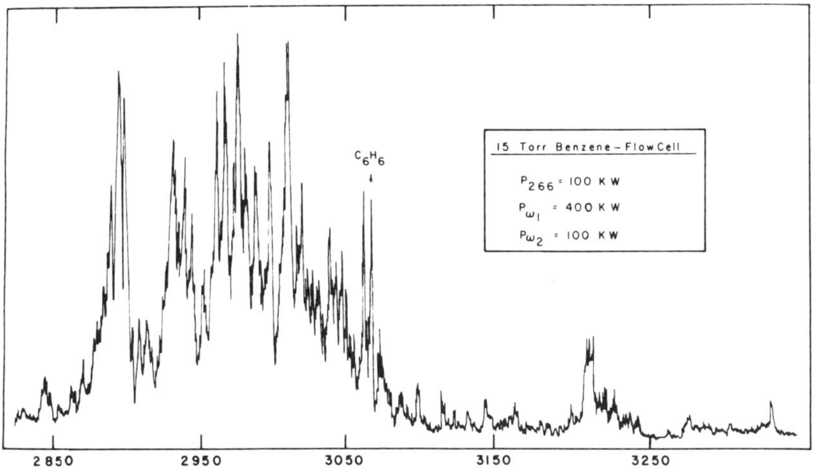

Figure 5.59. A spectrum observed for benzene vapor (15 torr) when subjected to a 266-nm laser beam simultaneous in time and space with the CARS beams. In the absence of the UV beam, only a single, strong CARS line is observed in the CH stretching region (ν_1 at 3073 cm^{-1}). On exposure to the 266-nm beam, the 3073-cm^{-1} feature is decreased tenfold and the complex spectrum shown in this figure results [Nibler and Knighten (1979)].

of benzene by its 992-cm^{-1} Raman line in benzene–toluene mixtures was possible only down to benzene concentrations of $\sim 1\%$. Below this, χ^{NR} exceed χ^R. A variety of approaches have been adopted [Nibler and Knighten (1979)] to minimize the effects of χ^{NR}. Resonance enhancement is possible when the source at ω_1 can be tuned near an electronic transition of the scattering molecule. A recent example of the enhancement possible with resonant CARS can be found in the report by Guthals *et al.* (1979).

CARS has been shown to be a novel and highly useful technique that is particularly well suited as a diagnostic tool in many areas where more conventional analytic methods are of little use. New applications for CARS and other coherent Raman spectroscopic techniques are likely to develop as instrumentation for the utilization of these processes continues to improve.

5.7. LASER MAGNETIC RESONANCE

The optical analogue of the familiar electron paramagnetic resonance effect has been widely used in the acquisition of fundamental molecular data. Laser magnetic resonance (LMR) can also be used in analysis for the detection of paramagnetic gas-phase molecules. As shown in Figure 5.60, light from a stable infrared laser is passed through a sample containing the molecule of interest inside a solenoid. The frequency of the laser line is chosen to be slightly off resonance for an absorption line in the molecule. By varying the magnetic field strength, a transition between Zeeman-split components of a rotation–vibration band of the molecule can be brought into resonance with the laser line. This increases absorption of the laser light. When the magnetic field is modulated, a similar modulation occurs in the intensity of laser radiation passed by the sample. This can be extracted from the CW background by synchronous detection.

LMR has been used in a quantitative detection of NO_2 in N_2 at the ppmv (parts per million-volume) level [Freund *et al.* (1976)]. A near coincidence is achieved between the $P(11)$ transition of the CO laser and a spin component of the $6_{16} \leftarrow 6_{15}$ transition of the ν_3 band of NO_2. It is noted that since this technique samples only paramagnetic species it is free from background absorption due to nonparamagnetic species such as H_2O. Thus the presence of a molecule such as NO_2 can be detected under conditions that might restrict detection limits via absorption spectroscopy.

Molecules that possess a permanent electric dipole moment can be detected via laser Stark spectroscopy (LSS). LSS utilizes the Stark effect to shift or split an infrared molecular absorption line into coincidence with a laser line. A summary of the capabilities of LSS in trace analysis of organic gases has been published by Sweger and Travis (1979). Table 5.20 gives detection limits for some gases using a CW CO_2 laser source.

Since the only requirements for detection using LSS are that the molecule have a permanent dipole moment and an absorption band near lines from a given laser source, this technique is widely applicable. Sweger and Travis note that no other process is likely to exceed LSS in terms of *selectivity* since measurements are made at or near the Doppler limit. The technique is not sensitive to saturated hydrocarbons or to H_2O given the large rotational constant of this molecule.

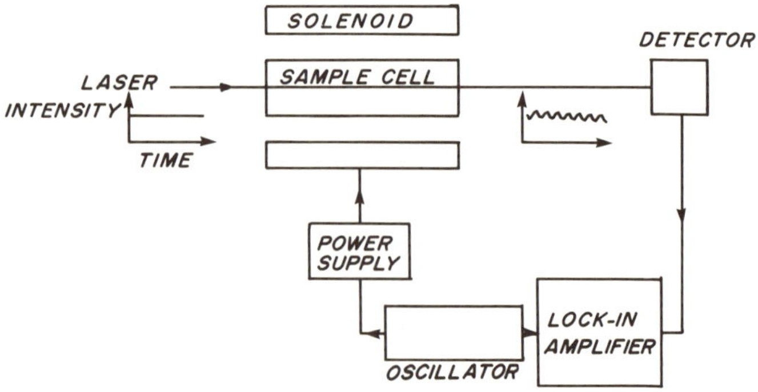

Figure 5.60. Laser magnetic resonance system for measurement of absorption by gas-phase paramagnetic molecules.

Table 5.20. Detection Limits from LSS of Gases[a]

Gas	Laser line	Band	Electric field (kV cm^{-1})	Detection limit (ppm)
C_2H_3Cl	*P*(42)	00°1–10°0	1.05	0.55
C_2H_3Cl	*P*(16)	00°1–10°0	0.35	5.4
C_2H_3CN	*P*(28)	00°1–10°0	0.14	0.03
C_2H_3Cl	*R*(24)	00°1–02°0	1.20	1.1
CH_3OH	*P*(34)	00°1–02°0	1.40	0.08

[a]From Sweger and Travis (1979).

5.8. LASER PHOTOACOUSTIC SPECTROSCOPY

While the optoacoustic effect was discovered in 1880 by A. G. Bell, only recently has this technique found application as a diagnostic and spectroscopic tool. In fact, it was some 90 years after Bell's initial report that the analytical possibilities of this technique were investigated by Kreuzer and his coworkers [Kreuzer and Patel (1971), Kreuzer (1971), and Kreuzer *et al.* (1972)].

The optoacoustic or photoacoustic effect occurs in all materials over certain ranges of photon energy. It arises because the absorption of electromagnetic radiation by a sample of solid, liquid, or gas does not lead to fluorescence or phosphorescence with 100% efficiency. Some fraction of absorbed radiation is retained by the sample and finally appears as heat. For a gas, this heat appears directly as increased translational energy. In a liquid or solid, the heat retained causes a temperature rise that can be detected directly or that heats a gas in contact with the sample.

Since the net effect of absorption in a gas is to produce acoustic energy, a microphone in contact with the gas in a closed container will measure a pressure rise when the sample absorbs radiation. A plot of the output of this microphone as a function of the wavelength of incident light is the photoacoustic spectrum of the absorbing sample.

One of the primary advantages of the laser photoacoustic spectroscopy (LPS) technique derives from the fact that only the power actually absorbed by a sample contributes to the photoacoustic spectrum. Thus scattered light does not contribute to signal levels and hence does not interfere with measurements at the limit of sensitivity. This makes the LPS approach useful for measurements on highly absorbing or powdered materials.

Figure 5.61 shows a block diagram of a simple photoacoustic spectrometer. The laser source should be tunable over the wavelength range of interest. However, in many applications it is not essential that the tuning be continuous and line-tunable lasers are satisfactory.

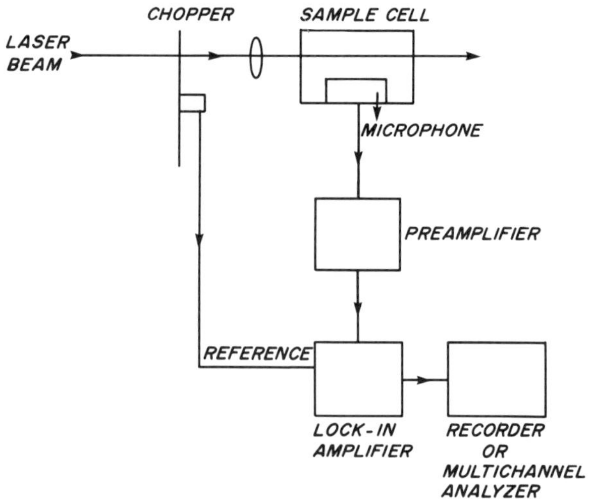

Figure 5.61. Block diagram of a simple photoacoustic spectrometer. The laser wavelength is tuned over the region of absorption of the sample.

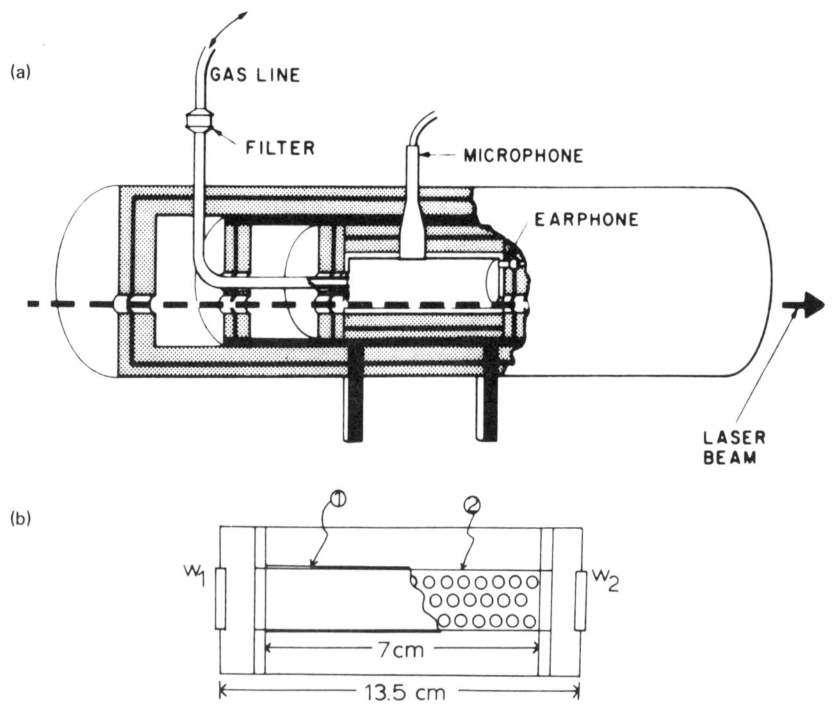

Figure 5.62. (a) Spectrophone developed by Terhune and Anderson (1977) for study of atmospheric aerosols; (b) cylindrical microphone–cell combination [Claspy *et al.* (1977)].

The optimum chopping frequency is chosen to maximize the signal-to-noise ratio. A discussion of criteria will be found in the review by Pao (1977). In a well designed system, the noise level is set by the Brownian motion of the diaphragm in the microphone. Acoustically tuning the cavity–microphone combination to provide a high Q resonance at the chopping frequency also contributes to ultimate sensitivity [Kamm (1976) and Max and Rosengren (1974)]. The distribution of the incident beam intensity within the cell is also related to sensitivity [Koch and Lahmann (1978) and Goldan and Goto (1974)].

A variety of cell geometries have been discussed [Pao (1977)]. For gaseous samples, multipass cells have been shown to be advantageous [Koch and Lahmann (1978), Shtrikman and Slatkine (1977), and Kamm (1976)]. In LPS of liquids, the cell may be a piezoelectric ceramic tube that also acts as a pressure sensor [Lahmann *et al.* (1977) and Sawada *et al.* (1979)]. In cells for gaseous species, the microphone can be a separate unit mounted on the side of the cell or can be an integral part of the cell. These two geometries are shown in Figure 5.62. Operation of a resonant cell inside the cavity of a CO_2 laser has been reported by Shtrikman and Slatkine (1977).

The development of the LPS equivalent to stimulated Raman scattering has been described by Barrett (1978), Barrett and Berry (1979), and West *et al.* (1980). This technique, called photoacoustic Raman spectroscopy (PARS), is feasible because stimulated Raman scattering results in the population of excited vibrational levels in gaseous scatterers. Vibrational–translational energy transfer then converts this excitation to acoustic energy which can be detected in the normal manner with a microphone.

West *et al.* (1980) have compared the capabilities of this technique with IR PAS and conclude that although detection limits with PARS are substantially higher than with IR PAS, the PARS technique offers the advantage of easy access to a wide range of chemical species given the wide tunability of dye laser sources. A quantitative comparison of detection limits for CH_4 in N_2 at 1 atm shows that PARS can detect CH_4 at the 1-ppm level. This is $\sim 10^3$ times the detection limit possible with IR PAS. It is suggested that the use of multipass cells may reduce the PARS detection limit substantially.

5.8.1. LPS of Gases

Emphasis in LPS studies of gaseous samples has been primarily on the measurement of gaseous pollutants. A summary of representative studies is given in Table 5.21. Most molecules of interest from an air pollution point of view can be detected at the 0.1–1.0 ppb level using either fixed-frequency (i.e., discrete line) or tunable laser excitation. Most fixed-frequency measurements have been made using lines from the $\sim$10-μm bands of the CW CO_2 laser or the 5-μm band of CO. Since many lines are available, wavelengths can often

Table 5.21. Detection of Trace Gases Using the Laser Photoacoustic Method

Gas	Detection limit (ppb)	Laser[a]	Reference
NO_2	4 ppb/w	FLDL	Claspy *et al.* (1977)
	10	CWDL	Angus *et al.* (1975)
	0.1	CO	Kreuzer *et al.* (1972)
NO	10	SFRL	Kreuzer and Patel (1971)
	0.4	CO	Kreuzer *et al.* (1972)
SO_2	0.1	Doubled CWDL	Koch and Lahmann (1978)
HF		Pressure tuned GaAs	Anzin *et al.* (1975)
CH_4	10	IR He–Ne	Kreuzer (1971)
n-butane	1	IR He–Ne	Kamm (1976)
NH_3	0.4	CO	Kreuzer *et al.* (1972)
Benzene	3	CO_2	
1,3 butadiene	1	CO	
	2	CO_2	
1 butene	2	CO, CO_2	
Ethylene	0.2	CO_2	
Methanol	0.3	CO_2	
Propylene	3	CO_2	
Trichloroethylene	0.7	CO_2	
H_2O	14	CO	
Nitroglycerine	0.28	CO_2	Claspy *et al.* (1976)
Ethylene glycol dinitrate	1500	CO_2	
Dinitrotoluene	220 ppm	CO	

[a]FLDL, flashlamp pumped dye laser; CWDL, CW dye laser; SFRL, spin flip Raman laser.

be chosen to be reasonably specific for individual molecules. This selectivity can be seen in the LPS of several organic compounds shown in Figure 5.63. These spectra were obtained using individual lines from a CW CO_2 laser.

Since more than one molecule may be present under practical measurement conditions, possible interferences can occur. Kreuzer *et al.* (1972) discuss this problem with regard to the detection of NO and NO_2 in automobile exhaust. The NO fundamental infrared absorption band occurs at $\sim 5.3\ \mu m$ which overlaps with bands of H_2O. Using line-tunable radiation from a CO laser to measure NO and H_2O concentrations, the rejection ratio for NO:H_2O was 10^6:1. Thus NO could be measured at the 0.1-ppm level in the presence of 10^5 ppm of H_2O vapor. In principle, continuous tunability in the infrared could improve on this result.

Detection of gaseous molecules at the ppb level is equivalent to a sensitivity of 10^{-8}–10^{-9} cm^{-1} W^{-1} of incident power. This compares favorably with some of the best results obtainable using techniques such as laser-induced fluorescence.

One advantage of the LPS technique is that absorption due to aerosol particles can also be detected using a gaseous sample cell [Terhune and Anderson (1977)]. Particles with sizes comparable to the wavelength of visible light are effective scatterers, and hence measurements were made in the presence of

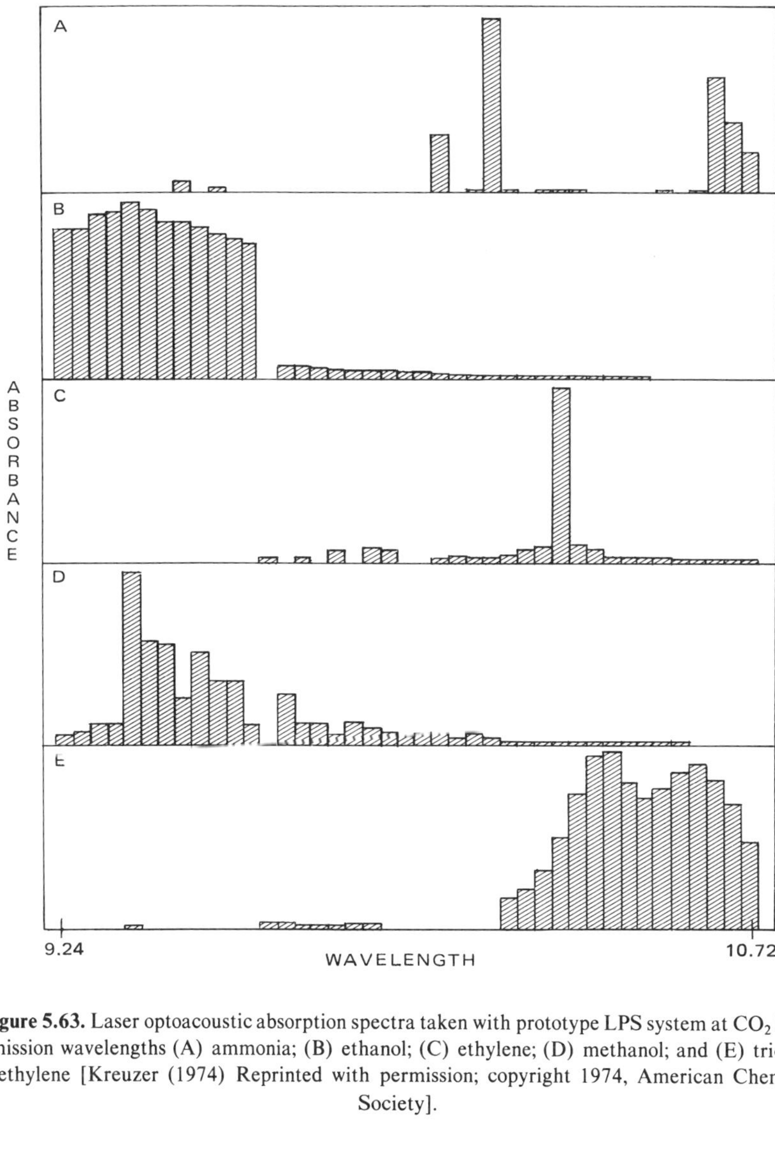

Figure 5.63. Laser optoacoustic absorption spectra taken with prototype LPS system at CO_2 laser emission wavelengths (A) ammonia; (B) ethanol; (C) ethylene; (D) methanol; and (E) trichloroethylene [Kreuzer (1974) Reprinted with permission; copyright 1974, American Chemical Society].

a large background of scattered light. For spherical particles of radius r, at density N cm^{-3} in a gas, the absorption coefficient α is

$$\alpha = Q_a \pi r^2 N$$

where Q_a is the efficiency factor for absorption. If we take r = 500 nm and Q_a = 0.1 then, with the detection limit of $\alpha = 2 \times 10^{-7}$ m^{-1} measured by Terhune and Anderson,

$$N(\text{detectable}) = 2.5 \times 10^6 \text{ particles m}^{-3}$$

This value is somewhat less than the normal ambient aerosol space density. The importance of LPS measurements on aerosols is that the absorption is measured directly while conventional instrumentation measures scattering. Independent estimates of the absorption efficiency of aerosol particles are useful in chemical characterization since for particles of a given size the ratio Q_a/Q_s, where Q_s is the efficiency factor for scattering, are highly dependent on particle composition.

Further work on LPS measurement of absorption by atmospheric aerosols and dust has been reported by Bruce and Pinnick (1977) and Schleuserner *et al.* (1976).

5.8.2. LPS of Liquids and Solids

Trace analysis in liquids by LPS has been discussed by Lahmann *et al.* (1977). A series of experimental investigations have been reported by Oda *et al.* (1978, 1979, 1980) and Sawada *et al.* (1979). These studies have all been carried out with lines from the CW Ar^+ laser or an Ar^+ pumped dye laser.

Lahmann *et al.* (1977) operate an Ar^+ laser on all lines and select out the 488- and 514-nm lines with a prism. These are then chopped sequentially and focused to irradiate the sample. When both beams have the same power and the background absorption of the sample is the same at both wavelengths, background absorption is reduced in the photoacoustic signal. As an example of the usefulness of this technique, Lahmann *et al.* report the detection of β-carotene is 1.02×10^5 liters mol^{-1} cm^{-1} at 488 nm and 3.7×10^4 liters mol^{-1} cm^{-1} at 514 nm while that of chloroform is relatively independent of wavelength over this range. The minimum absorption coefficient measured was $\sim 2.2 \times 10^{-5}$ cm^{-1}.

A study of the absorption by $BaSO_4$ particles in solution was carried out by Oda *et al.* (1980). The detection limit for particles was found to be ~100 times lower than that attainable with conventional turbidimetric analysis.

An improvement in detection limits for trace constituents in liquids via LPS using excitation with a pulsed tunable dye laser has been reported by Patel and Tam (1979). Detection of absorbers at the $\alpha = 10^{-7}\ \mathrm{cm}^{-1}$ level was found to be feasible.

Photoacoustic analysis of solids has been well reviewed by Rosencwaig and others [Rosencwaig (1975a, b), Harshbarger and Robin (1973), and Pao (1977)]. Fundamental work in this area has resulted in the development of techniques for the sampling of natural products and biological and medical systems.

5.8.3. Photoacoustic Imaging

A novel and potentially valuable innovation in the PAS technique has been reported by Busse (1980). Busse shows that the photoacoustic effect can be used to form images of absorbing samples. A schematic diagram of a photoacoustic imaging system is shown in Figure 5.64. To form an image, the laser beam is scanned over the sample surface. Changes in surface topology or absorptivity are then recorded as variations in the PAS signal. Since this measurement can be performed at different wavelengths, surface structures of different chemical composition can be detected. With *X*-*Y* scanning, a map of PAS signal vs. position is obtained. An example is shown in Figure 5.65. Detection is not limited to surface structures; Busse (1980) and Busse and Ograbeck (1980) showed that subsurface features may also be imaged.

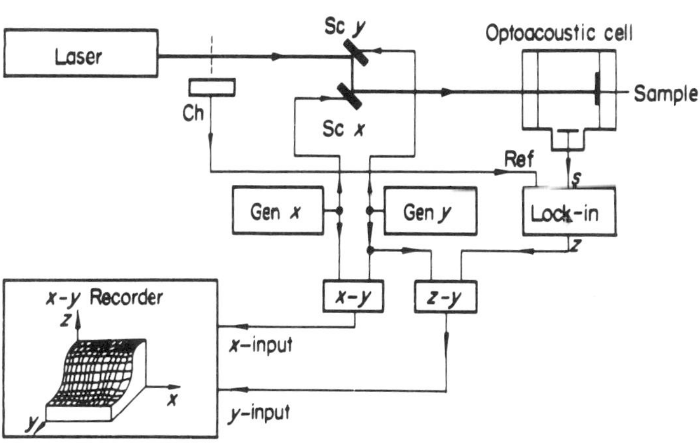

Figure 5.64. Experimental arrangement for optoacoustic imaging. The laser beam is deflected by a saw-tooth generator (Gen x and Gen y), driven rotable mirrors (Scx and Scy) and scanned across the sample. Chopper is indicated by Ch [Busse (1980)].

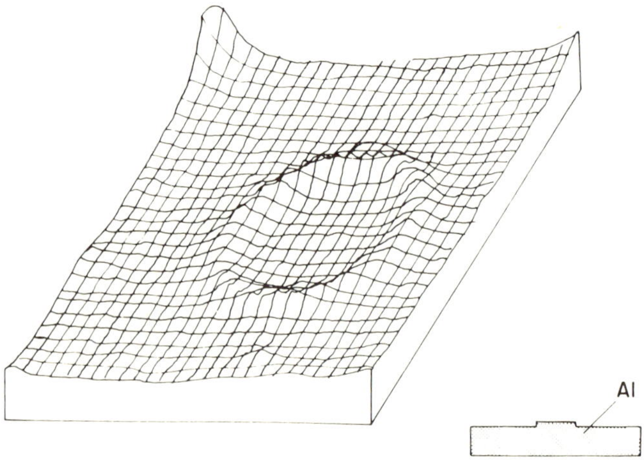

Figure 5.65. Surface imaging of an aluminum sample. Optoacoustic signal is enhanced at the circular step [Busse (1980)].

5.9. LASER MICROPROBE

Trace element detection with the laser microprobe combines the sensitivity of atomic emission spectrometry with the spatial selectivity obtainable with a tightly focused laser beam. The laser microprobe uses a pulsed laser to vaporize a minute volume of the sample to be studied. Material ejected from the laser focus is entrained in a supplementary arc that excites atomic emission from the elements present in the sample microvolume. Alternatively, the laser plasma itself may act as the source of spectroscopic emission.

A block diagram of the laser microprobe system developed by Marich *et al.* (1974) is shown in Figure 5.66. This system measures the emission from the laser-generated plasma using a six-channel polychromator and photoelectric detection. The polychromator format permits the emission from several different elements to be sampled simultaneously on a real time basis.

Kwong and Measures (1979) have provided a useful list of design criteria for laser microprobe systems:

1. Selectivity microsampling capability.
2. No sample preparation—*in situ* measurement.
3. Real time analysis.
4. High relative and absolute concentration sensitivity.
5. Freedom from matrix effects.
6. Highly selective for elements of interest.

7. Linearity of response over wide dynamic range of concentrations.
8. Minimum variation in sensitivity between elements.
9. Depth profiling capability.
10. Simultaneous multielement measurement possible.
11. Capability of distinguishing between isotopes.
12. Insensitivity to the nature of substrate material.

The system developed by Marich *et al.* (1974) and similar systems described by other groups [Morton *et al.* (1973) and Raspberry *et al.* (1967a,b)] satisfy many of these criteria. For example, all permit selective microsampling, require no sample preparation, and provide multielement measurement capability. Furthermore, with photoelectric detection real time analysis is possible.

Limitations to trace element analysis with a laser microprobe system are provided by sensitivity to matrix effects [Cerrai and Trucco (1968) and Marich *et al.* (1970)] and the contamination of the substrate when auxiliary arc excitation is used [Raspberry (1967a,b)]. The detection limit for metallic elements is typically 100 ppm but may be as low as 10 ppm. Background emission has been shown to be a significant factor in limiting detection efficiencies [Treytl *et al.* (1971) and Saffir *et al.* (1972)].

The matrix effect in sampling with the laser microprobe arises when the apparent concentration of an element homogeneously dispersed in a matrix is

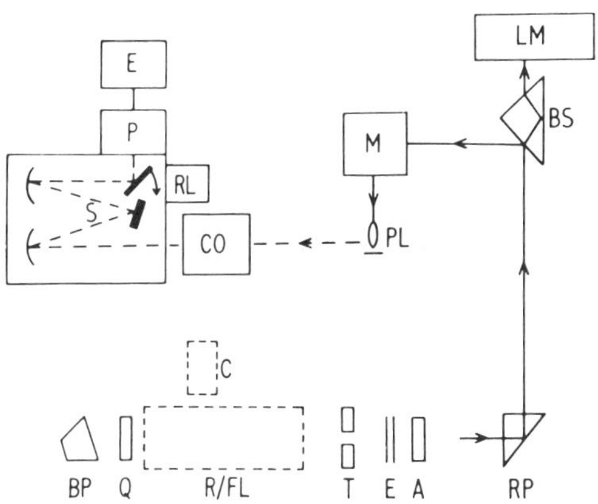

Figure 5.66. Diagram of laser microprobe apparatus; BP, Brewster angle roof prism; Q, liquid *Q*-switch; R/FL, rube and flashlamp; C, air cooling port: T, transverse mode selector; E, two-element sapphire etalon reflector; A, liquid attenuator; RP, right angle prism; BS, beam splitter prism; LM, laser monitor; M, microscope and TV viewing system; PL, optically emitting plasma; CO, coupling optics (mirrors or lens); S, spectrograph; P, six-channel polychromator; E, photoelectric detection electronics; and RL, He–Ne retroalignment laser [Marich *et al.* (1974) Reproduced with permission; copyright Institute of Physics].

found to depend on the type of matrix, surface preparation, grain size, or crystal orientation. A good example of this effect can be seen in laser microprobe studies of elemental metals [Kirchheim *et al.* (1976)]. It was found that the emitted intensity of lines from a given element depended on which crystal plane was excited by the laser spot. In particular, higher intensity was observed from planes with high atomic density. Since crater volume was found to be independent of crystal plane, this difference could not be attributed to different amounts of vaporized material. A detailed analysis showed that the "matrix" effect in this case was due to the dependence of the velocity of emitted atoms on the crystallographic plane from which these atoms were ejected. Velocities perpendicular to the surface of close-packed planes were preferred. This led to preferential excitation in the auxiliary spark source and enhanced emission intensity. It is evident that similar effects will occur when emission is from samples with a more heterogeneous structure as found, for example, in most samples of practical interest.

These considerations limit the overall accuracy of trace element determination using the laser microprobe. This can be overcome to some extent by the use of standards [Giovannini *et al.* (1976)] or by additional measurements on the sample [Morton *et al.* (1973)]. For example, Morton *et al.* report that a measurement of crater volume under conditions in which the laser energy is carefully adjusted to produce conical craters can be used to correct spectral line intensities for matrix effects that lead to differing crater volume.

An advantage of the laser microprobe technique is that samples can be irradiated in air. Two studies, however, have investigated the effect of ambient atmospheric conditions on solid sampling with the laser microprobe [Treytl *et al.* (1971) and Piepmeier and Osten (1971)]. Peipmeier and Osten found that crater diameter was independent of laser pulse energy over a wide range of energy at 760 torr. At 1 torr, however, a strong dependence was seen (Figure 5.67). The relative emission intensity of the background continuum was also found to show a strong dependence on pressure (Figure 5.68). Piepmeier and Osten conclude that both these phenomena are compatible with the creation of an absorbing plasma in front of the target. This plasma would tend to shield the substrate from further laser radiation under certain conditions and would prevent any increase in crater diameter.

Many of the sampling problems inherent in the conventional laser microprobe technique are removed when a second laser is used to selectively excite oblated material [Kwong and Measures (1979)]. A block diagram of this system is shown in Figure 5.69. In this system, material is evaporated with a pulse from a high-power ruby laser. The ejected material is then sampled with a light pulse from an N_2 pumped tunable dye laser. The output of this laser is tuned to the resonance line of an element present in the plume. This is absorbed by the element and reemitted as resonance fluorescence. The fluorescent signal is recorded on a pulse-by-pulse basis by a photomultiplier tube on a monochro-

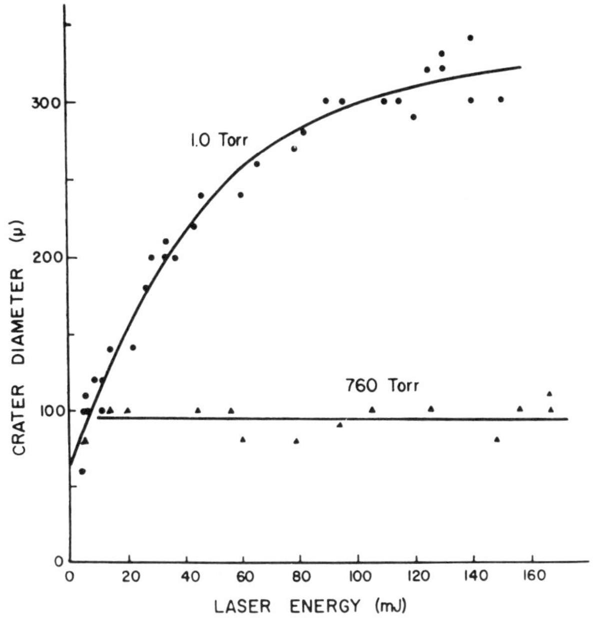

Figure 5.67. Crater diameter vs. laser energy for 1- and 760-torr ambient pressures in air [Piepmeier and Osten (1971)].

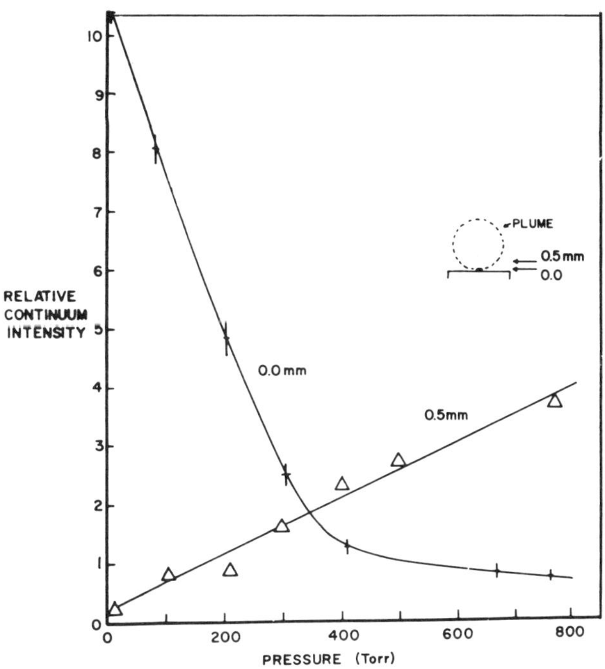

Figure 5.68. Atmospheric influences on relative continuum intensities at two positions above copper sample [Piepmeier and Osten (1971)].

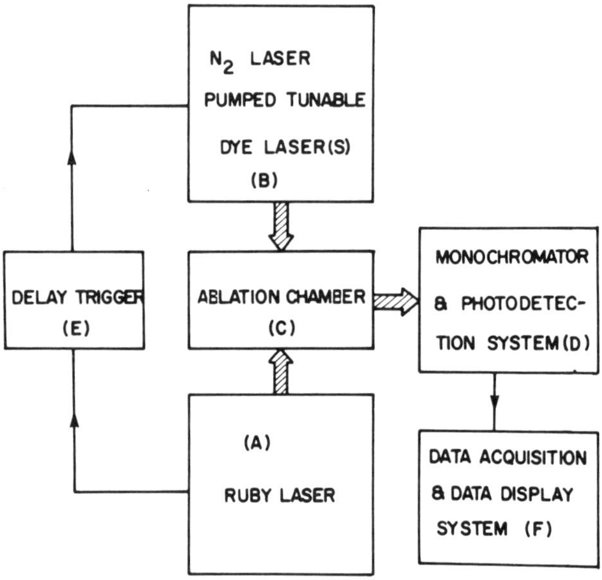

Figure 5.69. TABLASER block diagram [Kwong and Measures (1979)].

mator. Observed fluorescence signals were found to be relatively independent of matrix and were found to scale linearly with concentration in a particular matrix over several orders of magnitude.

5.10. ATOMIC ABSORPTION SPECTROMETRY

Vaporization of samples for atomic absorption analysis using pulsed laser radiation was first described by Mossotti *et al.* (1967). This system is similar to that developed in the laser microprobe with the exception that vaporized material is observed in absorption against a conventional light source such as a hollow cathode lamp or a Xe arc. As such, it retains several of the advantages of laser microprobe sampling without any complications due to arc excitation. Some disadvantages, however, are shared with the laser microprobe technique—notably the influence of matrix effects.

Mossotti *et al.* (1967) used a Xe flashlamp in a conventional atomic absorption spectrometer to record absorption spectra of a variety of elemental metals following cessation of emission from the laser plume. Osten and Piepmeier (1973) developed a pulsed hollow cathode lamp for recording spectra. The resulting increase in intensity available within spectral lines was advantageous in reversing lines from the laser plasma. Typical traces of the time-dependent emission from such lamps during laser excitation of metal targets are shown in Figure 5.70. Samples were vaporized by pulses from a Nd–YAG laser. A subsequent study by Manabe and Piepmeier (1979) showed that sample vaporization with 1-μs pulses from a dye laser minimized atmospheric absorption effects over the excited sample.

The need to directly sample plume species via absorption is eliminated in the system developed by Ishizuka *et al.* (1977), who have shown that the material evaporated from a sample with a laser pulse can be entrained in a flow of inert gas and passed through a separate absorption tube. This modification is shown in Figure 5.71. The absorption cell was placed between a hollow cathode spectral lamp and a monochromator. When material evaporated from the sample passed through the cell, reduction in the light transmitted by atomic species could be recorded with a photomultiplier tube mounted on the monochromator. A summary of detection limits for various metals evaporated from several alloys is given in Table 5.22. These limits were estimated by either measuring peak heights in oscillograms or total areas under absorption signals.

Matousek and Orr (1976) have investigated the use of a pulsed CO_2 laser to augment evaporation in a graphite tube furnace. A description of an atomic absorption system utilizing laser evaporation of biological samples is given by Sumino *et al.* (1980). An example is given of the detection of Cd in human kidney cortex at the 50–100 ppm level.

5.11. LASER MICROPROBE MASS SPECTROMETER

The laser microprobe mass spectrometer combines the spatial localization possible with a focused laser beam and the analytical capabilities of mass spectrometry. Devices have now been developed that provide analysis at the sub-ppm level with a spatial resolution of $\lesssim 1$ μm [Wechsung *et al.* (1978)].

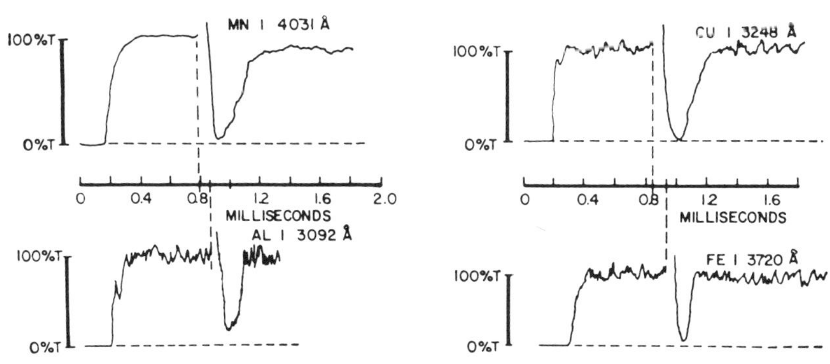

Figure 5.70. Typical oscilloscope tracings of atomic absorption pulse for four elements. Horizontal line indicates 0% T. Vertical dashed line indicates time of laser firing. Time zero indicates flashlamp firing [Osten and Piepmeier (1973)].

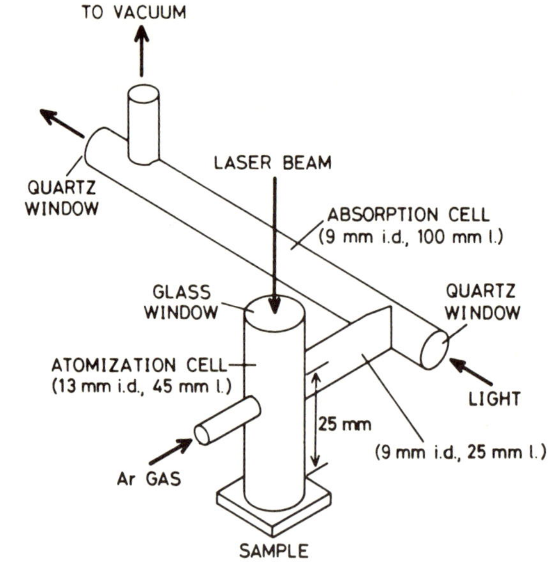

Figure 5.71. Schematic diagram of atomization–absorption cell. The cell is made with stainless steel walls and quartz or glass windows [Ishizuka *et al.* (1977) Reproduced with permission; copyright 1977, American Chemical Society].

Table 5.22. Detection Limits for Various Elements in Brass, Steel, and Aluminum Alloy[a]

			Detection limit			
			Peak height method		Integration method	
Element	Sample	Analytical line, (nm)[b]	Concentration (ppm)	Weight (g)	Concentration (ppm)	Weight (g)
Al	Steel	396.2	11	1.0×10^{-11}	39	$3.5 \times 10^{-[illegible]}$
		(309.3)	(7.3)	(6.7×10^{-12})	(26)	$(2.3 \times 10^{-[illegible]}$
Cr	Steel	429.0	20	1.8×10^{-11}	98	$8.8 \times 10^{-[illegible]}$
		(357.9)	(2.7)	(2.4×10^{-12})	(13)	$(1.2 \times 10^{-[illegible]}$
Cu	Steel	324.7	7.2	6.5×10^{-12}	33	$3.0 \times 10^{-[illegible]}$
	Al alloy	324.7	32	2.6×10^{-11}	120	$9.6 \times 10^{-[illegible]}$
Fe	Brass	372.0	17	1.7×10^{-11}	29	$2.9 \times 10^{-[illegible]}$
		(248.3)	(1.9)	(1.9×10^{-12})	(6.4)	$(6.4 \times 10^{-[illegible]}$
	Al alloy	372.0	100	8.0×10^{-11}	330	$2.6 \times 10^{-[illegible]}$
		(248.3)	(11)	(8.9×10^{-12})	(37)	$(3.0 \times 10^{-[illegible]}$
Mn	Steel	403.1	37	3.3×10^{-11}	93	$8.4 \times 10^{-[illegible]}$
		(279.5)	(2.1)	(1.9×10^{-12})	(7.4)	$(6.7 \times 10^{-[illegible]}$
	Al alloy	279.5	20	1.6×10^{-11}	42	$3.4 \times 10^{-[illegible]}$
Mo	Steel	313.3	10	9.0×10^{-12}	24	$2.2 \times 10^{-[illegible]}$
Ni	Brass	352.5	24	2.4×10^{-11}	38	$3.8 \times 10^{-[illegible]}$
		(232.0)	(4.7)	$(4.7 \times 10^{-12}$	(7.5)	$(7.5 \times 10^{-[illegible]}$
	Steel	352.5	29	2.6×10^{-11}	62	$5.6 \times 10^{-[illegible]}$
		(232.0)	(5.7)	(5.1×10^{-12})	(12)	$(1.1 \times 10^{-[illegible]}$
V	Steel	318.4	22	2.0×10^{-11}	30	$2.7 \times 10^{-[illegible]}$
		(318.5)	(16)	(1.4×10^{-11})	(21)	$(1.9 \times 10^{-[illegible]}$

[a]From Ishizuka *et al.* (1977); [b]The values calculated for the most sensitive lines are shown in parentheses.

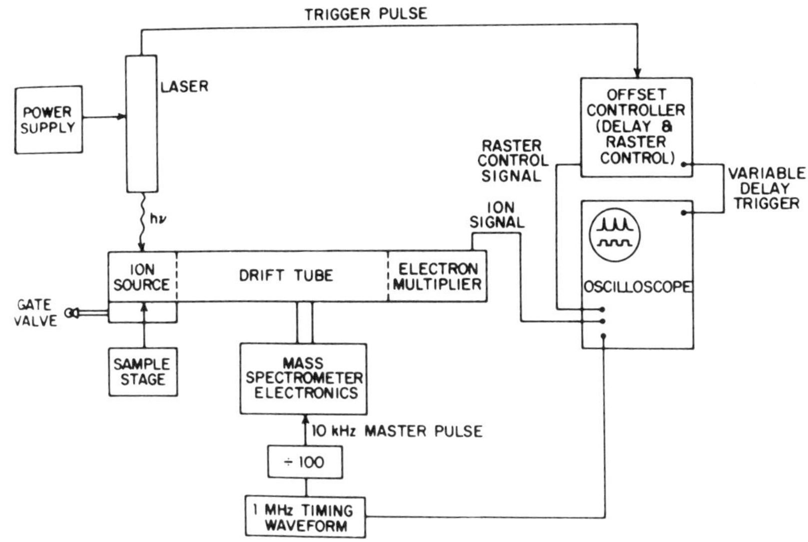

Figure 5.72. Schematic diagram of a laser mass spectrometer system [Thompson (1976) Reproduced with permission; copyright, 1976 American Chemical Society].

Early experiments on the composition of vapors ejected from solid surfaces subjected to focused laser radiation have been reviewed by Ready (1971) and Duley (1976). These experiments were usually carried out without regard to spatial resolution and tended to focus on the relationship between laser-induced and thermal evaporation. It was soon realized, however, that the laser probe mass spectrometric technique had considerable promise in chemical analysis [Bingham and Salter (1976a)].

While some early studies used photographic detection to record mass spectra [Bingham and Salter (1976a,b)], the greatest flexibility is obtained with time-resolved detection. Since sample excitation occurs in the form of a short pulse, time of flight (TOF) mass spectrometric technique are well suited to this method of analysis.

Figure 5.72 shows the TOF system designed by Thompson (1976) for laser microprobe studies of foreign particles on metals or in polymers. Laser pulses from a Nd–glass laser were focused to a $\sim$20-μm diameter spot on the target mounted in the ion source. The ejecta are mass analyzed in the TOF spectrometer and the output of mass spectrum vs. time appears on the oscilloscope. Statistical fluctuations were minimized by averaging several traces. It was possible to observe ions with m/e up to 500, making this system useful for the analysis of inclusions in polymeric solids.

A comprehensive study of capabilities of laser microprobe mass spectrometry has been published by Conzemius and Svec (1978). A summary of the parameters for the system used in this study is as follows:

Laser	Nd–YAG, acousto-optic Q switch
Pulse energy	10^{-3} J
Pulse length	~200 ns
Spot size	15 μm
Emitted charge	10^{-4} CJ^{-1}
Collected charge	10^{-10} CJ^{-1}
Spectrometer efficiency	10^{-6}
Mass resolution	2000
Time for 10^{-7} Coulomb	100 s

Ultimate limits to detectivity for concentration were 10^{-8} with photographic detection and 10^{-11} with electrical detection. An example of spatial resolution capability in a scanning mode can be seen from Figure 5.73. In this scan the level of Si in a Ta:Ta–Si junction was monitored as a function of distance across the junction.

Modification and development of the laser mass spectrometer microprobe for measurements on biological samples has been described by Hillenkamp *et al.* (1975) and Wechsung *et al.* (1978). Samples in the form of thin sections are mounted in vacuum on copper grids at the focus of an optical microscope. Visual acquisition of the spot to be irradiated is aided by centering the beam of a low-power He–Ne laser on the desired area. The laser output is focused through the same optical system on this region to produce evaporation. Sampling accuracy of ~0.5 μm was demonstrated.

The TOF mass spectrometer could be used to collect either positive or

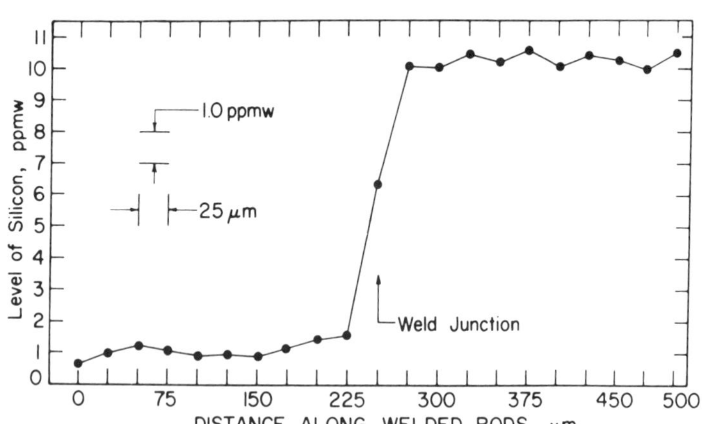

Figure 5.73. Scan of diffusion couple containing 1- and 10-ppm silicon in tantalum [Conzemius and Svec (1978) Reproduced with permission; copyright 1978, American Chemical Society].

Table 5.23. Detection Limits for Metals in Epoxy-Resin Films[a]

Metal	Absolute (g)	Relative (ppm)
Li	2×10^{-20}	0.2
Na	2×10^{-20}	0.2
K	1×10^{-20}	0.1
Ca	1×10^{-19}	1
Cu	2×10^{-18}	20
Rb	5×10^{-20}	0.5
Cs	3×10^{-20}	0.3
Ag	2×10^{-18}	20
Pb	2×10^{-18}	20

[a]After Wechsung *et al.* (1978).

negative ions. It was found that these two modes provided complementary information. For example, alkali ions were detected only in the positive ion spectrum while the halogens appeared only as negative ions. Detection limits for various metallic species in a epoxy-resin film are summarized in Table 5.23.

Increasing laser power was found to increase signal background due to decomposed organic species. It was found that the mass spectra of organic substrates produced with laser irradiation were less complex than those generated with electron impact, suggesting that this technique may be useful in chemical "fingerprinting." Related measurements of laser-induced decomposition of gas-phase organic molecules have shown that the mass spectrum may be wavelength dependent, particularly when resonance-enhanced multiphoton ionization processes dominate [Zandee and Bernstein (1979)].

5.12. LASER RAMAN MICROPROBE

A powerful new method for the chemical characterization of microscopic structures in chemical, geological, or biological samples by laser Raman spectroscopy has recently been developed by Lamelincourt *et al.* (1979). The laser Raman microprobe can be used to obtain Raman spectra of areas as small as several micrometers on the surface of solid samples or of individual particles as small as ~2 μm. In addition, by selecting the wavelength of an individual Raman line, the surface under study can be examined in light of that wavelength.

Figure 5.74 shows a schematic diagram of the Raman microprobe described by Lamelincourt *et al.* Two illumination schemes are shown: in the *punctual mode* the complete Raman spectrum of a microscopic volume can be obtained. The *global illumination mode* enables the surface to be scanned in

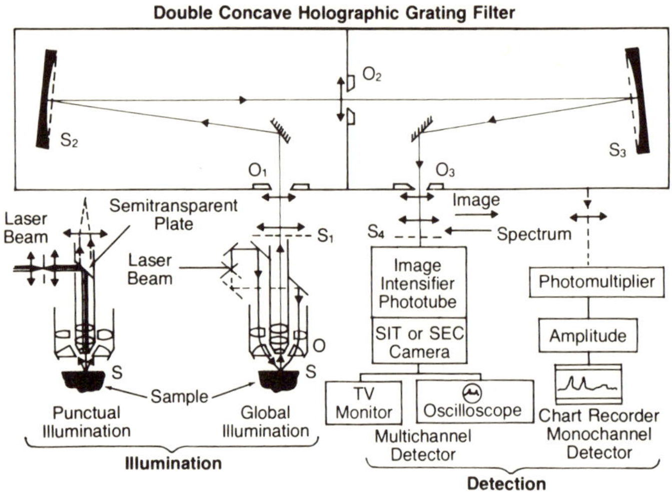

Figure 5.74. Optical scheme of Raman microprobe instrument [Dhamelincourt *et al.* (1979)].

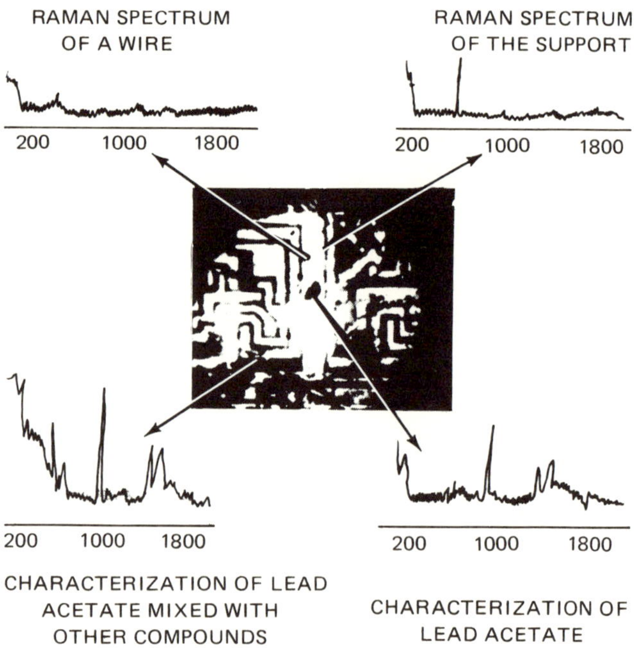

Figure 5.75. Analysis of pollutants found at surface of a commercial integrated circuit [Dhamelincourt *et al.* (1979) Reproduced with permission; copyright 1979, American Chemical Society].

the wavelength of a Raman line. As an example of the flexibility of this technique, Figure 5.75 shows the Raman spectra obtained at several points of an integrated circuit. These scans show the presence of lead acetate and organic contaminants at various points on the surface.

Since the laser probe radiation can be passed through transparent media, the Raman microprobe can be used to obtain spectra of inclusions or bubbles in materials such as glass. This permits information to be obtained on the composition of material found in these inclusions. Fluid inclusions in minerals have been examined in this way by Rosasco *et al.* (1975a).

Further discussion of techniques for obtaining Raman spectra of individual submicrometer diameter particles can be found in the article by Rosasco *et al.* (1975b).

5.13. LASERS IN CHROMATOGRAPHY

The use of lasers in the volatilization of compounds for analysis in gas chromatography has been described by Wiley and Vieravagu (1968), Guran *et al.* (1970), Ristau and Vanderborgh (1970, 1971), and Vanderborgh *et al.* (1975). Laser pyrolysis offers the advantage of rapid heating/cooling cycles minimizing the secondary reactions that are often present in conventional pyrolysis sources. Furthermore, compounds can be pyrolyzed while entrained in a carrier gas [Guran *et al.* (1970)]. It has been found that coating solid samples with carbon [Folmer (1971)] or metal powder [Vanderborgh *et al.* (1975)] is often required to enhance absorption by the sample to be degraded. While this method is simple and convenient, it may lead to spurious results due to reactions with the coating species (carbon powder) or catalyzation by a coating metal (nickel powder).

It has been shown that the use of laser excitation of fluorescence may offer some advantages in high-pressure liquid chromatography [Berman and Zare (1975) and Sepaniak and Yeung (1977)]. Berman and Zare used pulses from a N_2 laser to excite fluorescence from alfatoxins. Detection at the subnanogram level was reported.

Sepaniak and Yeung (1977) used two-photon excited fluorescence to detect several oxadiazoles in a high-pressure liquid chromatograph. The measurement involves the absorption of two photons from an argon laser by an organic solute. This is followed by fluorescence that can be detected with a filtered photomultiplier tube (Figure 5.76). Detection limits for the oxadiazoles PPD, PBD, and BBD compared favorably with those obtained with a conventional UV absorbance detector. Modification of this system to allow resonant two-photon excitation would greatly enhance sensitivity.

Resonant two-photon photoionization has been investigated as a detection mechanism in gas chromatography by Klimcak and Wessel (1980). A full

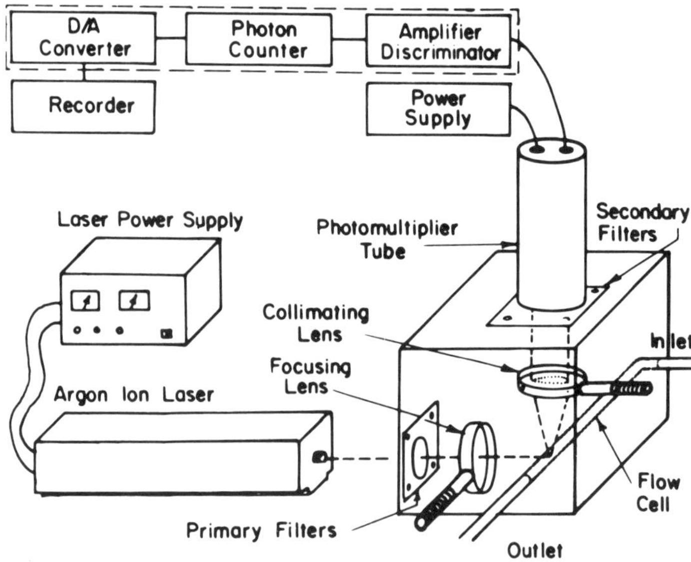

Figure 5.76. Schematic diagram of laser two-photon excited fluorometric detector [Sepaniak and Yeung (1977) Reproduced with permission; copyright 1977, American Chemical Society].

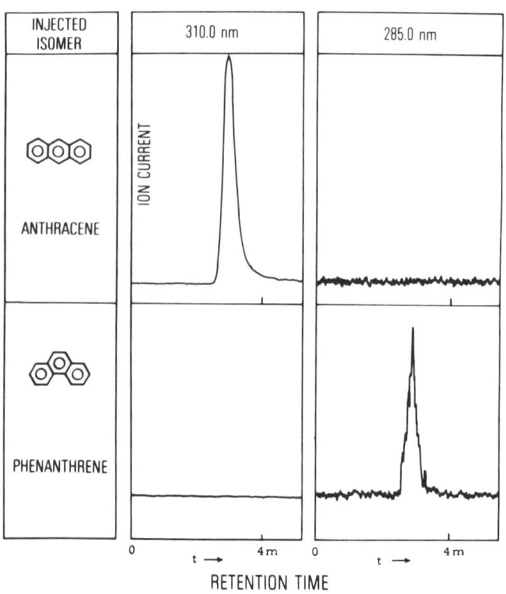

Figure 5.77. Resonant two-photon photoionization detection of a PAH pair [Klimcak and Wessel (1980) Reproduced with permission; copyright 1980, American Chemical Society].

Table 5.24. Detection Limits for Various PAH Species Using Resonant Two-Photon Photoionization[a]

	Detection limit (pg)
Anthracene	10
Napthalene	0.3
Phenanthrene	50
Acenapthene	10
Pyrene	6
1,2 benzanthracene	4

[a]From Klimcak and Wessel (1980).

description of the resonant two-photon photoionization process will be found in Section 5.5. Advantages in gas chromatography involve the ease with which the detector may be tuned to sample specific molecular species and the ability to sample individual chemical species on a real time basis. The specificity of detection can be seen in the separation of anthracene and phenanthrene shown in Figure 5.77. In the upper two scans, anthracene was injected and sampled with the laser tuned to 310 and 285 nm. Since 310 nm corresponds to a resonance in the two-photon photoionization cross section in anthracene, this molecule is detected only when the laser is tuned to this wavelength. This selectivity permitted quantitation of anthracene–phenanthrene mixtures with mixing ratios of $\sim 1:10^3$. Detection limits for several polynuclear aromatic hydrocarbons (PAH) are shown in Table 5.24. System response to the quantity of

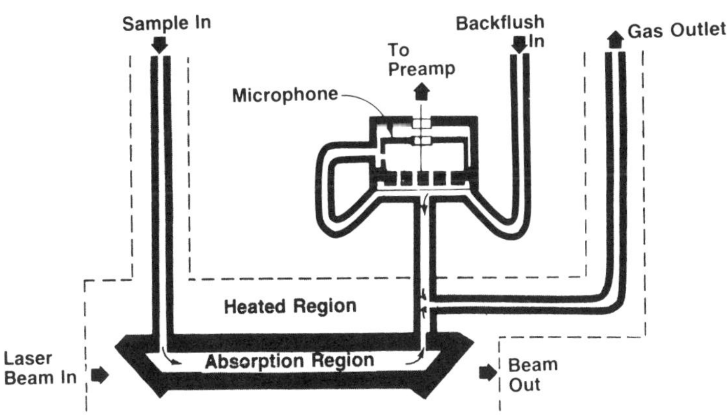

Figure 5.78. Cross section of optoacoustic detector. Tube connecting microphone and absorption region has inside diameter of 0.109 cm and length of 6 cm. Gas flow is about 10 ml min^{-1} [Kreuzer (1978) Reproduced with permission; copyright 1978, American Chemical Society].

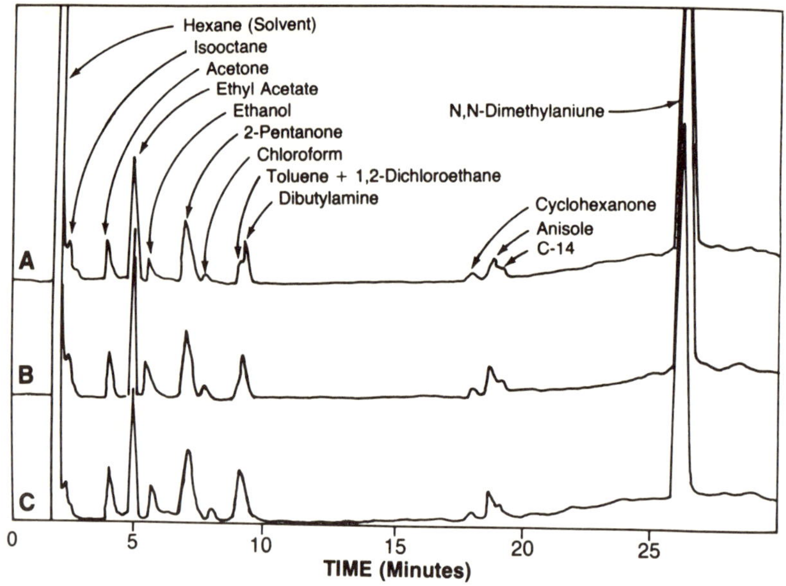

Figure 5.79. LPS chromatograms taken at 10.62 μM. Each chromatogram resulted from injection of 2 μl of sample mixture that contained about 2 μg of each of the compounds [Kreuzer (1978) Reproduced with permission; copyright 1978, American Chemical Society].

injected PAH was found to be linear over approximately five orders of magnitude.

The application of photoacoustic detection to gas chromatography has been discussed by Kreuzer (1978). A CW CO_2 laser was used in conjunction with the acoustic cell shown in Figure 5.78 to monitor the output of a gas chromatograph. The output spectrum taken with the laser tuned to the $P(22)$ line of the 10.4-μm branch is shown in Figure 5.79. This wavelength is not necessarily optimum for the detection of all the compounds appearing in the output of the chromatograph. However, the scan does show what can be measured at fixed laser frequency. The detectivity of this gas chromatograph with LPS detection was 0.2 pg s^{-1} for ethylene. Under similar conditions the detectivity with mass spectrometric detection was 0.05 pg^{-1}.

Chapter 6

Lasers in Environmental Analysis

The laser provides the atmospheric and environmental scientist with an instrument that can actively probe localized volumes of space at a large distance from a chosen location. This has resulted in widespread application of laser remote sensing to the monitoring and analysis of the environment. At the same time the sensitivity of point measurements of pollutant levels has been significantly improved with new laser-based spectroscopic techniques. Many of these new techniques are explored in this chapter.

6.1. PROPAGATION OF LASER RADIATION THROUGH THE ATMOSPHERE

The normal atmosphere contains a variety of gaseous, liquid, and solid contaminants. These absorb and scatter a propagating light beam, causing its intensity to decrease with distance from the emitter. In most cases the attenuation obeys Beer's law

$$I(x) = I_0 e^{-\alpha x}$$

where I_0 is the initial intensity, $I(x)$ the intensity after passing through a path length x, and α the absorption coefficient (cm^{-1}, m^{-1}, or km^{-1}).

The absorption coefficient, α, is determined by the concentration of absorbers and scatterers along the atmospheric path together with the absorption or scattering cross section for each attenuating species. If we take σ as the cross section for absorption or scattering on a per atom, per molecule, or per particle basis, then

$$\alpha = \sigma N$$

where N is the number of absorbers/scatterers per unit volume along the propagation path. If N is a function of x, then α also varies along the line of sight. In this case,

$$I(x) = I_0 \exp\left[-\int_0^x \sigma N(r)\,dr\right]$$

Generally, more than one species contributes to the attenuation of the incident beam. Then

$$I(x) = I_0 \exp\left[-\int_0^x \sum_i \sigma_i N_i(r)\,dr\right]$$

where σ_i is the cross section for ith absorber/scatterer present at density $N_i(r)$ per unit volume in the atmosphere. The aerosol component in the atmosphere contributes a continuous term to α due to classical, Mie, or Rayleigh scattering. This term usually shows no discrete spectral structure although σ(aerosol) varies in amplitude through the region IR $\rightarrow$ UV. Atomic and molecular species can contribute both continuous and discrete terms to α. As an example, O_3 absorbs continuously for $\lambda \lesssim 320$ nm and produces discrete rovibrational spectral structure near 10 μm.

The overall effect of these different terms can be included explicitly in α as follows

$$\alpha = \alpha(\text{ABS}) + \alpha(\text{EXT})$$

where α(ABS) contains terms due to absorption by atomic and molecular atmospheric constituents and α(EXT) includes the contribution of both particle absorption and scattering due to attenuation. If we neglect any possible variation in aerosol concentration with position along the absorption path, then

$$\alpha(\text{EXT}) = \sigma(\text{EXT})N$$

where N is the number of particles per unit volume. σ(EXT) contains terms in both absorption and scattering:

$$\sigma(\text{EXT}) = \sigma(\text{ABS}) + \sigma(\text{SCA})$$

If we consider that the absorbing/scattering particles are spherical with radius a, then

$$\begin{aligned}\sigma(\text{EXT}) &= Q(\text{ABS})\pi a^2 + Q(\text{SCA})\pi a^2 \\ &= [Q(\text{ABS}) + Q(\text{SCA})]\pi a^2 \\ &= Q(\text{EXT})\pi a^2\end{aligned}$$

where $Q(\text{ABS})$ is the efficiency factor for absorption by a sphere of radius a, $Q(\text{SCA})$ the efficiency factor for scattering, and $Q(\text{EXT})$ the efficiency factor for extinction. All Q's will depend on particle radius, refractive index, and wavelength of the incident light.

In most cases $Q(\text{ABS})$ and $Q(\text{SCA})$ must be obtained numerically from the Mie theory [van de Hulst (1957)]. However, when the particles are small compared to the wavelength (i.e., $a \ll \lambda$) and when $|m|\ 2\pi a/\lambda \ll 1$, where m is the particle complex refractive index, the Rayleigh approximation can be used. Then

$$Q(\text{ABS}) = -4\,\frac{2\pi a}{\lambda}\,Im\,\frac{m^2-1}{m^2+2}$$

$$Q(\text{SCA}) = \frac{8}{3}\left(\frac{2\pi a}{\lambda}\right)^4\left(\frac{m^2-1}{m^2+2}\right)^2$$

Figure 6.1 shows plots of $Q(\text{SCA})$ and $Q(\text{EXT})$ vs. particle radius for water droplets at two characteristic laser wavelengths. These were calculated with

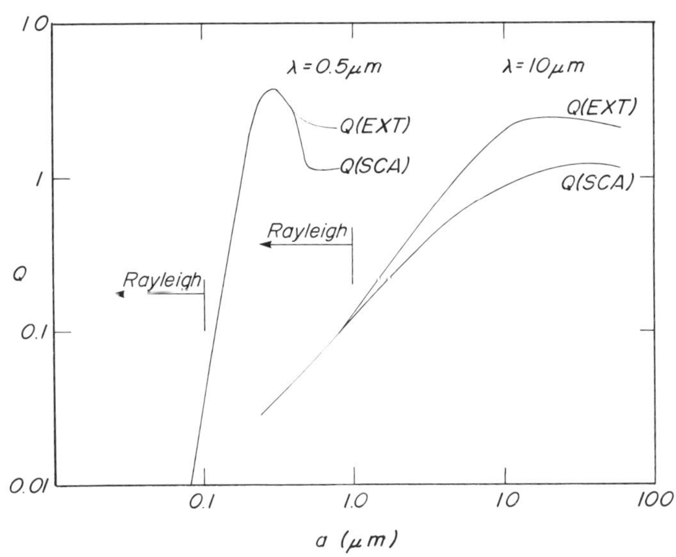

Figure 6.1. $Q(\text{EXT})$ and $Q(\text{SCA})$ vs. a for water droplets for $\lambda = 0.5$ and $10\ \mu\text{m}$. The region of validity of the approximate Rayleigh expressions is shown.

the full Mie theory. The region of agreement with the Rayleigh approximation is also shown. It is evident that the maximum value of Q is in the range of 1–5 and that maximum values are achieved when $a \sim \lambda$. This is a general characteristic of Mie scattering. The normal size distribution of atmospheric particles is shown in Figure 6.2 [Sutton (1978)].

As an example of the attenuation produced by aerosol particles, we take $a = 1.0\ \mu\text{m}$ and $Q(\text{EXT}) = 2$, then

$$\begin{aligned}\alpha(\text{aerosol}) &= \sigma(\text{EXT})N \\ &= Q\pi a^2 N \\ &= 6.3 \times 10^{-8}\ \text{N cm}^{-1}\end{aligned}$$

With $N = 10^3$ particles cm^{-3}, this gives $\alpha(\text{aerosol}) = 6.3 \times 10^{-5}\ \text{cm}^{-1}$ which corresponds to an attenuation of $e^{-6.3}$ per kilometer. Thus under these conditions $I = e^{-6.3} I_0 = 1.8 \times 10^{-3} I_0$. A 1-km atmospheric path would therefore transmit only $\sim$ 0.2% of the incident intensity. An empirical measure of α(aerosol) for visual wavelengths is

$$\alpha(\text{aerosol}) = \frac{3.91}{V}\left(\frac{0.55}{\lambda}\right)^{0.585V^{1/3}} \text{km}^{-1}$$

where V is the visual range in km and λ is the wavelength in μm.

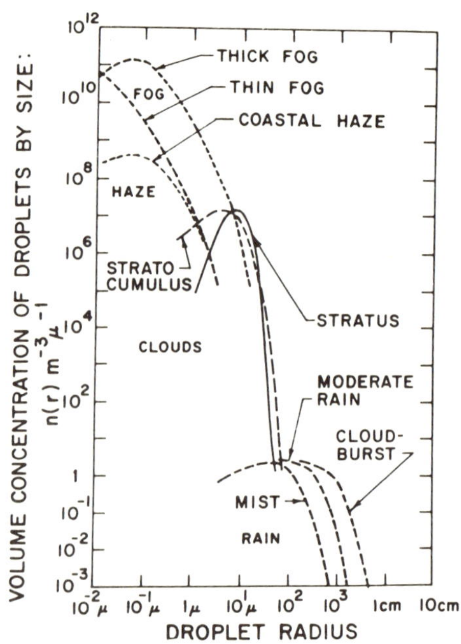

Figure 6.2. Measured water aerosol distribution functions [Sutton (1978)].

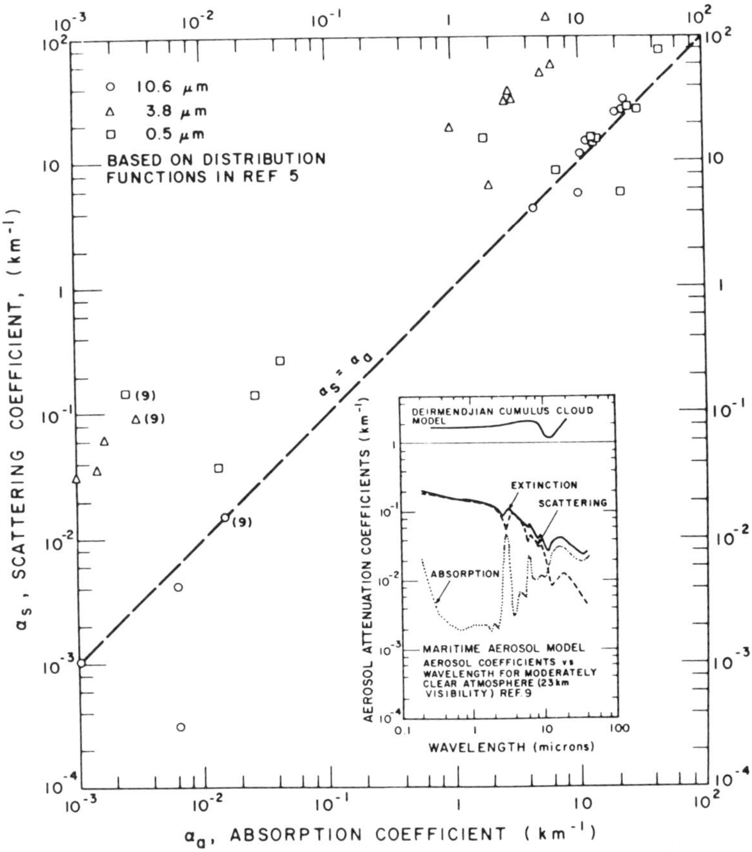

Figure 6.3. Absorption and scattering coefficients for fog, haze, and clouds [Sutton (1978)].

The interdependence of $\alpha(\mathrm{SCA})$ and $\alpha(\mathrm{ABS})$ for various wavelengths is shown in Figure 6.3 [Sutton (1978)]. Data points are shown for a variety of fogs, haze, and clouds as calculated from a model developed by Sutton. The dashed line corresponds to $\alpha(\mathrm{SCA}) = \alpha(\mathrm{ABS})$. This is the result expected for fog and haze particles at 10.6 μm, where $Q(\mathrm{SCA}) \sim Q(\mathrm{ABS})$. At shorter wavelengths, $Q(\mathrm{ABS}) \ll Q(\mathrm{SCA})$ and $\alpha(\mathrm{SCA}) \gg \alpha(\mathrm{ABS})$ as shown for many of the data points at 0.5 μm.

Radiation that is scattered out of the forward direction by aerosol particles is generally lost from the beam except when high-density aerosols are present. In this case, multiple scattering can occur and scattered light can find its way back into the forward direction.

The component scattered at 180° with respect to the forward direction is referred to as the backscattered beam. The detection of this beam with a telescope looking along the direction of propagation is exploited in the LIDAR technique. This is described in detail in Section 6.2.2.

The presence of molecular absorbers contributes a large term to α over certain spectral regions. The short-wavelength limit to the propagation of light in the troposhere is provided by O_2 which absorbs strongly for $\lambda \lesssim 190$ nm. The short-wavelength limit for light propagating in the stratosphere is due to O_3 absorption at $\lambda \lesssim 320$ nm.

Long-wavelength limits are due to the presence of strong absorption bands of H_2O, CO_2, and CH_4 in the infrared. These bands are localized to specific wavelength ranges, however, with the result that "windows" appear in the transmission of the atmosphere between these bands. Some notable windows are 1.6–1.8 μm, 2.1–2.3 μm, 3.5–4.0 μm, 4.5–4.8 μm, and 8–13.5 μm. A spectral scan of atmospheric transmission together with the wavelengths of absorption of common pollutant molecules is given in Figure 6.4. It is evident that the region between 9 and 13.5 μm is well suited to the detection of a variety of chemical contaminants because of the low background opacity of the atmosphere over this range.

Since the opacity of the atmosphere within the bands shown in Figure 6.4 is the result of the superposition of numerous rovibrational lines that are not resolved under low spectral resolution, the absorption experienced by a particular narrow spectral feature may vary considerably from that suggested by these low-resolution scans. This effect can be clearly seen in the comparison of high-resolution and low-resolution scans of the spectrum of SO_2 gas shown in Figure 6.5. Here it is apparent that a region appearing to be relatively opaque on the low-resolution scan actually shows regions of transparency between individual rotational lines.

When individual absorption lines are resolved spectroscopically, the probe laser can be alternately tuned on and off one of these lines in order to detect the presence of specific absorbing molecules. In such a case, the integrated pollutant concentration is given by

$$\int_0^x N(r)\, dr = -\frac{1}{\sigma} \ln \left| \frac{I(x)_{\text{on}}}{I(x)_{\text{off}}} \right|$$

where $I(x)_{\text{on}}$ is the laser intensity at distance x, recorded when the laser frequency is tuned to the absorption line, while $I(x)_{\text{off}}$ is that recorded when the laser is tuned off the absorption line.

The full spectral resolution available from laser sources is often not utilized in atmospheric studies because individual rotational lines usually have widths of $\sim$ 5 GHz due to the pressure broadening at atmospheric pressure. This is much greater than typical Doppler limited linewidths at 300 K ($\sim$ 50–100 MHz). Thus individual rotational lines may tend to overlap, precluding

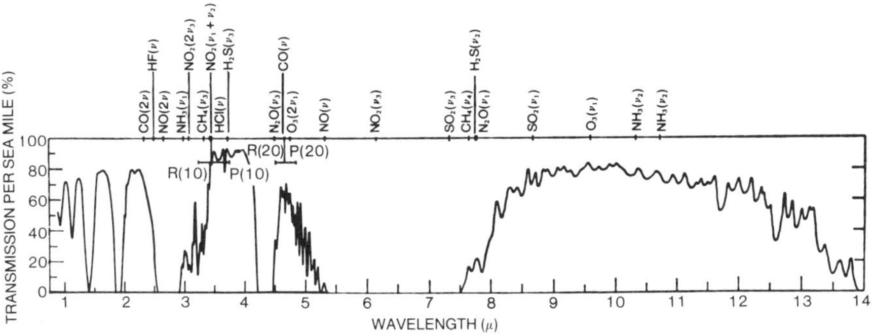

Figure 6.4. Atmospheric transmission in the 1–14 μm region of the infrared together with the important absorption bands of molecular pollutants (visual range is 14 km and water content is 17 mm/nmi which corresponds to a relative humidity of 70% at 60° F) [Kildal and Byer (1971) Reprinted with permission, copyright 1971, IEEE].

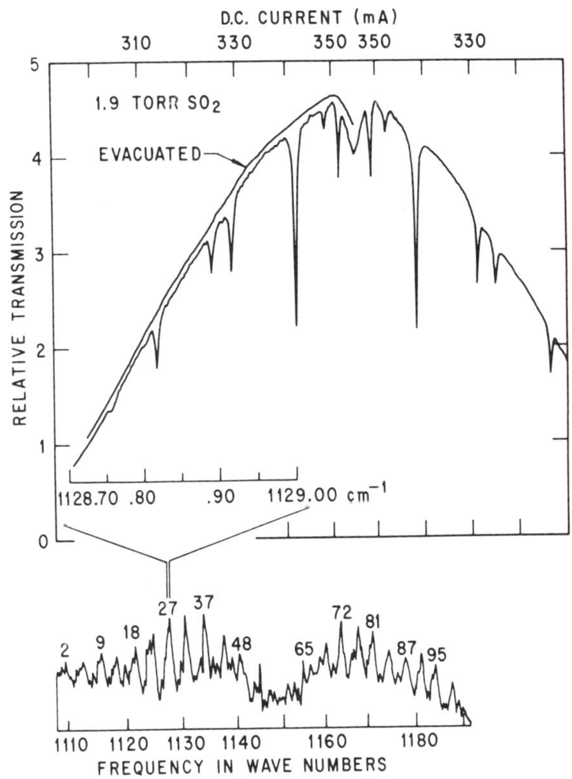

Figure 6.5. High-resolution spectral scan of SO_2 using a $Pb_{0.92}$ $Sn_{0.08}Te$ laser [Antcliffe and Wrobel (1972)].

Table 6.1. Summary of Infrared Molecular Linewidths and Resolution of Various Spectroscopic Sources

Resolution of a portable spectrometer	10^{11} Hz
Atmospheric pressure-broadened linewidth	5×10^9 Hz
Resolution of conventional laboratory spectrometer	10^9 Hz
Doppler width, 300 K, MW = 30	5×10^7 Hz
Semiconductor laser linewidth	10^5 Hz

the full utilization of the high resolution available from laser sources,* Table 6.1.

In addition to reducing spectral resolution, pressure broadening tends to reduce the value of σ at the center of an absorption line. This, in turn, increases the minimum detectable concentration of absorbers. With pressure broadening, a typical cross section for an infrared rotational line is 10^{-19} cm^2. Thus $I = I_0 \exp[-10^{-19}NL]$. If the exponent is small this can be approximated $I = I_0[1 - \sigma NL] = I_0[1 - 10^{-19}NL]$. Then

$$\frac{I - I_0}{I_0} = \sigma NL$$
$$= 10^{-19}NL$$

where $(I - I_0)/I_0$ is the fractional absorption. If we take this to be 0.01 over a path length of 1 km, we obtain N(detectable) $= 10^{-2}/(10^{-19} \times 10^5) = 10^{12}$ cm^{-3} $\simeq$ 30 ppb. Thus, molecular species present at levels exceeding 10–100 ppb in the atmosphere will contribute significantly to attenuation of a propagating beam when its frequency overlaps that of a molecular spectral feature.

The cross section at the center of a broadened atomic electronic transition in the visible or near UV is $\sim 10^{-16}$ cm^2. Then for a fractional absorption of 0.01 over a 1-km pathlength, N(detectable) $= 10^9$ cm^{-3}. An individual vibronic line of a molecular absorber in the visible or UV has $\sigma \sim 10^{-18}$ cm^2. Then N(detectable) $= 10^{+11}$ cm^{-3} under the same conditions.

The absorption of CO_2 laser radiation by atmospheric CO_2 and H_2O has been extensively studied [Yin and Long (1968), McCoy *et al.* (1969), Wood *et al.* (1971), and Peterson *et al.* (1979)]. Empirical expressions for the absorption due to CO_2 and H_2O at the frequency of the 10.59-μm $P(20)$ line is

$$\alpha(CO_2) = 1.44 \times 10^2 \left(\frac{295}{T}\right)^{1.5} \exp\left(\frac{-2233}{T}\right) \text{km}^{-1} \tag{6.1}$$

*For example, tunable diode lasers can have linewidths $\lesssim$ 100 kHz.

and

$$\alpha(H_2O) = 4.32 \times 10^{-6}p(P + 193p) \text{ km}^{-1}$$

where T is the temperature in K, p the partial pressure of H_2O, and P the total atmospheric pressure, with both pressures in torr. Some data on the variation of $\alpha(H_2O)$ with humidity at sea level are given in Table 6.2. Under these conditions $\alpha(CO_2) = 0.058 \text{ km}^{-1}$. Absorption by H_2O therefore dominates at 10.59 μm for relative humidities exceeding 30% at sea level. At higher altitudes, the contribution due to H_2O rapidly decreases and CO_2 becomes the dominant absorber.

Extensive measurements of H_2O absorption at the wavelengths of various CO_2 laser lines have been recently reported by Peterson *et al.* (1979). An example of their data for the $P(18)$ CO_2 laser line at 10.57 μm is given in Figure 6.6. These data were obtained with H_2O in N_2 at atmospheric pressure.

At high power levels a variety of nonlinear interactions occur between propagating laser beams and the atmosphere. With CW beams, vibrational–translational (V–T) transfer from excited molecules along the propagation path causes local heating and refractive index variations. This can result in self-focusing or defocusing. The latter effect is known as *thermal blooming* [Leite *et al.* (1964, 1967), Litvak (1966), Gebhardt and Smith (1969, 1971), Buser and Rhodes (1973), and Pridmore-Brown (1973)]. A review of this effect has been given by Akhamanov *et al.* (1972).

With high-intensity beams, air breakdown along the propagation path can result. The intensity threshold for breakdown is a strong function of the aerosol content of the air as well as the laser wavelength. The thermal interaction between high-power laser pulses and aerosol particles will be discussed in Section 6.3.3.

Table 6.2. Extinction Due to H_2O Absorption at 10.59-μm Laser Line[a]

$\alpha(H_2O)(\text{km}^{-1})$	Relative humidity
0.0125	10
0.0338	20
0.0653	30
0.107	40
0.157	50
0.215	60
0.284	70
0.363	80

[a] Sea level, P = 760 torr and T = 10.7°C [McCoy *et al.* (1969)].

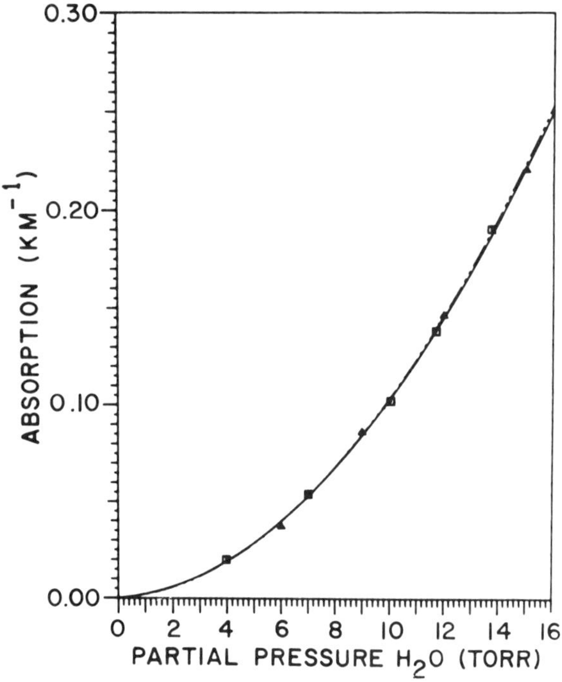

Figure 6.6. Measured H_2O and N_2 absorption coefficient at a total pressure of 760 torr for the $P(18)$ CO_2 laser lines at 945.980 cm^{-1}. The White cell data, represented by the symbol ▲, were obtained at an average temperature of 22.5 °C. The spectrophone data, represented by the symbol □, were obtained at an average temperature of 24.5 °C [Perterson *et al.* (1979].

6.2. LASER REMOTE SENSING OF THE ATMOSPHERE

The past 15 years has seen rapid development of a variety of laser-based techniques for the measurement of atmospheric properties and for the characterization of chemical species present at trace levels. These techniques are described in some detail in this chapter.

An overview of laser-based systems for atmospheric remote sensing is given in Table 6.3. Two basic categories can be defined: those systems that yield measurements of meteorological parameters and those that provide chemical information on atmospheric constituents.

6.2.1. Absorption Measurements

The measurement of air pollution via absorption can be made on a point basis or along a predefined atmospheric path. Some configurations are shown in Figure 6.7. The first two configurations involve the measurement of absorption due to pollutant molecules along an extended atmospheric path. In the first

Table 6.3. Summary of Laser Remote Sensing Techniques for Atmospheric Studies

Parameter	Method	Range	Precision	Reference
		Meterological data		
Cloud structure	LIDAR	100 km		Allen and Evans (1972)
Density	LIDAR	100	1–10%	Kent and Wright (1970)
Temperature	Raman LIDAR	2	1 K	Cooney (1972)
Humidity	Raman LIDAR	2	15%	Pourney *et al.* (1979)
Wind speed	Doppler LIDAR	5	0.2 m s^{-1}	Benedetti-Michalangeli (1972)
		Chemical Data		
Parameter	**Method**	**Range**	**Sensitivity**	**Reference**
Concentration	Long-path absorption (point monitor)	1 km	1 ppb	Reid *et al.* (1980)
	Remote path	400 m	4 ppm	Chaney *et al.* (1979)
	Laser Raman	10 m	30 ppm	Hirschfeld *et al.* (1973)
	DIAL	25 km	0.2 ppm	Uchino *et al.* (1979)
	Resonance fluorescence	1 km	50 ppb	Gelbwachs (1973)
	Heterodyne radiometer	Stratosphere		Frerking and Muehler (1977)

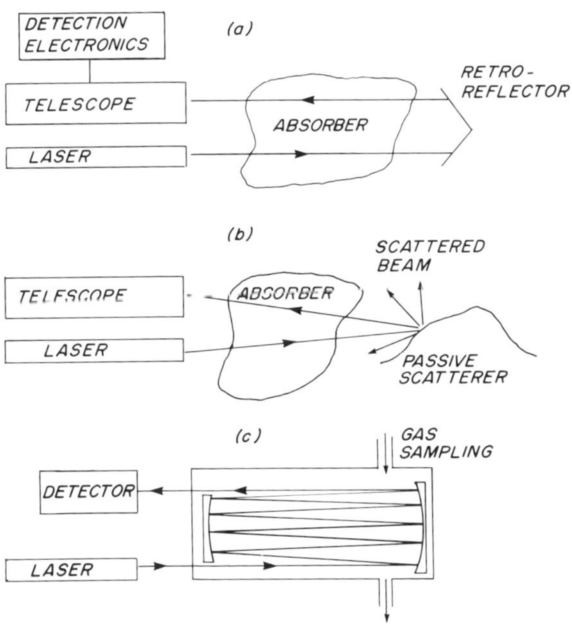

Figure 6.7. Pollution detection by absorption: (a) cooperative reflector, (b) passive scatterer, and (c) point measurement with White cell.

case, the beam is returned by a cooperative reflector; in the second case the backscatter from a suitable passive target is used. The third example shows a system used for a point measurement of pollutants. Here the requisite path length is provided by multiple reflections within a White cell.

With the exception of the DIAL technique, which will be described in Section 6.2.5, most absorption measurements have been made with infrared laser sources. The CO_2 laser is useful in this regard since both bands over which emission can be obtained lie within the 8–13.5 μm atmospheric window. However, when a fixed-frequency laser is used, detection of a particular pollutant relies on the chance coincidence between a laser line and a rotational line of the desired molecule. In addition, overlap with spectral lines from other cohabiting species can interfere with the detection of the desired molecule.

Tunable diode laser sources have the advantage that their output wavelength can be adjusted to correspond exactly to that of the desired pollutant. Overlap with other species can also be eliminated, improving the selectivity of the absorption measurement. A disadvantage, however, is that output power is low (typically 1–3 mW).

The presence of an absorption line at the chosen detection frequency can only be inferred if a measurement of background attenuation can be made near the detection frequency. Such a *dual-beam measurement* automatically compensates for system losses as well as for aerosol extinction and absorption due to the wings of other spectral lines that may overlap the desired line. Figure 6.8 shows how this measurement is carried out.

The measurement of absorption at ν_S yields an absorption coefficient $\alpha_S + \alpha_B$ that contains a contribution due to background absorption and other system losses. The difference, α_S, then gives the strength of the desired spectral line and hence the pollutant concentration since $\alpha = \sigma N$.

A thorough study of the detection of a wide variety of organic molecular species present at trace levels in the atmosphere using lines from the CO_2 laser has been published by Mayer *et al.* (1978). Their data, which includes optimum measurement and reference wavelengths, are given in Table 6.4. Absorp-

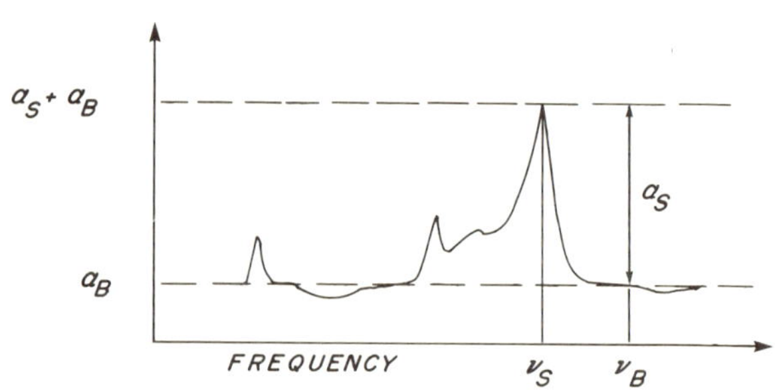

Figure 6.8. Compensation for absorption background by measurement at two frequencies: υ_S—frequency of absorption line due to pollutant and υ_B—nearby frequency showing only background absorption.

Table 6.4. Detection Limits for Various Atmospheric Species Using Lines from a CW CO_2 Laser[a]

	Measurement					Reference	
Gas	Line	Wavelength (μm)	Absorption coefficient (atm cm^{-1})	Line	Wavelength (μm)	Absorption coefficient (atm cm^{-1})	Sensitivity (ppb km^{-1})
Ethylene	*P*(14)	10.533	32.14	*P*(28)	10.675	1.27	2.5
Benzene	*P*(28)	9.621	2.0	*R*(20)	9.272	0.05	125
Ammonia	*R* (8)	10.333	25.8	*R*(12)	10.304	0.06	5
Sulfur dioxide	*R*(26)	9.240	0.105	*P*(18)	9.536	0.005	3×10^3
Ethylmercaptan	*R*(26)	10.208	0.56	*P*(20)	10.591	0.18	600
Vinyl chloride	*P*(22)	10.612	8.80	*R*(18)	9.282	0.05	20
Trichloroethylene	*P*(20)	10.591	12.60	*R*(20)	10.247	0.04	24
Monochloroethane	*R*(16)	10.275	3.3	*P*(20)	10.591	0.06	75
1,2 dichloroethane	*P*(20)	10.591	0.52	*R*(16)	10.275	0.01	550
Freon 12	*P*(32)	10.719	35.7	*P*(12)	10.514	0.08	6
Freon 11	*R*(22)	9.261	29.2	*P*(18)	9.536	0.10	7
Perchloroethylene	*P*(34)	10.742	4.9	*R*(24)	10.220	0.1	60
Water vapor	*R*(20)	10.247	0.002	*R*(16)	10.275	<0.002	85×10^3
Ozone	*P*(14)	9.505	12	*P*(24)	9.583	0.6	25

[a]All measurements made at atmospheric pressure. After Mayer *et al.* (1978).

tion coefficients are quoted in units of atm cm^{-1} and are defined from the equation $I = I_0 \exp(-\ \alpha\ px)$, where p is the partial pressure of absorber in atm units. Typical sensitivities for highly absorbing gases are $\sim$ 5 ppb km^{-1}.

Since CO_2 is present in the normal atmosphere, the use of a CO_2 laser to measure absorption may be subject to systematic errors. Although the laser lines correspond to transitions between excited levels of the CO_2 molecule, these levels are at low enough energy that appreciable thermal population occurs at 300 K. Using the expression (6.1) for $\alpha(CO_2)$ one obtains

$$\alpha(CO_2) = 0.82\ \text{km}^{-1} \qquad \text{at 300 K}$$

This corresponds to the absorption that would be produced by 100 ppb of an absorber with $\sigma = 10^{-19}\ cm^2$.

The effect of absorption by atmospheric CO_2 is compensated in part by the dual-beam method described above. However, it undoubtedly affects the ultimate sensitivity of pollution detection with fixed-frequency CO_2 lasers.

The effect of CO_2 absorption can be seen clearly in the high-resolution scan of atmospheric air absorption shown in Figure 6.9 [El-Sherbiny *et al.* (1979)]. The CO_2 line shown in this spectrum arises from ambient CO_2. This is the laser line listed in Table 6.4 as optimum for the detection of O_3. Note,

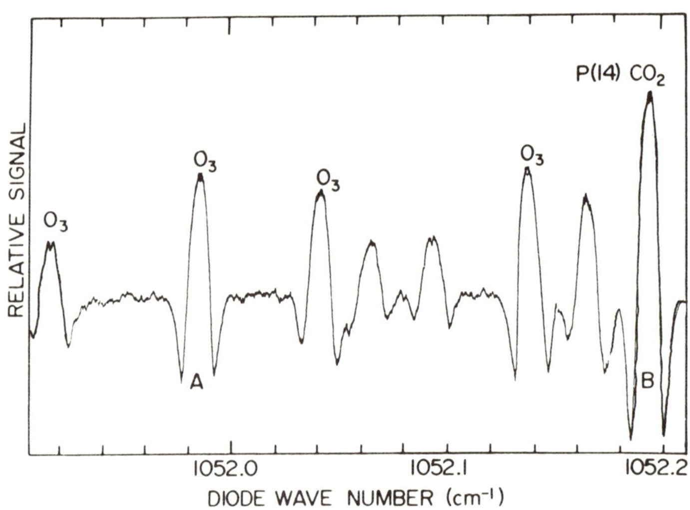

Figure 6.9. Detection of atmospheric ozone. This second harmonic scan was taken with 10-ppm O_2 in air at a pressure of 20 torr. The path length was 240 m. A 1-s time constant was used. The estimated ozone concentrations was $\sim$38 ppb. Note also that the instrumental sensitivity was reduced by a factor of 10 for line B [El-Sherbiny *et al.* (1979)].

however, that the scan shown in this figure was obtained at a pressure of 20 torr, so that the spectral lines are considerably narrower than they would be at 1 atm. In fact, at 1 atm the O_3 lines would each have widths of $\sim$ 0.1 cm^{-1} so that considerable overlap of individual O_3 lines with the CO_2 laser line would occur. This would tend to reduce the effect of CO_2 absorption on O_3 detection limits.

The use of tunable diode lasers eliminates the above problem. Some reports on the use of tunable diode lasers in the measurement of atmospheric pollution by absorption are those by Hinkley and Kelley (1971), Antcliffe and Wrobel (1972), Hinkley (1972), Ku *et al.* (1974), Hinkley *et al.* (1976), Reid *et al.* (1978a,b, 1980), El-Sherbiny *et al.* (1979), and Chaney *et al.* (1979).

An example of a tunable diode laser system used in the detection of atmospheric contaminants over a long atmospheric path is shown in Figure 6.10 [Chaney *et al.* (1979)]. The active element is a diode laser (30) mounted in a closed-cycle refrigerator (1). The output of this laser is frequency modulated at 1 kHz and amplitude modulated at 100 Hz prior to collimation with the off-axis paraboloid (16). The signal is detected at (10) and is processed with the lock-in system (23–28). Drifts in output frequency are minimized by passing part of the beam through a reference gas cell (filled with CO in this case) to the reference detector. This output is then processed in the reference feedback loop (5–8) using the second harmonic of the modulation frequency as a feedback signal. The laser output beam was passed through a scannable Fabry-Perot etalon (17–19) to ensure operation on only one mode. This etalon was also used to provide the 100-Hz modulation.

This instrument is designed to be portable and is housed in a mobile van. It is used to detect the presence of atmospheric CO. The full scale sensitivity for CO is reported as 4.6 ppm over a 400-m path length with an average peak-to-peak noise of 0.06 ppm.

High-sensitivity point monitoring of a number of pollutant gases has been reported by Reid *et al.* (1978a,b, 1980) and El-Sherbiny *et al.* (1979). A block diagram of the diode-laser spectrometer used is shown in Figure 6.11. The output of the laser is mechanically chopped at 150 Hz before focusing into the White cell. The beam exiting the White cell is passed through a 0.5-m spectrometer to separate the longitudinal modes of the diodes, and then into a HgCdTe detector. A Ge etalon or CO_2 laser discharge tube can be incorporated as shown in the beam to provide wavelength calibration.

With a path length of 300 m within the White cell, the noise level in the measurement of O_3 concentration was $\sim$ 0.5 ppb. It is reported that a sensitivity of $\sim$ 1 ppb for O_3 should be attainable with a 0.75-m long White cell in a portable instrument. A sensitivity of $\sim$ 0.1 ppb for NO_2 has also been reported [Reid *et al.* (1980)].

Atmospheric CO has also been detected at 2.3 μm using a parametric tunable laser [Hennington *et al.* (1974)]. The laser linewidth was 0.2 cm^{-1} and pulse energies were 2.0 μJ.

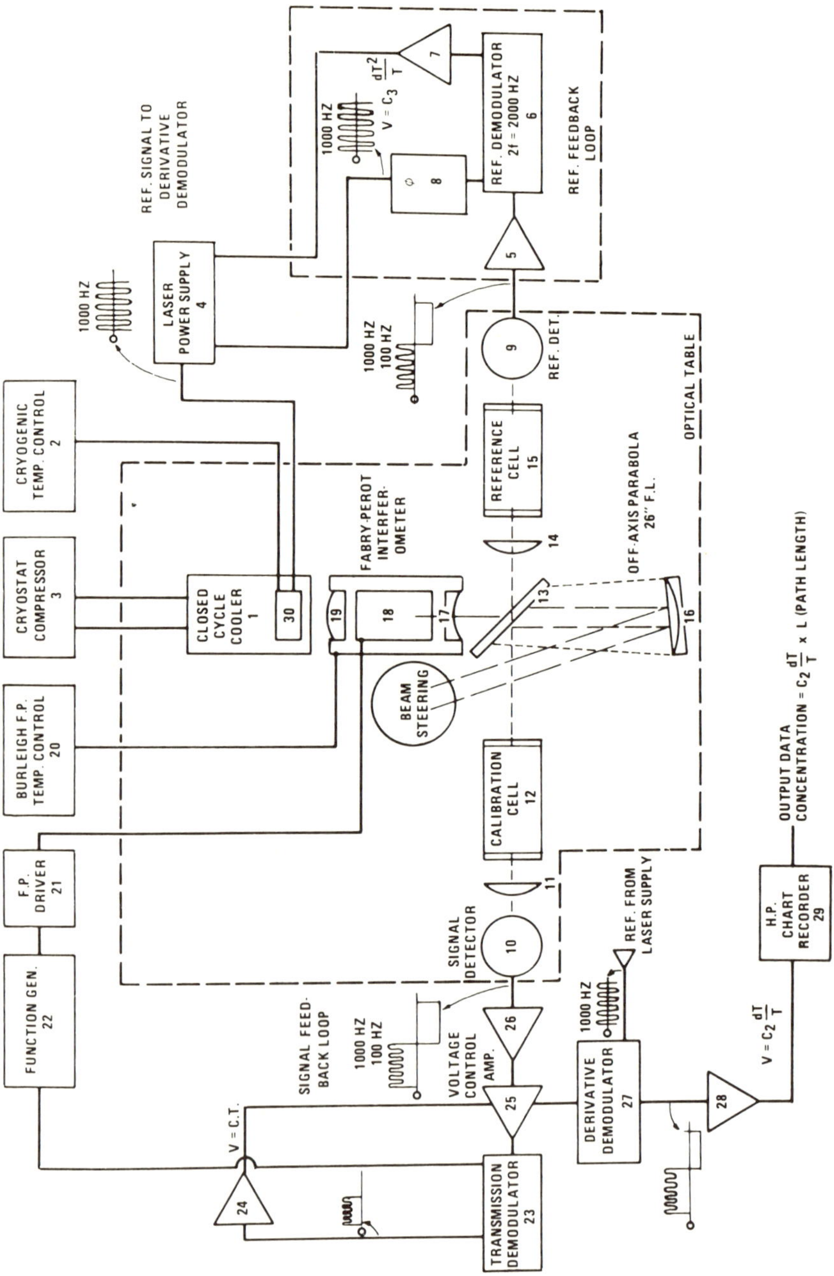

Figure 6.10. Long-path laser monitor block diagram: items are numbered and referenced in the text [Chaney *et al.* (1979)].

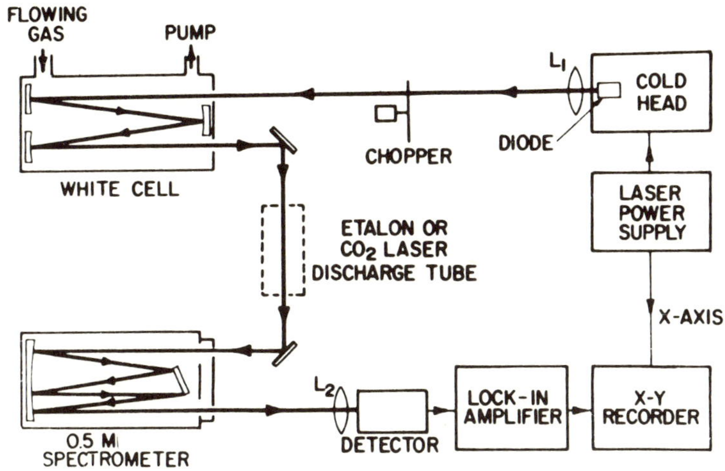

Figure 6.11. Schematic diagram of the laser absorption spectrometer (LAS). L_1 and L_2 are the laser output and detector input lenses, respectively [El-Sherbiny *et al.* (1979)].

6.2.2. LIDAR

Light detection and ranging (LIDAR) has revolutionized many aspects of atmospheric analysis since the first experiments of Fiocco and Smullin (1963). The propagation of a laser beam through the atmosphere is accompanied by a variety of processes that can lead to the attenuation of the beam. Many of these have been described in Section 6.1. In this section we discuss the interactions that occur with the atmospheric aerosol component and how these interactions can be used for the remote sensing of atmospheric properties.

A standard LIDAR measurement involves the transmission of a short laser pulse through the atmosphere. The degree to which photons from this pulse are scattered depends on the local density of atmospheric particles along the propagation path. Thus, if this scattering is monitored with a telescope the density of particles can be inferred. Since the laser pulse propagates at constant speed, the time dependence of the backscattered component is directly related to distance. The LIDAR geometry is shown in Figure 6.12.

The LIDAR equation—which relates the power received by the collecting telescope to that transmitted and can be expressed in a variety of forms [Collis (1970) and Derr and Little (1970)]—is, in simplified form,

$$P_R = P_T \frac{\Delta R\, \beta(R) A}{R^2} \exp\left[-2 \int_0^R \alpha(r)\, dr \right]$$

where P_R and P_T are received and transmitted powers, respectively, $\Delta R = c\tau/2$ is the range interval at range R, τ the pulse length, $\beta(R)$ the volume back-

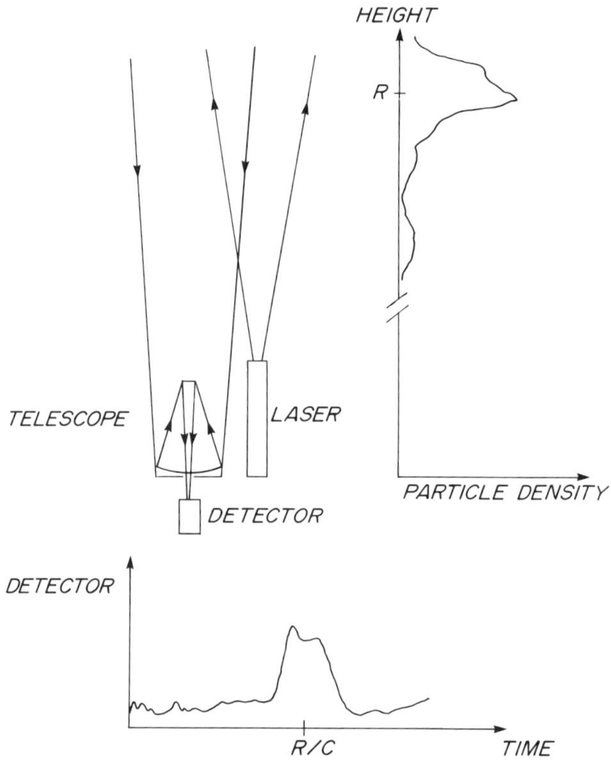

Figure 6.12. Basic LIDAR configuration: R—range, c—speed of light.

scatter coefficient (m^{-1} sr^{-1}), A the telescope area, and $\alpha(r)$ the atmospheric attenuation coefficient. This assumes a monostatic system, i.e., both receiver and telescope in the same location. Both β and α depend on the aerosol content of the atmosphere along the propagation path. Typical volume backscattering coefficients are given in Table 6.5.

The attenuation coefficient α is a function of absorption by contaminants, extinction by aerosols, and Rayleigh scattering. If we define a coefficient α (Ray), which measures aerosol extinction in the Rayleigh scattering regime, then $\beta(\text{Ray}) \sim \alpha(\text{Ray})/4\pi$. Thus under clear conditions $\beta = 1.5\alpha/4\pi$. When Mie scattering occurs, this simple relationship no longer holds although in general $\beta \sim \alpha$. Figure 6.13 shows a plot of P_R/P_T vs. range R for various values of β. The effect of atmospheric extinction with $\alpha \sim \beta$ is also shown when $\beta = 10^{-3}$ m^{-1} sr^{-1} (i.e., $\alpha = 10^{-3}$ m^{-1}). The effect of attenuation is significant only when β exceeds 10^{-4}–10^{-5}, that is, in a foggy or cloudy atmosphere.

Figures 6.14 and 6.15 show some data obtained from the vertical sounding

Table 6.5. Typical Volume Backscattering Coefficients

Conditions	β (m^{-1} sr^{-1})
Clear	1×10^{-6}
Haze	2×10^{-5}
Clouds	8×10^{-5}
Fog	2×10^{-4}

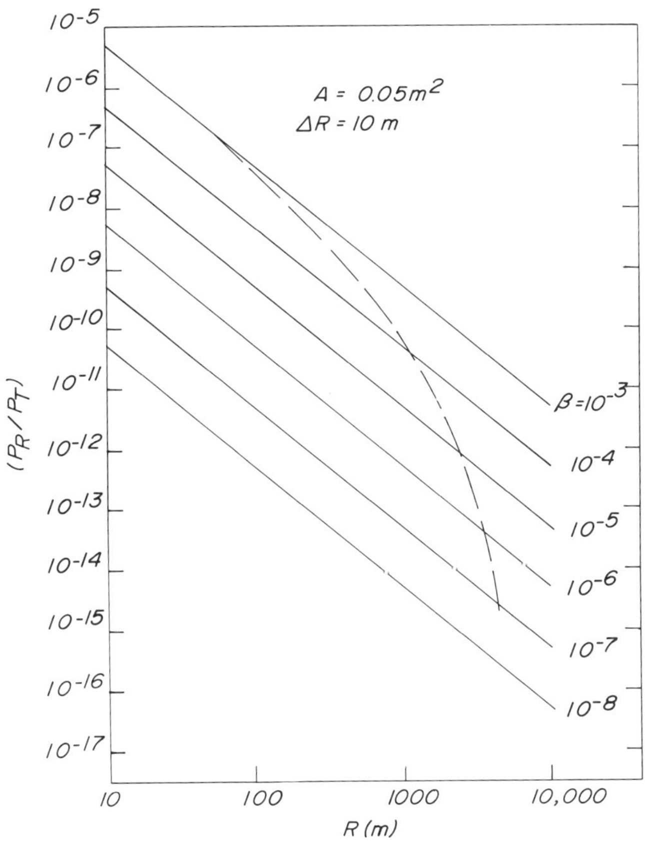

Figure 6.13. Ratio of received power to transmitted power for a LIDAR system with $A = 0.05$ m^2, $\Delta R = 10$ m, and various values of β (m^{-1} sr^{-1}). The dashed line shows the curve for P_R/P_T when an atmospheric extinction $\alpha = 4\pi\beta$ is assumed.

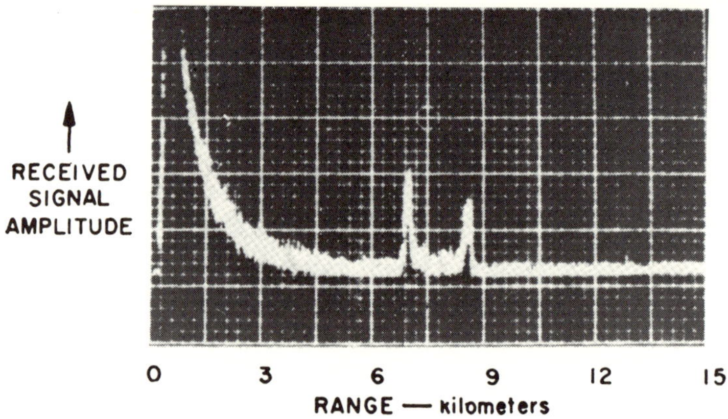

Figure 6.14. Return from two layers of overhead cirrus at 6.6 and 8.1 km [Northend *et al.* (1966)].

of cirrus clouds with an early ruby LIDAR [Northend *et al.* (1966)]. Clear returns are observed from clouds at the 6–11 km level. The power of the LIDAR technique in mapping cloud structure and pollution sources is shown in the LIDAR scans in Figure 6.16 [Allen and Evans (1972)]. This system incorporates automatic scanning and pulsing at equal angle increments or time intervals together with data storage on video disk. Instant playback provides real-time monitoring. The system is mounted in a van. Figure 6.16 shows examples of the use of such a system to detect point sources of pollution and to trace the time evolution of effluent from these sources.

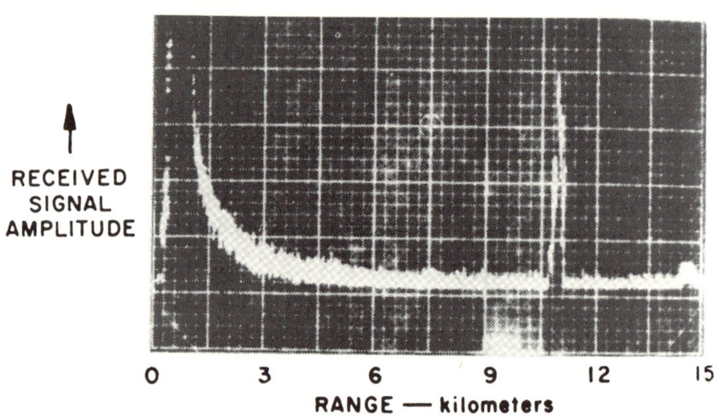

Figure 6.15. A strong return from overhead cirrus at 10.5 km [Northend *et al.* (1966)].

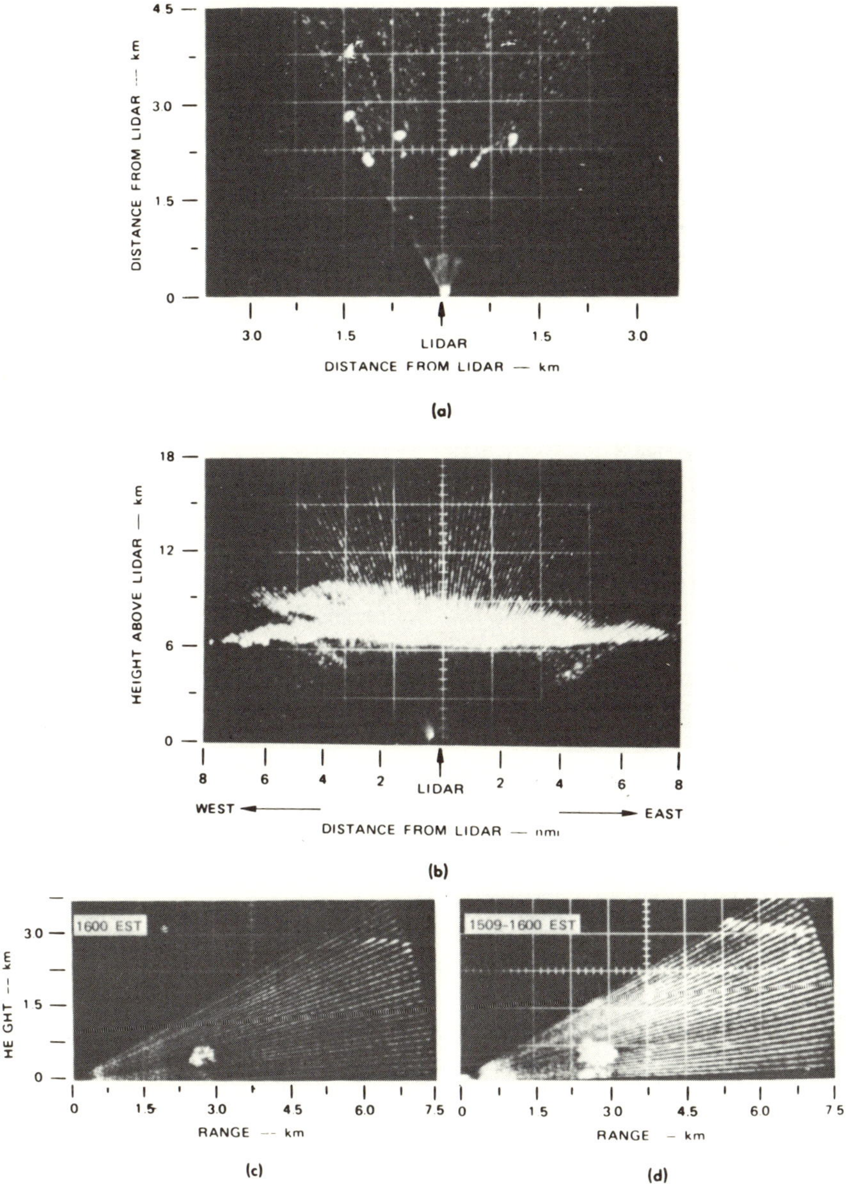

Figure 6.16. Disk playback using sin/cos sweep display: (a) horizontal fan above an industrial area showing location of particulate sources; (b) overhead vertical fan showing cloud structure; (c) low-angle verticle fan through plume, showing its cross section; (d) composite of a sequence of runs including (c) showing the plume spread and mixing. Q-switched ruby laser source [Allen and Evans (1972)].

Some features of LIDAR measurements of cloud characteristics are:

1. Precision measurement of cloud ceiling and its temporal and spatial variation.
2. Determination of the optical thickness of clouds and their morphology.
3. Measurement of high-level clouds.
4. Investigation of precipitation and nucleation within clouds.
5. Measurement of wave structure within clouds.
6. Measurement of wind velocity.

Summaries of the application of LIDAR to studies in these areas have been published by Collis (1970), Derr and Little (1970), and Kostko (1976).

Most LIDAR studies use ruby or Nd–YAG laser sources. This choice is made because the backscatter from atmospheric particles is greatest when $2\pi a/\lambda \sim 5$, where a is the particle radius and λ is the laser wavelength. Since $a \sim \lambda$ for maximum scattering, lasers with wavelengths in the visible or at 1 μm will detect particles with sizes in the range 1–2 μm. At 10 μm the Mie scattering maxima would be obtained only for $a \sim 10$ μm. Thus 10-μm LIDAR systems are of little use in cloud diagnostic studies. Conversely, 10-μm LIDAR beams experience significantly smaller attenuation in dusty, low-humidity atmospheres. A comparison of LIDAR measurements on artifically created dust clouds at 0.7 and 10.6 μm has been reported by Randhawa and van der Laan (1980).

The use of an IR LIDAR system to measure atmospheric IR extinction has been discussed by Murray *et al.* (1978). One-joule TEA laser pulses at several CO_2 laser wavelengths were transmitted through the atmosphere and the backscatter was monitored. The slope of the curve of backscatter intensity vs. range could then be related to α. Since CO_2 laser radiation is absorbed by a variety of atmospheric molecules as well as scattered by the atmospheric aerosol, α contains contributions from several sources of attenuation. Measured values of α were in the range 0.2–1.4 km^{-1}.

One of the most important applications of LIDAR has been in the monitoring of the aerosol component of effluent from point sources of pollution. These studies provide valuable information about the size distribution and dispersal of plumes generated as the result of industrial activity. When combined with DIAL or laser Raman sampling, the aerosol content can be examined in the context of the chemical composition of the plume. Some representative studies have been published by Johnson (1969), Collis and Uthe (1972), Allen and Evans (1972), Rothe *et al.* (1974), and Schuster and Kyle (1980).

The work of Carswell and his collaborators [Houston and Carswell (1978) and Pal *et al.* (1978)] has shown that much useful diagnostic information can be obtained from a study of the polarization of LIDAR returns. The depolarization of an initially polarized beam can be used to obtain information on the shape, size, and composition of cloud aerosols. Houston and Carswell show that in water droplet clouds linear polarization is strongly preserved but some rotation of plane of polarization and ellipticity are obtained.

6.2.3. Laser Remote Sensing of Wind Velocity

Several techniques for laser remote sensing of wind velocity have been reported [Lawrence *et al.* (1972), Benedetti-Michelangeli *et al.* (1972), Bartlett and She (1976), Schwiesow *et al.* (1977), Armstrong *et al.* (1976), and Eberhard and Schotland (1980)]. These are based on one or other of the following two basic methods:

Measurement of Doppler shift. The Doppler shift of light scattered from individual aerosol particles is detected by optical heterodyning with a beam that has not undergone scattering. Wind velocity can then be inferred from this Doppler shift.

Intensity modulation. Two beams from the same laser are superimposed in a volume of space at the remote location where the wind velocity is to be measured. This forms an interference pattern within this volume. Particles moving through this volume encounter a periodic light field and the scattered light is therefore modulated. Scattered light is detected incoherently.

Many modifications and combinations of these two techniques have been reported [see the review by Thompson and Stevenson (1979)]. In this section some of these techniques and their application to atmospheric diagnostics will be described.

A simple laser Doppler velocimeter is shown in Figure 6.17. The beam from a frequency-stabilized laser is brought to a focus at some remote location where it is scattered by atmospheric aerosols carried along by the wind at that point. The scattered radiation is Doppler shifted because of velocity components along the optic axis of the system. This is then returned to the laser where it is amplified. Both components are then mixed in a square-law detector where the Doppler frequency is detected. The relationship between Doppler shift $\Delta\nu$ and wind velocity, $\mathbf{v}$, is

$$\Delta\nu = \frac{2|\mathbf{v}| \cos\theta}{\lambda}$$

where θ is the angle between the wind direction and the optic axis. Defining a velocity component $|\mathbf{v}| \cos\theta = V$, one has

$$\begin{aligned} \Delta\nu &= 188.67V \text{ kHz}, \qquad \lambda = 10.6\ \mu\text{m} \\ &= 3887.3V \text{ kHz}, \qquad \lambda = 514.5\ \text{nm} \end{aligned}$$

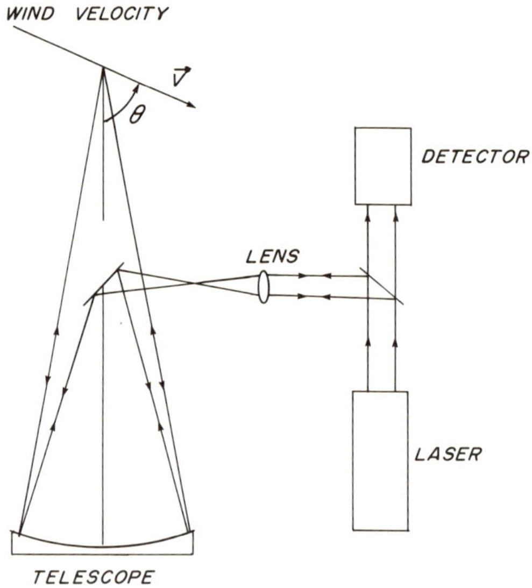

Figure 6.17. Laser Doppler velocimeter based on heterodyne detection.

where V is in m s^{-1}. At 10.6 μm it is evident that typical wind speeds will give $\Delta\nu$ in the range 10^5–10^7 Hz.

Lawrence *et al.* (1972) have used a system of this sort with a CW CO_2 laser to measure wind velocities at ranges of $\sim$ 300 m. Some of these data were compared to those obtained from conventional anemometers. A comparison of this sort is shown in Figure 6.18. Detailed analysis of power spectral densities for the two measurements give close agreement.

Benedetti-Michelangeli *et al.* (1972) have described a laser Doppler velocimeter based on an Ar^+ laser. A scanning Fabry-Perot interferometer was used to analyze the backscattered signal. Range resolution was obtained by chopping the laser beam and gating the signal photon counting system with an appropriate delay. Ranging to 750 m was reported. Figure 6.19 shows a sample of the data obtained with such a measurement. The RMS error on radial wind velocity was $\pm$ 0.27 m s^{-1}.

These measurements only provide information on the radial (i.e., parallel to optic axis) velocity component. Kennedy and Bilbro (1979) have described how intensity fluctuations in the Doppler-shifted backscattered component can be used to determine the transverse component of wind velocity. Measurements of both radial and transverse wind velocity components were made at a distance of 100 m using a CW CO_2 laser source.

A simple dual-beam laser Doppler velocimeter system is shown schematically in Figure 6.20. The two beams are obtained from the same laser source as shown. A series of interference fringes are generated when the beams are

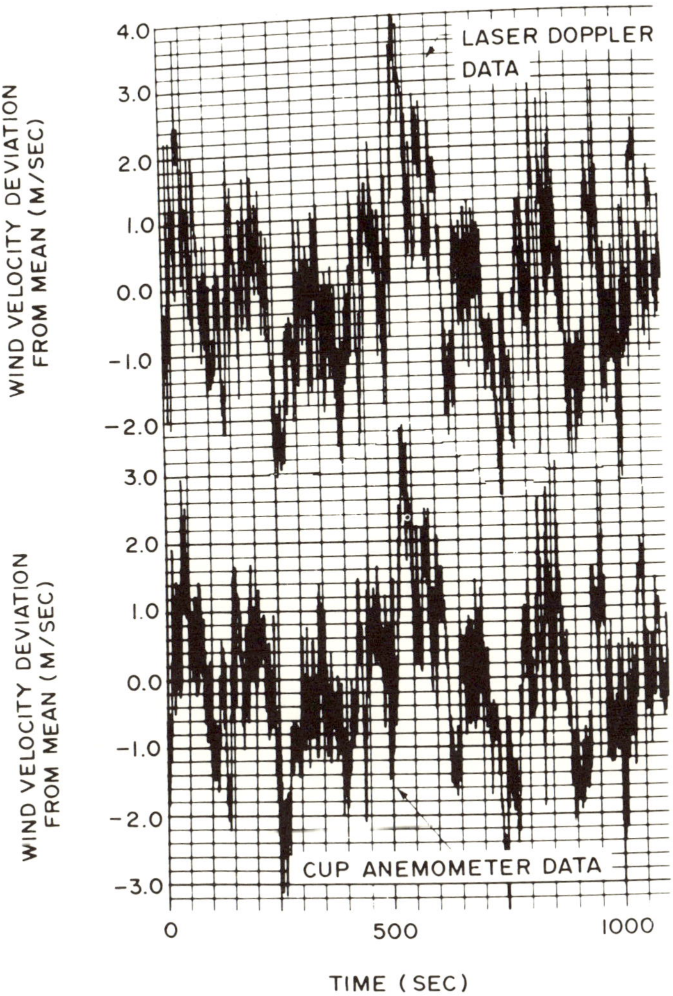

Figure 6.18. Laser Doppler derived wind velocity time histories compared with anemometer data. Comparisons are in terms of deviation from the mean (5.4 m s^{-1}). Anemometer data are resolved along laser line of sight [Lawrence *et al.* (1972)].

recombined within the focal volume at range R. The spacing between these fringes will be $\lambda/[2 \sin(\theta/2)]$, where θ is the angle between the two laser beams.

The remote sensing of wind velocities using this type of dual-beam instrument has been reported by Bourke and Brown (1971) and Bartlett and She (1976). A theoretical analysis of signal-to-noise ratios under a variety of conditions of laser power, range, wavelength, etc. has been published by Farmer and Brayton (1971). An example of correlation functions obtained from such a system operating at 5.4.5 nm is shown in Figure 6.21 [Bartlett and She (1976)]. These signals were recorded with a clipped digital correlator. They represent the passage of individual dust particles through the fringe pattern.

Schwiesow *et al.* (1977) have reported on the development of a dual-beam laser anemometer in which four scattered beams are combined with each other in a detector to provide simultaneous information on radial and transverse wind

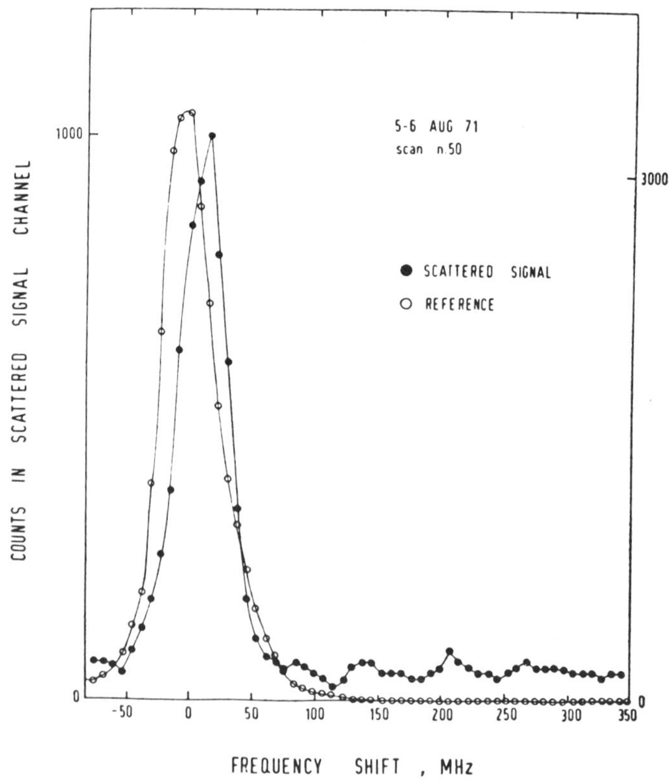

Figure 6.19. Spectrum of the signal backscattered from aerosols and the instrumental response (reference). The displacement is due to wind [Benedetti-Michalangeli *et al.* (1972)].

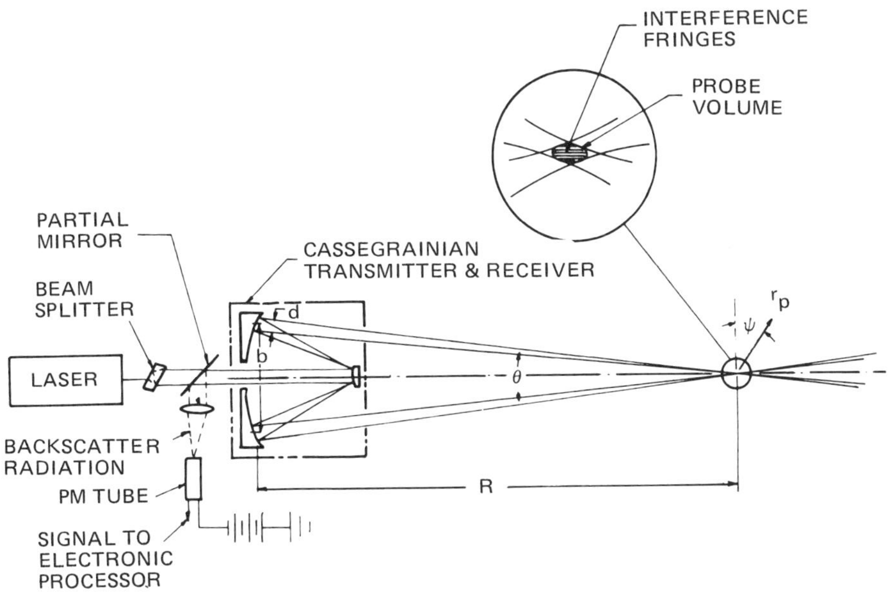

Figure 6.20. A one-velocity component long range, dual-scatter backscatter LDV system [Farmer and Brayton (1971)].

velocity components. The scattering geometry is shown in Figure 6.22. Frequency shifts from the transmitted beam frequency are

$$\Delta\nu_1 = -\frac{2V}{\lambda}\left(\cos\psi\cos\frac{\theta}{2} + \sin\psi\sin\frac{\theta}{2}\right)$$

$$\Delta\nu_2 = -\frac{2V}{\lambda}\left(\cos\psi\cos\frac{\theta}{2} - \sin\psi\sin\frac{\theta}{2}\right)$$

The frequency shift associated with the interbeam component, i.e., that component scattered from one beam to the other, is

$$\Delta\nu_0 = -\frac{2V}{\lambda}\cos\psi\cos\frac{\theta}{2}$$

These three components yield a triplet of output signals when heterodyned with the local oscillator output. The average frequency shift ν_L of this group is

$$\nu_L = \left|-\frac{2V}{\lambda}\cos\psi\right|$$

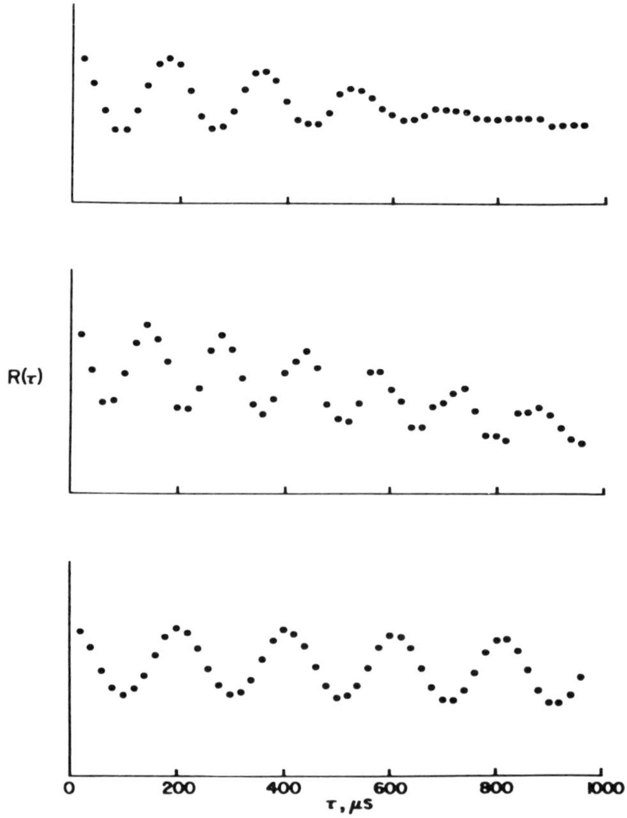

Figure 6.21. Three consecutive correlation functions, $R(r)$ obtained at a range of 20 m and with experimental parameters: 0.2-W laser power, 20 μs per channel, clipping level = 8, collection angle $\Delta = 5 \times 10^{-3}$ rad, D = 2.9 cm, and ρ = 1 cm. $D = 2R/\sin(\theta/2)$ and ρ is the beam radius at remote location [Bartlett and She (1976)].

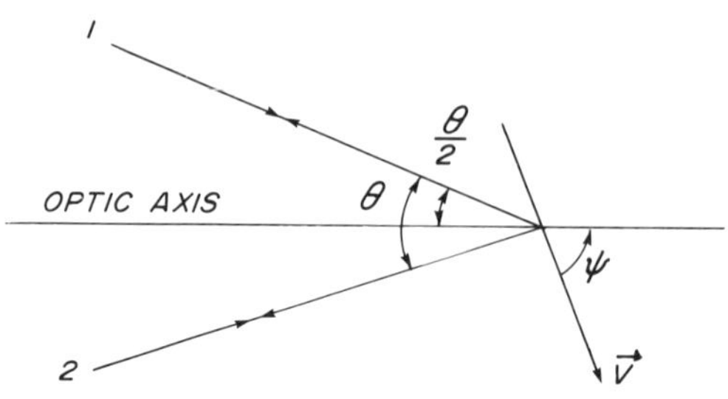

Figure 6.22. Scattering geometry used for remote sensing of transverse velocity.

This frequency is proportional to the radial velocity component. An intratriplet beat frequency

$$\nu_T = \left| -\frac{V\Theta}{\lambda} \sin \psi \right|$$

is shown to be proportional to the transverse wind velocity [Schwiesow *et al.* (1977)]. Typical values for these frequencies are $\nu_L \sim 500$ kHz and $\nu_T \sim 3$ kHz. Experimental verification of these predictions was given by Schwiesow *et al.* using test targets.

Eberhard and Schotland (1980) use a dual-frequency Doppler LIDAR system to measure wind velocity components on a remote basis. Two separate beams with frequencies ν_1 and ν_2, such that $\nu_1 - \nu_2 = \delta$ lies in the microwave region, are used as probes. Different Doppler shifts $\Delta\nu_1$ and $\Delta\nu_2$ are observed for scatterers in the two beams so that $\Delta\nu_1 - \Delta\nu_2 = \Delta\delta$. Processing of the backscattered signal provides the frequency $\Delta\delta$ and hence a component of the wind velocity. A prototype system based on an Ar^+ laser was described.

Laser wind sensing with a time of flight system has been reported by Armstrong *et al.* (1976). The geometry of this system is shown in Figure 6.23. Two target volumes are defined optically within the region excited by pulses from the laser. The correlation of signals from these two volumes can be related to wind velocity. Signals from each volume were averaged and then correlated. It was found that the prototype system which utilized an N_2 laser with pulse length and a beam divergence of 2.5×7.5 mrad could measure wind speeds in the range $1.25 \lesssim v \lesssim 75$ m s^{-1}. A comparison of particle velocity measurements with time of flight and laser Doppler anemometers has been published by Lading (1979). It was found that the laser Doppler anemometer was

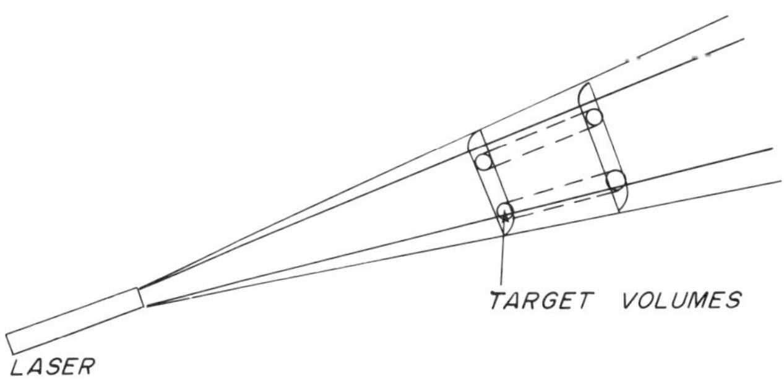

Figure 6.23. Geometry of system used by Armstrong *et al.* (1976) to measure wind velocity via time of flight measured. In the prototype system the target volumes were separated by 1.25 m and were at a distance of 250 m from the laser transmitter. The radial extent of the target volume is determined by the length of the laser pulse. It was $\sim \delta m$ in the system of Armstrong *et al.*

inferior to the time of flight instrument for remote sensing of wind velocity under certain conditions.

Laser anemometry has been widely applied in a variety of other systems in which particle flows are important. Some applications include the study of particle motion in combustion [Owen (1979)], in wind tunnels [Boutier *et al.* (1979)], in internal combustion engines [Witze (1979) and Rask (1979)], and in rocket exhausts [Farmer *et al.* (1979)].

6.2.4. Raman LIDAR

The Raman effect is extensively used in laboratory chemical analysis (cf. Section 5.6). The application of Raman spectroscopy to atmospheric studies was first described by Cooney (1968) and Inaba and Kobayasi (1969). A derivative of the standard LIDAR technique, its utility lies in the fact that a complete range-resolved chemical fingerprint is generated on a real-time basis. Disadvantages include low signal levels and the need to use high-energy laser pulses with wavelengths in the UV.

Raman shifts arise due to the excitation of Raman active vibrational transitions in gaseous molecules along the laser propagation path. Energy shifts are

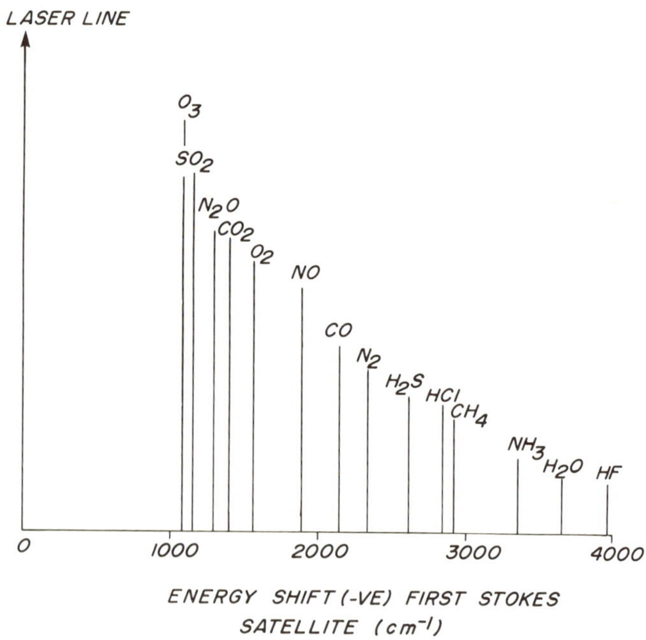

Figure 6.24. Energy shift of first Stokes satellite from laser line for various molecular species. The length of vertical lines has no physical significance.

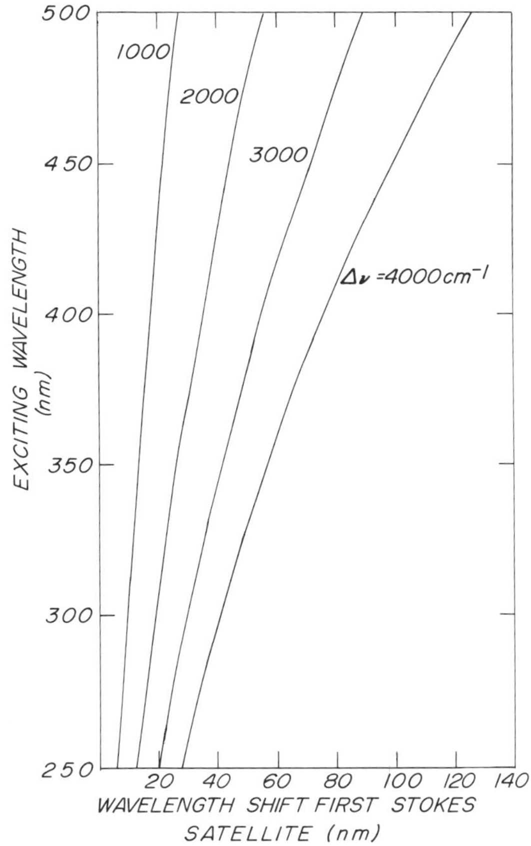

Figure 6.25. Variation of wavelength shift of first Stokes satellite with wavelength of exciting laser line.

typically 500–3500 cm^{-1}. Raman spectral features due to individual molecules are the superposition of a large number of rotational lines. This tends to produce relatively broad features although detection of the *Q*-branch reduces this effect. Figure 6.24 shows Raman spectral energy shifts observed for a variety of scatterers. Spectra obtained as the laser wavelength is varied are shown in Figure 6.25. Note that the wavelength shift between the exciting line and individual Raman lines decreases with decreasing wavelength.

The equation that relates the number of Raman-scattered photons to system parameters is

$$n(R) = \frac{E}{h\nu}\frac{c\tau}{2}\,T(\lambda_0,R)\,T(\lambda,R)\frac{A}{R^2}\,\xi\eta\left(\frac{d\sigma}{d\Omega}\right)_\nu N(R)$$

where $n(R)$ is the number of photons detected at frequency ν during integration time τ. $R = ct/2$ is the range, where t is the time elapsed from the start of the laser pulse. E is the laser pulse energy, T the atmospheric transmittance between 0 and R at wavelengths λ_0, the exciting line, and λ, the Raman line. A is the area of the receiver, ξ is a receiver transmission coefficient, η is the photomultiplier quantum efficiency, and $N(R)$ is the number of Raman scatterers per unit volume. $(d\sigma/d\Omega)_\nu$ is the Raman scattering cross section. Some values of $d\sigma/d\Omega$ for various molecules are listed in Table 6.6. These values refer to a wavelength of 514.5 nm for the exciting line. The resonance Raman effect (Section 5.6) can become important at shorter wavelengths when the wavelength of the exciting line becomes close to that of an absorption band in the scatterer. In this case, $d\sigma/d\Omega$ can increase by several orders of magnitude [Rosen *et al.* (1975)].

A simple Raman LIDAR system is shown schematically in Figure 6.26. Pulses of 20-ns duration from a Q-switched ruby laser were collimated by a telescope. The Raman backscatter was gathered by a 30-cm diameter, 150-cm focal length mirror. A monochromator together with a blocking filter were used to reject the Rayleigh component and select the Raman line required.

A comprehensive study by Pourny *et al.* (1979) has identified the following factors that limit the sensitivity of the Raman–LIDAR technique.

Table 6.6. Differential Raman Scattering Cross Sections for Various Gaseous Molecules[a]

Gas	Raman frequency (cm^{-1})	$d\sigma/d\Omega$ ($cm^2\ sr^{-1}$)
O_3	1103	1.8×10^{-30}
CO_2	1286	4.4×10^{-31}
N_2O	1287	1.2×10^{-30}
CO_2	1388	6.6×10^{-31}
O_2	1556	5.3×10^{-31}
NO	1877	2.4×10^{-31}
CO	2143	5.3×10^{-31}
N_2O	2223	2.3×10^{-31}
N_2	2330	4.4×10^{-31}
CH_4	2914	3.5×10^{-30}
	3020	3.5×10^{-31}
H_2O	3652	1.5×10^{-30}
H_2	4160	9.7×10^{-31}

[a]The exciting wavelength is 514.5 nm. They can be scaled to other wavelengths, λ, by multiplying by the factor $\left(\frac{514.5}{\lambda}\right)^4$

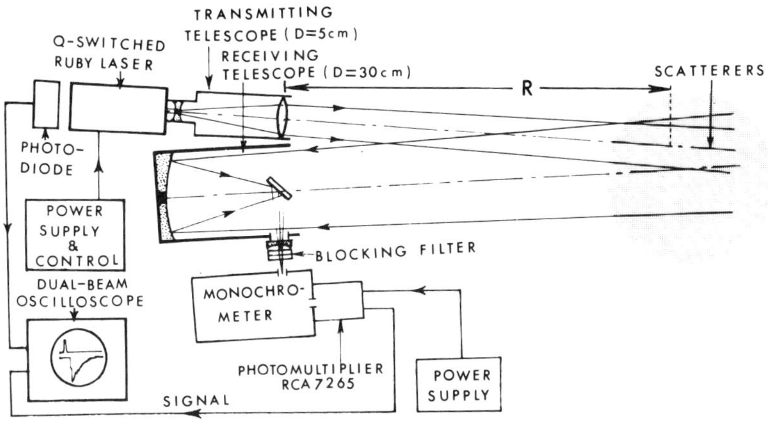

Figure 6.26. Schematic diagram of a laser–Raman radar system using a *Q*-switched ruby laser [Kobayasi and Inaba (1970b) Reprinted with permission; copyright 1970, IEEE].

Background sky light. The sky background contributes photons to the detection system. The number of photons detected in time τ is

$$n_b\tau = \frac{W}{h\nu} B\Omega A\xi\eta\tau$$

where W is the sky irradiance, B the optical bandwidth, and Ω the field of view solid angle. Pourny *et al.* estimate $n_b\tau(\text{day}) = 800$ and $n_b\tau(\text{night}) = 4 \times 10^{-5}$. The detection limit of their system (operating at 347.1 nm) was $n\tau = 0.5$ so that operation in daylight was precluded.

Elastic backscatterer. This arises from Rayleigh scattering by atmospheric molecules and Mie scattering from aerosols. Since it occurs at the wavelength of the exciting line it can be discriminated against by the use of interference filters or a monochromator.

Aerosol fluorescence. Absorption of the exciting laser line can lead to aerosol fluorescence. Little interference is expected from this effect in clear air.

Scattering from molecules in condensed state. This is usually important only when sampling H_2O. Since H_2O can be in liquid form as well as in vapor form, a Raman signal can arise from both phases. Pourny *et al.* estimate that the ratio of these two signals will be typically $I(\text{liquid})/I(\text{vapor}) \sim 10^{-2}$ when equal quantities of H_2O are in liquid and gaseous form. Similar effects would be expected for molecules such as SO_2 that can become incorporated in aerosol particles.

Dynamic range of signal. The large signals obtained from nearby scatterers together with the weak return from distant scatterers places stringent limits on the measurement accuracy.

Atmospheric changes. Variations in atmospheric content can limit accuracy by changing the atmospheric attenuation between laser pulses or over a series of laser pulses.

System electrical noise. Signal-to-noise ratio is given by

$$\frac{\mathrm{S}}{\mathrm{N}} = \frac{\bar{n}(p\tau)^{1/2}}{[\bar{n} + 2(\bar{n}_b + \bar{n}_d)]^{1/2}}$$

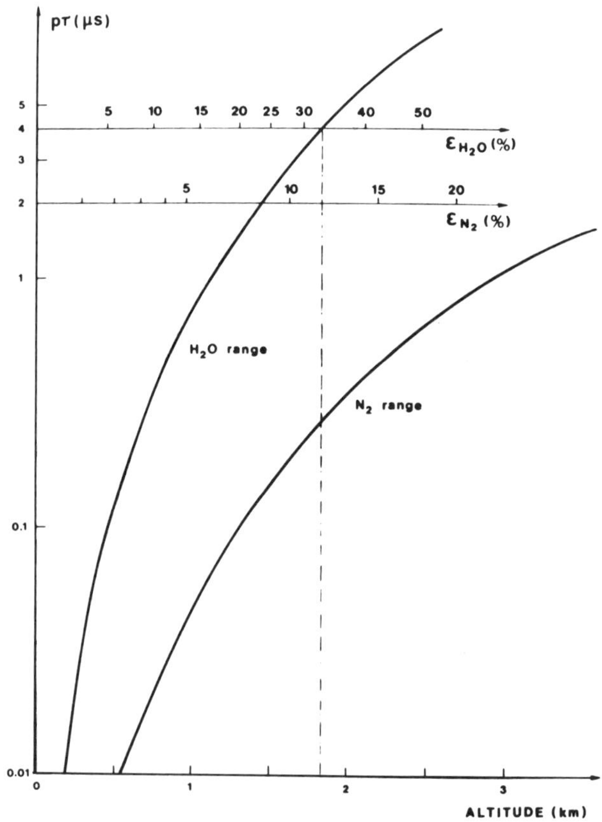

Figure 6.27. Raman LIDAR range (S/N = 3) for a vertical sounding as a function of product $p\tau$. Uncertainty ξ on H_2O signal for $p = 20\tau = 0.2\ \mu s$ and on N_2 signal for $p = 10\tau = 0.2\ \mu s$ as a function of altitude. System parameters $\lambda = 347.1\text{nm}$, $A = 0.05\ \text{m}^2$, $E = 0.2\ \text{J}$, $\xi = 0.29$, $\eta = 0.25$ [Pourny *et al.* (1979)].

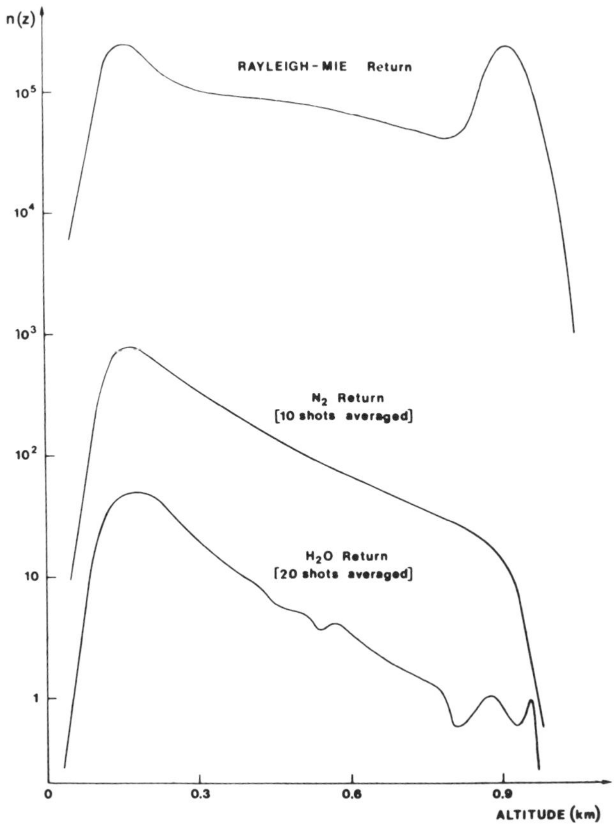

Figure 6.28. LIDAR and Raman LIDAR signals (photoelectrons count per 30-m range incremently) obtained in the presence of stable stratus layer on 25 November 1976 [Pourny *et al.* (1979)].

where $\bar{n} = n/\tau$ is the number of detected photons per unit time and $\overline{n_b}$ and $\overline{n_d}$ are similar expressions for the background optical noise and the photocathode dark noise, respectively. p is the number of laser pulses.

Some data obtained by Pourny *et al.* (1979) on the detection of H_2O via Raman LIDAR are shown in Figures 6.27 and 6.28. Figure 6.27 shows the sounding range obtained at a S/N of 3 for H_2O and N_2. The uncertainty in the measurement defined as $(S/N)^{-1}$ is also shown. It is evident that low noise data can only be obtained for ranges $\lesssim 1$ km.

A comparison between the height dependence of normal LIDAR and Raman LIDAR returns is shown in Figure 6.28. The independence of elastically scattered and Raman components is evident; note the absence of any excess emission at the 900-m level in the H_2O and N_2 returns.

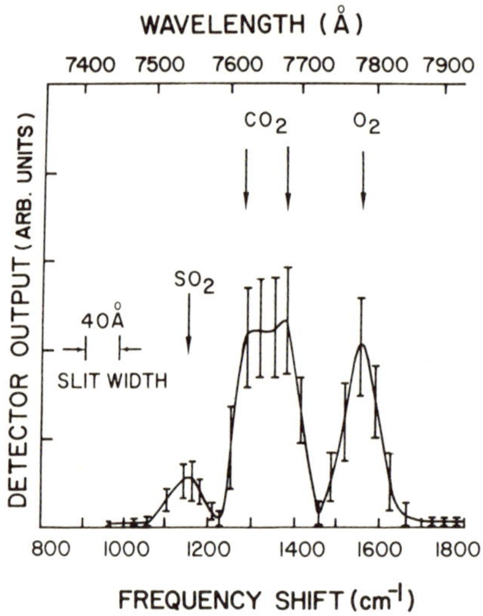

Figure 6.29. Observed spectra of Raman backscatter from a polluted atmosphere. This spectrum shows the relative detector output of Raman returns with an oil plume polluting the atmosphere. Raman peaks from CO_2 and SO_2 are from pollutants [Kobayasi and Inaba (1970)].

The limited range capability of Raman LIDAR has restricted applications to the remote sensing of localized sources. Several applications have been reported in which Raman LIDAR has been used to sample pollutants in stack plumes or other discreet sources of emission [Kobayasi and Inaba (1970a,b), Melfi *et al.* (1973), Hirschfeld *et al.* (1973), Delong (1974), Leonard (1975), and Rosen *et al.* (1975)].

An example of the Raman spectrum obtained under these conditions is shown in Figure 6.29. The detailed chemical information available from these measurements is apparent from this spectrum. An example of the range resolution attainable is shown in Figure 6.30.

The use of Raman LIDAR to obtain atmospheric temperature profiles has been described by Cooney (1972), Cooney and Pina (1976), and Cohen *et al.* (1976). Since the distribution of intensity within a rotational Raman spectrum is a function of the ambient temperature, a measurement of this distribution yields information about the ambient temperature. Cooney (1972) showed that such data can be obtained if the intensity of the rotational backscatter is measured at two wavelengths within the rotational band profiles. Development of this technique has resulted in a sensitive method of temperature sounding using LIDAR. Some experimental data obtained by Cooney and Pina (1976) are shown in Figure 6.31. Note the excellent temperature–range resolution.

6.2.5. *Differential Absorption LIDAR (DIAL)*

This useful remote sensing technique uses the LIDAR backscatter as a measure of the strength of laser radiation reaching a given point in space. If

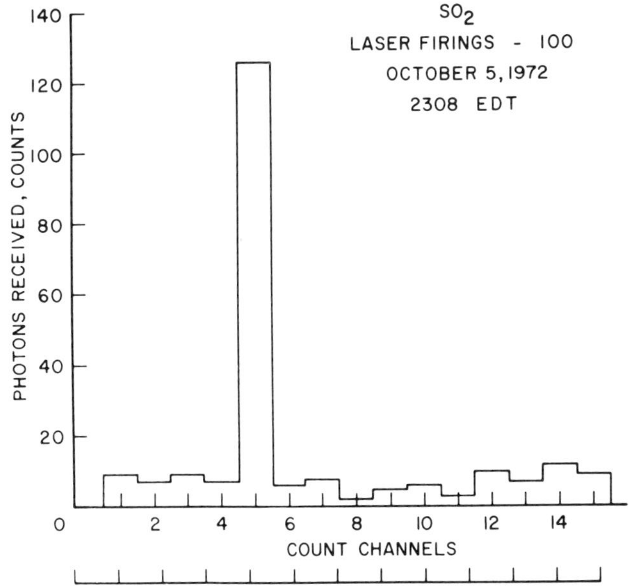

Figure 6.30. Number of photon counts received at 7546 Å (υ_1 lines of SO_2) for 100 laser firings. Excitation wavelength is 694.3 nm [Melfi *et al.* (1973)].

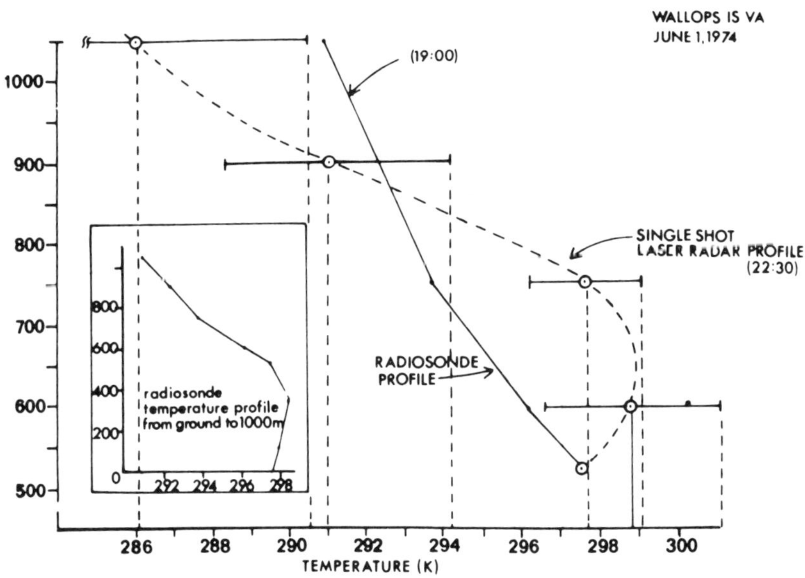

Figure 6.31. Comparison of laser radar temperature profile with radiosonde temperature profile [Cooney and Pina (1976)].

the wavelength of the probe beam is tuned across an absorption feature of a molecular pollutant, then a decrease in backscattered radiation will be observed. By comparing the LIDAR signal received from a given region of space at two laser frequencies—one resonant with the absorption line and the other off resonance—the concentration of absorbers along the path to that point in space can be ascertained.

The concentration of an absorber $\bar{n}(R)$ averaged over some small distance ΔR at distance R can be obtained from the LIDAR equation. One has for the number of photons backscattered at distance R

$$P(R) = \frac{KE}{h\nu} A \frac{\beta(R)\,\Delta R}{R^2} T^2$$

where K is the system counting efficiency, A the receiver area, E the pulse energy, $h\nu$ the photon energy, T the atmospheric transmission, and $\beta(R)$ the volume backscatter coefficient. β is the sum of contributions from molecules and particles

$$\beta(R) = \beta_m + \beta_p$$

where β_m is the coefficient for scattering from molecules and β_p is the coefficient for scattering from particles. The atmospheric transmission is

$$T(R) = \exp\left[-\int_0^R (\alpha_m + \alpha_p)\,dR\right]$$

where α_m and α_p are absorption coefficients for molecules and particles, respectively.

$P(R)$ is measured for two laser wavelengths, λ_1 and λ_2. The two values of $P(R)$ can then be combined to yield the following expression for $\bar{n}(R)$

$$\bar{n}(R) = \frac{1}{2(\sigma_1 - \sigma_2)R}\left\{\ln\left[\frac{P_1(R)P_2(R + \Delta R)}{P_2(R)P_1(R + \Delta R)}\right] + B' + T'\right\}$$

where

$$B' = \ln\left[\frac{\beta_1(R + \Delta R)}{\beta_1(R)}\,\frac{\beta_2(R)}{\beta_2(R + \Delta R)}\right]$$

and

$$T' = -2\,\Delta R(\bar{\alpha}_1 - \bar{\alpha}_2)$$

where σ_1 and σ_2 are absorption cross sections of the absorber at λ_1 and λ_2 and $\overline{\alpha}_1$ and $\overline{\alpha}_2$ are absorption coefficients at the two wavelengths averaged over the range ΔR. When B' and T' are zero, this expression simplifies to

$$\overline{n}(T) = \frac{1}{2(\sigma_1 - \sigma_2)\Delta R} \ln\left[\frac{P_1(R)P_2(R + \Delta R)}{P_2(R)P_1(R + \Delta R)}\right]$$

This expression can often be used when $\lambda_1 \sim \lambda_2$ and the two-wavelength measurement is made simultaneously. Uncertainties and errors in the use of the above expressions for $\overline{n}(R)$ have been estimated by Schotland (1974) and Byer and Garbuny (1973).

DIAL studies of atmospheric species have been carried out in the IR [Murray *et al.* (1976)], visible [Browell *et al.* (1979)], and UV [Uchino *et al.* (1979)]. It should be noted that the only requirement for the detection of a molecular species using DIAL is that it have different absorption coefficients at the two wavelengths used. This difference can be the result of variations in σ within a molecular continuum and need not necessarily arise from the presence of a discrete absorption line. For some measurements it is often sufficient to use a single laser line. This requires that the contribution due to background scattering be calculated from separate data [Uchino *et al.* (1979)].

The advantages and limitations of DIAL in the analysis of atmospheric species can be seen from several studies of stratospheric O_2 [Uchino *et al.* (1978, 1979), Grant and Hake (1975), Menzies and Shumate (1976), Gibson and Thomas (1975), Asai *et al.* (1979), and Bufton *et al.* (1979)]. System parameters for IR and UV DIAL are given in Table 6.7. The variation of $P(R)$

Table 6.7. Summary of System Parameters in Two DIAL Measurements of Atmospheric Ozone

Parameter	IR	UV[a]
Laser	CO_2 TEA	XeCl
Wavelength	$P(14)$ 9.505 μm $P(24)$ 9.586 μm	307.9, 308.2 nm
Pulse energy	5.5 J	0.05 J
Pulse length	100 ns	20 ns
Repetition rate	1 Hz	4 ppm
Beam divergence	3 mrad	2 × 5 mrad
Telescope diameter	30 cm	50 cm
Field of view	4.5 mrad	10 mrad
Detector	HgCdTe	PMT EMI 9558QB
Measured O_3 concentrations and range	0.04–0.02 ppm at 1.5 km	0.2 ppm at 2.5 km
Reference	Asai *et al.* (1979)	Uchino *et al.* (1979)

[a]Measurement uses both emission bands from XeCl laser as "single" wavelength within O_3 band.

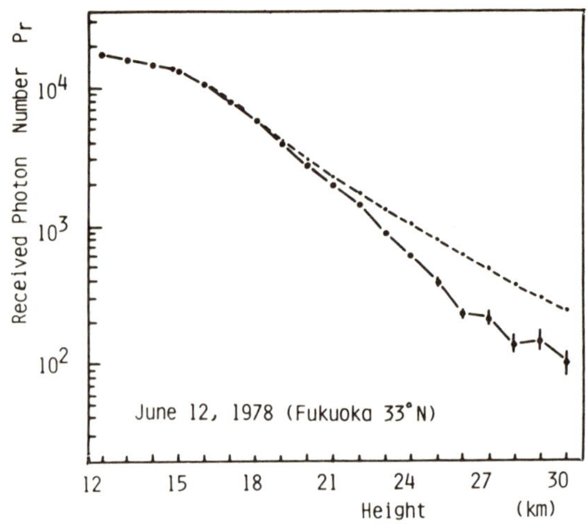

Figure 6.32. Total received photon number $P(R)$ in 1-km interval for 1050 XeCl laser shots. The error bars represent $\pm$ 1 standard deviation based on the receiving signal. Dashed line represents the expected value of $P(R)$ for absence of ozone [Uchino *et al.* (1979) Reprinted with permission; copyright 1979, IEEE].

with R for the UV DIAL system is shown in Figure 6.32. Note that the O_3 absorption is measured at essentially a single wavelength and that, in this case, deviations of O_3 concentration require that the background scattered component be estimated. This data can be obtained by separate LIDAR measurements at a wavelength that is not absorbed by O_3.

Some data on atmospheric O_3 concentrations obtained with a TEA-type CO_2 laser DIAL system are shown in Figure 6.33 [Asai *et al.* (1979)]. These data are compared with a point measurement made at a range of 1.5 km. An improved CO_2 laser DIAL system has been described by Bufton *et al* (1979). This system is shown in block diagram form in Figure 6.34.

The laser output energy is divided equally between lines at two separate wavelengths. A multielement array of HgCdTe photodiodes is used to provide simultaneous monitoring of the backscatter at both wavelengths. These signals are computer processed to yield data on O_3 concentrations. By monitoring both wavelengths simultaneously, wavelength-independent noise can be removed.

The detection of atmospheric NO_2 with DIAL has been reported by Grant *et al.* (1974), Rothe *et al.* (1974), Takeuchi *et al.* (1978), and Fredriksson *et al.* (1979). This molecule is well suited to remote sensing with visible laser radiation since it absorbs extensively in this region. An example of a DIAL probe of the effluent from an oil refinery smokestack is shown in Figure 6.35 [Fredriksson *et al.* (1979)]. The plume was located 2 km away from the transmitter. The two traces show that the beam whose wavelength is absorbed by NO_2 is decreased in amplitude at ranges exceeding 2 km. Similar returns were

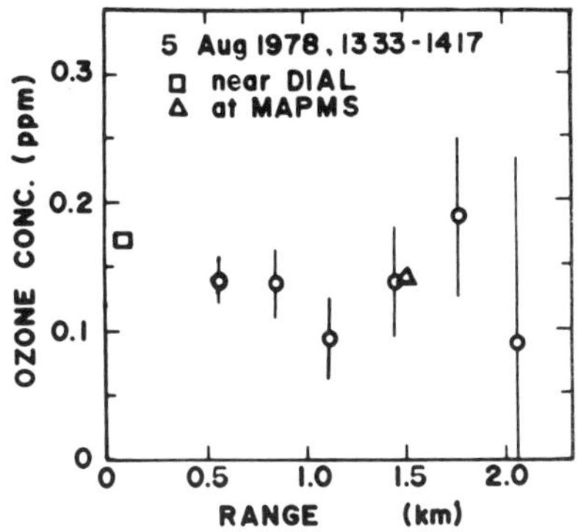

Figure 6.33. The range-resolved ozone concentration with the CO_2 laser radar (MAPMS = point measurement with separate instrumentation) [Asai *et al.* (1979)].

obtained by Grant *et al.* (1974) using a standard NO_2 cell at a distance of 365 m. A single pair of laser pulses (one on an absorption line, the other off the line) was sufficient to provide detection of NO_2 with an uncertainty equivalent to 0.05 km ppm.

Takeuchi *et al.* (1978) have discussed detection limits for NO_2 measured via DIAL. Daytime detection was assumed using a laser system with the following characteristics:

$$E = 100 \text{ mJ per pulse}, \quad A = 0.07 \text{ m}^2, \quad \xi = 0.2, \quad \eta = 0.1,$$
$$\lambda_1 = 448.1 \text{ nm}, \quad \lambda_2 = 446.8 \text{ nm},$$
$$\sigma_1 = 4.82 \times 10^{-19} \text{ cm}^2, \quad \sigma_2 = 2.59 \times 10^{-19} \text{ cm}^2$$

The efficiency of the transmitting optical system was taken to be 0.5. Two cases were treated: In the first, a uniform NO_2 distribution was assumed. The second assumed that the NO_2 pollutant was concentrated in a localized region of

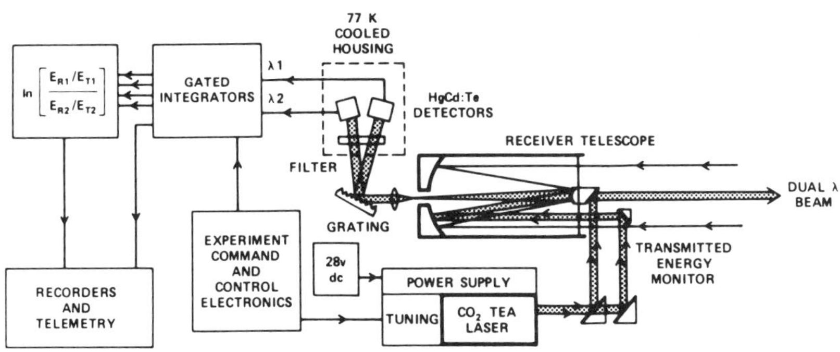

Figure 6.34. CO_2 DIAL system components [Bufton *et al.* (1979)].

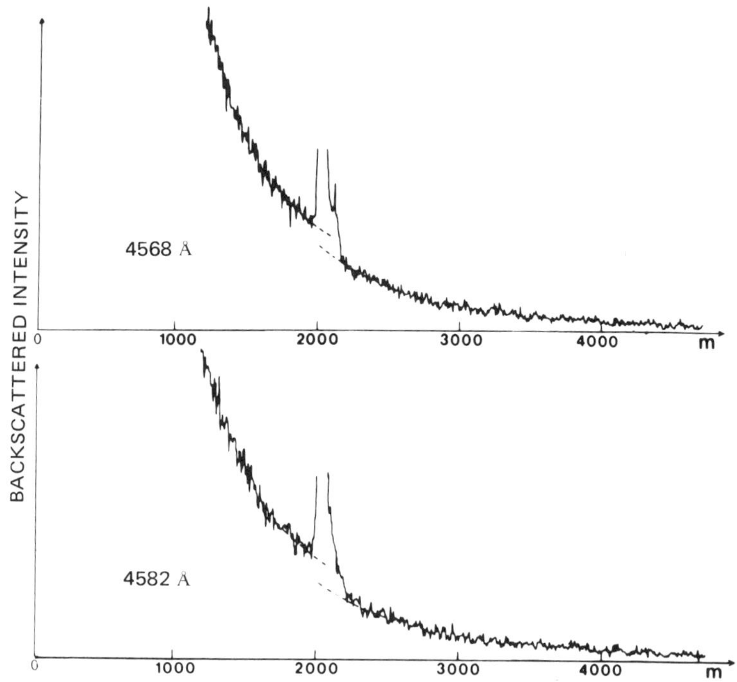

Figure 6.35. Measurement of NO_2 content in the plume from an oil-refinery stack situated 2 km from the remote sensing system. Two LIDAR curves for two close-lying wavelengths with different absorption cross sections are shown. A NO_2 concentration of 30 ± 10 mg m^{-3} through the 6-m thick plume is inferred [Fredriksson *et al.* (1979)].

length L at range R. This would correspond to the geometry obtained for emission from a source such as a smokestack. Detectivity was expressed in terms of the number of pulse pairs required for detection of NO_2 at the 1-ppm level with 10% error. A pulse pair consists of one pulse transmitted at λ_1 and another at λ_2. Their calculated results for both cases are shown in Figure 6.36. Detection limits were calculated under conditions of given visibility V. It was found that in all cases 1 ppm of NO_2 can be detected at $R = 1$ km in 10 pulse pairs.

Calculation of detection limits for NO_2 in ambient atmospheres using DIAL have also been reported by Byer and Garbuny (1973).

DIAL measurements of SO_2 concentrations have been discussed by Grant and Hake (1975), Baumgartner and Byer (1978a,b), and Fredriksson *et al.* (1979). Baumgartner and Byer used a parametric oscillator to generate tunable radiation within the 4-μm fundamental vibrational band of SO_2, while Grant and Hake and Fredriksson *et al.* used frequency-doubled dye laser radiation to sample SO_2 within its UV absorption band starting at 330 nm. A comparison of these two systems is given in Table 6.8. Detection limits were in the

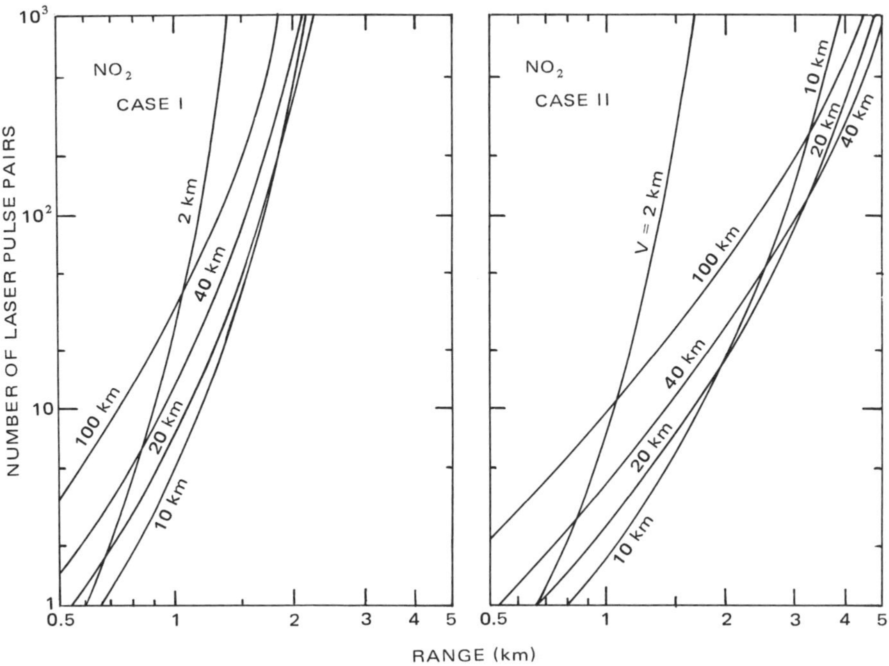

Figure 6.36. Number of laser pairs vs. the detectable range with a parameter of visibility under the conditions that the error ϵ = 10%, the range resolution L = 120 m, and the minimum concentration of 1 ppm for the laser energy of 100 mJ per shot in the daytime measurement: (a) in case I; (b) in case II. Case I—uniform NO_2 distribution; case II—NO_2 concentrated in distance L at range R [Takeuchi *et al.* (1978)].

Table 6.8. DIAL Systems for Measurement of SO_2 in IR and UV

	IR	UV
Laser	YAG pumped $LiNbO_3$	Flashlamp pumped Rh7G doubled in ADA
Wavelength	3.985, 3.998 μm	292.3, 293.3 nm
Pulse energy	0.6 mJ	0.8 mJ
Repetition rage	10 Hz	0.017 Hz
Beam divergence	0.63 mrad	0.7 mrad
Telescope diameter	40 cm	13 cm
Detector	InSb	RCA 7200
Detection sensitivity	0.9 ppm–km	0.6 ppm–m
Reference	Baumgartner and Byer (1978a)	Grant and Hake (1975)

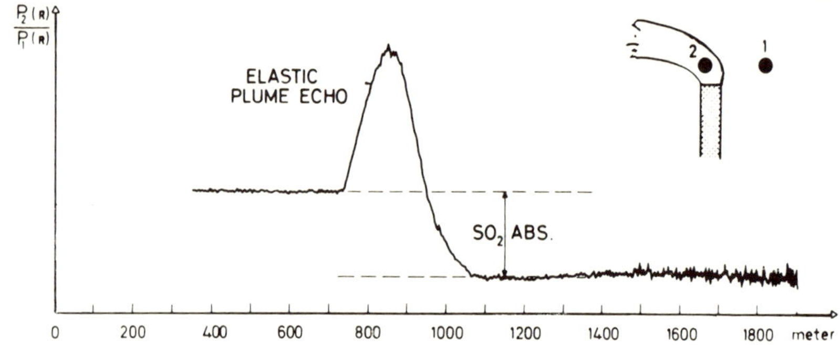

Figure 6.37. LIDAR measurement of SO_2 emission from an oil-burning power station. Two LIDAR curves, measured through and beside the smoke plume, have been divided displaying the smoke-particle scattering as well as the attenuation due to SO_2. A SO_2 concentration of 760 ± 150 mg m^{-3} was inferred across the 4.5-m diameter plume [Fredriksson *et al.* (1979)].

ppm–km range although both groups noted that relatively simple system improvements could reduce this limit by perhaps an order of magnitude.

Figure 6.37 shows a LIDAR probe of SO_2 emitted from a smokestack at a range of 800 m. The absorption produced by SO_2 is evident, as is the scattering due to the aerosol component in the ejecta. Spatial resolution was limited by the 1-μs duration of the probing laser pulse.

A summary of DIAL measurements made with IR laser sources is given in Table 6.9. In most of these studies the DIAL measurement was made using

Table 6.9. DIAL Detection of Atmospheric Species Using IR Laser Sources

Molecule	Laser	Wavelengths[a] (μm)	Detectivity	Reference
CO	Doubled CO_2	4.641 4.635	0.6 ppm–500 km	Killinger *et al.* (1980)
H_2O	CO_2	10.247 10.260		Murray *et al.* (1976a)
N_2O	DF	3.8903	0.24 ppm–km	Murray *et al.* (1976b)
N_2O	DF	3.8903 3.8548		Altmann *et al.* (1980)
C_2H_4	CO_2	10.532 10.551	1 ppb	Murray and van der Laan (1978)
CH_4	$LiNbO_3$ parametric oscillator	1.66	4 ppm–km	Baumgartner and Byer (1978)
CH_4	DF	3.7153	6 ppm–km	Murray *et al.* (1976b)
HCl	DF	3.6362	0.05 ppm–km	Murray *et al.* (1976b)
O_3	CO_2	9.505 9.586	0.1 ppm–km	Asai *et al.* (1979)

[a]First entry corresponds to line used for absorption; second entry is that used for background.

the return obtained from topographical features (buildings, trees, hillsides, clouds, etc.) rather than that backscattered directly from the atmosphere. As such, these experiments are a hybrid between DIAL and long-path absorption measurements. An example of an experimental system of this kind is shown in Figure 6.38. Using this system, Murray and van der Laan (1978) were able to detect ~ 1 ppb of ethylene in air over a path length of 5 km. Repetitive pulsing at wavelengths on resonance and off resonance with a line from ethylene produced the series of backscattered signals shown in Figure 6.39. Part of the observed difference between returns at *P*(14) and *P*(16) wavelengths arises from H_2O absorption. A correction for this effect was incorporated in the derivation of C_2H_4 concentrations.

Possible improvements in the measurement of atmospheric constituents with the DIAL technique using heterodyne detection have been discussed by Inaba and Kobayasi (1975) and Rye (1978). Significant advantages over direct detection methods appear to be possible in the infrared. These derive from the insensitivity of heterodyne detection systems to background radiation. This would be particularly important for daytime measurements. Inaba and Kobayasi conclude that improvements in sensitivity of up to several orders of mag-

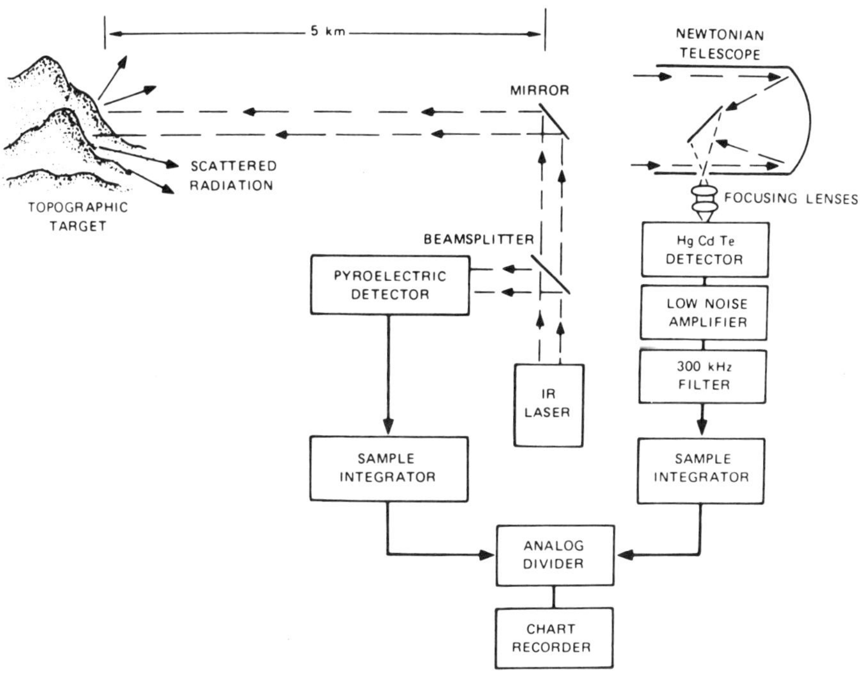

Figure 6.38. Infrared LIDAR system with topographical scatterer [Murray and van der Laan (1978)].

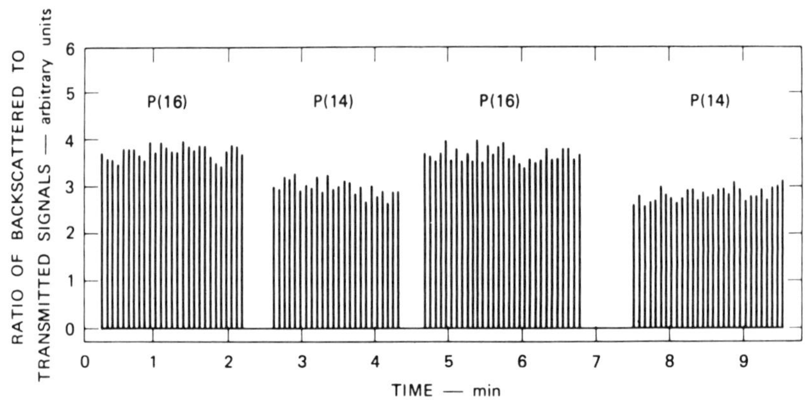

Figure 6.39. Ratios of backscattered to transmitted signal obtained on day of high ethylene concentration [Murray and van der Laan (1978)].

nitude may be attainable with heterodyne detection of DIAL backscatter. However, the heterodyne technique places stringent requirements on system alignment that may inhibit wide application.

6.2.6. Resonance Fluorescence

The excitation and detection of resonance fluorescence in laboratory systems has been described in some depth in Section 5.3. The same principles apply to the remote sensing of atmospheric species. A laser is tuned to the wavelength of an absorption feature in the atom or molecule of interest. As the beam propagates through the atmosphere it is absorbed by these species which then fluoresce. The time-dependent fluorescent backscatter can be monitored with a telescope and then used to infer the concentration profile of the absorber along the laser propagation path.

Differential cross sections for resonance fluorescence, $d\sigma/d\Omega$, can be as large as 10^{-13} cm^2 sr^{-1} for an electronic transition in an atom in the absence of collisional quenching. For a vibronic line in a polyatomic molecule $d\sigma/d\Omega \sim 10^{-18}$ cm^2 sr^{-1} under the same conditions. An infrared line would have $d\sigma/d\Omega \sim 10^{-19}$ cm^2 sr^{-1}. Of course, these cross sections are many orders of magnitude larger than those for Raman scattering.

Range resolution with the resonance fluorescence (RF) technique depends on the lifetime of the emitter. Spontaneous radiative lifetimes vary from $\tau_F \sim 10^{-9}$ s for an electronic transition in the near UV to 10^{-1}–10^{-2} s for IR transitions. In sampling atmospheric constituents, lifetimes exceeding the collisional relaxation time τ will not be seen. Collisional lifetimes may be as small as 10^{-9} s for electronic transitions. Collisional quenching of vibrational energy yields 10^{-4}–10^{-6}s as the lifetime of a vibrationally excited molecule at atmospheric pressure. The range resolution with RF then corresponds to $\Delta R = C(\tau + \tau_p)/2$ for infrared transitions and to $\Delta R = C(\tau + \tau_p + \tau_F)/2$ for electronic transitions, where τ_p is the pulse length. Thus RF gives poor range resolution when excitation occurs with IR lasers.

Excitation at a particular wavelength often leads to fluorescence over a range of wavelengths. It may often be advantageous to detect the backscattered radiation off resonance at one or more of these fluorescent wavelengths [Gelbwachs (1973), Solomon (1977), and Solomon and Silva (1976)]. The LIDAR equation for detection of fluorescence over a bandwidth ΔB is [Gelbwachs (1973)]

$$\phi_F = \frac{E}{h\nu} N\sigma \,\Delta R\, Q \frac{A}{4\pi R^2}\, \xi \frac{\Delta B}{B}\, T$$

where E is the laser pulse energy, $h\nu$ the photon energy (transmitted beam), N the concentration of fluorescent molecules, σ the absorption cross section at $h\nu$, ΔR the range resolution, Q the quenching factor for fluorescence at 1 atm, A the receiver area, R the range, ξ the efficiency of optical detection system, B the wavelength range of fluorescent emission, and T the atmospheric transmission. ϕ_F is the number of fluorescent photons received at the photomultiplier tube per laser pulse.

The corresponding expression for photons received from the sky background is

$$\phi_s = \frac{N\lambda}{h\nu} A\Omega\, \Delta B \frac{2\,\Delta R}{c}\, \xi T$$

where Ω is the solid angle of the receiver and c is the speed of light. Gelbwachs used the following values for system parameters to estimate detection limits for NO_2: $E/2 = 0.1$ J, $\lambda = 450$ nm, $\sigma = 3 \times 10^{-17}$ cm^2, $A = 0.1$ m^2, $\Omega = 3 \times 10^{-8}$ sr, $\xi = 0.1$, $\lambda_F = 650$ nm, $\Delta B = 100$ nm, $\Delta B/B = 0.2$, $N_\lambda = 10$ W m^{-2} sr^{-1} (day), and 10^{-2} W m^{-2} sr^{-1} (night). Atmospheric attenuation was neglected. The limits in Table 6.10 were obtained for NO_2 detection at the S/N = 3 level.

Detection of atomic Na in the atmosphere using a tunable laser and RF has been reported by Bowman *et al.* (1969). A Na column density of $\sim 6 \times 10^9$ cm^{-2} was derived from observation of RF by Na atoms at $\sim$ 100-km height.

McGee and McIlrath (1979) have discussed the use of laser-induced fluorescence to detect stratospheric OH. It is shown that the ambient pressure can be determined from a measurement of the ratio of fluorescence in the (0,0) band to that in the (1,0) band. Temperature profiles can be obtained by excitation of transitions from two separate rotational levels. This is possible since rotational equilibrium is established over a time scale that is significantly shorter than the lifetime of OH in the atmosphere. Figure 6.40 shows the ratio of fluorescence excited from two rotational levels as a function of ambient temperature.

The detection of atmospheric OH in a point measurement using RF was first reported by Wang and Davis (1974). The detection was made at a level of 5×10^{-12} by volume. Other point measurements of atmospheric species using RF have been reported by Tucker *et al.* (1975), Solomon and Silva (1976), and Halpern *et al.* (1979). Tucker *et al.* used a CW 10-mW He–Cd laser to monitor NO_2 at the 0.6-ppbv level in ambient air. Solomon and Silva studied the fluorescence produced by N_2^+ ions created in air by an x-ray source. N_2^+ was excited at 391.4 nm in the (0–0) $X \rightarrow B$ transition. Fluorescence in the $B \rightarrow X$ (0–1) transition was detected. Halpern *et al.* (1979) used an ArF laser to dissociate NH_3 via a two-photon transition and detected emission from the NH fragments at 336 nm.

McIlrath *et al.* (1979) have proposed the use of two-photon absorption to excite an energetic level in atomic O. Fluorescence from this level at 844.7 nm could be used for detection. The calculation indicates that a LIDAR system emitting 1 J per pulse at 2256 Å could be used to determine O atom concentrations down to 100 km from a spacecraft at an altitude of 200 km. A range resolution of 1 km could be expected.

An interesting application of laser-induced fluorescence in the detection of ^{238}U via its decay products Pb and Bi has been investigated by Solomon (1977). It is concluded that such a technique could be used to detect Pb at the level of 5–10 atoms cm^{-3} at a range of 200 m.

Table 6.10. Limits of NO_2 Detection at S/N = 3 via Resonance Fluorescence

Range (km)	Day (ppb)	Night (ppb)
0.3	36	5
0.4	60	8
0.5	100	13
0.6	145	18
0.7	195	24
0.8	250	31
0.9	320	40
1.0	400	50

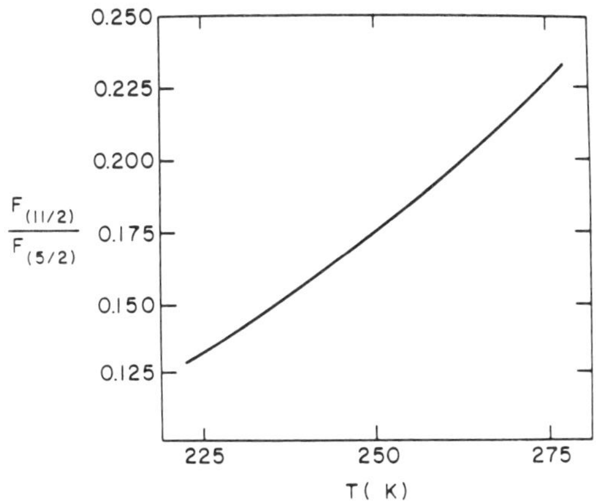

Figure 6.40. Ratio of OH fluorescence yield $F_{11/2}/F_{5/2}$ from J = 11/2 and J = 5/2 vs. temperature [McGee and McIlrath (1979)].

6.2.7. Heterodyne Detection

Several experiments have been reported on passive remote sensing of atmospheric pollutants using heterodyne detection [Menzies (1973), Menzies and Shumate (1974, 1976), and Frerking and Muehlner (1977)]. The advantage of heterodyne detection can be seen from the following simple analysis.

If two electromagnetic waves are incident on a square-law detector then the output of this detector is

$$i(t) \propto (E_1 \cos \omega_1 t)^2 + E_2 \cos(\omega_2 t + \psi) + 2E_1E_2 \cos \omega_1 t \cos(\omega_2 t + \psi)$$

where one field is $E_1 \cos \omega_1 t$ while the other has $E_2 \cos(\omega_2 t + \psi)$ and ψ is a phase angle. The component $2E_1E_2 \cos \omega_1 t \cos(\omega_2 t + \psi)$ contains terms in $\omega_1 - \omega_2$ and $\omega_1 + \omega_2$. If ω_1 and ω_2 are optical frequencies, the detector will not respond to the component at $\omega_1 + \omega_2$. However, when $\omega_1 \sim \omega_2$ the detector can respond at this frequency. The detector output at this intermediate frequency is

$$i_{IF}(t) \propto 2(\bar{I}_1\bar{I}_2)^{1/2}\cos[(\omega_1 - \omega_2)t - \psi]$$

where $\bar{I}_1$ and $\bar{I}_2$ are average intensities. Thus if I_2 is small (i.e., the signal for a remote emitter), the detected current can be increased if a strong local oscillator source is used at ω_1. In remote sensing systems a laser is used as a local oscillator to provide a local frequency close to that emitted from the remote source.

The minimum detectable power in the heterodyne detection regime is

$$P_H = 2\frac{h\nu}{\eta}B$$

where B is the intermediate frequency bandwidth and η is the detector quantum efficiency. This should be compared to the minimum detectable power for direct detection

$$P_D = (2B)^{1/2}\ \text{NEP}$$

where NEP is the noise equivalent power (W $\text{Hz}^{-1/2}$). The frequency dependence of P_H means that heterodyne detection is favored at low frequencies, i.e., in the infrared.

For example, assuming $B = 10$ MHz and a detector with an NEP of 10^{-11} W $\text{Hz}^{-1/2}$ at $\lambda = 10.6\ \mu\text{m}$ with a quantum efficiency of 0.2, one has $P_H = 6.2 \times 10^{-13}$ W and $P_D = 4.5 \times 10^{-8}$ W, both for a 1-s integration time. This situation is reversed for wavelengths $\lesssim 1\ \mu\text{m}$. The greatest advantage of heterodyne detection lies at wavelengths $\lambda \lesssim 5\ \mu\text{m}$.

The S/N ratio for a Dicke-type radiometer with heterodyne detection looking at a radiating slab of gas at temperature T with emissivity $\epsilon(\nu)$ is [Menzies and Shumate (1974)]

$$\frac{\text{S}}{\text{N}} = \frac{1}{4}\frac{\eta(\tau/B)^{1/2}}{\left[\exp\left(\dfrac{h\nu}{kT}\right)-1\right]}\int_{\nu_{\text{LO}}-B}^{\nu_{\text{LO}}+B}\epsilon(\nu)\,d\nu$$

where ν_{LO} is the local oscillator frequency and τ is the post intermediate frequency integration time.

Figure 6.41 shows a heterodyne radiometer used to detect emission from atmospheric ozone at 9.9 μm [Frerking and Muehlner (1977)]. A tunable diode laser acts as the local oscillator. The IF bandwidth $B = 100$ MHz and $\tau = 1$ s. The spectroscopic resolution was 0.0067 cm^{-1}.

Heterodyne radiometers based on lines from the CO and CO_2 lasers have been developed for monitoring the atmosphere [Menzies (1973) and Menzies and Shumate (1974, 1976)]. A summary of experimental sensitivities for various molecular species detected with a system of this type is given in Table 6.11. Sensitivities for gases such as SO_2 and NO are $\sim 10^{-2}$ atm cm or 3×10^{17} cm^{-2} while those for O_3, NH_3, and C_2H_4 are $\sim 10^{-4}$ atm cm or 3×10^{15} cm^{-2}. Note that a sensitivity of 3×10^{15} cm^{-2} is equivalent to 1 ppm over a path length of 1 m. These limits are easily obtained with integration times of $\sim$ 30 s and local oscillator powers of 10–50 mW. They are typically within a factor of $\sim$ 3 of theoretical sensitivities.

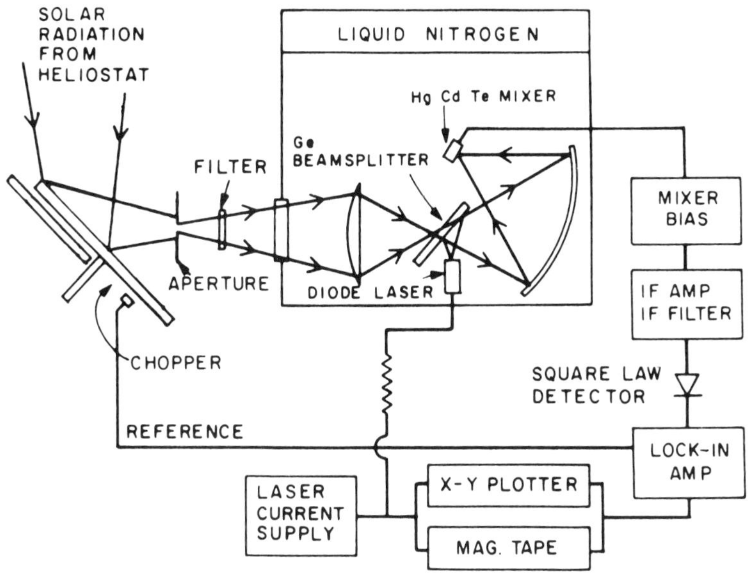

Figure 6.41. Apparatus used for atmospheric ozone measurements [Frerking and Muehler (1977)].

Table 6.11. Sensitivities for the Detection of Pollutant Molecules Using a Heterodyne Radiometer with Fixed Local Oscillator Frequency[a]

Gas	Laser	Line	Wavelength (μm)	Sensitivity (atm cm)
NO	$^{12}C^{16}O$	7-6, $P(15)$	5.19	10^{-2}
SO_2	$^{12}C^{18}O_2$	$R(40)$	9.024	10^{-2}
O_3	$^{12}C^{18}O_2$	$P(40)$	9.50	2×10^{-4}
	$^{12}C^{16}O_2$	$P(14)$	9.50	2×10^{-4}
C_2H_4	$^{12}C^{16}O_2$	$P(14)$	10.53	5×10^{-5}
NH_3	$^{12}C^{16}O_2$	$P(32)$	10.72	10^{-4}
SO_2[b]	$^{12}C^{18}O_2$	$R(40)$	9.024	10^{-2}
CO_2[b]	$^{12}C^{16}O_2$	$P(20)$	10.588	1

[a]All gases are at 290 K, except NO which was at 390 K. From Menzies and Shumate (1974); [b]Menzies (1973).

6.3. LASER SAMPLING OF AEROSOLS

In this section we discuss some of the laser-based systems that have been developed for the chemical analysis and characterization of aerosol particles. The discussion will be limited to techniques that are involved in point sampling; remote sensing methods have been discussed in the previous section.

In general, the complete description of an aerosol requires measurement of the following parameters:

Particle size and size distribution.
Chemical composition of individual particles.
Particle shape.
Particle refractive index and optical properties.

Much of this information can be obtained by studying the way in which light interacts with aerosol particles. It is this property that has ensured the widespread application of lasers in this area.

6.3.1. Particle Size and Distribution

The geometry shown in Figure 6.42 can be used to obtain particle size when the particle refractive index is known. The scattered light intensity is measured as a function of scattering angle and compared with that predicted from the Mie scattering theory for particles of the given refractive index. Usually a close fit to the experimental scattering curve can only be obtained over a limited size range providing the aerosol is monodisperse, i.e., all particles have the same size. When a range of particle sizes is present, derivation of the size

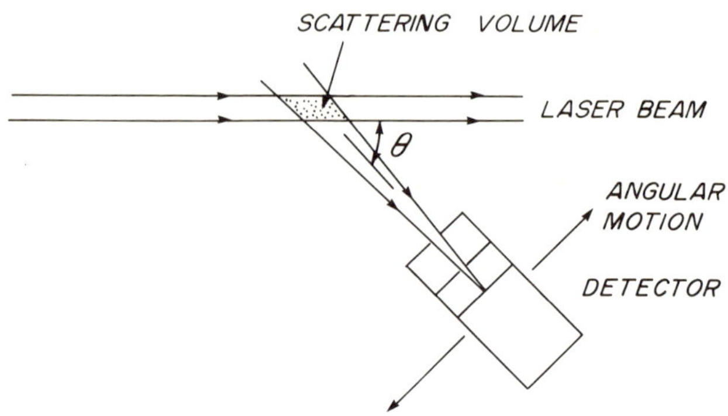

Figure 6.42. Basic laser scattering system for the determination of aerosol particle size.

distribution is possible, in principle, but the techniques for doing so are tedious at best.

The power scattered from a beam of intensity I_0 at angle θ is

$$P(\theta) = I_0 N V \sigma(\theta) \Delta\Omega$$

where N is the number of scatterers per unit volume, V the scattering volume seen by the detector, $\sigma(\theta)$ the differential scattering cross section at angle θ, and $\Delta\Omega$ the solid angle subtended by the detector. Burns *et al.* (1976) have reported some data on the scattering from a suspension of 2.02-μm diameter PVT latex spheres using the scattering geometry shown in Figure 6.42. Their plots of $\sigma(\theta)$ vs. θ at two wavelengths are shown in Figure 6.43 and compared with the results of Mie scattering calculations. Note the close agreement between theory and experiment, particularly in the reproduction of oscillatory behavior in $\sigma(\theta)$ for off-axis scattering. It is evident that an excellent estimate of particle size can be obtained under these rather controlled conditions.

Scattering of laser radiation from individual submicron particles has been reported by Phillips *et al.* (1970). Measurements were performed on particles whose positions were stabilized in the laser beam by use of an electric field. It was found that both particle refractive index and radius could be determined by fitting Mie calculations to the observed angular distribution of scattered radiation.

The effect of particle shape on the angular dependence of scattered light has been investigated by Pinnick *et al.* (1976). The system involved a measurement of the angular dependence of laser radiation scattered from a flowing sample of monodisperse aerosol. An example of the data obtained, together with best-fit Mie scattering calculations, are shown in Figure 6.44. Data are presented in terms of components of the Stokes scattering matrix [van de Hulst (1957)]. Defining a size parameter ρ equal to particle circumference divided by wavelength, it was found that irregular particles of NaCl or K_2SO_4 produced more forward scattering than predicted from the Mie theory when $\rho \gtrsim 5$. Better agreement with the Mie theory was obtained when $\rho < 5$. Angular scattering patterns obtained experimentally generally showed less structure than predicted from the Mie theory for spherical particles of equivalent size.

Carlon *et al.* (1980) have used a He–Ne laser to derive mass concentrations of monodisperse aerosols. This quantity is defined as

$$C = \frac{1}{\gamma_\lambda L} \ln \left[\frac{1}{T_\lambda} \right]$$

where C is the mass concentration (g m^{-3}), γ_λ the mass extinction coefficient (m^2 g^{-1}), L the optical path, and T_λ the optical transmittance of aerosol at wavelength λ. For liquid aerosols with particle diameters $D \gg \lambda$, the above expression simplifies to

$$C = \frac{\mathrm{D}d}{3L} \ln \left[\frac{1}{T_\lambda} \right]$$

where d is the particle density. Several applications of this technique to the real-time monitoring of aerosol properties were discussed.

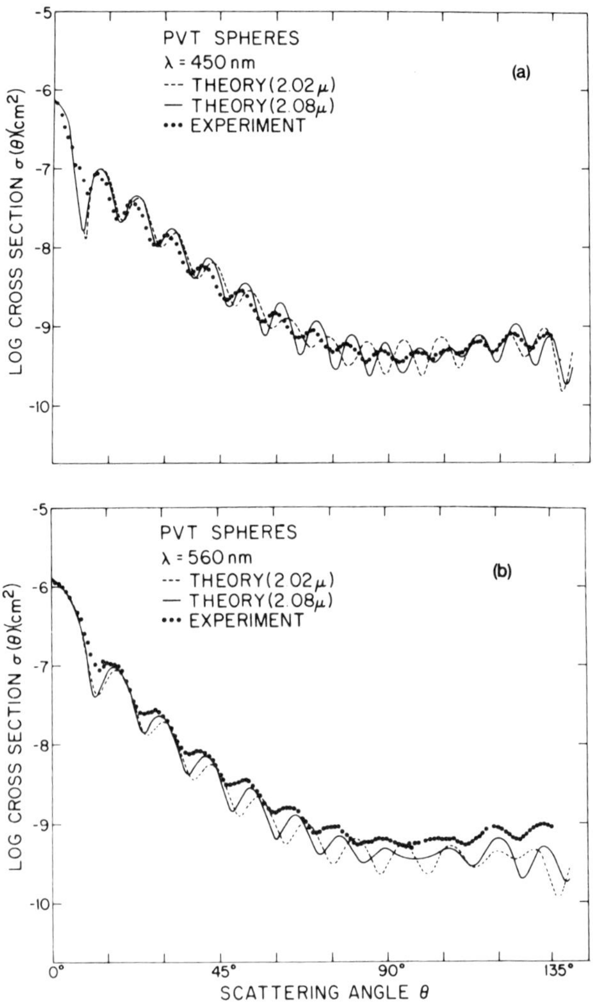

Figure 6.43. Measured differential scattering cross sections and comparison theory: (2) $\lambda = 450$ nm; (b) $\lambda = 560$ nm [Burns *et al.* (1976)].

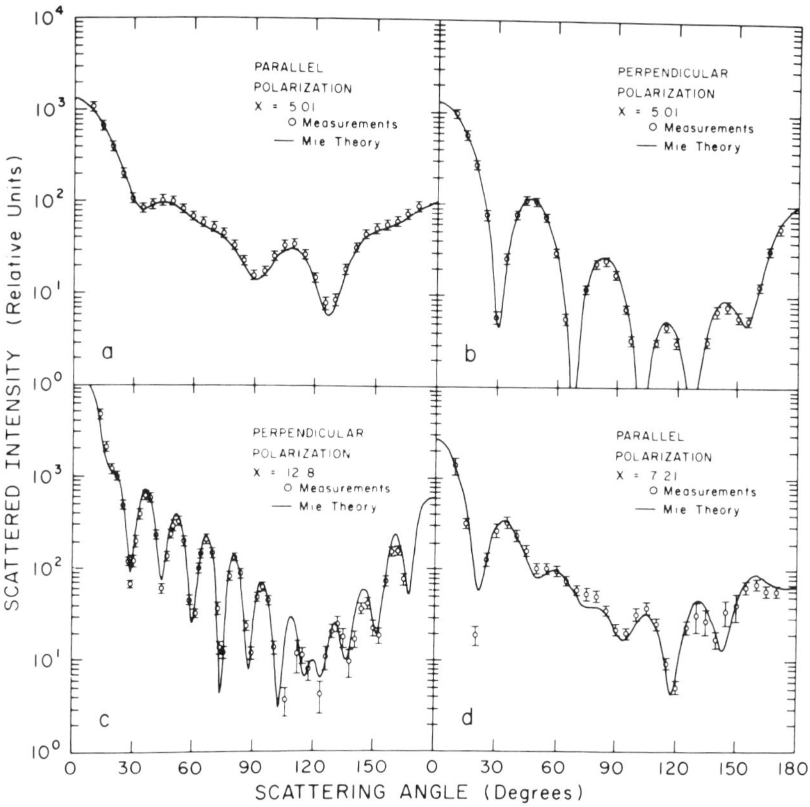

Figure 6.44. Measurements (circles) and Mie theory (smooth curves) for the matrix element S_{11} and S_{22} vs. scattering angle from the direction of forward scattering for uniform particles of potassium sulfate. Also shown in (b) are measurements of the particle depolarization corresponding to the matrix element S_{12}. Electron microscope micrographs of the particles are presented with the corresponding angular scattering measurements. The solid curves are Mie theoretical results for homogeneous spheres of equal cross-sectional areas as determined by electron microscope and for the refractive index of potassium sulfate: 1.49−0i. [Pinnick *et al.* (1976)].

There have been many studies of aerosol particle size using laser anemometer techniques [Farmer (1972), Robinson and Chu (1975), Kirsch and Mazumder (1975), Hong and Jones (1976), dos Santos and Stevenson (1977), and Adrian and Orloff (1977)]. Some basic laser anemometer methods have been described in Section 6.2.3. Information on particle velocity components can be obtained by analysis of the light scattered when an aerosol particle passes through a region of space in which interference fringes are present. Farmer (1972) showed that the light scattered by a particle under these conditions also can be used to derive particle size. Calculations showed that the fringe visibility in the scattered signal was determined by the particle size and space density as well as by particle shape. It was found that when fringe separation was comparable to particle diameter, the particle size could be estimated. When this separation was much greater than particle diameter, data could be obtained on the space density of scatterers. Further calculations on this effect and experimental verification were reported by Robinson and Chu (1975). Hong and Jones (1976) used an Ar^+ laser to sample a variety of artificially generated aerosols with particle sizes in the 1–10 μm range. Size distributions were measured with an accuracy that was comparable to that obtained via direct measurement with an optical microscope.

An experimental study of the visibility of signal fringes produced by particle scattering in a laser anemometer by Adrian and Orloff (1977) showed that visibility depended on the ratio of particle diameter to fringe spacing only under limited conditions. More generally, it was found that visibility depends on the Mie scattering parameter, the refractive index, beam polarization, and the size, shape, and location of the light-collecting aperture. Figure 6.45 shows some examples of signal fringes produced by sampling 9.8-μm diameter polystyrene particles with interferometer fringes of different spacing. These were produced with 488-nm radiation from an Ar^+ laser. Good agreement between

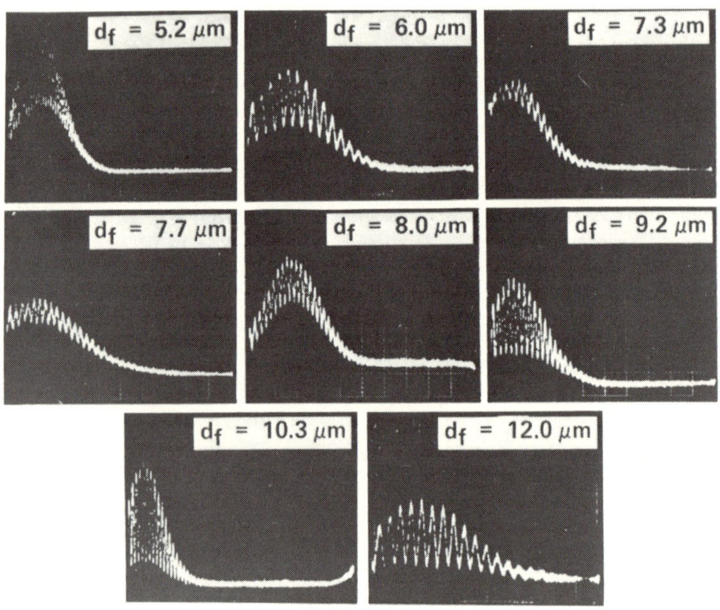

Figure 6.45. Burst signals produced by 9.8-μm particles for fringe spacings ranging from 5 to 12 μm [Adrian and Orloff (1977)].

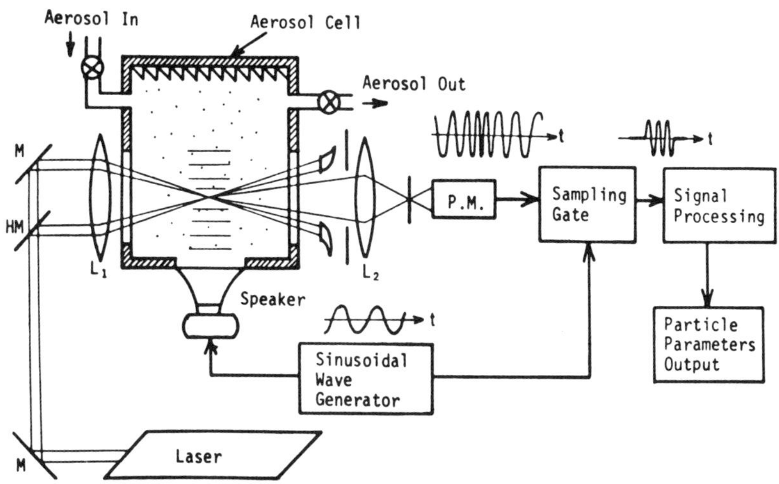

Figure 6.46. Schematic construction of laser Doppler velocimeter [Sato *et al.* (1978a)].

theoretical and empirical visibilities was found in this case for (particle diameter)/(fringe spacing) in the range 1–2.

A novel modification of the laser interferometer for sampling particle sizes in systems of fluorescent particles is described by dos Santos and Stevenson (1977). It is shown that monitoring the periodic fluorescence from particles moving through a series of interferometer fringes obviates many of the problems associated with measurements of scattered light intensity.

Acoustic modulation of particle velocity in a laser Doppler velocimeter has been utilized by Sato *et al.* (1978a) and Kirsch and Mazumder (1975) to derive particle sizes. The system used by Sato *et al.* is shown in Figure 6.46.

Kirsch and Mazumder show that the ratio of particle velocity v_p to that of the medium u_g can be written

$$\frac{v_p}{u_g} = \frac{1}{(1 + \omega^2\tau_p^2)^{1/2}}$$

where ω is the acoustic frequency and τ_p is the dynamic relaxation time of the particle due to the viscosity of the medium. When $\omega\tau_p \sim 0$, the particles oscillate in conjunction with the medium. v_p/u_g decreases as $\omega\tau_p$ increases until at high frequency the particles do not follow the motion of the medium. At intermediate frequencies the relative amplitude of oscillation of different particles can be related to their size since $\tau_p\ \alpha(\text{diameter})^2$. Sato *et al.* show that this information can be deduced from the power spectrum of the signal. This is

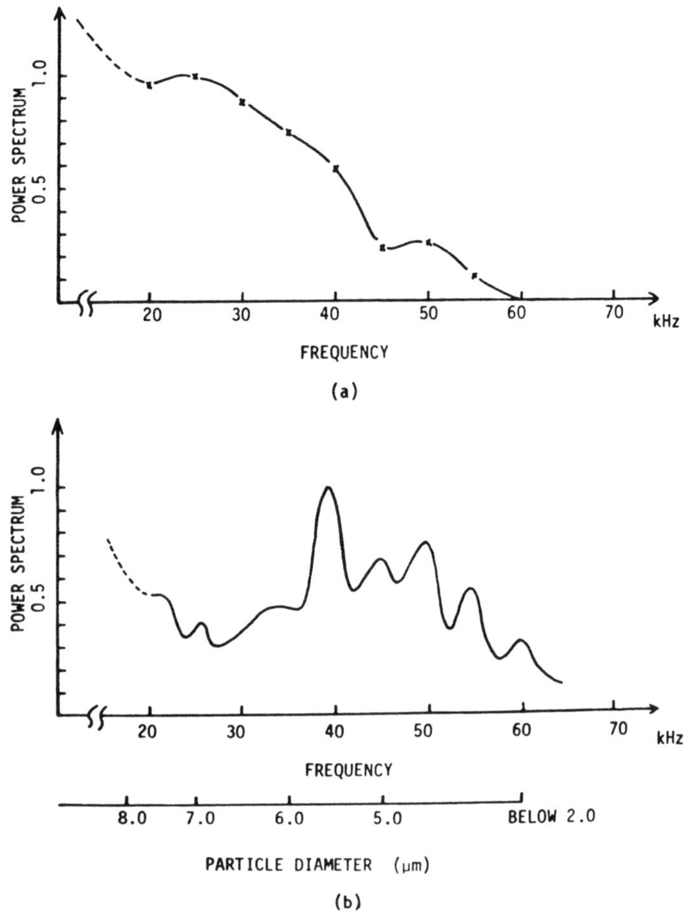

Figure 6.47. Experimental results for water droplets: (a) background-eliminated power spectrum; (b) restored power spectrum. The scale at the bottom of (b) shows the diameter of water droplet particle for corresponding Doppler frequency [Sato *et al.* (1978a)].

shown in Figure 6.47 for water droplets. Data for this spectrum were taken with the system shown in Figure 6.46. The acoustic oscillation was at 1 kHz and sound pressure was at 120 dB.

6.3.2. *Particle Composition*

Information on the chemical composition of aerosol particles can be obtained optically by measurement of Raman spectra [Rosen *et al.* (1978)] or infrared absorption spectra [Schleusener *et al.* (1975)]. These methods, of course, supplement conventional wet chemical or spectroanalytical techniques. Recently, considerable emphasis has been placed on the use of optoacoustic or spectrophone spectroscopy to determine aerosol composition by measurement of infrared absorption [Schleusener *et al.* (1975, 1976), Terhune and Anderson (1977), and Yasa *et al.* (1979)].

The advantages of optoacoustic detection include (i) real-time analysis and (ii) the ability to sample small quantities of material. With respect to point (ii), Schleusener *et al.* (1976) report on the detection of $\sim 3 \times 10^{-6}$ g of soot with a differential spectrophone. An example of the absorption spectrum of atmospheric dust measured with lines from a CW CO_2 laser and a differential spectrophone is shown in Figure 6.48 [Schleusener *et al.* (1975)].

6.3.3. Interaction of High-Power Laser Radiation with Aerosol Particles

At high intensity the energy absorbed by individual aerosol particles may be sufficient to induce normal or even explosive evaporation. More subtle changes, such as an alteration of particle charge, can also occur. The aerosol concentration present along an atmospheric path must therefore be considered as intensity dependent. The medium as a whole can be thought of as a saturable absorber with high absorption (or extinction) at low laser powers and low absorption above some critical intensity where aerosol removal occurs.

The question of aerosol evaporation initiated by the passage of high-intensity laser pulses has been extensively investigated. Interest centers on the effect of aerosol evaporation on communication links and on the possible use of high-

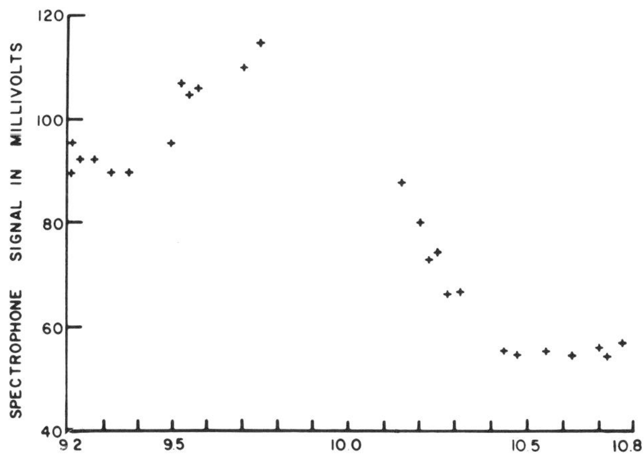

Figure 6.48. Spectrophone experimental results for atmospheric aerosol. The crosses (+) are the signals obtained at specific CO_2 laser wavelengths. The sample was present on an NaCl flat [Schleusener *et al.* (1975)].

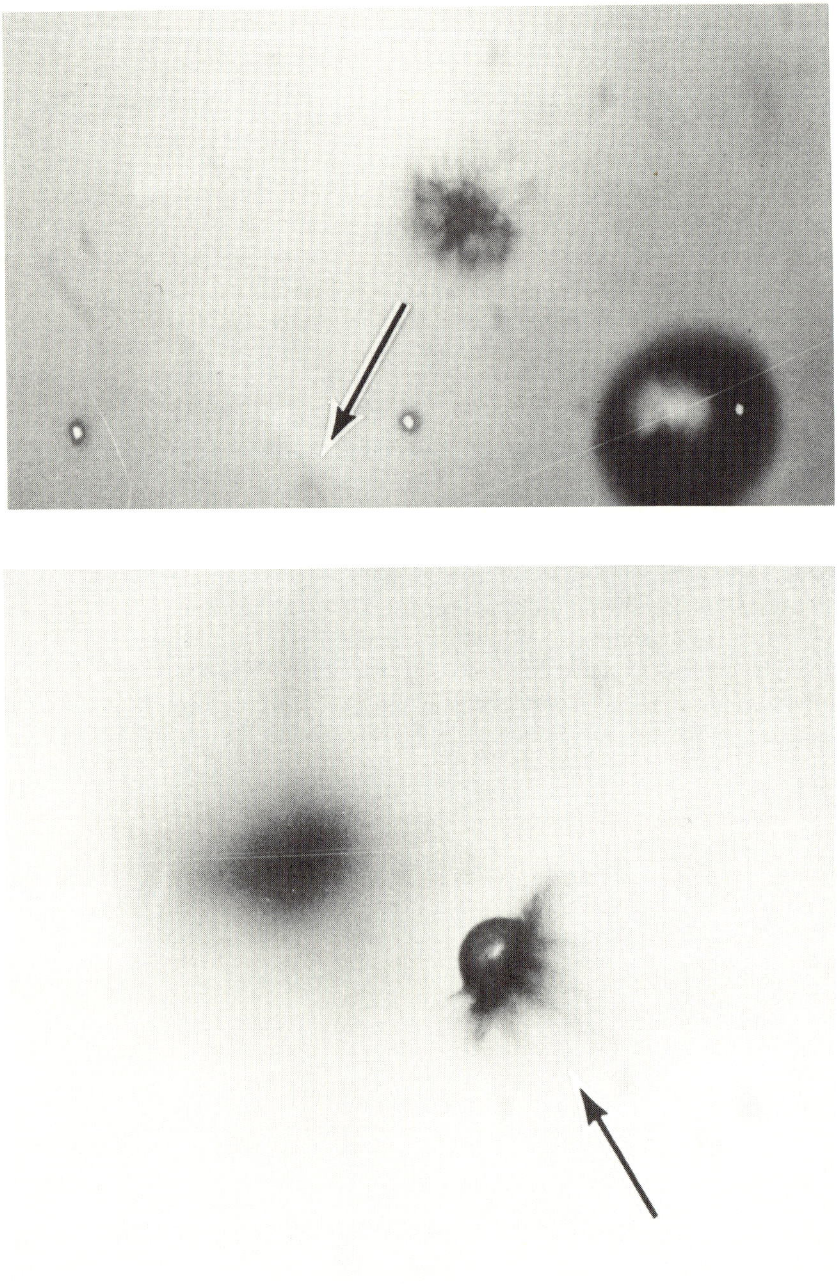

Figure 6.49. (a) A 12-μm (radius) droplet 2.0 μs after incidence of the laser pulse. (b) A 26-μm (radius) droplet 0.9 μs after incidence of the laser pulse [Kafalas and Ferdinand (1973)].

power laser precursor pulses to open up low-loss atmospheric paths [Mullaney *et al.* (1968), Lamb and Kinney (1969), Williams (1965), Lencioni and Lowder (1974), and Sutton (1978)]. Investigations of the evaporation of individual aerosol particles have been reported by Kafalas and Ferdinand (1973), Barinov and Sorokin (1973), and Weeks and Duley (1974, 1976).

Kafalas and Ferdinand found that fog particles with radii $\lesssim 10\ \mu m$ evaporated uniformly, but larger particles underwent asymmetric explosive evaporation when subjected to 300-ns pulses from a CO_2 laser. Some large particles were found to rocket away from the direction of the incident radiation driven by a jet of evaporating material. This behavior can be seen in Figure 6.49. A subsequent study showed that evaporation was accompanied by the creation of a shock wave in the ambient medium [Kafalas and Herrmann (1973)]. This shock propagates at supersonic speed during the first 2 μs after the laser pulse is absorbed.

A comprehensive study of fog hole boring with lasers by Sutton (1978) has shown that the laser energy required to burn a hole through fog is 50% greater than estimated from thermodynamic considerations when $\lambda = 10.6$ μm. This is due to scattering. Calculations of the length of path cleared by a single pulse from a CO_2 laser under various atmospheric conditions yield the data shown in Figure 6.50. It is evident that a collimated 10.6-μm beam with a fluence of 0.4 J cm^{-2} clears a 1-km path in haze. Clearing the same path length in thick fog requires $\sim$ 40 J cm^{-2}. The cleared path length is a sensitive

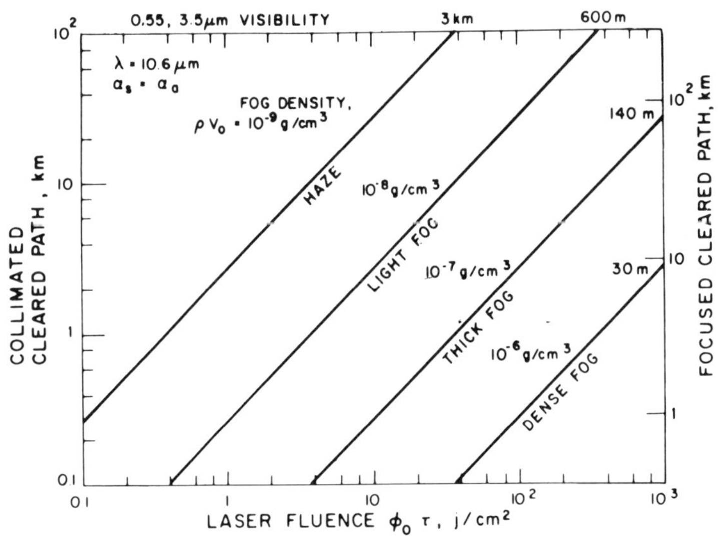

Figure 6.50. Length of fog cleared by pulsed laser for 10.6-μm wavelength and equal initial scattering and absorption coefficients [Sutton (1978)].

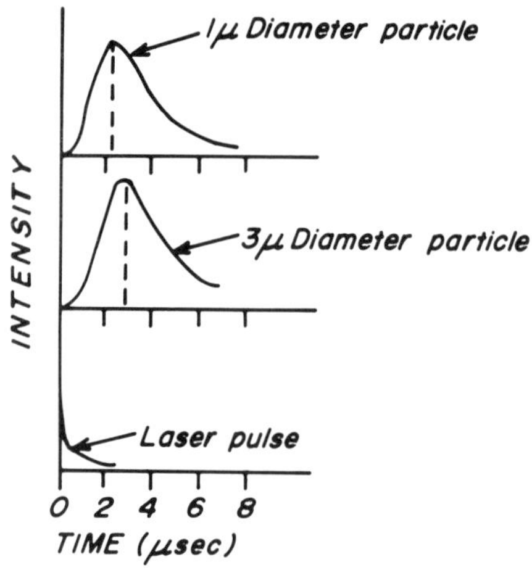

Figure 6.51. Variation of visible emission with particle radius in two monodisperse alumina aerosols [Weeks and Duley (1976)].

function of the laser wavelength since the absorption efficiency of water vapor droplets decreases considerably for $\lambda \lesssim 3.5\ \mu m$.

Sutton (1978) also shows that material evaporated from fog or haze droplets has little effect on beam attenuation due to molecular absorption. It was concluded that the material produced by particle evaporation has little effect on thermal blooming.

The interaction of high-power laser radiation with refractory aerosol particles has been discussed by Weeks and Duley (1974, 1976), Lencioni and Lowder (1974), Lencioni and Pettingill (1977), and Smith and Brown (1975).

Weeks and Duley showed that information on particle size could be obtained from an analysis of the time dependence of visible light emission from micron-sized particles heated by a pulse from a TEA-type CO_2 laser. An example of this effect is shown in Figure 6.51 for two monodisperse Al_2O_3 aerosols. By using an experimental configuration similar to that of the Millikan oil drop

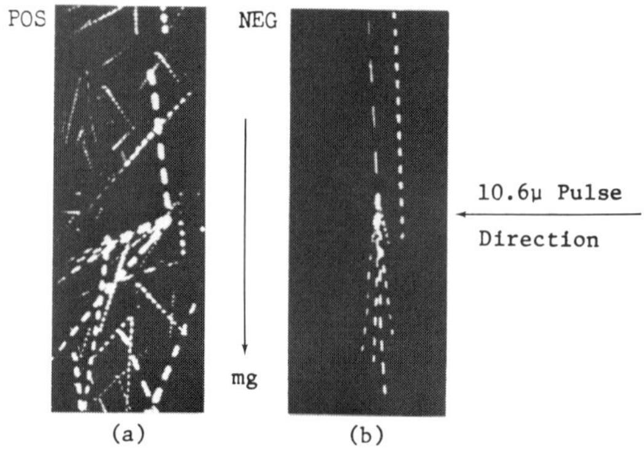

Figure 6.52. (a) Positively charged 1.6-μm radius alumina particle splits into several negative components when heated with CO_2 laser pulse. (b) Negatively charged 2.0-μm radius particle splits into negatively and positively charged particles. Particles fall between charged plates maintained at the polarity shown. Particle traces are segmented since a chopped He–Ne laser beam was used as a light source [Weeks and Duley (1976)].

experiment, the charge on individual aerosol particles before and after absorption of 10.6-μm laser radiation could be determined. It was found that the predominant effect at laser intensities $I \lesssim 10^5$ W cm^{-2} was a loss of positive charge by refractory oxide or carbon particles. This was often accompanied by particle fragmentation and mass loss (Figure 6.52).

The presence of particles along the propagation path of a high-power laser can significantly lower the threshold for air breakdown [Lencioni and Lowder (1974)]. The effect of individual ~ 50-μm particles on breakdown with CO_2 laser radiation has been investigated quantitatively by Lencioni and Pettingill (1977). It was found that 50-μm carbon and NaCl particles lowered the breakdown threshold to ~ 5×10^7 and ~ 8×10^7 W cm^{-2}, respectively. This is to be compared with ~ 3×10^9 W cm^{-2} for clean air. Smith and Brown (1975) found that the breakdown threshold intensity was relatively independent of particle size over the diameter range 1–70 μm. This threshold was also relatively independent of particle material and of laser beam diameter under conditions in which one particle was always in the beam. Focusing in ordinary atmospheric air containing aerosol particles produced the breakdown threshold dependence on beam diameter shown in Figure 6.53. The decreased threshold at large beam diameter was attributed to the presence of aerosol particles

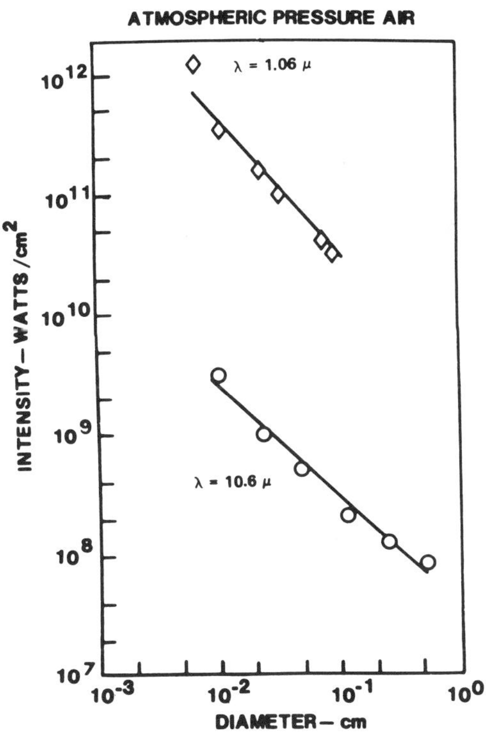

Figure 6.53. Aerosol-induced breakdown threshold vs. beam diameter for 10.6-μm wavelength and 1.06-μm wavelength [Smith and Brown (1975)].

within the beam. The dependence of breakdown on laser wavelength is not expected for particle-induced breakdown but can be understood if aerosol particles serve only to provide precursor electrons that then participate in normal cascade ionization.

6.4. LASER REMOTE SENSING OF WATER QUALITY

The high peak power and irradiance obtainable from laser sources has opened up a variety of active remote sensing applications in monitoring water quality. These properties of laser radiation have made it possible to excite and measure fluorescence and Raman spectra with portable instrumentation under ambient conditions.

Some early experiments on active remote sensing with laser sources were reported by Measures *et al.* (1973, 1974). The system shown in Figure 6.54 was used to excite fluorescence from oil spills on water. The active element in this system was an N_2 laser with a peak output power of 100 kW per pulse at a pulse repetition rate of up to 100 Hz. The pulse duration was 10 ns. Fluorescence was monitored through a broad band filter centered at 460 nm. Excitation and measurement from a single remote station overlooking a large body of water was reported at ranges of up to 1450 m (corresponding to a slant angle of 79° from the fixed station). It was shown that while a single measurement of fluorescent intensity was sufficient to indicate the presence of an oil layer, identification of the emitting species requires more sophisticated measurements. This can be seen from the fluorescent spectra of some commercial oil samples shown in Figure 6.55.

Measures *et al.* suggest that the use of dual-wavelength excitation may enable this information to be obtained. With this method the fluorescent intensity at a given wavelength is monitored as the wavelength of excitation is varied. Visser (1979) has shown that the measurement of fluorescent intensity at these wavelengths while exciting at 337 nm can be related to the composition and thickness of oil films on water.

Capelle and Franks (1979) have compared the performance of laser fluorosensors based on N_2 and KrF lasers. Performance rating was based on the ability to detect polycyclic aromatic hydrocarbons in water. It was found that the KrF system operating at 249 nm could be used for both daytime and nighttime measurements. The N_2 laser system was only effective in nighttime operation because solar radiation interferes with detection of the fluorescent signal when excitation occurs at 337.1 nm. With 249-nm excitation strong fluorescence from polynuclear aromatic hydrocarbons produced in the effluent from a coal plant occurs at 290 nm. This band lies in a spectral region where the

Figure 6.54. (a) Schematic diagram of laser fluorosensor system; (b) module in operating position [Measures *et al.* (1973)].

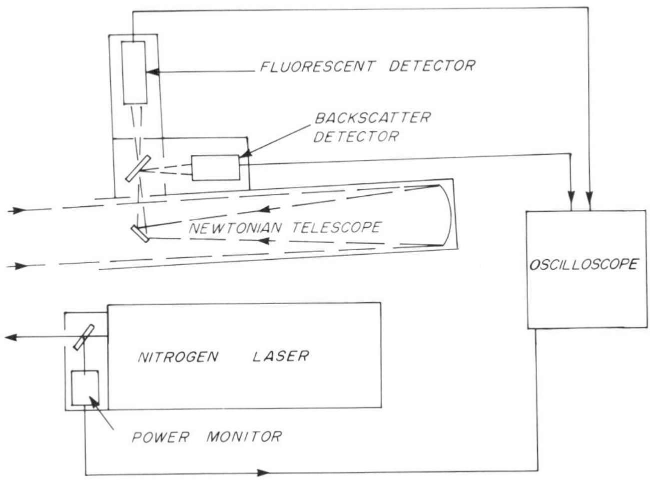
FLUORESCENT DETECTOR
BACKSCATTER DETECTOR
NEWTONIAN TELESCOPE
OSCILLOSCOPE
NITROGEN LASER
POWER MONITOR

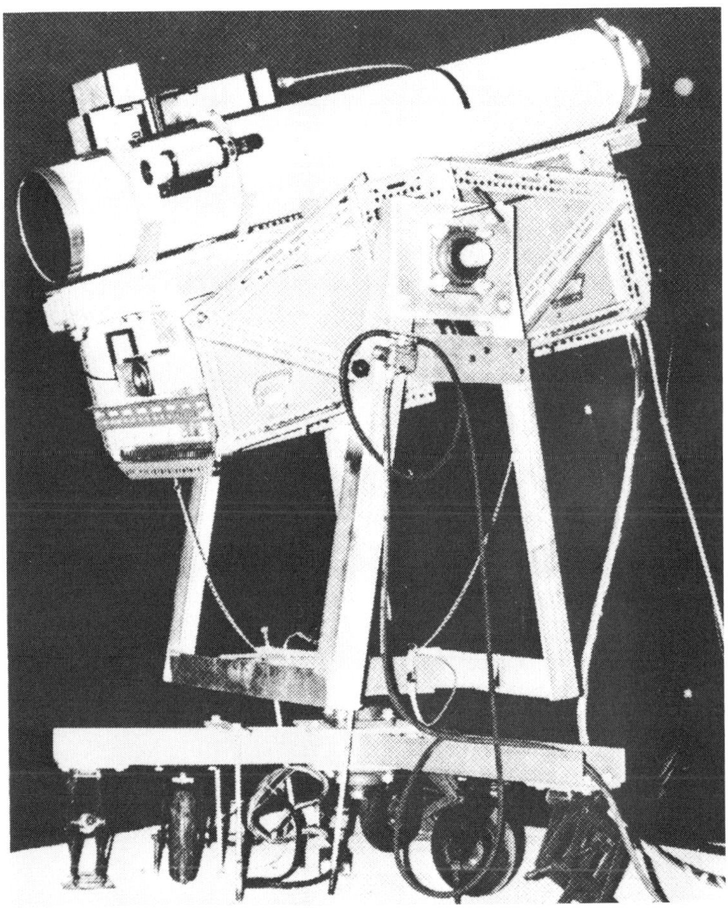

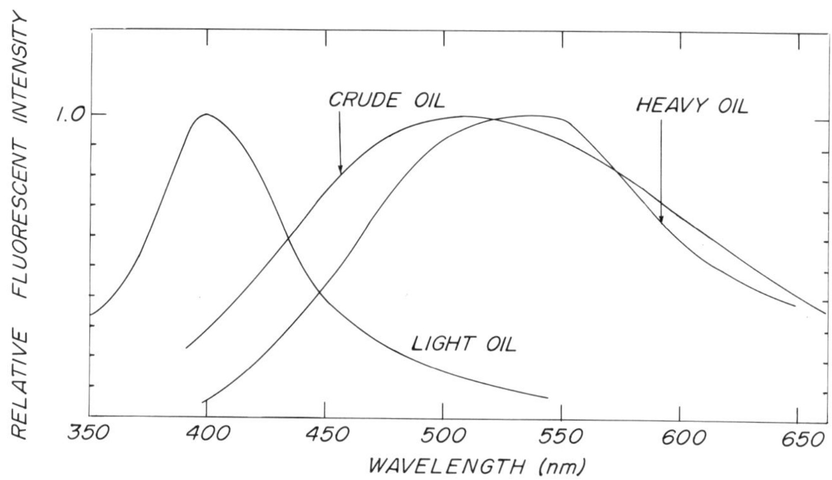

Figure 6.55. Relative fluorescent intensity from light oil, crude oil, and heavy oil excited at 337.1 nm [data from Rayner and Szabo (1978)]

solar background is low. Hence the KrF system is "solar blind" allowing daytime operation.

Another parameter that can be monitored in a remote sensing laser fluorosensing system is the fluorescent lifetime. Rayner and Szabo (1978) have studied the time dependence of the fluorescence of a variety of oils using 337-nm excitation. They find that the fluorescent decay can be characterized by a single lifetime only for some heavy oils. The higher oils are characterized by two exponential lifetimes. Some data are summarized in Table 6.12. The heavier oils have lifetimes that are generally $\lesssim 1$ ns while the light oils and crude oils have two components in their fluorescence curves with lifetimes differing by about one order of magnitude. Some experimental studies in remote sensing of oil slick fluorescence using time-resolved fluorescence have been reported by Measures *et al.* (1975b). Kung and Itzkan (1976) and Rayner *et al.* (1978)

Table 6.12. Fluorescent Lifetimes for Commercial Oils Excited at 337 nm[a]

Oil	Exponential time constants		
Medium bunker C		0.23 ± 0.07 ns	
Bunker fuel		0.22 ± 0.02	
Esso fuel Int.2		0.63 ± 0.05	
Venezuelan Largo medium crude	0.77 ± 0.10		7.3 ± 0.8
Transmountain crude	1.42 ± 0.08		8.2 ± 0.5
Marine gas/oil	3.35 ± 0.40		23.1 ± 2.0
Furnace fuel 2B (winter)	4.38 ± 0.50		20.5 ± 0.7

[a]From Rayner and Szabo (1978).

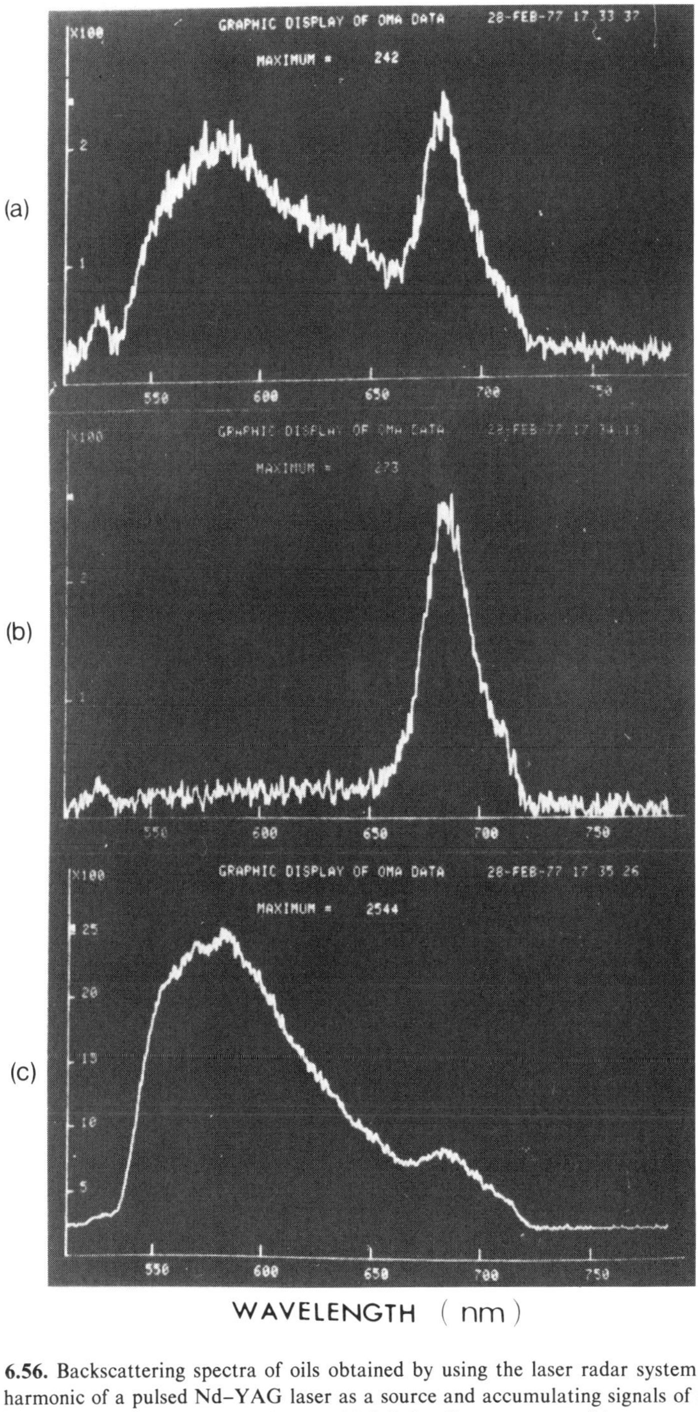

Figure 6.56. Backscattering spectra of oils obtained by using the laser radar system with the second harmonic of a pulsed Nd–YAG laser as a source and accumulating signals of OMA 10 scans: (a) backscattering spectrum of 1-cm thick light oil on seawater; (b) seawater; and (c) 1-mm thick heavy oil on seawater. Spectra have been stored in an optical multichannel analyzer (OMA) [Sato *et al.* (1978b)].

have reported further theoretical work on the remote sensing of oils using the laser fluorosensor.

Ahmadjian and Brown (1973) and Sato *et al.* (1978b) have reported on the remote detection of oil layers on water via laser-excited Raman spectra. This method offers the advantage of providing additional information on chemical constitution. However, the Raman spectrum is often obscurred by sample or water fluorescence. Figure 6.56 shows the Raman and fluorescence spectrum of two oil–water layers together with the Raman spectrum of seawater. These spectra were excited with light from a radiation of a frequency-doubled Nd–YAG laser. The combination of temporal and Raman spectral information would constitute a powerful remote sensing technique.

Hoge and Swift (1980) have developed a novel remote sensing system in which an N_2 laser excites the Raman band of OH from seawater. The presence of an oil slick is accompanied by a decrease in the strength of the Raman feature. The degree by which the intensity of the Raman line is diminished can be related to the thickness of the oil layer. Remote sensing of oil slicks from an aircraft was reported at a height of 150 m. Oil slicks with thicknesses of 0.05–5 μm could be detected.

The remote detection of bio-organisms in water with laser fluorosensors has been reported by Kim (1973) and Fadeev (1979). The system developed by Kim utilized a dye laser to excite at 590 nm. The chlorophyl fluorescence at 685 nm was detected. Chlorophyl-a concentrations were measured down to a limit of < 1 mg m^{-3}.

References

Chapter 1

R. L. ABRAMS (1974), *Appl. Phys. Lett.* **25,** 304.

J. D. ANDERSON (1976), *Gas Dynamic Lasers,* Academic Press, New York.

A. J. BEAULIEU (1970), *Appl. Phys. Lett.* **16,** 504.

W. B. BRIDGES (1979), in *Quantum Electronics,* Part A, C. L. Tang, Ed. Academic Press, New York, p. 33.

R. L. BYER (1980), *Electro-Opt. Syst. Des.* February, p. 24.

T. Y. CHANG (1970), *Opt. Commun.* **2,** 77.

C. C. DAVIS and T. A. KING (1975), in *Advances in Quantum Electronics,* Vol. 3, D. W. Goodwin, Ed. Academic Press, New York, p. 170.

A. DIENES (1975), in *Laser Applications to Optics and Spectroscopy,* S. F. Jacobs *et al.,* Eds., Addison-Wesley, Reading, Mass., p. 53.

W. W. DULEY (1976), *CO_2 Lasers: Effects and Applications,* Academic Press, New York.

G. T. FORREST (1980), *Appl. Opt.* **19,** 2094.

C. FREED, A. H. M. ROSS, and R. G. O'DONNELL (1974), *J. Mol. Spectrosc.* **49,** 439.

D. FRÖLICH, L. STEIN, H. W. SCHRÖDER, and H. WELLING (1976), *Appl. Phys.* **11,** 97.

E. T. GERRY (1970), *IEEE Spectrum* **7,** 51.

D. H. GILL and B. E. NEWMAN (1979), *Laser Focus,* September, p. 76.

L. S. GOLDBERG (1975), *Appl. Opt.* **14,** 653.

J. P. GOLDSBOROUGH (1972), in *Laser Handbook,* Vol. 1, F. T. Arecchi and E. O. Schulz-Dubois, Eds., North-Holland, Amsterdam.

A. HERTZBERG, W. H. CHRISTIANSEN, E. W. JOHNSTON, and H. G. AHLSTROM (1972), *AIAA J.* **10,** 394.

E. D. HINKLEY, K. W. NILL, and F. A. BLUM (1976), in *Laser Spectroscopy of Atoms and Molecules,* H. Walther, Ed., Springer-Verlag, Heidelberg.

E. P. IPPEN and C. V. SHANK (1979), in *Quantum Electronics,* Part B, C. L. Tang, Ed., Academic Press, New York.

M. G. LITMAN and H. J. METCALF (1978), *Appl. Opt.* **17,** 2224.

F. M. LUSSIER (1976), *Laser Focus,* December, p. 47.

I. MELNGAILIS and A. MOORADIAN (1975), in *Laser Applications to Optics and Spectroscopy,* S. F. Jacobs *et al.,* Eds., Addison-Wesley, Reading, Mass., p. 1.

O. G. PETERSON (1979), in *Quantum Electronics,* Part A, C. L. Tang, Ed., Academic Press, New York, p. 251.

M. S. PILTCH, J. RINK, and C. TALLMAN (1975), *Opt. Commun.* **15,** 112.

R. W. RALSTON, I. MELNGAILIS, W. T. LINDLEY, and A. R. CALAWA (1973), *IEEE J. Quantum Electron.* **QE-9,** 350.
R. J. SEYMOUR and F. ZERNIKE (1976), *Appl. Phys. Lett.* **29,** 76.
A. E. SIEGMAN (1971), *Laser Focus, July, p. 42.*
D. R. SKINNER and R. E. WHITCHER (1972), *J. Phys. E* **5,** 237.
D. H. SLINEY and W. J. MARSHALL (1979), *Electro-Opt. Syst. Des.* November, p. 31.
W. J. SPAWR and R. L. PIERCE (1975), *Laser Focus,* March, p. 37.
M. J. WEBER (1979), in *Quantum Electronics,* Part A, C. L. Tang, Ed., Academic Press, New York, p. 167.
B. WELLEGEHAUSEN, D. FRIEDE, H. VOGT, and S. SHAHDIN (1976), *Appl. Phys.* **11,** 363.
V. WILKE and W. SCHMIDT (1979), in *Laser-Induced Processes in Molecules,* K. L. Kompa and S. D. Smith, Eds., Springer-Verlag, New York, p. 12.
F. ZERNIKE (1979), in *Quantum Electronics* Part B, C. L. Tang, Ed., Academic Press, New York, p. 143.

Chapter 2

H. J. ALLELEIN, R. HECKER, A. MAX, TH. OVERHOFF, and D. STOVER (1979), *J. Appl. Phys.* **50,** 6162.
T. R. ANTHONY and H. E. CLINE (1977), *J. Appl. Phys.* **48,** 3888.
T. R. ANTHONY and H. E. CLINE (1978), *J. Appl. Phys.* **49,** 1248.
M. I. ARZUOV, A. I. BARCHUKOV, F. V. BUNKIN, V. I. KUNOV, and A. M. PROKHOROV (1975), *Sov. J. Quantum Electron.* **5,** 523.
M. I. ARZUOV, M. E. KARASEV, V. I. KONOV, V. V. KOSTIN, S. M. METEV, A. S. SILENOK, and N. I. CHAPLIEV (1979), *Sov. J. Quantum Electron.* **8,** 892.
J. F. ASMUS (1979), *Chem. Abstr.* **90,** 38086.
V. P. BABENKO and V. P. TYCHINSKII (1973), *Sov. J. Quantum Electron.* **2,** 399.
J. M. BALDWIN (1970), *Appl. Spectrosc.* **24,** 429.
J. M. BALDWIN (1973), *J. Appl. Phys.* **44,** 3362.
W. C. BALL and C. M. BANAS (1974), Soc. Auto. Eng., Report 740863.
C. M. BANAS, A. P. WALCH, and C. O. BROWN (1971), *Recordings of the 11th Symposium on Electron Ion, Laser Beam Technology,* IEEE, San Francisco Press, San Francisco.
V. BARANAUSKAS, C. I. Z. MAMMANA, R. E. KLINGER, and J. E. GREENE (1980), *Appl. Phys.* **36,** 930.
B. T. BARNES (1966), *J. Opt. Soc. Am.* **56,** 1546.
S. M. BEDAIR and H. P. SMITH (1969), *J. Appl. Phys.* **40,** 4776.
A. L. BERG, D. E. LOOD (1968), *Solid State Electron.* **11,** 773.
I. M. BETEROV, V. P. CHEBOTAEV, N. I. YURSHINA, and B. YA YURSHIN (1978), *Sov. J. Quantum Electron.* **8,** 1310.
V. M. BOYAKOV, V. M. EPIKHIN, B. A. KALIN, M. K. MAKLATOV, I. N. MIKOLAEV, and G. N. SHISHKIN (1978), *Sov. J. Quantum Electron.* **8,** 902.
L. BRAUN and D. BREUER (1969), *Solid State Technol.* **12,** 56.
E. M. BREINAN and C. M. BANAS (1981), in *Advances in Metal Processing* J. J. Burke, R. Mehrabian, and V. Weiss, Eds., Plenum Press, New York.
E. M. BREINAN, C. M. BANAS, and M. A. GREENFIELD (1975), United Technol. Res. Ctr. Report R75-111 087-3.
E. M. BREINAN, B. H. KEAR, and C. M. BANAS (1976), *Phys. Today,* November, p. 44.
A. D. BRISBANE and T. M. JACKSON (1968), *Electron. Compon.* **9,** 73.
K. BRUGGER (1972), *J. Appl. Phys.* **53,** 577.
K. R. BUBE (1975), *Am. Ceram. Soc. Bull.* **54,** 528.
L. BUENE, J. M. POATE, D. C. JACOBSON, C. W. DRAPER, and J. K. HIRVONEN (1980), *Appl. Phys. Lett.* **37,** 385.

L. L. BURNS (1980), *Laser Focus,* July, p. 48.
C. A. BURRUS and L. A. GOLDREN (1977), *Appl. Phys. Lett.* **31,** 383.
C. CALI, V. DANEN, A. ORIOLI, and R. RIVA-SANSEVERINO (1976), *Appl. Opt.* **15,** 1327.
D. K. CAMPBELL and D. W. SWEENEY (1978), *Appl. Opt.* **17,** 3727.
C. O. CARLSON, E. STONE, H. L. BERNSTEIN, W. K. TOMITA, and W. C. MYERS (1966), *Science* **154,** 1550.
H. S. CARSLAW and J. C. JAEGER (1959), *Conduction of Heat in Solids,* 2nd ed., Oxford University Press (Clarendon), London and New York.
D. CHEN and J. D. ZOOK (1975), *Proc. IEEE* **63,** 1207.
D. CHEN, B. KOEPKE, J. D. ZOOK, and E. BERNAL G. (1976), *Appl. Phys. Lett.* **29,** 657.
C. P. CHRISTENSEN and K. M. LAKIN (1978), *Appl. Phys. Lett.* **32,** 254.
J. CLARKE and W. M. STEEN (1979), Proceedings of the Laser '79 Conference, Munich.
H. E. CLINE and T. R. ANTHONY (1977), *J. Appl. Phys.* **48,** 3895.
B. COCKAYNE and D. B. GASSON (1971), *J. Mater. Sci.* **6,** 126.
M. I. COHEN (1972) in *Laser Handbook,* F. T. Arecchi and E. O. Schulz-Dubois, Eds., North Holland, Amsterdam, p. 1577.
M. I. COHEN and J. P. EPPERSON (1968a), *Adv. Electron. Electron Phys. Suppl.* **4,** 139.
M. I. COHEN and J. P. EPPERSON (1968b), in *Electron Beam and Laser Beam Technology,* L. Marton and A. B. El-Kareh, Eds., Academic Press, New York.
M. I. COHEN, B. A. UNGER, and J. F. MILKOWSKY (1968), *Bell Syst. Tech. J.* **47,** 385.
S. M. COPLEY, D. BECK, O. ESQUIVEL, and M. BASS (1979), in *Laser Solid Interactions and Laser Processing,* S. D. Ferris *et al.,* Eds., AIP Conference Proceedings, 50, American Institute of Physics, New York.
C. COURTNEY and W. M. STEEN (1978), Proceedings of Advances in Surface Coating Technology, London, p. 219.
R. C. CRAFER (1976), *Weld. Inst. Res. Bull.* **17,** 28.
F. W. DABBY and U. C. PAEK (1972), *IEEE J. Quantum Electron.* **QE-8,** 106.
K. DAREE and W. KAISER (1978), *Opt. Laser Technol.,* April, p. 65.
T. F. DEUTSCH, D. J. EHRLICH, and R. M. OSGOOD (1979), *Appl. Phys. Lett.* **35,** 175.
W. W. DULEY (1976), *CO_2 Lasers: Effects and Applications,* Academic Press, New York.
W. W. DULEY and J. N. GONSALVES (1974), *Opt. Laser Technol.* **6,** 78.
W. W. DULEY, D. J. SEMPLE, J.-P. MORENCY, and M. GRAVEL (1979), *Opt. Laser Technol.* **11,** 281.
M. EBOO, W. M. STEEN, and J. CLARKE (1978), *Advances in Welding, Proceedings,* Harrogate.
K. EGASHIRA and M. KOBAYASHI (1977a), *Appl. Opt.* **16,** 1636.
K. EGASHIRA and M. KOBAYASHI (1977b), *Appl. Opt.* **16,** 2743.
D. J. EHRICH, R. M. OSGOOD, and T. F. DEUTSCH (1980), *Appl. Phys. Lett.* **36,** 916.
K. EICKHOFF and K. GÜRS (1969), *J. Crystal. Growth* **6,** 21.
W. A. ELLIOTT, F. P. GAGLIANO and G. KRAUSS (1973), *Met. Trans.* **4,** 2031.
S. L. ENGEL (1976), *Am. Mach.* May, p. 107.
S. L. ENGEL (1976), *Laser Focus,* February, p. 44.
S. L. ENGEL (1978), *Laser Focus,* December, p. 66.
A. ENGEL, J. STEFFEN and G. HERZIGER (1974), *Appl. Opt.* **13,** 269.
J. P. EPPERSON, R. W. DYER, and J. C. GRZYWA (1966), *West Elec. Eng.* **10,** 2.
N. N. EVTIKHIEV, G. R. LEVINSON, K. P. TSVETAEV, and O. V. GOLOSNOI (1974), *Sov. J. Quantum Electron.* **4,** 527.
M. L. FABER (1974), *IEEE Trans. Parts. Hybrids, Packag.* **10,** 169.
J. A. FOX (1975), *Appl. Phys. Lett.* **26,** 682.
H. FUJITA, Y. SUZAKI, and A. TACHIBANA (1976), *Appl. Opt.* **15,** 230.
F. P. GAGLIANO, R. M. LUMLEY, and L. S. WATKINS (1969), *Proc. IEEE* **57,** 114.
F. P. GAGLIANO and V. J. ZALECKAS (1972), in *Lasers in Industry,* S. S. Charschan, Ed., P. 139, Van Nostrand Reinhold, Princeton.

D. B. GASSON and B. COCKAYNE (1970), *J. Mater. Sci.* **5,** 100.

D. S. GNANAMUTHU, C. B. SHAW, W. E. LAWRENCE, and M. R. MITCHELL (1979), in *Laser Solid Interactions and Laser Processing—1978,* S. D. Ferris *et al.,* Eds., AIP Conference Proceedings, Vol. 50, American Institute of Physics, New York, p. 173.

E. E. GRAY (1972), *IEEE Trans. Magn.* **MAG-8,** 416.

L. E. GREENWALD, E. M. BREINAN, and B. H. KEAR (1979), in *Laser Solid Interactions and Laser Processing,* S. D. Ferris *et al.,* Eds., AIP Conference Proceedings, Vol. 50, American Institute of Physics, New York, p. 189.

D. C. HAMILTON and D. J. JAMES (1976), *J. Phys. D* **9,** L41.

D. C. HAMILTON and I. R. PASHBY (1979), *Opt. Laser Technol.,* August, p. 183.

M. HANABUSA, A. NAMIKI, and K. YOSHIHARA (1979), *Appl. Phys. Lett.* **35,** 626.

J. E. HARRY and F. W. LUNAU (1972), *IEEE Trans. Ind. Appl.* **IA-8,** 418.

G. HASS and J. B. RAMSEY (1969), *Appl. Opt.* **8,** 1115.

R. C. HEADLEY (1973), *Electron.* **46,** 121.

J. W. HILL, M. J. LEE, I. J. SPALDING (1974), *Opt. Laser Technol.,* December, p. 276.

G. HOLZINGER, K. KOSANKE, and W. MENZ (1973), *Opt. Laser Technol.,* December, p. 257.

E. INOUE, H. KOKADO, I. SHIMIZU, and S. OOTSUKA (1973), *Photogr. Sci. Eng.* **17,** 405.

R. E. JAEGER (1976), *Am. Ceram. Soc. Bull.* **55,** 270.

R. E. JAEGER and W. LOGAN (1973), *Opt. Spectra,* November, p. 40.

A. KESTENBAUM (1979), *J. Appl. Phys.* **50,** 5012.

P. H. KIM, K. TAKI, and S. NAMBA (1971), Recordings of the 11th Symposium on Electron, Ion, Laser Beam Technology, IEEE, San Francisco Press, San Francisco, p. 375.

Y-W KIM, P. R. STRUTT, and H. NOWOTNY (1979), *Met. Trans.* **10A,** 881.

K. KINOSHITA and K. EGASHIRA (1978), *Appl. Opt.* **17,** 1210.

J. K. KLIWER (1973), *J. Appl. Phys.* **44,** 490.

F. KUMMER and I. TAITL (1977), *Chem. Abstr.* **89,** 98291.

V. I. LITTLE, P. Y. KEY, R. WILTSCHER, and D. M. ROWLEY (1971), *Nature, Phys. Sci.* **232,** 165.

E. V. LOCKE and R. A. HELLA (1974), *IEEE J. Quantum Electron* **QE-10,** 179.

E. V. LOCKE, E. D. HOAG, and R. A. HELLA (1972), *IEEE J. Quantum Electron.* **QE-8,** 132.

J. LONGFELLOW (1970). *Rev. Sci. Instrum.* **41,** 1485.

J. LONGFELLOW (1971), *Am. Ceram. Soc. Bull.* **50,** 251.

D. Y. LOU (1977), *J. Appl. Phys.* **48,** 2015.

R. M. LUMLEY (1969), *Am. Ceram. Soc. Bull* **48,** 850.

F. W. LUNAU (1970), Conference on Electronic Methods of Machining, Forming, and Coating (reprint).

H. LYDTIN (1972), 3rd International Conference on Chemical Vapor Deposition, Salt Lake City, April 24.

M. E. MAHER and R. B. HALL (1978), *J. Appl. Phys.* **49,** 2254.

S. MARCUS, J. E. LOWDER and D. L. MOONEY (1976), *J. Appl. Phys.* **47,** 2966.

A. J. MARTIN, R. H. BUCK, R. G. LOASBY, and J. SAVAGE (1968), *Thin Solid Films* **2,** 253.

D. MAYDAN (1971), *Bell Syst. Tech. J.* **50,** 1761.

J. MAZUMDER (1977), Ph.D. thesis, University of London.

J. MAZUMDER and W. M. STEEN (1977), *Proceedings of the Laser '77 Conference,* Munich, p. 307.

J. MAZUMDER and W. M. STEEN (1980), *J. Appl. Phys.* **51,** 941.

J. A. MCKAY and J. T. SCHRIEMPF (1979), *Appl. Phys. Lett.* **35,** 433.

R. MEHRABIAN, S. KOU, S. C. HSU, and M. MUNITZ (1979), *AIP Conf. Proc.* **50**, 129.

C. H. MILLER and T. A. OSIAL (1969), Status Report on 250-W CO_2 Laser for Applications in the Pulp and Paper Industry, IEEE 1969, Annual Pulp Paper Industry Technical Conference, Atlanta.

C. J. MOHR, J. A. RING, E. H. STEVENS, and R. A. BAKER (1977), *Proc. IEEE* **65,** 269.

P. Moore, C. Kim, and L. S. Weinman (1979), *AIP Conf. Proc.* **50,** 221.
A. J. Moorhead (1971), *Weld. J.* **50,** 97.
J. M. Moran (1971), *Appl. Opt.* **10,** 412.
M. Mori and H. Kumehara (1976), *Annal. CIRP* **25,** 115.
Y. Nakada and M. A. Giles (1971), *J. Am. Ceram. Soc.* **54,** 354.
S. L. Narasimhau, S. M. Copley, E. W. van Stryland, and M. Bass (1979), *Met. Trans.* **10A,** 654.
G. Nath (1974), *Opt. Laser Technol.,* October, p. 233.
L. S. Nelson and N. L. Richardson (1972), *Mater. Res. Bull.* **7,** 971.
L. S. Nelson, N. L. Richardson, K. Keil, and S. R. Skaggs (1973), *High Temp. Sci.* **5,** 138.
L. S. Nelson, S. R. Skaggs, and N. L. Richardson (1970), *J. Am. Ceram. Soc. Bull.* **53,** 113.
L. S. Nelson, A. G. Whittaker, and B. Tooper (1972), *High Temp. Sci.* **4,** 445.
J. C. North (1977), *J. Appl. Phys.* **48,** 2419.
M. Oakes (1978), *Opt. Eng.* **17,** 217.
U. C. Paek (1974), *Appl. Opt.* **13,** 1383.
U. C. Paek and A. Kestenbaum (1973), *J. Appl. Phys.* **44,** 2260.
U. C. Paek and A. L. Weaver (1975), *Appl. Opt.* **14,** 294.
T. G. Pavlopoulos and K. Crabtree (1974), *J. Appl. Phys.* **45,** 4964.
J. F. Ready (1971), *Effects of High Power Laser Radiation,* Academic Press, New York.
J. F. Ready (1978), *Industrial Applications of Lasers,* Academic Press, New York.
J. E. Robin (1978), *J. Appl. Phys.* **49,** 5306.
J. E. Robin and P. Nordin (1975a), *Appl. Phys. Lett.* **26,** 289.
J. E. Robin and P. Nordin (1975b), *Appl. Phys. Lett.* **27,** 593.
J. E. Robin and P. Nordin (1976), *Appl. Phys. Lett.* **29,** 3.
D. M. Roessler and V. G. Gregson (1978), *Appl. Opt.* **17,** 992.
H. Schonhorn, C. R. Kurkjian, R. E. Jaeger, H. N. Vazirani, R. V. Albarino, and F. V. Di Marcello (1976), *Appl. Phys. Lett.* **29,** 712.
H. Schonhorn, H. N. Vazirani, and H. L. Frisch (1978), *J. Appl. Phys.* **49,** 3703.
H. Schwarz and H. A. Tourtellote (1969), *J. Vac. Sci. Technol.* **6,** 373.
F. Schwirzke, H. Brinkschulte, and M. Hashmi (1974), *J. Appl. Phys.* **46,** 4891.
F. Schwirzke, L. Oren, S. Talmadge, and R. J. Taylor (1978), *Phys. Rev. Lett.* **40,** 1181.
J. G. Seigman (1968), *Microelectron. Reliability* **7,** 305.
J. G. Seigman (1969), *Microelectron. Reliability* **8,** 87.
J. Shewell (1977), *Weld. Des. Fab.,* June, p. 100.
I. P. Shkarofsky (1975), *RCA Rev.* **36,** 336.
S. H. Slivinsky and N. E. Ogle (1977), *J. Appl. Phys.* **48,** 3660.
H. M. Smith and A. F. Turner (1965), *Appl. Opt.* **4,** 147.
M. Sparks (1976), *J. Appl. Phys.* **47,** 837.
W. M. Steen (1977), *Lett. Heat Mass Trans.* **4,** 167.
W. M. Steen (1978), Advances in Surface Coating Technology International Conference, London, p. 175.
W. M. Steen (1979), private communication.
W. M. Steen and M. Eboo (1979), *Met. Const.* **11,** 332.
F. Stern (1973), *J. Appl. Phys.* **44,** 4204.
B. Steverding, R. W. Conrad, and H. P. Dudel (1979), *J. Appl. Phys.* **50**, 6713.
P. R. Strutt, D. A. Gilbert, H. Nowotny, and Y-M Kim (1979), in *Laser Solid Interactions and Laser Processing,* S. D. Ferris *et al.,* Eds., AIP Conference Proceedings, Vol. 50, American Institute of Physics, New York, p. 232.
E. Stürmer and M. von Allmen (1978), *J. Appl. Phys.* **49,** 5648.
L. L. Sveshnikova, V. I. Donin, and S. M. Repinskii (1977), *Sov. Phys. Tech. Phys. Lett.* **3,** 223.
D. T. Swift-Hook and A. E. F. Gick (1973), *Weld. J.* **52**, 492S.

K. Takagi and M. Ishii (1977), *J. Cryst. Growth* **40,** 1.
K. Takei, S. Fujimoni, and K. Nagai (1980), *J. Appl. Phys.* **51,** 2903.
M. Terao, K. Shigematsu, M. Ojima, Y. Taniguchi, S. Horigome, and S. Yonezawa (1979), *J. Appl. Phys.* **50,** 6881.
S. A. Thompson (1970), *Electron. Eng.* **29,** 48.
L. E. Topol and R. A. Happe (1974), *J. Non-Cryst. Solids* **15,** 116.
L. C. Towle, J. A. McKay, and J. T. Schriempf (1979), *J. Appl. Phys.* **50,** 4391.
B. A. Tozer (1976), *Opt. Laser Technol.,* April, p. 57.
W-T Tsang and S. Wang (1975), *Appl. Phys. Lett.* **27,** 79.
A. A. Uglov and A. N. Kokora (1977), *Sov. J. Quantum Electron.* **7,** 671.
A. A. Uglov, A. N. Kotova, and N. V. Berlin (1978), *Sov. J. Quantum Electron.* **8,** 884.
M. von Allmen (1976), *J. Appl. Phys.* **47,** 5460.
R. J. von Gutfeld, E. E. Tynan, R. L. Melcher, and S. E. Blum (1979), *Appl. Phys. Lett.* **35,** 651.
C. von Osenbruggen and Th. P. J. Botden (1975), Proceedings of the 2nd International Symposium of the Japanese Welding Society, *Adv. Weld. Technol.,* p. 597.
R. E. Wagner (1974), *J. Appl. Phys.* **45,** 4631.
J. M. Webster (1970), *Met. Progr.* **98,** 59.
W. W. Weick (1972), *IEEE J. Quantum Electron.* **QE-8,** 126.
L. S. Weinman and J. N. Devant (1979), in *Laser Solid Interactions and Laser Processing,* S. D. Ferris *et al.,* Eds. AIP Conference Proceedings, Vol. 50, American Institute of Physics, New York.
L. S. Weinman, C. Kim, T. R. Tucker, and E. A. Metzbower (1978), *Appl. Opt.* **17,** 906.
D. M. Wick (1975), Paper MR-75-491, Soc. Man. Engineers.
T. J. Wieting and J. L. de Rosa (1979), *J. Appl. Phys.* **50,** 1071.
T. J. Wieting and J. T. Schriempf (1976), *J. Appl. Phys.* **47,** 4009.
R. A. Willgoss, J. H. P. C. Megaw, and J. M. Clark (1979), *Opt. Laser Technol.,* April, p. 73.
J. R. Williamson (1976), in *Industrial Applications of High Power Laser Technology,* Vol. 86, J. F. Ready, Ed., Proceedings of the Society of Photo-Optical Instrument Engineers, Palo Verdes Estates, California.
J. H. Wood (1977), *Microelectron. Symp.,* p. 81.
V. J. Zaleckas and J. C. Koo (1977), *Appl. Phys. Lett.* **31,** 615.
D. M. Zehner, C. W. White, and G. W. Ownby (1980), *Appl. Phys. Lett.* **36,** 56.

Chapter 3

K. Affolter, W. Lüthy, and M. von Allmen (1978), *Appl. Phys. Lett.* **33,** 185.
K. Affolter, W. Lüthy, and M. Wittmer (1980), *Appl. Phys. Lett.* **36,** 559.
R. Andrew and M. Lovato (1979), *J. Appl. Phys.* **50,** 1142.
A. K. Antonenko, N. N. Gerasimenko, A. V. Dvureihenskii, L. S. Smirnov, and G. M. Tseitlin (1976), *Sov. Phys. Semicond.* **10,** 81.
D. H. Auston, J. A. Golovchenko, P. R. Smith, C. M. Surko, and T. N. C. Venkatesan, (1978), *Appl. Phys. Lett.* **33,** 539.
D. H. Auston, J. A. Golovchenko, and T. N. C. Venkatesan (1979), *Appl. Phys. Lett.* **34,** 558.
G. Bahir, T. Bernstein, and R. Kalish (1979), *Appl. Phys. Lett.* **34,** 486.
P. A. Barnes, H. J. Leamy, J. M. Poate, S. D. Ferris, J. S. Williams, and G. K. Celler (1978), *Appl. Phys. Lett.* **33,** 965.

J. C. Bean, H. J. Leamy, J. M. Poate, G. A. Rozgonyi, J. P. van der Ziel, J. S. Williams, and G. K. Celler (1979), *J. Appl. Phys.* **50,** 881.

A. Bontemps, S. U. Campisano, G. Foti, G. Mondio, and G. Saitta (1980), *Appl Phys. Lett.* **36,** 542.

G. K. Celler, J. M. Poate, and L. C. Kimerling (1978), *Appl. Phys. Lett.* **32,** 464.

G. K. Celler, J. M. Poate, G. A. Rozgonyi, and T. T. Sheng (1979), *J. Appl. Phys.* **50,** 7264.

P. W. Cook, S. E. Schuster, and R. J. von Gutfeld (1975), *Appl. Phys. Lett.* **26,** 125.

P. D. Dapkus, W. W. Weick, and R. W. Dixon (1974), *Appl. Phys. Lett.* **24,** 292.

T. F. Deutsch, D. J. Ehrlich, R. M. Osgood, Z. L. Liau (1980), *Appl. Phys. Lett.* **36,** 847.

J. C. C. Fan, R. L. Chapman, J. P. Donnelly, G. W. Turner, and C. O. Bozler (1979), *Appl. Phys. Lett.* **34,** 780.

J. C. C. Fan and H. J. Zeiger (1975), *Appl. Phys. Lett.* **27,** 224.

J. C. C. Fan, H. J. Zeiger, R. P. Gale, and R. L. Chapman (1980), *Appl. Phys. Lett.* **36,** 158.

J. M. Fairfield and G. H. Schwuttke (1968), *Solid State Electron.* **11,** 1175.

G. Foti, S. U. Campisano, E. Rimini, and G. Vatali (1978a), *J. Appl. Phys.* **49,** 2569.

G. Foti, G. Della Mea, E. Jannitti, and G. Majni (1978b), *Phys. Lett. A,* **68,** 368.

A. B. Fowler and R. T. Hodgson (1980), *Appl. Phys. Lett.* **36,** 914.

A. Gat, L. Gerzberg, J. F. Gibbons, T. J. Magee, J. Peng, and J. D. Hong (1978), *Appl. Phys. Lett.* **33,** 775.

A. Gat and J. F. Gibbons (1978), *Appl. Phys. Lett.* **32,** 143.

A. Gat, J. F. Gibbons, T. J. Magee, J. Peng. V. R. Deline, P. Williams, and C. A. Evans (1978a), *Appl. Phys. Lett.* **32,** 776.

A. Gat, J. F. Gibbons, T. J. Magee, J. Peng, P. Williams, V. R. Deline, and C. A. Evans (1978b), *Appl. Phys. Lett.* **33,** 389.

A. Gat, A. Lietoila and J. F. Gibbons (1979), *J. Appl. Phys.* **50,** 2926.

J. A. Golovchenko and T. N. C. Venkatesan (1978), *Appl. Phys. Lett.* **32,** 147.

F. I. Harper and M. I. Cohen (1970), *Solid State Electron.* **13,** 1103.

G. A. Kachurin, N. B. Pridachin, and L. S. Smirnov (1976), *Sov. Phys. Semicond.* **9,** 946.

R. A. Kaplan, M. G. Cohen and K. C. Liu (1979), from A. R. Tebo, *Electro-Opt. Syst. Des.,* December, p. 29.

A. G. Klimenko, E. A. Klimenko, and V. I. Donin (1976), *Sov. J. Quantum Electron.* **5,** 1289.

M. Koyanagi, H. Tamura, M. Miyao, N. Hashimoto, and T. Tokuyama (1979), *Appl. Phys. Lett.* **35,** 621.

R. A. Laff and G. L. Hutchins (1974), *IEEE Trans. Electron. Devices* **E-21,** 743.

B. C. Larson and J. F. Barhorst (1980), *J. Appl. Phys.* **51,** 3181.

B. C. Larson, C. W. White, and B. R. Appleton (1978), *Appl. Phys. Lett.* **32,** 801.

H. J. Leamy, G. A. Rozgonyi, T. T. Sheng, and G. K. Celler (1978), *Appl. Phys. Lett.* **32,** 535.

K. F. Lee, J. F. Gibbons, K. C. Saraswat, and T. I. Kamins (1979), *Appl. Phys. Lett.* **35,** 173.

A. Lietoila and J. F. Gibbons (1979), *Appl. Phys. Lett.* **34,** 332.

W. Lüthy, K. Affolter, H. P. Weber, M. E. Roulet, M. Fallavier, J. P. Thomas, and J. Mackowski (1979), *Appl. Phys. Lett.* **35,** 873.

V. V. Makarov, T. Tuomi, K. Naukkarinen, M. Luomajärvi, and M. Riihonen (1979), *Appl. Phys. Lett.* **35,** 922.

Y. Makita, M. Yokata, H. Nojiri, T. Tsurushima, H. Tanoue, T. Kanayama, and T. I. Shimada (1979), *Appl. Phys. Lett.* **35,** 633.

S. Margalit, D. Fekete, D. M. Pepper, C-P. Lee, and A. Yariv (1978), *Appl. Phys. Lett.* **33,** 346.

M. Miyao, M. Tamura, and T. Tokuyama (1978), *Appl. Phys. Lett.* **33,** 828.

J. C. Muller, A. Grob, J. J. Grob, R. Stuck, and P. Siffert (1978), *Appl. Phys. Lett.* **33,** 287.

K. MURAKAMI, K. GAMO, S. NAMBA, M. KAWABE, and Y. AOYAGI (1979), *Appl. Phys. Lett.* **35,** 628.

J. NARAYAN (1980), *Appl. Phys. Lett.* **37,** 66.

J. NARAYAN, R. T. YOUNG, and C. W. WHITE (1978a), *J. Appl. Phys.* **49,** 3912.

J. NARAYAN, R. T. YOUNG, R. F. WOOD, and W. H. CHRISTIE (1978b), *Appl Phys. Lett.* **33,** 338.

H. OKABAYASKI, M. YOSHIDA, K. ISHIDA, and T. YAMANE (1980), *Appl.. Phys. Lett.* **36,** 202.

M. ORON and A. SØRENSEN (1979), *Appl. Phys. Lett.* **35,** 782.

C. W. PEARCE and V. J. ZALECKAS (1979), *J. Electrochem. Soc.* **126,** 1436.

N. S. PLATAKIS (1976), *J. Appl. Phys.* **47,** 2120.

J. M. POATE, H. J. LEAMY, T. T. SHENG, and G. K. CELLER (1978), *Appl. Phys. Lett.* **33,** 918.

S. PRUSSIN and W. VON DER OHE (1980), *J. Appl. Phys.* **51,** 3853.

M. E. ROULET, P. SCHWOB, K. AFFOLTER, W. LÜTHY, M. VON ALLMEN, M. FALLAVIER, J. M. MACKOWSKI, M. A. NICOLET, and J. P. THOMAS (1979), *J. Appl. Phys.* **50,** 5536.

G. A. SAI-HALASZ, F. F. FANG, T. O. SEDGWICK, and A. SEGMÜLLER (1980), *Appl. Phys. Lett.* **36,** 419.

T. SHIBATA, J. F. GIBBONS, and T. W. SIGMON (1980), *Appl. Phys. Lett.* **36,** 566.

E. I. SHTYRKOV, I. B. KHAIBULLIN, M. M. ZARIPOV, M. F. GALYATUDINOV, and R. M. BAYAZITOV (1976), *Sov. Phys. Semicond.* **9,** 1309.

M. R. T. SIREGAR, W. LUTHY, and K. AFFOLTER (1980), *Appl. Phys. Lett.* **36,** 787.

M. TAKAI and H. RYSSEL (1979), *Appl. Phys. Lett.* **35,** 696.

H. TAMURA, M. MIYAO, and T. TOKUYAMA (1979), *J. Appl. Phys.* **50,** 3783.

G. J. VAN GURP, G. E. J. EGGERMONT, Y. TAMMINGA, W. T. STACY, and J. P. M. GIJSBERS (1979), *Appl. Phys. Lett.* **35,** 273.

T. N. C. VENKATESAN, J. A. BOLVCHENKO, J. M. POATE, P. COWAN, and G. K. CELLER (1978), *Appl. Phys. Lett.* **33,** 429.

G. VITALI, M. BERTOLOTTI, G. FOTI, and E. RIMINI (1977), *Phys. Lett. A* **63,** 351.

M. VON ALLMEN, S. S. LAU, M. MÄENPÄÄ, and B. Y. TSAUR (1980a), *Appl. Phys. Lett.* **36,** 205.

M. VON ALLMEN, S. S. LAU, M. MÄENPÄÄ, and B. Y. TSAUR (1980b), *Appl. Phys. Lett.* **37,** 84.

M. VON ALLMEN, W. LÜTHY, J. P. THOMAS, M. FALLAVIER, J. M. MACKOWSKI, R. KIRSCH, M.-A. NICOLET, and M. E. ROULET (1979), *Appl. Phys. Lett.* **34,** 82.

M. VON ALLMEN and M. WITTMER (1979), *Appl. Phys. Lett.* **34,** 68.

C. W. WHITE, W. H. CHRISTIE, B. R. APPLETON, S. R. WILSON, P. P. PRONKO, and C. W. MAGEE (1978), *Appl. Phys. Lett.* **33,** 662.

C. W. WHITE, J. NARAYAN, B. R. APPLETON, and S. R. WILDON (1979a), *J. Appl. Phys.* **50,** 2967.

C. W. WHITE, P. P. PRONKO, S. R. WILSON, B. R. APPLETON, J. NARAYAN, and R. T. YOUNG (1979b), *J. Appl. Phys.* **50,** 3261.

J. Z. WILCOX (1980), *J. Appl. Phys.* **51,** 2866.

M. WITTMER and M. VON ALLMEN (1979), *J. Appl. Phys.* **50,** 4786.

G. YARON and L. D. HESS (1980), *Appl. Phys. Lett.* **36,** 220.

R. T. YOUNG and J. NARAYAN (1978), *Appl. Phys. Lett.* **33,** 14.

Chapter 4

A. P. ALESKSANDROV, V. N. GENKIN, M. S. KITAI, I. M. SMIRNOVA, and V. V. SOKOLOV (1977), *Sov. J. Quantum Electron.* **7,** 547.

R. V. AMBARTZUMYAN *et al.* (1974), *Chem. Phys. Lett.* **25,** 515.

R. V. AMBARTZUMYAN *et al.* (1975), *Chem. Phys. Lett.* **36,** 301.

R. V. AMBARTZUMYAN *et al.* (1975), *Zh. Eksp. Teor. Fiz. Pis'ma Red.* **21,** 375.

R. V. AMBARTZUMYAN *et al.* (1976), *Opt. Commun.* **18,** 220.

R. V. Ambartzumyan *et al.* (1976a), *JETP Lett.* **23,** 22.
R. V. Ambartzumyan *et al.* (1976b), *Zh. Eksp. Teor. Fiz.* **71,** 440.
R. V. Ambartzumyan *et al.* (1976c), *Opt. Commun.* **18,** 517.
R. V. Ambartzumyan *et al.* (1976d), *Sov. J. Quantum Electron.* **5,** 1196.
R. V. Ambartzumyan *et al.* (1978a), *Opt. Commun.* **25,** 69.
R. V. Ambartzumyan *et al.* (1978b), *Opt. Commun.* **27,** 79.
R. V. Ambartzumyan, N. V. Chekalin, V. S. Doljikov, V. S. Letokhov, and V. N. Lokhman (1976e,) *Opt. Commun.* **18,** 400.
R. V. Ambartzumyan and V. S. Letokhov (1972), *Appl. Opt.* **11,** 354.
R. V. Ambartzumyan and V. S. Letokhov (1977), in *Chemical and Biochemical Applications of Lasers,* Vol. 3, C. B. Moore, Ed., Academic Press, New York, p. 167.
R. V. Ambartzumyan, V. S. Letokhov, G. N. Makarov, and A. A. Puretskii, (1972), *Zh. Eksp. Teor. Fiz. Pis'ma Red.* **15,** 709.
V. Arkhipov, N. G. Basov, E. M. Belenov, B. N. Duvanov, E. P. Markin, and A. N. Oreavsky (1972), *Zh. Eksp. Teor. Fiz. Pis'ma Red.* **16,** 469.
D. Arnoldi, K. Kaufmann, and J. Wolfrum (1975), *Phys. Rev. Lett.* **34,** 1597.
D. Arnoldi and J. Wolfrum (1974), *Chem. Phys. Lett.* **24,** 234.
A. Ashkin (1970), *Phys. Rev. Lett.* **24,** 156.
G. A. Askaryan and V. A. Namiot (1975a), *JETP Lett.* **21,** 305.
G. A. Askaryan and V. A. Namiot (1975b), *JETP Lett.* **42,** 1009.
J. F. Asmus (1971), Recordings of the 11th Symposium on Electron, Ion, Laser Beam Technology, IEEE, San Francisco Press, San Francisco, p. 367.
P. Avouris, M. M. T. Loy, and I. Y. Chan (1979), *Chem. Phys. Lett.* **63,** 624.
H. R. Bachmann, N. Nöth, R. Rinck, and K. L. Kompa (1974), *Chem. Phys. Lett.* **29,** 627.
H. R. Bachmann, R. Rinck, H. Nöth, and K. L. Kompa (1977), *Chem. Phys. Lett.* **45,** 169.
V. I. Balykin, V. S. Letokhov, V. I. Mishin, and W. A. Semchishen (1976), *Chem. Phys.* **17,** 111.
N. G. Basov, E. M. Belenov, V. A. Isakov, E. P. Markin, A. N. Oraevskii, V. I. Romanenko, and N. B. Ferapantov (1974a), *Zh. Eksp. Teor. Fiz. Pis'ma Red.* **20,** 607.
N. G. Basov, E. M. Belenov, E. P. Markin, A. N. Oraevsky, and A. V. Pankratov (1973), *Zh. Eksp. Teor. Fiz.* **64,** 485.
N. G. Basov, A. N. Oraevsky, and A. V. Pankratov (1974b), in *Chemical and Biochemical Applications of Lasers,* Vol. 1, C. B. Moore, Ed., Academic Press, New York, p. 203.
H. E. Bass and J. R. Fanchi (1976), *J. Chem. Phys.* **64,** 4417.
S. H. Bauer, E. Bar-Ziv, and J. A. Haberman (1978), *IEEE J. Quantum Electron.* **QE-14,** 237.
S. H. Bauer and J. A. Haberman (1978), *IEEE J. Quantum Electron.* **QE-14,** 233.
S. H. Bauer, D. M. Lederman, E. L. Resler, and E. R. Fisher (1973), *Int. J. Chem. Kinet.* **5,** 93.
A. F. Bernhardt, D. E. Duerre, J. R. Simpson, and L. L. Wood (1974), *Appl. Phys. Lett.* **25,** 617.
A. F. Bernhardt, D. E. Duerre, J. R. Simpson, and L. L. Wood (1976a), *J. Opt. Soc. Am.* **66,** 420.
A. F. Bernhardt, D. E. Duerre, J. R. Simpson, and L. L. Wood (1976b), *J. Opt. Soc. Am.* **66,** 416.
A. F. Bernhardt, D. E. Duerre, J. R. Simpson, and L. L. Wood (1976c), *Opt. Commun.* **16,** 166.
A. F. Bernhardt, D. E. Duerre, J. R. Simpson, and L. L. Wood (1976d), *Opt. Commun.* **16,** 169.
M. J. Berry (1974), *Ann. Rev. Phys. Chem.* **26,** 259.
D. R. Betteridge and J. T. Yardley (1979), *Chem. Phys. Lett.* **62,** 570.

R. Bhatnagar, P. E. Dyer, and G. A. Oldershaw (1979), *Chem. Phys. Lett.* **61,** 339.
J. H. Birely and J. F. Lyman (1975), *J. Photochem.* **4,** 269.
J. P. Biscar (1971), *J. Chromatogr.* **56,** 348.
J. E. Bjorkholm, A. Ashkin, and D. B. Pearson (1975), *Appl. Phys. Lett.* **27,** 534.
D. Blaxell, R. Peterson, D. J. Malcolme-Lawes, and R. Wolfgang (1972), *J. Chem. Soc. Chem. Commun.* **2,** 110.
U. Boesl, H. J. Neusser, and E. W. Schlag (1979), *Chem. Phys. Lett.* **61,** 62.
H. -D. V. Böhm, W. Michaelis, and C. Weitkamp (1978), *Opt. Commun.* **26,** 177.
E. Borsella, R. Fantoni, A. GiardiniGuidoni, and G. Sanna (1980), *Chem. Phys. Lett.* **72,** 25.
W. Braun, M. J. Kurylo, A. Kaldor, and R. P. Wayne (1974), *J. Chem. Phys.* **61,** 461.
W. Braun and W. Tsang (1976), *Chem. Phys. Lett.* **44,** 354.
D. M. Brenner (1978), *Clem. Phys. Lett.* **57,** 357.
D. M. Brenner, S. Datta, and R. N. Zare (1977), *J. Am. Chem. Soc.* **99,** 4554.
U. Brinkmann, W. Hartig, H. Telle, and H. Walther (1974), *Appl. Phys.* **5,** 109.
R. D. H. Brown, G. P. Glass, and I. W. M. Smith (1975), *Chem. Phys. Lett.* **32,** 517.
R. D. H. Brown, I. W. M. Smith, and S. W. J. van derMerwe (1976), *Chem. Phys.* **15,** 143.
J. L. Buechele, E. Weitz, and F. D. Lewis (1979), *J. Am. Chem. Soc.* **101,** 3700.
I. I. Bukin and V. I. Kosyahov (1967), *Chem. Abstr.* **67,** 117483.
J. G. Calvert and J. N. Pitts (1966), *Photochemistry,* Wiley, New York.
J. D. Campbell, G. Hancock, J. B. Halpern, and K. H. Welge (1976a), *Opt. Commun.* **17,** 38.
J. D. Campbell, G. Hancock, J. B. Halpern, and K. H. Welge (1976b), *Chem. Phys. Lett.* **44,** 404.
C. Cheng and P. Keehn (1977), *J. Am. Chem. Soc.* **99,** 5809.
K. R. Chien and S. H. Bauer (1976), *J. Phys. Chem.* **80,** 1405.
S. L. Chin and G. Bedard (1971), *Phys. Lett.* **36A,** 271.
J. H. Clark and R. G. Anderson (1978), *Appl. Phys. Lett.* **32,** 46.
J. H. Clark, Y. Haas, P. L. Houston, and C. B. Moore (1975), *Chem. Phys. Lett.* **35,** 82.
D. M. Cox, R. B. Hall, J. A. Horsley, G. M. Kramer, P. Rabinowitz, and A. Kaldor (1979), *Science* **205,** 390.
D. M. Cox and E. T. Maas (1980), *Chem. Phys. Lett.* **71,** 330.
W. C. Danen, D. F. Koster, and R. N. Zitter (1979), *J. Am. Chem. Soc.* **101,** 4281.
B. Davies, A. McNeish, M. Poliakoff, and J. J. Turner (1977), *J. Am. Chem. Soc.* **99,** 7573.
B. Dellinger, D. S. King, R. M. Hochstrasser, and A. B. Smith (1977), *J. Am. Chem. Soc.* **99,** 3197.
M. Desnoyer, G. Nief, and E. Roth (1963), *J. Chem. Phys.* **60,** 14.
D. G. Dever and E. Grunwald (1976), *J. Am. Chem. Soc.* **98,** 5055.
M. Drouin, M. Gauthier, R. Pilon, P. A. Hackett, and C. Willis (1978), *Chem. Phys. Lett.* **60,** 16.
B. L. Earl and A. M. Ronn (1976), *Chem. Phys. Lett.* **41,** 29.
D. K. Evans, R. D. McAlpine, and F. K. McClusky (1978), *Chem. Phys.* **32,** 81.
D. K. Evans, R. D. McAlpine, and F. K. McClusky (1979), *Chem. Phys. Lett.* **65,** 226.
W. E. Farneth, G. Flynn, R. Slater, and N. J. Turro (1976), *J. Am. Chem. Soc.* **98,** 7877.
W. E. Farneth, M. W. Thomsen, and M. A. Berg (1979), *J. Am. Chem. Soc.* **101,** 6468.
S. V. Filseth, J. Danon, D. Feldmann, J. D. Campbell, and K. H. Welge, (1979), *Chem. Phys. Lett.* **63,** 615.
M. P. Freeman, D. N. Travis, and M. F. Goodman (1974), *J. Chem. Phys.* **60,** 231.
S. M. Freund and J. L. Lyman (1978), *Chem. Phys. Lett.* **55,** 435.
S. M. Freund and J. J. Ritter (1975), *Chem. Phys. Lett.* **47,** 595.
G. E. Frigerio and A. Stefanini (1971), *Nuovo Cimento Lett.* **2,** 810.
T. E. Gangwer (1978), *Opt. Laser Technol.,* April, p. 81.

M. GAUTHIER, W. S. NIP, P. A. HACKETT, and C. WILLIS (1980a), *Chem. Phys. Lett.* **69,** 372.
M. GAUTHIER, P. A. HACKETT, and C. WILLIS (1980b), *Chem. Phys.* **45,** 39.
J. GELBWACHS and T. S. HARTWICK (1975), *IEEE J. Quantum Electron.* **QE-11,** 52.
I. GLATT and A. YOGEV (1976), *J. Am. Chem. Soc.* **98,** 7087.
K. S. GOCHELASHVILI, N. V. KARLOV, A. N. ORLOV, R. P. PETROV, YU. N. PETROV, and A. M. PROKHOROV (1975), *JETP Lett.* **21,** 302.
V. I. GOLDANSKII, V. A. NAMIOT, and R. V. KHOKHLOV (1976), *Sov. Phys. JETP* **43,** 1226.
R. J. GORDON and M. C. LIN (1973), *Chem. Phys. Lett.* **22,** 262.
R. J. GORDON and M. C. LIN (1976), *J. Chem. Phys.* **64,** 1058.
M. C. GOWER and K. W. BILLMAN (1977), *Opt. Commun.* **20,** 123.
U. HAAS and G. YAHAV (1977), *Chem. Phys. Lett.* **48,** 63.
R. L. HANSON, D. BROOKINS, and N. E. VANDERBORGH (1976), *Anal. Chem.* **48,** 2210.
R. L. HANSON, N. E. VANDERBORGH, and D. G. BROOKINS (1975), *Anal. Chem.* **47,** 335.
A. HARTFORD JR. (1978), *Chem. Phys. Lett.* **53,** 503.
A. HARTFORD and S. A. TUCCIO (1979), *Chem. Phys. Lett.* **60,** 431.
I. P. HERMAN, R. P. MARIELLA JR., and A. JAVAN (1976), *J. Chem. Phys.* **65,** 3792.
I. P. HERMAN and J. B. MARLING (1979), *Chem. Phys. Lett.* **64,** 75.
R. M. HOCHSTRASSER and D. S. KING (1975), *J. Am. Chem. Soc.* **97,** 4760.
R. M. HOCHSTRASSER and D. S. KING (1976), *J. Am. Chem. Soc.* **98,** 5443.
D. S. Y. HSU and T. J. MANUCCIA (1978), *Appl. Phys. Lett.* **33,** 915.
R. E. HUIE, J. T. HERRON, R. BRAUN, and W. TSANG (1978), *Chem. Phys. Lett.* **56,** 193.
N. R. ISENOR and M. C. RICHARDSON (1971), *Appl. Phys. Lett.* **18,** 224.
W. A. JALENAK and N. S. NOGAR (1979), *Chem. Phys.* **41,** 407.
W. A. JALENAK, D. SCHULTZ, M. FISHER, and N. S. NOGAR (1979), *Chem. Phys. Lett.* **64,** 457.
A. S. JAMES, I. ITZKAN, C. T. PIKE, R. H. LEVY, and L. LEVIN (1976), *IEEE J. Quantum Electron.* **QE-12,** 111.
W. K. JOY, W. R. LADNER, and E. PRITCHARD (1968), *Nature (London)* **217,** 640.
W. K. JOY, W. R. LADNER, and E. PRITCHARD (1970), *Fuel* **49,** 26.
A. KALDOR, R. B. HALL, D. M. COX, J. A. HORSLEY, D. RABINOWITZ, and G. M. KRAMER (1979), *J. Am. Chem. Soc.* **101,** 4465.
I. N. KALVINA, V. F. MOSKALENKO, E. O. OSTAPCHENKO, L. L. PAVOVSKII, T. V. PROTSENKO, and V. I. RYCHKOV (1975), *Sov. J. Quantum Electron* **4,** 1285.
R. R. KARL and K. K. INNES (1975), *Chem. Phys. Lett.* **36,** 275.
N. V. KARLOV, B. B. KRYNETSKII, V. A. MISHIN, and A. M. PROKHOROV (1978), *Appl. Opt.* **17,** 856.
N. V. KARLOV, B. B. KRYNETSKII, V. A. MISHIN, A. M. PROKHOROV, A. D. SALEBEV, and V. V. SMIRNOV (1977a), *Opt. Comm.* **21,** 384.
N. V. KARLOV, R. P. PETROV, YU. N. PETROV, and A. M. PROKHOROV (1977b), *JETP Lett.* **24,** 259.
F. S. KARN, R. A. FRIEDEL, and A. G. SHARKEY JR. (1967), *Carbon* **5,** 25.
F. S. KARN, R. A. FRIEDEL, and A. G. SHARKEY JR. (1969), *Fuel* **48,** 297.
F. S. KARN, R. A. FRIEDEL, and A. G. SHARKEY JR. (1970), *Chem. Ind. (London)* **7,** 239.
F. S. KARN and A. G. SHARKEY JR. (1968), *Fuel* **47,** 193.
F. S. KARN and J. M. SINGER (1968), *Fuel* **47,** 235.
Z. KARNY and B. KATZ (1976), *Chem. Phys. Lett.* **38,** 382.
Z. KARNY, B. KATZ, and A. SZOKE (1975), *Chem. Phys. Lett.* **35,** 100.
Z. KARNY and R. N. ZARE (1977), *Chem. Phys.* **23,** 321.
K. C. KIM, R. FLEMING, D. SEITZ, and M. REISFELD (1979), *Chem. Phys. Lett.* **62,** 61.
S. KIMEL and S. SPEISER (1977), *Chem. Rev.* **77,** 437.
P. KOLODNER, C. WINTERFIELD, and E. YABLONOVITCH (1977), *Opt. Commun.* **20,** 119.
G. KOREN, M. OKON, and U. P. OPPENHEIM (1977), *Opt. Commun.* **22,** 351.
G. KOREN and U. P. OPPENHEIM (1978), *Opt. Commun.* **26,** 449.

G. Koren and U. P. Oppenheim (1979), *Chem. Phys. Lett.* **67,** 289.

G. Koren, U. P. Oppenheim, D. Tal, M. Okon, and R. Weil (1976), *Appl. Phys. Lett.* **29,** 40.

L. N. Krasnopyorov, E. N. Chesnokov, and V. N. Panfilov (1979), *Chem. Phys.* **42,** 345.

R. Kuhlmann *et al.* (1977), *Chem. Abstr.* **86,** 90368d.

M. J. Kurylo, W. Braun, A. Kaldor, S. M. Freund and R. P. Wayne (1974), *J. Photochem.* **3,** 71.

M. J. Kurylo, W. Braun, C. N. Xaun, and A. Kaldor (1975), *J. Chem. Phys.* **62,** 2065.

M. J. Kurylo, W. Braun, S. Abramowitz, and M. Krauss (1976), *J. Res. Natl. Bur. Stand. A* **80,** 167.

V. S. Kushawaha (1980), *Chem. Phys. Lett.* **72,** 451.

M. Lamotte, H. J. Dewey, R. A. Keller, and J. R. Ritter (1975), *Chem. Phys. Lett.* **30,** 165.

S. R. Leone, R. G. McDonald, and C. B. Moore (1975), *J. Chem. Phys.* **63,** 4735.

S. R. Leone and C. V. Moore (1974), *Phys. Rev. Lett.* **33,** 269.

M. L. Lesiecki and W. A. Guillary (1977), *J. Chem. Phys.* **66,** 4317.

V. S. Letokhov (1977), *Physics Today,* May, p. 23.

V. S. Letokhov and C. B. Moore (1977), in *Chemical and Biochemical Applications of Lasers,* Vol. III, C. B. Moore, Ed., Academic Press, New York.

V. S. Letokhov, E. A. Ryabov, and O. A. Tumanov (1972), *Zh. Eksp. Teor. Fiz.* **63,** 2025.

C. T. Lin, T. D. Z. Atvars, and F. B. T. Pessine (1977), *J. Appl. Phys.* **48,** 1720.

S. T. Lin, S. M. Lee, and A. M. Ronn (1978), *Chem. Phys. Lett.* **53,** 260.

D. D.-S. Liu, S. Datta, and R. N. Zare (1975), *J. Am. Chem. Soc.* **97,** 2557.

E. R. Lory, S. H. Bauer, and T. Manuccia (1975), *J. Phys. Chem.* **79,** 545.

J. L. Lyman and R. J. Jensen (1972), *Chem. Phys. Lett.* **13,** 421.

J. L. Lyman and R. J. Jensen (1973), *J. Phys. Chem.* **77,** 883.

J. L. Lyman, R. J. Jensen, J. Rink, C. P. Robinson, and S. D. Rockwood (1975), *Appl. Phys. Lett.* **27,** 87.

J. L. Lyman and S. D. Rockwood (1976), *J. Appl. Phys.* **47,** 595.

L. Mannik, G. M. Keyser, and K. B. Woodall (1979), *Chem. Phys. Lett.* **65,** 231.

R. G. Manning, W. Braun, and M. J. Kurylo (1976), *J. Chem. Phys.* **65,** 2609.

T. J. Manuccia, M. D. Clark, and E. R. Lory (1976), *Opt. Commun.* **18,** 219.

J. B. Marling (1975), *Chem. Phys. Lett.* **34,** 84.

J. B. Marling (1976), *Opt. Commun.* **18,** 36.

J. B. Marling (1977), *J. Chem. Phys.* **66,** 4200.

S. W. Mayer, M. A. Kwok, R. W. F. Gross, and D. J. Spencer (1970), *Appl. Phys. Lett.* **17,** 516.

K. B. McAfee, R. M. Lum, and R. S. Hozack (1976), *J. Chem. Phys.* **64,** 5073.

R. D. McAlpine, D. K. Evans, and F. K. McClusky (1979), *Chem. Phys.* **39,** 263.

R. S. McDowell, H. W. Galbraith, B. J. Krohn, C. D. Cantrell, and E. D. Hinkley (1976), *Opt. Commun.* **17,** 178.

F. P. Miknis and J. P. Bisnar (1971), *J. Phys. Chem.* **75,** 725.

S. S. Miller, D. D. DeFord, T. J. Marks, and E. Weitz (1979), *J. Am. Chem. Soc.* **101,** 1036.

Y. N. Molin, V. N. Panifilov, and V. P. Strunin (1978), *Chem. Phys. Lett.* **56,** 557.

C. B. Moore (1973), *Acc. Chem. Res.* **6,** 323.

J. P. Morand and G. Nief (1968), *J. Chim. Phys.* **65,** 2058.

S. Mukamel and J. Jortner (1976a), *Chem. Phys. Lett.* **40,** 150.

S. Mukamel and J. Jortner (1976b), *J. Chem. Phys.* **65,** 5204.

C. H. Muller, K. Schofield, and M. Steinberg (1978), *Chem. Phys. Lett.* **57,** 364.

K. Nagai and M. Katayama (1977), *Chem. Phys. Lett.* **51,** 329.

K. Nagai, M. Katayama, H. Mikuni, and M. Takahasi (1979), *Chem. Phys. Lett.* **62,** 499.

S. Namba, P. H. Kim, and K. Taki (1969), Recordings of the 10th Symposium on Electron, Ion, Laser Beam Technology, L. Marton, Ed., San Francisco Press, San Francisco, p. 493.

N. Nishi, H. Shinohara, and I. Hanazaki (1980), *Chem. Phys. Lett.* **73,** 473.

T. J. ODIORNE, P. R. BROOKS, and J. V. V. KASPER (1971), *J. Chem. Phys.* **55,** 1980.
A. E. OREL and W. H. MILLER (1978), *Chem. Phys. Lett.* **57,** 362.
V. N. PANFILOV and Y. N. MOLIN (1978), *Russ. Chem. Rev.* **47,** 503.
J. M. PRESES, R. E. WESTON, JR., and C. G. FLYNN (1977), *Chem. Phys. Lett.* **46,** 69.
D. PROCH and H. SCHRÖDER (1979), *Chem. Phys. Lett.* **61,** 426.
C. R. QUICK JR., A. B. HORWITZ, R. E. WESTON, JR., and A. W. FLYNN (1980), *Chem. Phys. Lett.* **72,** 352.
P. RABINOWITZ, A. STEIN, and A. KALDOR (1978), *Opt. Commun.* **27,** 381.
L. J. RADZIEMSKI, S. GERSTENKORN, and P. LUE (1975), *Opt. Commun.* **15,** 273.
K. V. REDDY and M. J. BERRY (1980), *Chem. Phys. Lett.* **72,** 29.
E. A. RINEHART, J. H. RICHARDSON, D. C. JOHNSON, and L. W. HRUBESH (1976), *Appl. Phys. Lett.* **28,** 131.
J. J. RITTER (1978), *J. Am. Chem. Soc.* **100,** 2441.
S. ROCKWOOD and S. W. RABIDEAU (1974), *IEEE J. Quantum Electron.* **QE-10,** 789.
C. K. ROFER-DEPOORTER and G. L. DEPOORTER (1979), *Chem. Phys. Lett.* **61,** 605.
C. RILEY and R. SHATAS (1979), *J. Phys. Chem.* **83,** 1679.
C. RILEY, S. SHATAS, and V. ARKLE (1978), *J. Am. Chem. Soc.* **100,** 658.
C. RILEY, R. SHATAS, and L. OPP (1979), *Inorg. Chem.* **18,** 460.
M. ROSSI, J. R. BARKER, and D. M. GOLDEN (1979), *Chem. Phys. Lett.* **65,** 523.
E. W. ROTHE, B. P. MATHUR, and G. P. RECK (1978), *Chem. Phys. Lett.* **53,** 74.
R. K. SANDER, T. R. LOREE, S. D. ROCKWOOD, and S. M. FREUND (1977), *Appl. Phys. Lett.* **30,** 150.
R. SCHAEFFER and R. K. PEARSON (1969), *J. Am. Chem. Soc.* **91,** 2153.
A. G. SHARKEY JR., J. L. SHULTZ, and R. A. FRIEDEL (1964), *Nature (London)* **202,** 968.
S. SHATAS, D. GREGORY, R. SHATAS, and C. RILEY (1978), *Inorg. Chem.* **17,** 163.
W. M. SHAUB and S. H. BAUER (1975), *Int. J. Chem. Kinet.* **7,** 509.
J. L. SHULTZ and A. G. SHARKEY JR. (1967), *Carbon* **5,** 57.
R. C. SLATER and J. H. PARKS (1979), *Chem. Phys. Lett.* **60,** 275.
V. SLEZAK, J. CABELLERO, A. BURGOS, and E. QUEL (1978), *Chem. Phys. Lett.* **54,** 105.
I. W. M. SMITH (1977), *Chem. Phys.* **20,** 437.
B. B. SNAVELY, R. W. SOLARZ, and S. A. TUCCIO (1975), in *Laser Spectroscopy* H. Walther, Ed., Springer-Verlag, Berlin, p. 268.
J. C. STEPHENSON and S. M. FREUND (1976), *J. Chem. Phys.* **65,** 4303.
M. STUKE and E. E. MARINERO (1979), *Chem Phys. Lett.* **68,** 28.
K. Y. TANG and E. K. C. LEE (1976), *J. Phys. Chem.* **80,** 1833.
J. TARDIEU DE MALEISSYE, F. LEMPEREUR, C. MARSAL, and R. I. BEN-AÏM (1976), *Chem. Phys. Lett.* **42,** 46.
M. TASUMI, H. TAKEUCHI, and H. NAKANO (1979), *Chem. Phys. Lett.* **68,** 44.
J. J. TIEE, F. B. WAMPLER, and W. W. RICE (1979), *Chem. Phys. Lett.* **68,** 403.
J. J. TIEE and C. WITTIG (1978), *Opt. Commun.* **27,** 377.
W. B. TIFFANY (1968), *J. Chem. Phys.* **48,** 3019.
S. A. TUCCIO, J. W. DURBIN, O. G. PETERSON, and B. B. SNAVELY (1974), *IEEE J. Quant. Electron.* **QE–10,** 790.
S. A. TUCCIO and A. HARTFORD (1979), *Chem. Phys. Lett.* **65,** 234.
M. E. UMSTEAD and M. C. LIN (1978), *J. Phys. Chem.* **82,** 2047.
E. WEITZ and C. W. FLYNN (1974), *Ann. Rev. Phys. Chem.* **25,** 275.
G. A. WEST, R. E. WESTON, JR., and G. W. FLYNN (1978), *Chem. Phys.* **35,** 275.
M. YAMASHITA, M. KASAMATSU, H. KASHIWAGI, and K. MACHIDA (1978), *Opt. Commun.* **26,** 343.
J. T. YARDLEY (1979), in *Quantum Electronics,* Part B, C. L. Tang, Ed., Academic Press, New York, p. 269.
E. S. YEUNG and C. B. MOORE (1972), *Appl. Phys. Lett.* **21,** 109.

A. YOGEV and R. M. J. BENMAIR (1975), *J. Am. Chem. Soc.* **97,** 4430.
A. YOGEV and R. M. J. BENMAIR (1977), *Chem. Phys. Lett.* **46,** 290.
A. YOGEV and R. M. J. BENMAIR (1979), *Chem. Phys. Lett.* **63,** 558.
A. YOGEV and R. M. J. LOEWENSTEIN-BENMAIR (1973), *J. Am. Chem. Soc.* **95,** 8487.
A. YOGEV, R. M. J. LOEWENSTEIN-BENMAIR, and D. AMAR (1972), *J. Am. Chem. Soc.* **94,** 1091.
R. N. ZITTER and D. F. KOSTER (1976), *J. Am. Chem. Soc.* **98,** 1613.
R. N. ZITTER and D. F. KOSTER (1978), *J. Am. Chem. Soc.* **100,** 2265.
R. N. ZITTER, R. A. LAU, and K. S. WILLS (1975), *J. Am. Chem. Soc.* **97,** 2578.

Chapter 5

A. ANDERSON (1971), *The Raman Effect,* Vol. 1, Marcel Dekker, New York.
A. ANDERSON (1973), *The Raman Effect,* Vol. 2, Marcel Dekker, New York.
A. M. ANGUS, E. E. MARINERO, and M. J. COLLES (1975), *Opt. Commun.* **14,** 223.
E. N. ANTONOV, V. G. KOLOSHNIKOV, and V. R. MIRONENKO (1975a), *Sov. J. Quantum Electron.* **5,** 105.
E. N. ANTONOV, V. G. KOLOSHNIKOV, and V. R. MIRONENKO (1975b), *Opt. Commun.* **15,** 99.
V. S. ANTONOV, I. N. KRYAZEV, V. S. LETOKHOV, V. M. MATINK, V. G. MOVSHEV, and V. K. POTAPOV (1978), *Opt. Lett.* **3,** 37.
V. B. ANZIN, M. V. GLUSHKOV, V. P. ZHAROV, YU. V. KOSICHKIN, V. O. SHAIDUROV, and A. M. SHIROKOV (1975), *Sov. J. Quantum Electron.* **5,** 754.
G. H. ATKINSON, T. N. HEIMLICK, and M. W. SCHUYLER (1977), *J. Chem. Phys.* **66,** 5005.
G. H. ATKINSON, A. H. LAUFER, and M. J. KURYLO (1973), *J. Chem. Phys.* **59,** 350.
R. H. BARNES and J. F. KIRCHER (1978), *Appl. Opt.* **17,** 1099.
R. H. BARNES, C. E. MOELLER, J. F. KIRCHER, and C. M. VERBER (1973), *Appl. Opt.* **12,** 2531.
A. P. BARONAVSKI and J. R. MCDONALD (1977), *Appl. Opt.* **16,** 1897.
J. J. BARRETT (1978), *J. Opt. Soc. Am.* **68,** 1433.
J. J. BARRETT and M. J. BERRY (1979), *Appl. Phys. Lett.* **34,** 144.
R. F. BEGLEY, A. B. HARVEY, and R. L. BYER (1974), *Appl. Phys. Lett.* **25,** 387.
M. R. BERMAN and R. N. ZARE (1975), *Anal. Chem.* **47,** 1201.
R. A. BINGHAM and P. L. SALTER (1976a), *Int. J. Mass Spectrom. Ion Phys.* **21,** 133.
R. A. BINGHAM and P. L. SALTER (1976b), *Anal. Chem.* **48,** 1735.
D. C. BLAZEJ and W. L. PETICOLAS (1977), *Proc. Natl. Acad. Sci. USA* **74,** 2639.
M. A. BOLSHOV, A. V. ZYBIN, L. A. ZYBINA, V. G. KOLOSHRIKOV, and I. A. MAJOROV (1976), *Spectrochim. Acta B* **31,** 493.
G. BOUTILIER and J. D. WINEFORDNER (1979), *Anal. Chem.* **51,** 1384.
R. G. BRAY, W. HENKE, S. K. LIN, K. V. REDDY, and M. J. BERRY (1977), *Chem. Phys. Lett.* **47,** 213.
E. J. BREFORD and F. ENGELKE (1979), *J. Chem. Phys.* **71,** 1994.
W. B. BRIDGES (1978), *J. Opt. Soc. Am.* **68,** 352.
G. O. BRINK (1980), *Opt. Commun.* **32,** 123.
H. L. BROD and E. S. YEUNG (1976), *Anal. Chem.* **48,** 344.
J. H. BROPHY and C. T. RETTNER (1979a), *Opt. Lett.* **4,** 337.
J. H. BROPHY and C. T. RETTNER (1979b), *Opt. Lett.* **67,** 351.
R. E. BROWN, K. D. LEGG, M. W. WOLF, L. SINGER, and J. H. PARKS (1974), *Anal. Chem.* **46,** 1690.
C. W. BRUCE and R. G. PINNICK (1977), *Appl. Opt.* **16,** 1762.
W. BRUNNER and H. PAUL (1974), *Opt. Commun.* **12,** 252.
W. BRUNNER and H. PAUL (1976), *Opt. Commun.* **19,** 253.
W. BRUNNER and H. PAUL (1978), *Opt. Commun.* **24,** 11.
G. BUSSE (1980), *Opt. Laser Technol.,* June, p. 150.
G. BUSSE and A. OGRABECK (1980), *J. Appl. Phys.* **51,** 3576.

E. Cerrai and R. Trucco (1968), *Energ. Nucl. (Italy)* **15,** 581.
W. S. Childs, M. S. Freds, and L. S. Goodman (1974), *Appl. Opt.* **13,** 2297.
P. C. Claspy, C. Ha, and Y-H. Pao (1977), *Appl. Opt.* **16,** 2972.
P. C. Claspy, Y-H. Pao, S. Kwong, and E. Nodov (1976), *Appl. Opt.* **15,** 1506.
R. J. Conzemius and H. J. Svec (1978), *Anal. Chem.* **50,** 1854.
F. Crescenzi and J. S. Shirk (1977), *Chem. Phys.* **20,** 331.
P. J. Dagdigian (1976), *J. Chem. Phys.* **64,** 2609.
P. J. Dagdigian, H. W. Cruse, A. Schultj, and R. N. Zare (1974), *J. Chem. Phys.* **61,** 4450.
J. W. Daily (1977), *Appl. Opt.* **16,** 568.
J. W. Daily (1978), *Appl. Opt.* **17,** 225.
S. Datta, R. W. Anderson, and R. N. Zare (1975), *J. Chem. Phys.* **63,** 5503.
M. B. Denton and H. V. Malmstadt (1971), *Appl. Phys. Lett.* **18,** 485.
P. D. Dhamelincourt, F. Wallart, M. Leclereq, A. T. N'Guyen, and D. O. Landon (1979), *Anal. Chem.* **51,** 414A.
R. Dirscherl and H. U. Lee (1979), in *Laser-Induced Processes in Molecules,* K. L. Kompa and S. D. Smith, Eds., Springer-Verlag, Berlin, p. 239.
S. Druet and J-P Taran (1979), in *Chemical and Biochemical Applications of Lasers,* C. B. Moore, Ed., Academic Press, New York, p. 187.
M. Dubs and H. H. Günthard (1978), *Appl. Opt.* **17,** 3593.
W. W. Duley (1976), *CO_2 Lasers: Effects and Applications,* Academic Press, New York.
A. C. Eckbreth, P. A. Bonczyk, and J. F. Verdieck (1977), *Appl. Spectros. Rev.* **13,** 15.
D. L. Feldman, R. K. Lengel, and R. N. Zare (1977), *Chem. Phys. Lett.* **52,** 413.
P. S. H. Fitch, L. Wharton, and D. H. Levy (1979), *J. Chem. Phys.* **70,** 2018.
O. F. Folmer, Jr. (1971), *Anal. Chem.* **43,** 1057.
G. T. Forrest (1980), *Appl. Opt.* **19,** 2094.
L. M. Fraser and J. D. Winefordner (1972), *Anal. Chem.* **44,** 1444.
S. K. Freeman (1974), *Applications of Laser Raman Spectroscopy,* Wiley, New York.
S. M. Freund, D. M. Sweger, and J. C. Travis (1976), *Anal. Chem.* **48,** 1944.
R. Frueholz, J. Wessel, and E. Wheatley (1980), *Anal. Chem.* **52,** 281.
J. A. Gelbwachs, C. F. Klein, and J. E. Wessel (1977), *Appl. Phys. Lett.* **30,** 489.
K. P. Ghiggino, A. J. Roberts and D. Phillips (1979), in *Laser-Induced Processes in Molecules,* K. L. Kompa and S. D. Smith, Eds., Springer-Verlag, Berlin, p. 98.
T. R. Gilson and P. J. Hendra (1970), *Laser Raman Spectroscopy,* Wiley, London.
E. Giovannini, G. B. Principato, and F. Rondelli (1976), *Anal. Chem.* **48,** 1517.
D. A. Goff and E. S. Yeung (1978), *Anal. Chem.* **50,** 625.
P. D. Goldan and K. Goto (1974), *J. Appl. Phys.* **45,** 4350.
R. B. Green, R. A. Keller, G. G. Luther, P. K. Schenck, and J. C. Travis (1976b), *Appl. Phys. Lett.* **29,** 726.
R. B. Green, R. A. Keller, P. K. Schenck, J. C. Travis, and G. G. Luther (1976a), *J. Am. Chem. Soc.* **98,** 8517.
R. B. Green and H. W. Latz (1974), *Spectros. Lett.* **7,** 419.
R. B. Green, J. C. Travis, and R. A. Keller (1976), *Anal. Chem.* **48,** 1954.
G. G. Guilbault (1973), *Practical Fluorescence. Theory, Methods and Techniques,* Marcel Dekker, New York.
B. T. Guran, R. J. O'Brien, and D. H. Anderson (1970), *Anal. Chem.* **42,** 115.
F. J. Gustafson and J. C. Wright (1977), *Anal. Chem.* **49,** 1680.
F. J. Gustafson and J. C. Wright (1979), *Anal. Chem.* **51,** 1762.
D. M. Guthals, K. P. Gross, and J. W. Nibler (1979), *J. Chem. Phys.* **70,** 2393.
T. W. Hänsch, A. L. Schawlow, and P. E. Toschek (1972), *IEEE J. Quantum Electron.* **QE-8,** 802.
W. R. Harshbarger and M. B. Robin (1973), *Acc. Chem. Res.* **6,** 329.
A. B. Harvey and J. W. Nibler (1978), *Appl. Spectros. Rev.* **14,** 101.

P. J. HENDRA (1975), in *Vibrational Spectra and Structure,* Vol. 2, J. R. Durig, Ed., Marcel Dekker, New York.

G. HERZBERG (1945), *Infrared and Raman Spectra of Polyatomic Molecules* Van Nostrand, Princeton.

F. HILLENKAMP, R. KAUFMANN, R. NITSCHE, and E. UNSÖLD (1975), *Appl. Phys.* **8,** 341.

E. D. HINKLEY and G. FREED (1969), *Phys. Rev. Lett* **23,** 277.

E. D. HINKLEY, R. T. KU, K. W. NILL, and J. F. BUTLER (1976), *Appl. Opt.* **15,** 1653.

J. P. HOHIMER and P. J. HARGIS (1979), *Anal. Chem.* **51,** 930.

H. HOLT (1975), *Phys. Rev. A* **11,** 625.

G. HORLICK and E. G. CODDING (1974), *Anal. Chem.* **46,** 133.

T. IMASAKA, H. KADONE, T. OGAWA, and N. ISHIBASHI (1977), *Anal. Chem.* **49,** 667.

H. INABA (1976), in *Laser Monitoring of the Atmosphere,* D. E. Hinkley, Ed., Springer-Verlag, Berlin.

N. ISHIBASHI, T. OGAWA, T. IMASAKA, and M. KUNITAKE (1979), *Anal. Chem.* **51,** 2096.

T. ISHIZUKA, Y. UWAMINO, and H. SUNAHARA (1977), *Anal. Chem.* **49,** 1339.

M. V. JOHNSTON and J. C. WRIGHT (1979), *Anal. Chem.* **51,** 1774.

D. G. JONES and C. J. MACKIE (1976), *Combust. Flame* **27,** 143.

R. D. KAMM (1976), *J. Appl. Phys.* **47,** 3550.

R. A. KELLER, E. F. ZALEWSKI, and N. C. PETERSON (1972), *J. Opt. Soc. Am.* **62,** 319.

J. L. KINSEY (1977), in *Annual Reviews of Physical Chemistry,* Vol. 28, B. S. Rabinovitch, J. M. Schurr, and H. L. Strauss, Eds., Annual Rev. Inc., Palo Alto, Calif., p. 349.

R. KIRCHHEIM, U. NAGORNY, K. MAIER, and G. TÖLG (1976), *Anal. Chem.* **48,** 1505.

V. J. KLEINSCHMIDT, W. TOTTLEBEN, and S. REUTSCH (1974), *Exp. Tech. Phys.* **22,** 191.

C. M. KLIMCAK and J. E. WESSEL (1980), *Anal. Chem.* **52,** 1233.

K. P. KOCH and W. LAHMANN (1978), *Appl. Phys. Lett.* **32,**289.

J. A. KONINGSTEIN (1972), *Introduction to the Theory of the Raman Effect,* D. Reidel, Dordrecht.

L. B. KREUZER (1971), *J. Appl. Phys.* **42,** 2934.

L. B. KREUZER (1974), *Anal. Chem.* **46,** 235A.

L. B. KREUZER (1978), *Anal. Chem.* **50,** 597A.

L. B. KREUZER, N. D. KENYON, and C. K. N. DATEL (1972), *Science* **177,** 347.

L. B. KREUZER and C. K. N. PATEL (1971), *Science* **173,** 45.

J. KUHL and G. MAROWSKY (1971), *Opt. Commun.* **4,** 125.

J. KUHL, S. NEUMANN, and M. KRIESE (1973), *Z. Naturforsch. A* **28,** 273.

J. KUHL and H. SPITSCHAN (1973), *Opt. Commun.* **7,** 256.

H. S. KWONG and R. M. MEASURES (1979), *Anal. Chem.* **51,** 428.

W. LAHMANN, J. J. LUDEWIG, and H. WELLING (1977), *Anal. Chem.* **49,** 549.

M. D. LEVENSON (1977), *Physics Today,* May, p. 44.

L. R. LEWELLEN and G. O. BRINK (1981), *Opt. Commun.* **39,** 159.

H. -B. LIN and M. R. TOPP (1977), *Chem. Phys. Lett.* **47,** 442.

F. E. LYTLE and M. S. KELSEY (1974), *Anal. Chem.* **46,** 855.

M. MAEDA, F. ISHITSUKA, M. MATSUMOTO, and Y. MIYAZOE (1977), *Appl. Opt.* **16,** 403.

M. MAEDA, F. ISHITSUKA, and Y. MIYAZOE (1975), *Opt. Commun.* **13,** 314.

R. M. MANABE and E. H. PIEPMEIER (1979), *Anal. Chem.* **51,** 2066.

J. R. MAPLE, E. L. WEHRY, and G. MAMANTOV (1980), *Anal. Chem.* **52,** 920.

K. W. MARICH, P. W. CARR, W. J. TRETYL, and D. GLICK (1970), *Anal. Chem.* **42,** 1775.

K. W. MARICH, W. J. TRETYL, J. G. HAWLEY, N. A. PEPPER, R. E. MYERS, and D. GLICK (1974), *J. Phys. E* **7,** 830.

J. P. MATOUSEK and B. J. ORR (1976), *Spectrochim. Acta B* **31,** 475.

E. MAX and L. G. ROSENGREN (1974), *Opt. Commun.* **11,** 422.

S. MAYO, R. A. KELLER, J. C. TRAVIS, and R. B. GREEN (1976), *J. Appl. Phys.* **47,** 4012.

B. Meyer (1971), *Low Temperature Spectroscopy,* Elsevier, New York.
M. P. Miller, D. R. Tallant, F. J. Gustafson, and J. C. Wright (1977), *Anal. Chem.* **49,** 1474.
V. R. Mironenko and I. Pack (1979), *Sov. J. Quantum Electron.* **8,** 1394.
F. J. Morgan, C. H. Dugan, and A. G. Lee (1978), *Opt. Commun.* **27,** 451.
K. L. Morton, J. D. Nohe, and B. S. Madsen (1973), *Appl. Spectrosc.* **27,** 109.
V. G. Mossotti, K. Laqua, and W.-D. Hagenah (1967), *Spectrochim. B* **23,** 197.
S. Neumann and M. Kriese (1974), *Spectrochim. Acta B* **29,** 127.
J. W. Nibler and G. V. Knighten (1979), in *Raman Spectroscopy of Gases and Liquids,* A. Weber, Ed., Springer-Verlag, Berlin, p. 253.
J. W. Nibler, J. R. MacDonald, and A. B. Harvey (1976), in Proceedings of the 5th International Conference on Raman Spectroscopy, E. D. Schmidt *et al.,* Eds., Schulz Verlag, Freilberg, p. 717.
S. Oda, T. Sawada, and H. Kamada (1978), *Opt. Commun.* **50,** 865.
S. Oda, T. Sawada, M. Nomura, and H. Kamada (1979), *Anal. Chem.* **51,** 686.
S. Oda, T. Sawada, T. Moriguchi, and H. Kamada (1980), *Anal. Chem.* **52,** 650.
N. Omenetto, N. N. Hatch, L. M. Fraser, and J. D. Winefordner (1973), *Spectrochim. Acta B* **28,** 65.
G. C. Orner and M. R. Topp (1975), *Chem. Phys. Lett.* **36,** 295.
D. E. Osten and E. H. Piepmeier (1973), *Appl. Spectrosc.* **27,** 165.
Y.-H Pao (1977), *Optoacoustic Spectroscopy and Detection,* Academic Press, New York.
L. Pasternack and P. J. Dagdigian (1978), *Chem. Phys.* **33,** 1.
C. K. N. Patel and A. C. Tam (1979), *Appl. Phys. Lett.* **34,** 467.
J. A. Perry, M. F. Bryant, and H. V. Malmstadt (1977), *Anal. Chem.* **49,** 1702.
N. C. Peterson, M. J. Kurylo, W. Braun, A. M. Bass, and R. A. Keller (1971), *J. Opt. Soc. Am.* **61,** 746.
E. H. Piepmeier (1972), *Spectrochim Acta B***27,** 431.
E. H. Piepmeier and D. E. Osten (1971), *Appl. Spectrosc.* **25,** 642.
J. G. Pruett and R. N. Zare (1975), *J. Chem. Phys.* **62,** 2050.
J. G. Pruett and R. N. Zare (1976), *J. Chem. Phys.* **64,** 1774.
S. D. Rasberry, B. F. Scribner, and M. Margoshes (1967a), *Appl. Opt.* **6,** 81.
S. D. Rasberry, B. F. Scribner, and M. Margoshes (1967b), *Appl. Opt.* **6,** 87.
J. F. Ready (1971), *Effects of High Power Laser Radiation,* Academic Press, New York.
P. Regnier and J-P. E. Taran (1973), *Appl. Phys. Lett.* **23,** 240.
J. Reid, J. Shewchun, B. K. Garside, and E. A. Ballik (1978), *Appl. Opt.* **17,** 300.
J. P. Reilly and G. C. Pimentel (1976), *Appl. Opt.* **15,** 2372.
J. H. Richardson (1977), *Anal. Biochem.* **83,** 754.
J. H. Richardson and M. E. Ando (1977), *Anal. Chem.* **49,** 955.
J. H. Richardson and S. M. George (1978), *Anal. Chem.* **50,** 616.
J. H. Richardson, B. W. Wallin, D. C. Johnson, and L. W. Houbesh (1976), *Anal. Chim. Acta* **86,** 263.
W. T. Ristau and N. E. Vanderborgh (1970), *Anal. Chem.* **42,** 1848.
W. T. Ristau and N. E. Vanderborgh (1971), *Anal. Chem.* **43,** 702.
A. B. Rodrigo and R. M. Measures (1973), *IEEE J. Quantum Electron.* **QE-9,** 972.
G. J. Rosasco, E. Roedder, and J. H. Simmons (1975a), *Science* **190,** 557.
G. J. Rosasco, E. S. Etz, and W. A. Cassatt (1975b), *Appl. Spectrosc.* **29,** 396.
A. Rosencwaig (1975a), *Anal. Chem.* **47,** 593A.
A. Rosencwaig (1975b), *Physics Today,* September, p. 23.
D. L. Rousseau, J. M. Friedman, and P. F. Williams (1979), in *Raman Spectroscopy of Gases and Liquids,* A. Weber, Ed., Springer-Verlag, New York.
A. J. Saffir, K. W. Marich, and J. R. Orenburg (1972), *Appl. Spectrosc.* **26,** 429.

R. K. Sander, B. Soep, and R. N. Zare (1976), *J. Chem. Phys.* **64,** 1242.
T. Sawada, S. Oda, H. Shimizu, and H. Kamada (1979), *Anal. Chem.* **51,** 688.
P. K. Schenck, W. G. Mallard, J. C. Travis, and K. C. Smyth (1978), *J. Chem. Phys.* **69,** 5147.
S. A. Schleuserner, J. D. Lindberg, K. O. White, and R. L. Johnson (1976), *Appl. Opt.* **15,** 2546.
W. Schmidt, A. Siegel, and A. Schultz (1976), *Chem. Phys.* **16,** 161.
H. Schröder, H. J. Neusser, and E. W. Schlag (1975), *Opt. Commun.* **14,** 395.
A. Schultz, H. W. Cruse, and R. N. Zare (1972), *J. Chem. Phys.* **57,** 1354.
M. J. Sepaniak and E. S. Yeung (1977), *Anal. Chem.* **49,** 1554.
W. Sharfin, K. E. Johnson, L. Wharton, and D. H. Levy (1979), *J. Chem. Phys.* **71,** 1292.
B. L. Sharp and A. Goldwasser (1976), *Spectrochim. Acta B* **31,** 431.
S. Shtrikman and M. Slatkine (1977), *Appl. Phys. Lett.* **31,** 830.
M. P. Sinha, A. Schultz, and R. N. Zare (1973), *J. Chem. Phys.* **58,** 549.
R. E. Smalley, D. H. Levy, and L. Wharton (1976), *J. Chem. Phys.* **64,** 3266.
R. E. Smalley, B. L. Ramakrishna, D. H. Levy, and L. Wharton (1974), *J. Chem. Phys.* **61,** 4363.
R. E. Smalley, L. Wharton, and D. H. Levy (1975), *J. Chem. Phys.* **63,** 4977.
B. W. Smith, M. B. Blackburn, and J. D. Winefordner (1977), *Can. J. Spectrosc.* **22,** 57.
B. W. Smith, F. W. Plankey, N. Omenetto, L. P. Hart, and J. Winefordner (1974), *Spectrochim. Acta A* **30,** 1459.
G. P. Smith and R. N. Zare (1976), *J. Chem. Phys.* **64,** 2632.
K. C. Smyth and P. K. Schenck (1978), *Chem. Phys. Lett.* **55,** 466.
R. W. Solarz and S. A. Johnson (1979), *J. Chem. Phys.* **70,** 3592.
R. C. Spiker and J. S. Shirk (1974), *Anal. Chem.* **46,** 572.
T. G. Spiro (1974), in *Chemical and Biochemical Applications of Lasers* Vol. 1, C. B. Moore, Ed., Academic, New York.
B. I. Stepanov and A. N. Rubinov (1975), *Spectros. Lett.* **8,** 621.
M. Stuke and F. P. Schäffer (1977), *Chem. Phys. Lett.* **48,** 271.
K. Sumino, R. Yamamoto, F. Hatayama, S. Kitamura, and H. Itoh (1980), *Anal. Chem.* **52,** 1064.
D. M. Sweger and J. C. Travis (1979), *Appl. Spectrosc.* **33,** 46.
R. W. Terhune and J. E. Anderson (1977), *Opt. Lett.* **1,** 70.
K. R. Thompson (1976), *Anal. Chem.* **48,** 696.
R. J. Thrash, H. von Wessenhoff, and J. S. Shirk (1971), *J. Chem. Phys.* **55,** 4659.
K. Tohma (1975), *Opt. Commun.* **15,** 17.
K. Tohma (1976), *J. Appl. Phys.* **47,** 1422.
W. M. Tolles, J. W. Nibler, J. R. McDonald, and A. B. Harvey (1977), *Appl. Spectros.* **31,** 253.
M. R. Topp (1978), *Appl. Spectros. Rev.* **14,** 1.
M. R. Topp and H. B. Lin (1977), *Chem. Phys. Lett.* **50,** 412.
J. C. Travis, P. K. Schenck, G. C. Turk, and W. G. Mallard (1979), *Anal. Chem.* **51,** 1516.
W. J. Treytl, K. W. Marich, J. B. Orenburg, P. W. Carr, D. C. Miller, and D. Glick (1971), *Anal. Chem.* **43,** 1452.
G. C. Turk, W. G. Mallard, P. K. Schenck, and K. C. Smyth (1979b), *Anal. Chem.* **51,** 2408.
G. C. Turk, J. C. Travis, J. R. DeVoe, and T. C. O'Haver (1979a), *Anal. Chem.* **51,** 1890.
C. P. Wang (1976), *Combust. Sci. Technol.* **13,** 211.
C. C. Wang and L. I. Davis (1974a), *Phys. Rev. Lett.* **32,** 349.
C. C. Wang and L. I. Davis (1974b), *Appl. Phys. Lett.* **25,** 34.

A. WEBER (1979), *Raman Spectroscopy of Gases and Liquids,* Springer-Verlag, New York.
R. WECHSUNG, F. HILLENKAMP, R. KAUFMANN, R. NITSCHE, and H. VOGT (1978), *Scan Electron. Microsc.* **1,** 611.
S. J. WEEKS, H. HARAGUCHI, and J. D. WINEFORDNER (1978), *Anal. Chem.* **50,** 360.
G. A. WEST, D. R. SIEBERT, and J. J. BARRETT (1980), *J. Appl. Phys.* **51,** 2823.
R. H. WILEY and P. VIERAVAGU (1968), *J. Phys. Chem.* **72,** 2417.
R. M. WILSON and T. L. MILLER (1975), *Anal. Chem.* **47,** 256.
J. D. WINEFORDNER and T. J. VICKERS (1970), *Anal. Chem.* **42,** 206R.
J. C. WRIGHT (1977), *Anal. Chem.* **49,** 1690.
J. C. WRIGHT and F. J. GUSTAFSON (1978), *Anal. Chem.* **50,** 1147A.
N. E. VANDERBROUGH, R. L. HANSON, and C. BROWER (1975), *Anal. Chem.* **47,** 2277.
C. A. VAN DIJK, C. TH. J. ALKEMADE, and P. J. ZEEGERS (1978a), *Appl. Spectrosc.* **32,** 189.
C. A. VAN DIJK, P. J. TH. ZEEGERS, G. NIENHUIS, and C. TH. J. ALKEMADE (1978b), *J. Quantum Spectros. Rad. Transfer* **20,** 55.
R. P. VAN DUYNE (1979), in *Chemical and Biochemical Applications of Lasers,* Vol. 4, C. B. Moore, Ed., Academic Press, New York.
T. F. VAN GEEL and J. D. WINEFORDNER (1976), *Anal. Chem.* **48,** 335.
H. VON WESSENHOFF and U. REHLING (1974), *Z. Naturforsch A* **29,** 256.
M. YAMASHITA, M. KASAMATSU, H. KASHIWAGI, and K. MACHIDA (1978), *Opt. Commun.* **26,** 343.
E. F. ZALEWSKI, R. A. ZELLER, and R. ENGLEMAN (1979), *J. Chem. Phys.* **70,** 1015.
L. ZANDEE, R. B. BERNSTEIN, and D. A. LICHTIN (1978), *J. Chem. Phys.* **69,** 3427.
L. ZANDEE and R. B. BERNSTEIN (1979), *J. Chem. Phys.* **70,** 2574.
R. N. ZARE (1979), in *Laser-Induced Processes in Molecules,* K. L. Kompa and S. D. Smith, Eds., Springer-Verlag, Berlin, p. 225.

Chapter 6

R. J. ADRIAN and K. L. ORLOFF (1977), *Appl. Opt.* **16,** 677.
M. AHMADJIAN and C. W. BROWN (1973), *Env. Sci. Technol.* **7,** 453.
S. A. AKHMANOV, R. V. KHOKHLOV, and A. P. SUKHORUKOV (1972), in *Laser Handbook,* Vol. 2, F. T. Arecchi and E. O. Shulz-Dubois, Eds., North Holland Publ., Amsterdam, p. 115.
R. J. ALLEN and W. E. EVANS (1972), *Rev. Sci. Instrum.* **43,** 1422.
J. ALTMANN, W. LAHMANN, and C. WEITKAMP, (1980), *Appl. Opt.* **18,** 3453.
G. A. ANTCLIFFE and J. F. WROBEL (1972), *Appl. Opt.* **11,** 1548.
R. L. ARMSTRONG, J. B. MASON, and T. BARBER (1976), *Appl. Opt.* **15,** 2891.
K. ASAI, T. ITABE and T. IGARASHI (1979), *Appl. Phys. Lett.* **35,** 60.
V. V. BARINOV and S. A. SOROKIN (1973), *Sov. J. Quantum Electron.* **3,** 89.
K. G. BARTLETT and C. Y. SHE (1976), *Appl. Opt.* **15,** 1980.
R. A. BAUMGARTNER and R. L. BYER (1978a), *Opt. Lett.* **2,** 163.
R. A. BAUMGARTNER and R. L. BYER (1978b), *Appl. Opt.* **17,** 3555.
G. BENEDETTI-MICHANGELI, F. CONGEDUTI, and G. FIOCCO (1972), *J. Atmos. Sci.* **29,** 906.
P. J. BOURKE and C. G. BROWN (1971), *Opt. Laser Technol.,* February, p. 23.
A. BOUTIER, J. LÉFEVRE, C. PÈROUZE, and O. PAPIRNIK (1979), in *Laser Velocimetry and Particle Sizing,* H. D. Thompson and W. H. Stevenson, Eds., Hemisphere Publishing, Washington, D. C., p. 15.
M. R. BOWMAN, A. J. GIBSON, and M. C. W. SANDFORD (1969), *Nature* **221,** 456.
E. V. BROWELL, T. D. WILKERSON, and T. J. MCILRATH (1979), *Appl. Opt.* **18,** 3474.
J. L. BUFTON, R. W. STEWART, and C. WENG (1979), *Appl. Opt.* **18,** 3363.
M. BURNS, R. FRANCKE, M. AUSTIN, L. W. RYAN JR., and M. S. FELD (1976), *Appl. Opt.* **15,** 672.

R. G. BUSER and R. S. RHODES (1973), *Appl. Opt.* **12,** 205.
R. L. BYER and M. GARBUNY (1973), *Appl. Opt.* **12,** 1496.
G. A. CAPELLE and L. A. FRANKS (1979), *Appl. Opt.* **18,** 3579.
R. H. CARLON, D. V. KIMBALL, and R. J. WRIGHT (1980), *Appl. Opt.* **19,** 2366.
L. W. CHANEY, D. G. RICKEL, G. M. RUSSWARM, and W. A. MCCLENNY (1979), *Appl. Opt.* **18,** 3004.
A. COHEN, J. A. COONEY, and K. N. GELLER (1976), *Appl. Opt.* **15,** 2896.
R. T. H. COLLIS (1970), *Appl. Opt.* **9,** 1782.
R. T. H. COLLIS and E. E. UTHE (1972), *Opt. Electron.* **4,** 87.
J. COONEY (1968), *Appl. Phys. Lett.* **14,** 210.
J. COONEY (1972), *J. Appl. Meteor.* **11,** 108.
J. COONEY and M. PINA (1976), *Appl. Opt.* **15,** 602.
H. P. DELONG (1974), *Opt. Eng.* **13,** 5.
V. E. DERR and C. G. LITTLE (1970), *Appl Opt.* **9,** 1976.
R. DOS SANTOS and W. H. STEVENSON (1977), *Appl. Phys. Lett.* **30,** 236.
W. L. EBERHARD and R. M. SCHOTLAND (1980), *Appl. Opt.* **19,** 2967.
M. EL-SHERBINY, E. M. BALLIK, J. SHEWCHUN, B. K. GARSIDE, and J. REID (1979), *Appl. Opt.* **18,** 1198.
V. V. FADEEV (1979), *Sov. J. Quantum Electron.* **8,** 1251.
W. M. FARMER (1972), *Appl. Opt.* **11,** 2603.
W. M. FARMER and D. B. BRAYTON (1971), *Appl. Opt.* **10,** 2319.
W. M. FARMER, J. O. HORNKOHL, G. J. BRAND, and J. MEIER (1979), in *Laser Velocimetry and Particle Sizing,* H. D. Thompson and W. H. Stevenson, Eds., Hemisphere Publishing, Washington, D.C., p. 428.
G. FIOCCO and L. D. SMULLIN (1963), *Nature* **199,** 1275.
K. FREDRIKSSON, B. GALLE, K. NYSTRÖM, and S. SVANBERG (1979), *Appl. Opt.* **18,** 2998.
M. A. FRERKING and D. J. MUEHLNER (1977), *Appl. Opt.* **16,** 526.
F. G. GEBHARDT and C. D. SMITH (1969), *Appl. Phys. Lett.* **14,** 52.
F. G. GEBHARDT and D. C. SMITH (1971), *IEEE J. Quantum Electron.* **QE-7,** 63.
J. GELBWACHS (1973), *Appl. Opt.* **12,** 2812.
A. J. GIBSON and L. THOMAS (1975), *Nature* **256,** 561.
W. B. GRANT and R. D. HAKE, JR. (1975), *J. Appl. Phys.* **46,** 3019.
W. B. GRANT, R. D. HAKE, JR., E. M. LISTON, R. C. ROBBINS, and E. K. PROCTOR, JR. (1974), *Appl. Phys. Lett.* **24,** 550.
J. B. HALPERN, W. M. JACKSON, and V. MCCRARY (1979), *Appl. Opt.* **18,** 590.
T. HENNINGSEN, M. GARBUNY, and R. L. BYER (1974), *Appl. Phys. Lett.* **24,** 242.
E. D. HINKLEY and P. L. KELLEY (1971), *Science* **171,** 635.
E. D. HINKLEY (1972), *J. Opt. Soc. Am.* **62,** 736.
E. D. HINKLEY, R. T. KU, K. W. NILL and J. F. BUTLER (1976), *Appl. Opt.* **15,** 1653.
T. HIRSCHFELD, E. R. SCHILDKRAUT, H. I. TANNENBAUM, and D. TANNENBAUM (1973), *Appl Phys. Lett.* **22,** 28.
F. E. HOGE and R. N. SWIFT (1980), *Appl. Opt.* **19,** 3269.
N. S. HONG and A. R. JONES (1976), *J. Phys. D* **9,** 1839.
J. D. HOUSTON and A. E. CARSWELL (1978), *Appl. Opt.* **17,** 614.
H. INABA and T. KOBAYASI (1969), *Nature* **224,** 170.
H. INABA and T. KOBAYASI (1975), *Opt. Commun.* **14,** 119.
W. B. JOHNSON (1969), *J. Air Pollution Control Assoc.* **19,** 176.
P. KAFALAS and A. P. FERDINAND, JR. (1973), *Appl. Opt.* **12,** 29.
P. KAFALAS and J. HERRMANN (1973), *Appl. Opt.* **12**, 772.
L. Z. KENNEDY and J. W. BILBRO (1979), *Appl. Opt.* **18,** 3010.
G. S. KENT and R. W. H. WRIGHT (1970), *J. Atmos. Terr. Phys.* **32,** 917.

H. KILDAL and R. L. BYER (1971), *Proc. IEEE* **59,** 1644.
D. K. KILLINGER, N. MENYUK, and W. E. DEFEO (1980), *Appl. Phys. Lett.* **36,** 402.
H. H. KIM (1973), *Appl. Opt.* **12,** 1454.
K. J. KIRSCH and M. K. MAZUMDER (1975), *Appl. Phys. Lett.* **26,** 193.
T. KOBAYASI and H. INABA (1970a), *Appl. Phys. Lett.* **17,** 139.
T. KOBAYASI and H. INABA (1970b), *Proc. IEEE* **58,** 1568.
O. K. KOSTKO (1976), *Sov. J. Quantum Electron.* **5,** 1161.
R. T. KU, E. D. HINKLEY, and J. O. SAMPLE (1975), *Appl. Opt.* **14,** 854.
R. T. V. KUNG and I. ITZKAN (1976), *Appl. Opt.* **15,** 409.
L. LADING (1979), in *Laser Velocimetry and Particle Sizing,* H. D. Thompson and W. H. Stevenson, Eds., Hemisphere Publishing, Washington, D.C., p. 26.
G. L. LAMB, JR., and R. B. KINNEY (1969), *J. Appl. Phys.* **40,** 416.
T. R. LAWRENCE, D. J. WILSON, C. E. CRAVEN, I. P. JONES, R. M. HUFFAKER, and J. A. L. THOMSON (1972), *Rev. Sci. Instrum.* **43,** 512.
R. C. C. LEITE, R. S. MOORE, and J. R. WHINNERY (1964), *Appl. Phys. Lett.* **5,** 141.
R. C. C. LEITE, S. P. S. PORTO, and T. C. DAMEN (1967), *Appl. Phys. Lett.* **10,** 100.
D. E. LENCIONI and J. E. LOWDER (1974), *IEEE J. Quantum Electron.* **QE-10,** 235.
D. E. LENCIONI and L. C. PETTINGILL (1977), *J. Appl. Phys.* **48,** 1848.
D. A. LEONARD (1975), *Opt. Quantum Electron.* **7,** 197.
A. G. LITVAK (1966), *JETP Lett.* **4,** 341.
A. MAYER, J. COMERA, H. CHARPENTIER, C. JAUSSAUD (1978), *Appl. Opt.* **17,** 391.
J. H. MCCOY, D. B. RENSCH, and R. K. LONG (1969), *Appl. Opt.* **8,** 1471.
T. J. MCGEE and T. J. MCILRATH (1979), *Appl. Opt.* **18,** 1710.
T. J. MCILRATH, R. HUDSON, A. AIKIN, and T. D. WILKERSON (1979), *Appl. Opt.* **18,** 316.
S. H. MELFI, M. L. BRUMFIELD, and R. W. STOREY, JR. (1973), *Appl. Phys. Lett.* **22,** 402.
R. M. MEASURES, W. HOUSTON, and M. BRISTOW (1973), *Can. Aeronaut. Space J.* **19,** 501.
R. M. MEASURES, W. R. HOUSTON, and D. G. STEPHENSON (1974), *Opt. Eng.* **13,** 494.
I. MELNGAILIS (1972), *IEEE Trans. Geosci. Electron.* **GE-10,** 7.
R. T. MENZIES (1971), *Appl. Opt.* **10,** 1532.
R. T. MENZIES (1973), *Appl. Phys. Lett.* **22,** 592.
R. T. MENZIES and M. S. SHUMATE (1974), *Science* **184,** 570.
R. T. MENZIES and M. S. SHUMATE (1976), *Appl. Opt.* **15,** 2080.
G. J. MULLANEY, W. H. CHRISTIANSEN, and D. A. RUSSELL (1968), *Appl. Phys. Lett.* **13,** 145.
E. R. MURRAY, R. D. HAKE, JR., J. E. VAN DER LAAN, and J. G. HAWLEY (1976a), *Appl. Phys. Lett.* **28,** 542.
E. R. MURRAY and J. E. VAN DER LAAN (1978), *Appl. Opt.* **17,** 814.
E. R. MURRAY, J. E. VAN DER LAAN, and J. G. HAWLEY (1976b), *Appl. Opt.* **15,** 3140.
E. R. MURRAY, M. F. WILLIAMS, and J. E. VAN DER LAAN (1978), *Appl. Opt.* **17,** 296.
C. A. NORTHEND, R. C. HONEY, and W. E. EVANS (1966), *Rev. Sci. Instrum.* **37,** 393.
F. K. OWEN (1979), in *Laser Velocimetry and Particle Sizing,* H. D. Thompson and W. H. Stevenson, Eds., Hemisphere Publishing, Washington, D.C., p. 123.
S. R. PAL, J. S. RYAN, and A. I. CARSWELL (1978), *Appl. Opt.* **17,** 2257.
J. C. PETERSON, M. E. THOMAS, R. J. NORDSTROM, E. K. DAMON, and R. K. LONG (1979), *Appl. Opt.* **18,** 834.
D. T. PHILLIPS, P. J. WYATT, and R. M. BERKMAN (1970), *J. Coll. Interface Sci.* **34,** 159.
R. G. PINNICK, D. E. CARROLL, and D. J. HOFFMANN (1976), *Appl. Opt.* **15,** 384.
J. C. POURNY, D. RENAUT, and A. ORSZAG (1979), *Appl. Opt.* **18,** 1141.
D. C. PRIDMORE-BROWN (1973), *Appl. Opt.* **12,** 2188.
J. S. RANDHAWA and J. E. VAN DER LAAN (1980), *Appl. Opt.* **19,** 2291.
R. B. RASK (1979), in *Laser Velocimetry and Particle Sizing,* H. D. Thompson and W. H. Stevenson, Eds., Hemisphere Publishing, Washington, D.C., p. 251.

D. M. RAYNER, M. LEE, and A. G. SZABO (1978), *Appl. Opt.* **17,** 2730.

D. M. RAYNER and A. G. SZABO (1978), *Appl. Opt.* **17,** 1624.

J. REID, M. EL-SHERBINY, B. K. GARSIDE, and E. A. BALLIK (1980), *Appl. Opt.* **19,** 3349.

J. REID, B. K. GARSIDE, J. SHEWCHUN, M. EL-SHERBINY, and E. A. BALLIK (1978), *Appl. Opt.* **17,** 1806.

J. REID, J. SHEWCHUN, B. K. GARSIDE, and E. A. BALLIK (1978b), *Appl. Opt.* **17,** 300.

D. M. ROBINSON and W. P. CHU (1975), *Appl. Opt.* **14,** 2177.

H. ROSEN, A. D. A. HANSEN, L. GUNDEL, and T. NOVAKOV (1978), *Appl. Opt.* **17,** 3859.

H. ROSEN, P. ROBRISH, and O. CHAMBERLAIN (1975), *Appl. Opt.* **14,** 2703.

K. W. ROTHE, U. BRINKMANN, and H. WALTHER (1974), *Appl. Phys.* **4,** 181.

B. J. RYE (1978), *Appl. Opt.* **17,** 3862.

T. SATO, T. KISHIMOTO, and K. SASAKI (1978a), *Appl. Opt.* **17,** 230.

T. SATO, Y. SUZUKI, H. KASHIWAGI, M. NANJO, and Y. KAKUÏ (1978b), *Appl. Opt.* **17,** 3798.

S. A. SCHLEUSENER, J. D. LINDBERG, and K. O. WHITE (1975), *Appl. Opt.* **14,** 2564.

S. A. SCHLEUSENER, J. D. LINDBERG, K. O. WHITE, and R. J. JOHNSON (1976), *Appl. Opt.* **15,** 2546.

R. M. SCHOTLAND (1974), *J. Appl. Meteor.* **13,** 72.

B. G. SCHUSTER and T. G. KYLE (1980), *Appl. Opt.* **18,** 2524.

R. L. SCHWIESOW, R. E. CUPP, M. J. POST, and R. F. CALFEE (1977), *Appl. Opt.* **16,** 1145.

D. C. SMITH and R. T. BROWN (1975), *J. Appl. Phys.* **46,** 1146.

J. E. SOLOMON (1977), *J. Appl. Phys.* **48,** 3588.

J. E. SOLOMON and D. M. SILVA (1976), *J. Appl. Phys.* **47,** 1519.

G. W. SUTTON (1978), *Appl. Opt.* **17,** 3424.

N. TAKEUCHI, H. SHIMIZU, and M. OKUDA (1978), *Appl. Opt.* **17,** 2734.

R. W. TERHUNE and J. E. ANDERSON (1977), *Opt. Lett.* **1,** 70.

H. D. THOMPSON and W. H. STEVENSON, Eds. (1979), *Laser Velocimetry and Particle Sizing,* Hemisphere Publishing, Washington, D.C.

A. W. TUCKER, M. BIRNBAUM, and C. L. FINCHER (1975), *Appl. Opt.* **14,** 1418.

O. UCHINO, M. MAEDA, and M. HIRONO (1979), *IEEE J. Quantum Electron.* **QE-15,** 1094.

O. UCHINO, M. MAEDA, J. KOHNO, T. SHIBATA, C. NAGASAWA, and M. HIRONO (1978), *J. Appl. Phys. Lett.* **33,** 807.

J. C. VAN DEHULST (1957), *Light Scattering by Small Particles,* Wiley, New York.

H. VISSER (1979), *Appl. Opt.* **18,** 1746.

C. C. WANG and L. I. DAVIS, JR. (1974), *Phys. Rev. Lett.* **32,** 349.

R. W. WEEKS and W. W. DULEY (1974), *J. Appl. Phys.* **45,** 4661.

R. W. WEEKS and W. W. DULEY (1976), *Appl. Opt.* **15,** 2917.

F. A. WILLIAMS (1965), *Int. J. Heat Mass Transfer* **8,** 575.

P. O. WITZE (1979), in *Laser Velocimetry and Particle Sizing,* H. D. Thompson and W. H. Stevenson, Eds., Hemisphere Publishing, Washington, D.C., p. 239.

O. R. WOOD, E. G. BURKHARDT, M. A. POLLACK, and T. J. BRIDGES (1971), *Appl. Phys. Lett.* **18,** 261.

Z. YASA, N. M. AMER, H. ROSEN, A. D. A. HANSEN, and T. NOVAKOV (1979), *Appl. Opt.* **18,** 2528.

P. R. L. YIN and R. L. LONG (1968), *Appl. Opt.* **7,** 1551.

Subject Index

Materials Index